Crop Nutrient Requirements and Advanced Fertilizer Management Strategies

Crop Nutrient Requirements and Advanced Fertilizer Management Strategies

Editors

Christos Noulas
Shahram Torabian
Ruijun Qin

Basel • Beijing • Wuhan • Barcelona • Belgrade • Novi Sad • Cluj • Manchester

Editors

Christos Noulas
Institute of Industrial and
Forage Crops
Hellenic Agricultural
Organization
Larissa
Greece

Shahram Torabian
College of Agriculture
Virginia State University
Petersburg
United States

Ruijun Qin
Department of Crop and
Soil Science
Oregon State University
Hermiston
United States

Editorial Office
MDPI
St. Alban-Anlage 66
4052 Basel, Switzerland

This is a reprint of articles from the Special Issue published online in the open access journal *Agronomy* (ISSN 2073-4395) (available at: www.mdpi.com/journal/agronomy/special_issues/Crop_Fertilizer).

For citation purposes, cite each article independently as indicated on the article page online and as indicated below:

Lastname, A.A.; Lastname, B.B. Article Title. *Journal Name* **Year**, *Volume Number*, Page Range.

ISBN 978-3-7258-0632-4 (Hbk)
ISBN 978-3-7258-0631-7 (PDF)
doi.org/10.3390/books978-3-7258-0631-7

Contents

About the Editors

Christos Noulas

Dr. Christos Noulas is an agronomist and principal researcher at the Institute of Industrial and Forage Crops of the Hellenic Agricultural Organization, "DIMITRA". His research interests include topics on soil fertility and sustainable nutrient management, improved nitrogen (N) use efficiency, soil-plant relations, plant nutritional physiology, land use and soil quality, crop production, and root morphological traits in relation to crop nutrition. From 2004 to 2010, he also acted as an external teaching assistant at the School of Agricultural Technology, Thessaly University of Applied Sciences, Greece. He received his diploma (major in Soil Science, MSc equivalent) in 1996 from the Agricultural University of Athens. In 2002, he completed his PhD studies at the Swiss Federal Institute of Technology in Zurich (ETH) (Institute of Agricultural Sciences, IAS). His research focused on topics related to parameters of N use efficiency in cereals, soil–plant relations, and cereal root morphological traits in relation to crop nutrition and production in the temperate region. He has participated in several research projects at international, national, and regional levels and in seven European Concerted Actions (COST Actions), dealing with soil–plant–water relationships, nutrient use efficiency, soil fertility, and sustainable crop production. He has published 91 peer reviewed journal articles (SCI and SCIE), 2 book chapters, and more than 50 contributions to national and international conferences. His articles include 700 (Scopus, h-index 13) and 1077 citations (in Google Scholar, h-index 16). He acted as a reviewer for 44 research articles in more than 18 peer-reviewed scientific journals. He is a member of the ongoing United Nations Environment Project "Towards INMS" on Nitrogen Management and the Collaborative Network on Quinoa (FAO, TCP). He is also a member of the Hellenic Society of Agricultural Engineers and the European Society for Soil Conservation (ESSC).

Shahram Torabian

Shahram Torabian, Ph.D., is an accomplished crop physiologist currently serving as an assistant research professor at the Agricultural Research Station of Virginia State University. He continues to lead research initiatives to identify and develop new, high-value crops suitable for Virginia. Moreover, he works on new cropping systems and studies how crops react to different farming methods and the environment. Dr. Torabian holds a Ph.D. in Agronomy with a specialization in Crop Physiology from the University of Tabriz, achieved in 2018. His Ph.D. thesis investigated the response of common bean varieties to drought stress, exploring the role of polyamines in mitigating drought impacts on these crops. Prior to his current position, Dr. Torabian served as a postdoctoral scholar at Oregon State University's Hermiston Agricultural Research and Extension Center (OSU-HAREC), where he spearheaded research and extension programs focused on nutrient management and cropping systems in irrigated fields in the Columbia Basin. Optimizing fertilization strategies for potatoes, enhancing agricultural sustainability through diversification of cropping systems, and evaluating the impacts of various factors on wheat, grass seed crops, and alfalfa were his research interests. Before joining Oregon State University, Dr. Torabian worked as a breeder and researcher at the Plant Improvement and Seed Production Research Center in Isfahan, Iran, where he managed field trials and greenhouse studies for hybrid seed production of various vegetables. He was a visiting scholar at the University of Adelaide, Australia, where he focused on evaluating rhizobia isolated from Australian soil and its effects on legume crops' root growth.

He has authored numerous peer-reviewed papers in esteemed journals, contributed to extension papers, and served as an associate editor for the *Agronomy Journal*. Additionally, he has actively engaged in teaching soil science courses (SOSC-242) and mentoring students.

Ruijun Qin

Dr. Ruijun Qin is an extension agronomist at the Oregon State University (OSU), Hermiston Agricultural Research and Extension Center, and an associate professor at the Department of Crop and Soil Science at OSU. He plays a vital role in organizing growers' meetings and field days in Northeastern Oregon and Southeastern Washington. His extension and research activities focus on sustainable soil/plant nutrition and water management, crop production, cropping systems, environmental quality, and soil health. From 2006 to 2016, he worked for the San Joaquin Agricultural Research Center of the USDA-ARS and the University of California Davis, and his research was mainly focused on soil fumigation, high-value crop production, plastic mulching, and water management. He received his Ph.D. (2003) at the Swiss Federal Institute of Technology (ETH-Zurich, Institute of Agricultural Sciences). His research topics included the study of cereal root morphological traits in relation to crop production and nutrition, tillage systems, and crop rotation in the temperate region. He received his MSc in Soil Science from the Chinese Academy of Agricultural Sciences in 1995, and he successfully carried out projects concerning soil reclamation, the use of soil organic amendments, and the productivity of cropping systems in the subtropical zone. He is an associate editor of the *Agronomy Journal* and was chair of the Soil Health Community and the Nutrient Management Professionals Community of the American Society of Agronomy. He has published 81 peer-reviewed research articles, given 92 presentations at various international conferences, and served as a reviewer for more than 20 peer-reviewed scientific journals. His articles include 1256 citations (Scopus, h-index 20) and 1814 citations (Google Scholar, h-index 23). He is a program organizing committee member for the Pacific Northwest Vegetable Association Conference and Hermiston Farm Fair in Oregon and Washington, USA.

Preface

In order to attain significant improvements in enhancing nutrient-use efficiency (NUE) in production agriculture, accurate assessments of plant-available nutrients and advanced fertilizer management strategies are required. This is indispensable in order to meet the increasing global food demands in the present situation of the ever-increasing world population, the availability of limited resources (i.e., land and water), and climate change. Elevated NUE comes from nutrient increases in the root zone, enhanced crop response to applied nutrients, and reduced nutrient loss to the soil–plant–water–atmosphere continuum. To help readers keep up with the latest progress in the field, this Special Issue (SI) gathers 19 original research articles and 1 communication on topics related to enhanced NUE-related parameters in cropping systems, from the agronomic perspective to environmental considerations. Moreover, this SI focuses on the physiological basis of genotypic differences in the uptake and utilization of key nutrients and provides demonstrated experimental data in order to optimize fertilizer management. The studies have been carried out under both field and laboratory conditions, as well as modeling studies, and a wide range of geographic regions are also covered. The collection of these manuscripts presented in this SI update provides a relevant knowledge contribution for crop nutrient requirements and advanced fertilizer management strategies. We express our sincere thanks to all contributing authors, and we greatly appreciate the constructive support of the editorial staff for the development of this Special Issue, making it a great success.

Christos Noulas, Shahram Torabian, and Ruijun Qin
Editors

agronomy

Editorial

Crop Nutrient Requirements and Advanced Fertilizer Management Strategies

Christos Noulas [1,*], Shahram Torabian [2,3] and Ruijun Qin [2]

1 Institute of Industrial and Forage Crops, Hellenic Agricultural Organization "Dimitra", 41335 Larissa, Greece
2 Hermiston Agricultural Research and Extension Center, Oregon State University, Hermiston, OR 97838, USA; storabian@vsu.edu (S.T.); ruijun.qin@oregonstate.edu (R.Q.)
3 Agricultural Research Station, Virginia State University, Petersburg, VA 23806, USA
* Correspondence: noulaschristos@gmail.com

1. Introduction

From an estimated 7.6 billion people worldwide in 2021, projections of the United Nations (UN) indicate global population growth to around 8.6 billion by 2030, 9.8 billion by 2050, and more than 11.0 billion in 2100 [1]. As a consequence, the global total demand for all agricultural products is expected to increase by 1.1% per year until 2050. Therefore, challenge for the coming decades will be to ensure long-term food security of the ever-growing world population by increasing crop productivity using sustainable agricultural practices while, at the same time, maintaining soil health and preserving the quality of the environment [2,3]. This must be accomplished in the context of the shrinking availability of arable land and shortage of fossil fuels since many of the resources needed for crop production are limited (mainly agricultural land, water, and nutrients), making it indispensable that must be used responsibly [4,5]. The UN has set ending hunger, achieving food security and improved nutrition, and promoting sustainable agriculture among the 17 Sustainable Development Goals (SDGs) by the year 2030. Improving nutrient-use efficiency (NUE) and crop yield through improved nutrient management practices also ensures SDG 1 (no poverty), SDG 3 (good health and wellbeing), and SDG 15 (life on land). However, crop production depends on several interrelated agronomic factors, such as soil (e.g., pH, texture, organic matter content, water holding capacity, mineral composition, and nutrient availability, etc.), plant genetic material, crop management, and several other biotic and abiotic factors. Apart from soil testing and nutrient removal by harvestable products that are traditionally used to derive the amounts of nutrients required by the crop [6], the role of roots, which has often been neglected, should also be taken into consideration for better resource acquisition [7]. Crop nutrition and balanced fertilization (both from inorganic and organic sources) are considered among the primary actions towards satisfactory crop growth and production while decreasing production costs. Nutrient elements are essential resources for food, feed, and biofuel production, next to energy, water, carbon dioxide (CO_2), biodiversity, labor, capital, and management.

Sustainable nutrient management is critical to increase or maintain crop yields, and soil fertility must be consistently high in order to meet crop needs throughout a growing season [5]. To increase crop yields, elevated levels of nitrogen (N), potassium (K), and phosphorus (P)-containing fertilizers as well as other macro and micronutrients have been applied in croplands since the end of World War II, have prevented soil nutrient depletion, and, in some cases, have even built-up soil fertility (maintenance fertility) [8]. However, fertilizer recommendations are regularly at the fore of production and environmental concerns related to agriculture. At the same time, worldwide fertilizer use is forecasted to decline up to 7% (in a pessimistic scenario) before partial recovery, with food security implications a reflection of significant uncertainty in market conditions due to the war in Ukraine [9]. Balanced fertilization refers to the application of plant nutrients in optimum quantities and in the right proportions through appropriate methods and at the right times for a specific crop's needs and agroclimatic conditions [10–12]. In this context, the development of novel and sophisticated fertilization practices is the challenge for future nutrient management that helps to improve crop NUE, maintain adequate levels of soil nutrients, and prevent deficiencies or the imbalance or overuse of fertilizers, leading to economic and environmental benefits [13]. However, crop-specific information on nutrient management, including diverse nutrient sources as part of an integrated nutrient management as well as improving NUE by developing novel and practical fertilizer recommendations for farmers, needs to be further explored under diverse pedoclimatic environments. The role of plant roots should also be taken into account as key parameters for improving NUE, which is central but still under debate. It will provide a better understanding of how crop plants acquire water and nutrients through their roots and maintain growth and performance under diverse pedoclimatic conditions.

This Special Issue (SI) provides a base for revealing the principal mechanisms of enhanced NUE-related parameters in cropping systems, from the agronomic perspective to environmental considerations. Moreover, it focuses on the physiological basis of genotypic differences in the uptake and utilization of key nutrients, including the primary macronutrients nitrogen (N), phosphorus (P), and potassium (K); the secondary macronutrients calcium (Ca), magnesium (Mg), and sulfur (S); and the micronutrients iron (Fe), copper (Cu), manganese (Mn), zinc (Zn), boron (B), and silicon (Si) and provide demonstrated experimental data in order to optimize fertilizer management. It tries to identify the barriers that exist to the improvement of nutrient management and which interventions can lead farmers along pathways towards the adoption of novel and more profitable and sustainable fertilization strategies.

2. Overview of This SI

The Special Issue (SI) comprises 18 original research articles and one communication on various topics of rational crop nutrient management, reporting novel scientific finding updates and recent developments on fertilization strategies of crops with quite diverse utilizations: from primary arable crops used for both food and fodder, like wheat (*Triticum aestivum* L.), maize (*Zea mays* L.), and faba beans (*Vicia faba* L.), to forage crops, such as marandu grass (*Urochloa brizantha* cv. Marandu), to oil crops with food or non-food usage, like soybean (*Glycine max* L.), olives (*Olea europaea* L.), oil palm (*Elaeis guineensis* Jacq.), and oilseed rape (*Brassica napus* L.). It also includes crops for the brewing industry (malting barley, *Hordeum vulgare* L.) and the perfume and pastry industry (Rose scented geranium, *Pelargonium graveolens* L.). Moreover, dual-purpose crops such as cotton (*Gossypium spp.* L.) grown for fiber and oil purposes or multipurpose crops are also included in this SI, like hemp (*Cannabis sativa* L.), which is cultivated worldwide for fiber, oil, and cannabinoids for medical purposes; flax (linseed) (*Linum usitatissimum* L.) for human nutrition, cosmetics, and the pharmaceutical industry; and sugarcane (*Saccharum officinarum* L.) for agro-industrial uses and the pharmaceutical and the chemical energy sectors. The studies have been carried out under both field and laboratory conditions, as well as modelling studies, and a wide range of geographic regions are also covered: six studies originated from China; three from Egypt and Brazil; two from Poland, the Czech Republic, and the United States; and one from Tailand.

The aim of the study by Yang et al. [14] was to investigate N accumulation, assimilation, and utilization in four commercial domestic hemp cultivars, as well as the growth and physiological response of hemp to N concentrations in a pot experiment conducted in a greenhouse. Those aims were well covered by the results and provide more precise answers about hemp responses to N in controlled conditions, which could serve for field recommendations of N for hemp production in the future. The study suggests that N application up to 6.0 mmol/L (NO_3-N in the nutrient solution) is sufficient to regulate morpho-physiological attributes, antioxidant capacities, and N accumulation to achieve the optimal growth of hemp. The study also investigated root parameters (root weight, root to shoot ratio, root N) exploring the role of roots as key parameters for improving NUE.

Even though the use of imaging and Near-Infrared (NIR) applications to detect N status in cereal crops is not all that new, the main objective by Klem et al. [15] in their study was to improve prediction of N status in malting barley using multiple spectral reflectance wavelengths, by selecting vegetation indices, using N status indicators, and employing artificial neural networks. The employment of artificial neural networks in remote sensing provides a number of advantages in comparison to regression models. Increasing the accuracy of N status estimation in barley aboveground biomass by combining indirect N status indicators, such as N nutrition index (NNI) or N uptake, and an artificial neural network is expected to advance the potential impact to improve N nutrition of malting barley and avoid over fertilization. Combining NNI or N uptake and a neural network increased the accuracy of N status estimation to up 94%, compared to less than 60% for N concentration.

Sulfur (S) is an essential secondary macronutrient involved in the growth and development of plants. After N, P, and K, it is increasingly seen as the fourth major nutrient in plants. The role of sulfur (S) in plant growth and development, the functions of which include both being a structural component of macrobiomolecules and modulating several physiological processes and tolerating abiotic stresses, is still under debate. The topic of the article by Stepaniuk et al. [16] covers the dose and method of S application for winter oilseed rape, which is an important crop for edible oil and biofuel. Moreover, the optimization of mineral S fertilization is considered to be particularly important among agricultural practices to boost oilseed rape yields grown in a monoculture. In this respect, soil fertilization must be supplemented with foliar fertilizers, and their doses and dates should be defined correctly. The impact of S on winter oilseed rape yield depended significantly on both the dose and the application method. Even at the lowest dose (20 kg·ha^{-1}), S increased seed yield, regardless of the application method. Fertilization with S increased the mineral composition of rapeseeds, whereas the contents of macroelements in the straw were more variable than in the seeds. Each of the S fertilization treatments reduced the S harvest index. The findings of this study seem to be interesting since a fertilization scheme of winter oilseed rape plants growing in a monoculture could be suggested.

The inhibitors of nitrification and urease play an important role in sustainable fertilization strategies and the influence of nitrification inhibitors (NIs) on soil N losses are widely known. However, there is no solid information on the fate of fertilizers containing N-transformation inhibitors (NIs) in soil. It is still not clear how long the effectiveness of the NIs can last and what factors can affect their efficiency. More studies are required about factors that affect NI efficiency, which can help growers to use NIs in fields correctly. These issues were well addressed by three original articles [17–19] and one communication [20] in this SI and contribute to our better understanding on N management affecting the economic and environmental aspects of fertilization.

The study by Školníková et al. [17] compared the effect of conventional N fertilizers with those containing N-transformation inhibitors and evaluated the timings of their applications on the wheat-grain yield and quality. A single application of urea with NI and/or urease inhibitors resulted in a relatively average increase in the wheat grain yield, whereas grain protein content, and the Zeleny test values were significantly increased compared to the split N application. Significant increases in the grain yield (by 6.3%) and the Zeleny test value (by 16.5%) were observed after inhibited urea applications compared to the control treatment (without inhibitors).

The study by Torabian et al. [18] evaluated the impact of two different types of "nitrate stabilizers" (NSs), in combination with urea and urea ammonium nitrate as N fertilizer sources under two N application methods and rates (single and split applications; 100 and 85% of fertilizers) on grain yield, SPAD (flag leaf greenness), protein concentrations of wheat, and mineral soil N contents. The results demonstrated that selecting effective NSs, suitable N sources, reducing N rates, and splitting N fertilizers during the growing season could be regarded as practical strategies to reduce NO_3^--N leaching while not compromising the wheat yield. However, they highlighted the importance to carrying out trials based on multiple years and locations to draw solid conclusions on the effects of NSs and N management on potential yield benefits and the N dynamics of soils.

The objective of the study by Cassimiro et al. [19] was to evaluate ammonia volatilization and dry matter production of *Urochloa brizantha* cv. Marandu (a pasture crop) in response to rates (100 and 200 kg ha^{-1} year) and four N sources of enhanced-efficiency N fertilizers (Urea—UrConv; Ammonium nitrate—AN; Urea+NBPT—UrNBPT; Urea+Duromide—UrDuromide). When urea or UAN are applied to a soil, several factors, such as soil moisture content, high temperature, soil acidity, soil organic C and N, and high crop residues, can contribute to N loss (mainly by ammonia volatilization). These losses can reduce N availability, and, therefore, crop dry matter yield and quality. The findings of the study would inform N management strategies by incorporating urease inhibitors and reducing N losses by volatilization and contribute to the advancement of the knowledge of pasture production and quality.

NIs were originally intended to improve N retention in soil by blocking the microbial oxidation of ammonium (NH_4^+) to nitrate (NO_3^-). However, NIs also have the potential to alter other components of the N cycle, such as denitrification. The outcome of the communication by Li et al. [20] could provide information as to how improved N use efficiency through the use of NIs promotes crop growth and decreases N losses in soil and the atmosphere. They studied the effects of the inhibitors on denitrification rates, which remain largely unclarified. The study monitored the dynamics in annual denitrification rates affected by NIs from a maize field. Their results showed that the denitrification rates and denitrifying enzyme activities were highly variable in different growing periods but were not affected by the applications of inhibitors. Partial inhibition of the nitrification process was observed in the inhibitor treatments compared with the urea- or manure-only treatments. The formation of NO_3^--N and the nitrification rates could be markedly reduced by DMPP (3,4-dimethylpyrazole phosphate), whereas NO_3^--N availability did not affect the denitrification rates. To provide insightful information for our understanding of the achievement of inhibitors on the mitigation of N losses in arable soil under field conditions, more studies are needed under different sites to explore additional mechanisms driving changes over longer time periods.

The soybean (*Glycine max* L. Merr.) is an important legume crop and is widely grown as an oilseed crop and a protein source worldwide. Effective fertilization regimes to achieve a balance between grain yield, plant biomass, and quality traits, as well as to contribute to the promotion of the large-scale cultivation of soybeans under drip irrigation in arid areas worldwide, are provided in the work by Li et al. [21]. Such information will help in fine-tuned soybean fertilization management practices to increase yield, resource-use efficiency, and to minimize environmental risk. They confirmed that N fertilizer significantly affects grain yield, whereas P and K fertilizers influence harvest index and biomass, respectively. They have also gone one step further by describing the optimized combination of fertilizers for high yield, as well as biological and quality traits, by a quadratic polynomial regression analysis. They found that a fertilization combination of 411.6–418.4 kg ha^{-1} N, 154.0–251.0 kg ha^{-1} P_2O_5, and 117.8–144.7 kg ha^{-1} K_2O was required in order to obtain a theoretical grain yield and plant biomass of more than 7.21 tons ha^{-1} and 16.38 tons ha^{-1} with 300,000 plants ha^{-1}, respectively. They also proposed an economical fertilizer combination that could promote the use of profitable fertilizer in the future production of soybean.

The study by Duangpan et al. [22] contributed to the enhancement of our knowledge on improved Si fertilization and on the growth and physiological responses of oil palm seedlings and nursery production under non-stress conditions. Oil palm could be considered as an intermediate Si accumulator, and, therefore, Si-improved management is crucial for vigorous oil palm plantations. Overall, Si fertilization provided beneficial effects on growth and physiological responses in oil palm seedlings. Correlation analysis revealed a highly significant and positive association among Si accumulation, chlorophyll *a* content, photosynthetic rate, total fresh weight, total dry weight, and N content of seedlings, indicating that Si fertilization enhanced the performances of these attributes. It is important, however, to notice that Si application can be more effective in particular soil than others since Si solubility is dependent on soil pH, redox potential, particle size, and organic matter; therefore, the soil type should be concerned when applying Si fertilizer.

In view of the cost for the producer, harm to humans, and the environmental use of chemical fertilizers, nanofertilizers are emerging as a promising alternative to conventional fertilizers through their positive roles in slow releases of nutrients into soils and enhanced nutrient use efficiencies. The use of nanofertilizers as new technology fertilizers was addressed by the study by El-Sonbaty et al. [23], who evaluated the effectiveness of innovative iron (Fe) oxide nanoparticle fertilizer formulations (NPs) against traditional Fe compounds (sulfate or chelate) and on rose-scented geranium herbs in terms of plant growth, biochemical attributes, essential oil, and its constituents. The effects of Fe NPs on growth, secondary metabolites, and essential oil components of rose-scented geranium herbs are of great value for the application of nanomaterials in agriculture. Therefore, the manuscript contributes to increasing and improving our knowledge on new technological fertilizers, ways to alleviate Fe deficiency, and increasing Fe-use efficiency, providing a solid confirmation of the high effectiveness of a nanofertilizer on plant productivity and product quality over conventional Fe sources. Iron deficiency is commonly found in sensitive crop species grown in arid and semiarid regions with calcareous soils, which have been estimated to comprise over one-third of the total world's land area. Therefore, the study has impacts on several regions of the world, challenging similar issues and demonstrating the significance of using Fe NPs for commercial purposes while also being environmentally preferred in alkaline soils.

However, despite the fact that nanofertilizers are undoubtedly opening new approaches towards sustainable agriculture, one should consider the potential limitations of the commercial use of these fertilizers (i.e., interaction of nanomaterials with the environment, potential effects on human health, toxicities of different nanoparticles, evaluations of different soil physio-chemical properties before their uses, market considerations, etc.).

Balanced fertilization is the best agronomic practice for soil management in plants grown under stressed conditions. The study by Hussein et al. [24] evaluated the potential performances of two types of highly soluble phosphorus fertilizers (HSPFs), namely, mono-ammonium phosphate (MAP) and urea phosphate (UP), in comparison to the most widely used phosphate fertilizer, granular calcium super-phosphate (GCSP with a high pH > 7.0), in an attempt to overcome the problem of P fixation and the unavailability of micronutrients under some abiotic stresses in olive trees (*Olea europaea* L.), trees grown under multi-stress conditions (calcareous alkaline soils), which, in turn, affect the growth and productivity characteristics. In short, the application of HSPFs under these conditions might be an alternative surrogate to improve nutrient efficiency and thus improve olive productivity.

Of interest is the extensive work by Korzeniowska and Stanisławska-Glubiak [25], who presented results of micronutrient (B, Cu, Fe, Mn, and Zn) concentsationd in the shoots of 12 wheat, 10 maize, and 12 rape varieties obtained on the bases of plant and soil samples collected from 950 fields in Poland. Such research material is undoubtedly credential of the results obtained, and yet the literature provides little information on the variations in micronutrient concentrations in staple crop cultivars. They tested the hypothesis that the variations in the micronutrient contents in plants between varieties of the same species may be similar or even greater than the differences between species. These varieties may also show significant differences in micronutrient concentrations and thus require different fertilization techniques. This is also more relevant than ever nowadays, where breeders are constantly creating new varieties in search of crops that will have a better yield and be resistant to stresses, with better quality traits for the consumers. Differences were found in micronutrient concentrations between crop species and also between varieties. Even though these observations were not surprising, as different crop species and varieties have unique genetic backgrounds, the study concluded that the cultivar should be taken into account when assessing the need to fertilize wheat, maize, and rape with Cu, Fe, and Mn, whereas the assessment of the need for fertilization of these species with B and Zn could be carried out independently of the cultivar used. When fertilizing certain crops with micronutrients, it would be advisable to take into account not only the nutritional needs of the individual species but also the adaptation of micronutrient doses to the requirements of the cultivars within the species. Such a measure could contribute to a more efficient use of fertilizers in line with sustainable agriculture. Moreover, further research should confirm to what extent the concentrations of micronutrients in the early stage of growth affect the size of the final crop yield.

As an initial step to (i) determine crop nutrient demand corresponding to different target yields, (ii) estimate soil nutrient supply dynamics, and (iii) determine corresponding nutrient application rates and timings, theoretical models are needed for relating plant growth dynamics and crop nutrient uptakes, soil nutrient supplies and climates, and crop nutrient uptakes and yield component formations. Phosphorus, an essential macronutrient for plants, is often available at insufficient levels, limiting crop yield and productivity. The critical dilution curve (CDC) for phosphorus (Pc) was proposed as a suitable analytical tool to assess the flax (*Linum usitatissimum* L.) P nutrition status using four field experiments, with five P applications in the study by Xie et al. [26]. The Pc dilution curve could be useful as a reference curve to assess flax's nutritional status through the P nutrition index (PNI). Although the Pc dilution curve as a simple, accurate, and more rapid tool to diagnose crop P status tools has been used in crop production worldwide, the Pc concentration curve on the capsule dry matter of flax has not been reported. Curves of Pc have been established for a range of crops, such as potato, wheat, timothy, mungbean, urdbean, rapeseed, and maize, defining scenarios of luxury (excess), sufficiency, and deficiency for plant nutrient statuses, but work on the Pc dilution curve for flax for optimizing seed yield, grower profits, P-use efficiencies, and reducing environmental risks is meagre in the literature. Moreover, Pc dilution curves vary among different regions, species, genotypes within species, and practice managements. The results by Xie et al. [26] validated that the capsule Pc dilution curve could be an alternative and more rapid tool to diagnose flax P statuses to support the precise decisions of P fertilization during the reproductive growth of flax in a semi-arid to arid continental climate (Köppen *BSk* or *BWk*).

Under the same climatic conditions, Xie et al. [27] extended their research and studied the relationship between the increase in soil P fertility and the P and N contents in flax to build the model for critical P concentration in this plant as a function of N concentration in a shoot of flax for diagnostic purposes. This work provided a diagnosis tool that used the relationship between P and N concentrations for the entire growth period to estimate the critical P concentration for quantifying the degree of P deficiency. This tool could be used to adjust P fertilization in the following growing seasons for the species-specific conditions of an approximate soil pH of 8.

The need to increase the efficiencies of phosphate fertilizers in tropical soils, and the lack of information about the issue, motivated the hypothesis by Oliveira et al. [28] that the application of a polymer-coated fertilizer raised phosphate fertilization efficiencies and crop yields. The aim of the study was the evaluation of the effects of phosphate fertilizers with (Policote coating—fixation inhibitors) and without polymer coatings on the productivity and nutritional status of sugarcane ratoon and its effects on soil phosphorus availability for tropical soils with low P agronomic efficiencies. This was in light of the global importance of sugarcane, the crop requirements during the cycle, and low P levels in highly weathered soils. Increasing the longevity of sugarcane ratoons is of utmost importance; however, it is necessary to understand the best way to reapply P fertilizers. This original research paper was interesting because it showed sugarcane ratoon's yield, nutritional status, technological quality, and soil phosphorus availability in response to an enhanced efficiency phosphate fertilizer. Their results indicated that fertilizers with or without a Policote coating induced positive responses in soil P. The P contents varied with the applied doses. The treatments did not influence the concentration of P in the leaf. The technological qualities of cane stalks varied between the studied growing seasons, with better results in the second year. They suggested that further research should be encouraged to understand the dynamics between polymers, the availability of P in soil, and the possible effect on the physiology and production of enzymes that may contribute to nutrient-use efficiency.

Legumes have traditionally been used in cropping systems as part of crop rotations and are also intercropped with other crops (especially cereals). Cereal and legume intercropping for cereal yield and grain protein improvements is common worldwide practice, but the role of intercropping in grain quality has not yet fully understood. The study by Zhu et al. [29] quantified the effect of intercropping (wheat and faba bean intercropping) on wheat grain protein and amino acids under different N input conditions, and they identified the impact of intercropping on the relationship between GY and quality. The research study provided a unique contribution of intercropping technology on grain quality with specific references to proteins and associated amino acids that are essential for human health. Although intercropping has central role in this paper, it also recognizes that intercropping yield advantages can be modified by different N levels. It concludes that N management should be taken into account to achieve both intercropping yield and quality advantages.

Cotton is a dual-purpose crop grown for fiber and oil purposes. Under field conditions, various factors such as environmental conditions and agricultural management practices can significantly influence cotton growth and productivity. Among them, the optimization of crop nutrition by synchronizing nutrient availability with crop demand are key elements for sustainable nutrient management in cotton production. The aim of the study by Amissah et al. [30] was to assess the impact on the productivities and fiber qualities of modern cotton varieties to varying degrees of nutrient stresses (early (E) and late (L) season) under different production conditions. Late stress (30–40% of the full nutrient rates, only at the initial stage of planting) decreased the lint and cotton seed yields by 34.4% and 36.2%, respectively, across all production conditions. Compared to the full nutrient rate, the E-stress (no nutrient application early in the season, but the full rates were split-applied equally at the initiation of squares and the second week of bloom stages) did not adversely impact cotton yield. Significant nutrient stress effects on fiber quality were observed, but the magnitude of the differences was small, and it did not affect the grading class. The minimal impact of E-stress on cotton yield and quality in this study suggested that the rates of nutrients often applied in the early season could be reduced. The study concluded that soil and plant tissue analyses could assist in applying tailored nutrient application rates shortly before the reproductive phase of the crop synchronizing nutrient availability with crop demand.

Soil salinity and alkalinity are among the major challenges that threaten food security globally. Climate change will have a negative impact on agriculture, particularly in arid and semi-arid regions. In semi-arid and arid regions, soil salinity is a major and widespread threat to crop yields, food security, and the environment. Soils in arid and semi-arid regions are commonly alkaline, with high pH values as a result of water scarcity, in addition to low precipitation and high potential evapotranspiration. Therefore, studies related to soil alkalinity modifications are indispensable since alkaline soils cover more than one-fourth of Earth's surface. Soil pH is an important chemical property because it affects plant growth and nutrient availability in many different and complex ways. Soil pH affects plant growth both directly and also indirectly by affecting the availability of essential nutrients, levels of phytotoxic elements, and microbial activities. A pH, either far above neutral (alkaline) or far below neutral (acidic), makes essential plant nutrients less available. For a high soil pH (alkaline soils), limited solutions exist for reducing pH because they are impractical or uneconomical. In this context, the study by Beheiry et al. [31] exhibited particular interest since they investigated the potential impacts of some acidifying agents (acetic acid, citric acid, and sulfuric acid) applied in an attempt to adjust the high soil pH values in olive orchards, which are the main problem with Egyptian soils, whose values vary between neutral and extreme alkaline. The study concluded that significant improvements in total olive yield and their attributes, as well as the olive oil contents, resulted from the positive effect of acidifying agents on reducing soil pH, which, in turn, improved the availability of nutrients in the soil, enhancing their absorption, as mirrored from the leaf nutrient contents. The study also provided an evaluation of the effectiveness of the treatments applied as a valuable practical tool to be used by the farmers to correct soil alkalinity problems, which, in turn, influence physiological and growth parameters, the yield of table and oil olives, and the fruit's physical attributes.

On the other hand, soil acidity significantly decreases the availability of nutrients to plants, such as P and molybdenum (Mo), and increases the availability of aluminum (Al) and manganese (Mn) even to toxic levels. Moreover, other essential plant nutrients can also be leached below the rooting zone. Soil acidity affects approximately 30% of the world's potential food production area. These problems are particularly severe in humid tropical regions that have highly weathered soils. Liming is the most common practice to mitigate soil acidity, but the low solubility of lime and its application on a soil's surface, especially in no-till systems, restrict its reaction on the first soil layers. In this respect, the study by De Souza et al. [32] investigated the effect of the joint application of lime and gypsum to enrich a subsoil with calcium (Ca) and to alleviate Al acidity in an intercropping system with soybean, followed by maize–guinea grass. Their work also investigated the synergistic effects of subsoil Ca associated with N on root growth and yield of maize and soybean. Liming resulted in greater root growth for both crops; however, when lime was associated with gypsum, root growth was further enhanced. Moreover, soil acidity correction and N supply resulted in better distribution of the soybean and maize root systems in the soil's profile, increasing soil exploration, which favored water extraction in periods of scarcity and nutrient absorption in deeper layers of the soil, resulting, eventually, in higher yields. Nitrogen fertilization increased total maize grain yield by 36%, with a more expressive increase when applying 160 kg ha^{-1} or more, and, despite a positive effect on soybean grain yields in the long term, this response seemed not to be a direct effect of the N applied to maize. Overall, benefits resulting from the combination of lime and gypsum include greater plant biomass production, a denser root system, higher crop yield, and, eventually, a positive impact on soil C and N.

3. Conclusions

The collection of these manuscripts presented in this Special Issue (SI) updates and provides a relevant knowledge contribution for the usefulness of improving the fertilizer-use efficiency of crops, thus ensuring enough food for the rising world population of acceptable quality, taking into account environmental considerations. Plant nutrient requirements and nutrition are complex issues, starting from the 17 known and necessary nutrients for plant growth and merging a group of sciences, namely, soil science, plant physiology, chemistry, circular economy, environmental science, etc. Plant nutrition is one of the most important elements on which the yield and the quality of agricultural products depend. For about a century, significant yield increases were the result of the introduced revolutionary method by Nobel Laureate Norman Borlaug on the use of chemical fertilizers by crops who covered the nutritional needs of the world. But, what one should we expect today? The global tendency is to adjust to the actual nutritional needs of the plants, maximizing yields and improving quality, with special attention paid to the environment and the grower, with respect to the consumer. This Special Issue provides nutrient management strategies and advanced knowledge on fertilizer-use efficiency as one of the primary inputs to match the quality and quantity of crops for contributing to the smooth and healthy characteristics of the food chain.

Author Contributions: Conceptualization, C.N., S.T. and R.Q.; writing—original draft preparation, C.N.; writing—review and editing, C.N., S.T. and R.Q. All authors have read and agreed to the published version of the manuscript.

Acknowledgments: We would like to thank all contributing authors in this Special Issue on "Crop Nutrient Requirements and Advanced Fertilizer Management Strategies" and all anonymous reviewers who dedicated their time and constructive efforts to improve the quality of science during the review process. Special thanks go to Rita Ren the managing Editor for their guidance and continuous support.

Conflicts of Interest: The authors declare no conflict of interest.

References

1. United Nations, Department of Economic and Social Affairs, Population Division. World Population Prospects 2022, Data Sources. UN DESA/POP/2022/DC/NO. 9. 2022. Available online: https://population.un.org/wpp/publications/Files/WPP2022_Data_Sources.pdf (accessed on 11 November 2022).
2. Beltran-Peña, A.; Rosa, L.; D' Odorico, P. Global food self-sufficiency in the 21st century under sustainable intensification of agriculture. *Environ. Res. Lett.* **2020**, *15*, 095004. [CrossRef]
3. Li, C.; Hoffland, E.; Kuyper, T.W.; Yu, Y.; Zhang, C.; Li, H.; Zhang, F.; van der Werf, W. Syndromes of production in intercropping impact yield gains. *Nat. Plants* **2020**, *6*, 1–8. [CrossRef] [PubMed]
4. Arora, N.K. Agricultural sustainability and food security. *Env. Sustain.* **2018**, *1*, 217–219. [CrossRef]
5. Zhang, X.; Davidson, E.A.; Zou, T.; Lassaletta, L.; Quan, Z.; Li, T.; Zhang, W. Quantifying nutrient budgets for sustainable nutrient management. *Glob. Biogeochem. Cycles.* **2020**, *34*, e2018GB006060. [CrossRef]
6. Slaton, N.A.; Lyons, S.E.; Osmond, D.L.; Brouder, S.M.; Culman, S.W.; Drescher, G.; Gatiboni, L.C.; Hoben, J.; Kleinman, P.J.A.; McGrath, J.M.; et al. Minimum dataset and metadata guidelines for soil-test correlation and calibration research. *Soil Sci. Soc. Am. J.* [illegible]
7. Lynch, J.P.; Strock, C.F.; Schneider, H.M.; Sidhu, J.S.; Ajmera, I.; Galindo-Castañeda, T.; Hanlon, M.T. Root anatomy and soil resource capture. *Plant Soil* **2021**, *466*, 21–63. [CrossRef]
8. Bouwman, A.F.; Beusen, A.H.W.; Lassaletta, L.; Van Apeldoorn, D.F.; Van Grinsven, H.J.M.; Zhang, J. Lessons from temporal and spatial patterns in global use of N and P fertilizer on cropland. *Sci. Rep.* **2017**, *7*, 1–11. [CrossRef]
9. IFA, International Fertilizer Association. Public Summary Medium-Term Fertilizer Outlook 2022–2026, IFA Market Intelligence Service. A/22/82. 2022. Available online: https://www.fertilizer.org/public/resources/publication_detail.aspx?SEQN=6198&PUBKEY=C5D3054A-4F40-4FFD-8F1A-24AD36D4087D (accessed on 14 November 2022).
10. IFA. *The Global "4r" Nutrient Stewardship Framework–Developing Fertilizer Best Management Practices for Delivering Economic, Social and Environmental Benefits*; International Fertilizer Industry Association: Paris, France, 2009.
11. IPNI. *4R Plant Nutrition Manual: A Manual for Improving the Management of Plant Nutrition, Metric Version*; Bruulsema, T.W., Fixen, P.E., Sulewski, G.D., Eds.; International Plant Nutrition Institute: Norcross, GA, USA, 2012.
12. Johnston, A.M.; Bruulsema, T.W. 4R Nutrient Stewardship for Improved Nutrient Use Efficiency. *Procedia Eng.* **2014**, *83*, 365–370. [CrossRef]
13. Bruulsema, T. Climate-Smart Fertilizers in 4R Nutrient Stewardship. *Crops Soils Mag.* **2022**, *55*, 32–35. [CrossRef]
14. Yang, Y.; Zha, W.; Tang, K.; Deng, G.; Du, G.; Liu, F. Effect of Nitrogen Supply on Growth and Nitrogen Utilization in Hemp (*Cannabis sativa* L.). *Agronomy* **2021**, *11*, 2310. [CrossRef]
15. Klem, K.; Křen, J.; Šimor, J.; Kováč, D.; Holub, P.; Míša, P.; Svobodová, I.; Lukas, V.; Lukeš, P.; Findurová, H.; et al. Improving Nitrogen Status Estimation in Malting Barley Based on Hyperspectral Reflectance and Artificial Neural Networks. *Agronomy* **2021**, *11*, 2592. [CrossRef]
16. Stepaniuk, M.; Głowacka, A. Yield of Winter Oilseed Rape (*Brassica napus* L. var. napus) in a Short-Term Monoculture and the Macronutrient Accumulation in Relation to the Dose and Method of Sulphur Application. *Agronomy* **2022**, *12*, 68. [CrossRef]
17. Školníková, M.; Škarpa, P.; Ryant, P.; Kozáková, Z.; Antošovský, J. Response of Winter Wheat (*Triticum aestivum* L.) to Fertilizers with Nitrogen-Transformation Inhibitors and Timing of Their Application under Field Conditions. *Agronomy* **2022**, *12*, 223. [CrossRef]
18. Torabian, S.; Farhangi-Abriz, S.; Qin, R.; Noulas, C.; Wang, G. Performance of Nitrogen Fertilization and Nitrification Inhibitors in the Irrigated Wheat Fields. *Agronomy* **2023**, *13*, 366. [CrossRef]
19. Cassimiro, J.B.; de Oliveira, C.L.B.; Boni, A.d.S.; Donato, N.d.L.; Meirelles, G.C.; da Silva, J.F.; Ribeiro, I.V.; Heinrichs, R. Ammonia Volatilization and Marandu Grass Production in Response to Enhanced-Efficiency Nitrogen Fertilizers. *Agronomy* **2023**, *13*, 837. [CrossRef]
20. Li, J.; Wang, W.; Wang, W.; Li, Y. The Ability of Nitrification Inhibitors to Decrease Denitrification Rates in an Arable Soil. *Agronomy* **2022**, *12*, 2749. [CrossRef]
21. Li, J.; Luo, G.; Shaibu, A.S.; Li, B.; Zhang, S.; Sun, J. Optimal Fertilization Level for Yield, Biological and Quality Traits of Soybean under Drip Irrigation System in the Arid Region of Northwest China. *Agronomy* **2022**, *12*, 291. [CrossRef]
22. Duangpan, S.; Tongchu, Y.; Hussain, T.; Eksomtramage, T.; Onthong, J. Beneficial Effects of Silicon Fertilizer on Growth and Physiological Responses in Oil Palm. *Agronomy* **2022**, *12*, 413. [CrossRef]

23. El-Sonbaty, A.E.; Farouk, S.; Al-Yasi, H.M.; Ali, E.F.; Abdel-Kader, A.A.S.; El-Gamal, S.M.A. Enhancement of Rose Scented Geranium Plant Growth, Secondary Metabolites, and Essential Oil Components through Foliar Applications of Iron (Nano, Sulfur and Chelate) in Alkaline Soils. *Agronomy* **2022**, *12*, 2164. [CrossRef]
24. Hussein, H.A.Z.; Awad, A.A.M.; Beheiry, H.R. Improving Nutrients Uptake and Productivity of Stressed Olive Trees with Mono-Ammonium Phosphate and Urea Phosphate Application. *Agronomy* **2022**, *12*, 2390. [CrossRef]
25. Korzeniowska, J.; Stanisławska-Glubiak, E. Differences in the Concentration of Micronutrients in Young Shoots of Numerous Cultivars of Wheat, Maize and Oilseed Rape. *Agronomy* **2022**, *12*, 2639. [CrossRef]
26. Xie, Y.; Li, Y.; Wang, L.; Nizamani, M.M.; Lv, Z.; Dang, Z.; Li, W.; Qi, Y.; Zhao, W.; Zhang, J.; et al. Determination of Critical Phosphorus Dilution Curve Based on Capsule Dry Matter for Flax in Northwest China. *Agronomy* **2022**, *12*, 2819. [CrossRef]
27. Xie, Y.; Li, L.; Wang, L.; Zhang, J.; Dang, Z.; Li, W.; Qi, Y.; Zhao, W.; Dong, K.; Wang, X.; et al. Relationship between Phosphorus and Nitrogen Concentrations of Flax. *Agronomy* **2023**, *13*, 856. [CrossRef]
28. de Oliveira, C.L.B.; Cassimiro, J.B.; Lira, M.V.D.S.; Boni, A.D.S.; Donato, N.D.L.; Reis, R.D.A.; Heinrichs, R. Sugarcane Ratoon Yield and Soil Phosphorus Availability in Response to Enhanced Efficiency Phosphate Fertilizer. *Agronomy* **2022**, *12*, 2817. [CrossRef]
29. Zhu, Y.-A.; He, J.; Yu, Z.; Zhou, D.; Li, H.; Wu, X.; Dong, Y.; Tang, L.; Zheng, Y.; Xiao, J. Wheat and Faba Bean Intercropping Together with Nitrogen Modulation Is a Good Option for Balancing the Trade-Off Relationship between Grain Yield and Quality in the Southwest of China. *Agronomy* **2022**, *12*, 2984. [CrossRef]
30. Amissah, S.; Baidoo, M.; Agyei, B.K.; Ankomah, G.; Black, R.A.; Perry, C.D.; Hollifield, S.; Kusi, N.Y.; Harris, G.H.; Sintim, H.Y. Early and Late Season Nutrient Stress Conditions: Impact on Cotton Productivity and Quality. *Agronomy* **2023**, *13*, 64. [CrossRef]
31. Beheiry, H.R.; Awad, A.A.M.; Hussein, H.A.Z. Response of Multi-Stressed Olea europaea Trees to the Adjustment of Soil pH by Acidifying Agents: Impacts on Nutrient Uptake and Productivity. *Agronomy* **2023**, *13*, 539. [CrossRef]
32. De Souza, M.; Barcelos, J.P.d.Q.; Rosolem, C.A. Synergistic Effects of Subsoil Calcium in Conjunction with Nitrogen on the Root Growth and Yields of Maize and Soybeans in a Tropical Cropping System. *Agronomy* **2023**, *13*, 1547. [CrossRef]

 *agronomy*

Article

Dynamics of Mineral Uptake and Plant Function during Development of Drug-Type Medical Cannabis Plants

Avia Saloner and Nirit Bernstein *

Institute of Soil Water and Environmental Sciences, Volcani Center, Rishon LeZion 7505101, Israel
* Correspondence: nirit@agri.gov.il

Abstract: Recent studies have demonstrated dose-responses of the cannabis plant to supply of macronutrients. However, further development of precision nutrition requires a high-resolution understanding of temporal trends of plant requirements for nutrients throughout the developmental progression, which is currently not available. As plant function changes during development, temporal information on nutrient uptake should be considered in relation to gradients in developmental-related physiological activity. Therefore, the present study investigated tempo-developmental trends of nutritional demands in cannabis plants, and in relation to physiological performance. Three cultivars differing in phenotype and chemotype were analyzed to evaluate genotypic variability. The results demonstrate that nutrient acquisition and deposition rates change dramatically during plant development. Uptake of individual minerals generally increased with the progression of both vegetative and reproductive development and the increase in plant biomass, while the deposition rates into the plant demonstrated nutrient specificity. The average concentrations of N, P, and K in the shoots of the different cultivars were 2.33, 4.90, and 3.32 times higher, respectively, at the termination of the reproductive growth phase, compared to the termination of the vegetative growth phase. Surprisingly, the uptake of Ca was very limited during the second part of the reproductive growth phase for two cultivars, revealing a decrease in Ca demand at this late developmental stage. Root-to-shoot translocation of most nutrients, including P, K, Mg, Mn, and Zn, as well as Na, is higher during the reproductive than the vegetative growth phase, and Fe, Mn, Zn, Cu, and Na displayed very little root-to-shoot translocation. The physiological characteristics of the plants, including gas exchange parameters, membrane leakage, osmotic potential, and water use efficiency, changed over time between the vegetative and the reproductive phases and with plant maturation, demonstrating a plant-age effect. The revealed tempo-developmental changes in nutritional requirements of the cannabis plant are a powerful tool required for development of a nutritional protocol for an optimal ionome.

Keywords: cannabis; development; fertilizer; growth; nutrition; uptake; deposition; translocation

Citation: Saloner, A.; Bernstein, N. Dynamics of Mineral Uptake and Plant Function during Development of Drug-Type Medical Cannabis Plants. *Agronomy* **2023**, *13*, 2865. https://doi.org/10.3390/agronomy13122865

Academic Editors: Christos Noulas, Shahram Torabian and Ruijun Qin

Received: 7 November 2023
Revised: 14 November 2023
Accepted: 15 November 2023
Published: 21 November 2023

1. Introduction

Considerable progress has recently been made in our understanding of the 'drug-type' (medical) cannabis plant ionome. Ample information is already available concerning the plant responses to fertilization [1–6], and numerous studies reported effects of additional cultivation conditions including exposure to light [7–11], salinity [12], root zone systems [13], planting density [14], and plant architecture manipulations [15,16].

Cannabis sativa is a dioecious, annual, short-photoperiod plant [17,18]. In short-day plants, the photoperiod controls steps of the flowering mechanism such as flower induction or inflorescence elongation. These plants thereby develop vegetatively under long photoperiod and require nights longer than a critical threshold for reproductive development [19]. The physiological performance and hence agronomic requirements often vary between the vegetative and the reproductive phases of development, and with plant maturation [20–22]. In *C. sativa*, a short photoperiod is required for inflorescence development, and therefore, the cultivation cycle of cannabis as well is composed of two distinct developmental phases: A vegetative growth-phase under long-photoperiod establishes the vegetative infrastructure of the plants, which is followed by a reproductive growth phase under short-photoperiod for development of the inflorescence yield.

The environmental conditions suitable for optimal plant development and function are known to be phase-specific for many crops, and optimal cultivation requires phase-specific adjustments of agronomic inputs such as light intensity and quality [23–25] and mineral supply [1,26]. Furthermore, as the metabolic, physiological, and developmental activity during the flowering (reproductive) phase in cannabis is not uniform over time, inputs may need to vary within this phase to facilitate optimal plant function. Medical cannabis flower maturation under short photoperiod is composed of two main sub-phases: 1. At the first 1–3 weeks, vegetative growth is accompanied by initiation of inflorescences, with relatively low secondary metabolism. 2. At the following 4–6 weeks, the vegetative growth ceases and is replaced by intensive inflorescence production and secondary metabolism [3,27–29]. These temporal changes in plant growth, development, and metabolic activity throughout the reproductive phase are no doubt accompanied by changes to physiological function, and dictate corresponding variations in requirements for exogenous inputs such as mineral nutrients. Indeed, gas-exchange parameters were shown to vary during the reproductive growth phase in cannabis [3,10], and the photosynthetic ability of the leaves of industrial hemp were shown to change with leaf age [30].

Plants require mineral elements as nutrients for their growth and reproduction throughout their life cycle. It is reasonable that the demand and, therefore, the uptake of nutrients by the plant will increase in times of rapid growth or flower/fruit development and decrease during growth deceleration, leaf senescence, and dormancy. It is well established that nutrient uptake and the concentration of nutrients in the plant change with plant age and development [31–33]. Moreover, the accumulation of nutrients in the plant differs between plant organs and changes over time to meet the plant's demands [34–37]. The nutrients taken up by the root are translocated to the shoot, under a rate of translocation that changes over time according to the plant's physiological state, and demonstrate mineral specificity that is affected by environmental factors such as root temperature and leaf transpiration [38–41]. As medical cannabis plants tend to have a rapid vegetative growth under long photoperiod [26,42,43], which proceeds to intensive inflorescence production under a short photoperiod [44,45], it is likely that the plant's nutritional demands will alter during plant development.

An increasing body of information is available on the impact of nutritional regimes on medical cannabis plants. Optimal supply concentrations were determined for N at the vegetative [42] and the reproductive [5] phases, and for NH_4/NO_3 ratios [4]; for K at the vegetative [26] and the reproductive [1] phases; for P at the vegetative [43] and the reproductive phases [3]; for Mg at the vegetative phase [6]; and for combinations of macronutrients [2,46]. A field experiment with fiber hemp showed that the accumulation of most nutrients was higher in the leaves > stem and that their uptake and partitioning to plant organs were affected by cultivar characteristics and plant yield [47]. Despite the substantial progress in understanding the nutritional requirements of 'drug-type' cannabis, the available studies were performed for one growth phase (vegetative or reproductive); and plant response was demonstrated for only one or two time points. No study has elaborated on, and compared responses of the cannabis crop plant throughout the growth cycle, or investigated the nutritional requirement of the plant over time.

The present study was therefore set forth to examine how the nutritional demands of 'drug-type' cannabis change during plant development throughout the vegetative and the reproductive phases, and in relation to the physiological performance of the plants. The hypothesis guiding the work plan was that nutritional demands and physiological functions change during plant development, between the two growth phases, and during the flowering phase. To this end, we have analyzed changes in nutrient uptake, nutrient deposition rate, and root-to-shoot nutrient translocation throughout plant development, in parallel to changes in plant morphology (root:shoot ratio), biomass accumulation, and plant function traits. Responses of three genotypes of 'drug-type' medical cannabis were analyzed to compare genotypic variability. The information gained on temporal rates of mineral uptake into the cannabis plant, in conjunction with the understanding of changes in plant function throughout the cultivation cycle can guide the development of precision mineral nutrition regimes. Development of an optimal fertigation practice will improve the ability of cannabis growers to stabilize plant cultivation, optimize the ionome, minimize agricultural inputs, and improve yield quality and quantity.

2. Materials and Methods

2.1. Plant Material and Growing Conditions

Three genotypes of medical cannabis (*Cannabis sativa* L.) were used as a model system in this study: 'Royal medic' (RM) and 'Desert queen' (DQ) (Teva Adir Ltd., Israel), and Annapurna (ANP) (Canndoc Ltd., Herzliya, Israel). The plants were propagated from cuttings that were rooted in coconut fiber plugs (Jiffy International AS, Kristiansand, Norway). After rooting, the plants were transferred to 3 L plastic pots with a perlite 2-1-2 cultivation media (Agrekal, Habonim, Israel). In order to examine responses at the vegetative growth phase, the plants were grown for one month under 18/6 h light/dark photoperiod in a controlled environment growing room. Light was supplied by Metal Halide bulbs (380 $\mu mol \cdot m^{-2} \cdot s^{-1}$; Solis Tek Inc., Carson, CA, USA). In order to examine the reproductive growth phase, a parallel set of plants was propagated and grown as is described above, and after one week of vegetative growth under long photoperiod (two weeks for ANP), they were transferred to a short photoperiod for the induction of inflorescence development. For the remainder of the experiment and until flower maturation, the plants were grown under 12:12 light/dark photoperiod using High-Pressure Sodium bulbs (860 $\mu mol \cdot m^{-2} \cdot s^{-1}$, Greenlab by Hydrogarden, Petah Tikva, Israel). Light intensity, light quality, and the photoperiod at the various phases of plant development were designed to follow conventional practices for cannabis cultivation [11,48]. Flower maturation occurred 51, 57, and 74 days after the transfer to the short photoperiod for the DQ, ANP, and RM genotypes, respectively. Flower maturation, and the time of the final harvest, was determined following the conventional agronomic practice for these varieties, as the stage at which ~50% of the trichomes of the inflorescences were of amber color. Temperatures in the cultivation rooms were 27/25 °C during the day/night, respectively; relative humidity was 58/48%, respectively; and CO_2 was at ambient levels. Irrigation was supplied via 1 L h^{-1} discharge-regulated drippers (Netafim, Tel-Aviv, Israel), one dripper per pot, to allow 30% drainage. Mineral nutrients were supplied dissolved in the irrigation solution, from final (pre-mixed) solutions, which were monitored throughout the experiment duration. During the vegetative growth phase, nitrogen (N) phosphorus (P), and potassium (K) concentrations were 16.3, 2.0, and 2.3 mM for RM and DQ, and 11.1, 1.9, and 4.6 mM for ANP, respectively. During the reproductive growth phase, N, P, and K concentrations were 16, 1.8, and 2.3 mM for RM and DQ, and 11.4, 2.0, and 2.7 mM for ANP, respectively. The complete composition of the irrigation solutions, including their pH and electric conductivity (EC), is detailed in Table S1. Zinc, Cu, and Mn were supplied chelated with EDTA, and Fe as chelated with EDDHSA. Mo and B were added as a part of the fertilizers Bar-Koret and B-7000, respectively (Israel chemicals, Tel-Aviv, Israel). During the last week before harvest, the plants were irrigated with distilled water without fertilizers as is routinely practiced in the commercial cultivation of medical cannabis. The experiment was arranged in a complete randomized design; all measurements were conducted with five replicates per genotype following the experimental design; and results are presented as averages $\pm$ standard errors (S.E.).

2.2. Plant Biomass and Inorganic Mineral Analysis

Biomass of the plant organs, i.e., inflorescences, inflorescence leaves, fan leaves, stem, and root biomass, was evaluated several times throughout plant development by destructive sampling of the plants. During the vegetative growth phase, the cultivars RM and DQ were sampled five times: 0, 7, 14, 21, and 29 days after the beginning of the vegetative growth phase; and ANP was sampled three times: 0, 14, and 31 days after the beginning of vegetative growth. During the reproductive growth phase, all three cultivars were sampled destructively three times (days are counted from the transfer to the short-photoperiod): 0, 30, and 74 days for RM; 0, 30, and 51 days for DQ; and 0, 29, and 57 days for ANP. Shoot biomass was calculated as the integration of the biomass of the above-ground organs, i.e., the inflorescences, inflorescence leaves, fan leaves, and stem biomass. Root biomass was evaluated for all samplings of the reproductive growth phase and for the last destructive sampling of the vegetative growth phase. Root:shoot ratio was calculated by dividing the root dry biomass by the shoot dry biomass. Dry weights were determined after drying for 48 h at 64 °C (128 h for the inflorescences). The results are averages $\pm$ SE for five replicated plants per cultivar. The plant material that was destructively sampled for biomass determination was used for the analyses of inorganic mineral contents in the plant organs, as is described by Saloner et al. [26]. In short, the plant samples were analyzed for concentrations of N, P, K, Ca, Mg, Fe, Mn, Zn, Cu, and Na. Two different procedures were applied for extraction of the various inorganic mineral elements from ground plant tissue. For the analysis of Ca, Mg, Fe, Zn, Cu, and Mn, the ground tissue was digested with HNO3 (65%) and HClO4 (70%), and the elements were analyzed with an atomic absorption spectrophotometer, AAnalyst 400 AA Spectrometer (PerkinElmer, Waltham, MA, USA). For the analysis of N, P, K, and Na, the dry tissue was digested with H_2SO_4 (98%) and H_2O_2 (70–72%). Na and K were analyzed by flame photometer (410 Flame Photometer Range, Sherwood Scientific Limited, The Paddocks, UK), and N and P were analyzed by an autoanalyzer (Lachat Instruments, Milwaukee, WI, USA). Mineral analysis of the irrigation solution was performed as described for the plant extraction and digestion solutions [26,49].

2.3. Physiological Parameters

The physiological activity of the plants was analyzed three times, once at the vegetative growth phase and twice during the reproductive growth phase. The timing of the analyses was chosen to represent three developmental stages: 1. Vegetative growth; 2. Early flowering stage; 3. Late flowering stage (near harvest). Royal Medic plants were analyzed 24 days after the beginning of the vegetative growth phase, and 67 days after the beginning of the reproductive growth phase. Desert Queen plants were analyzed 24 days after the beginning of the vegetative growth phase, and 24 and 45 days after the beginning of the reproductive growth phase. Annapurna plants were analyzed 32 days after the beginning of the vegetative growth phase, and 24 and 45 days after the beginning of the reproductive growth phase.

All measurements were conducted with five replicates each from a different plant, following the experimental design. In both growth phases, the youngest fully developed fan leaf on the main stem, located at the fourth node from the plant's top, was analyzed. Photosynthetic pigments and membrane leakage analyses were conducted following Saloner et al. [26] and pigment concentrations were calculated following Lichtenthaler and Wellburn [50]. The two most peripheral leaflets were used for the analysis of osmotic potential, as was previously described [26], and relative water content (RWC) was analyzed and calculated following Bernstein et al. [51]. Leaf gas exchange, i.e., photosynthesis, transpiration, stomatal conductance, and intercellular CO_2 concentration, was measured using a Licor 6400 XT system (LI-COR, Lincoln, NE, USA). Leaf water use efficiency (WUEi) was calculated using the photosynthesis and stomatal conductance results, as the net photosynthetic rate divided by the stomatal conductance [42].

2.4. Calculation of Nutrient Uptake, Deposition, and Translocation

For characterization of the uptake, deposition, and translocation of the mineral nutrients in the plant, three calculations were conducted. (i) **Uptake Curves**: An uptake curve of a mineral presents the changes over time of the total amount of the mineral in the shoot throughout plant development. It was calculated by multiplying the concentration of the mineral in each shoot's organs (fan leaves, stem, inflorescence leaves, and inflorescence) by the organ's dry weight, and summing the amounts in the various organs to receive the shoot mineral uptake. (ii) **Deposition rate Curves**: A deposition curve of a mineral presents the daily rate of deposition of a mineral into the shoot. The deposition rates were calculated as the differentials of the shoot concentrations over time periods throughout plant development, as per Equation (1). (iii). **Translocation Factor (TF)**: The translocation factor is the ratio between a mineral concentration in the shoot and the root, which reflects on root-to-shoot translocation. Translocation factor (TF) >1 means that the concentration of the specific nutrient is greater in the shoot than in the root, marking a higher accumulation of the nutrient in the shoot, and vice versa [43]. It was analyzed twice, at the termination of the vegetative growth phase, and at the termination of the reproductive growth phase, using Equation (2), following Shiponi and Bernstein [43].

$$
Deposition\ rate\ \left[\frac{g}{day}\right] = \frac{\left(\dfrac{Shoot\ mineral\ content\ on\ day\ (X)}{Shoot\ mineral\ content\ on\ day\ (X+n)}\right)}{No.\ of\ days\ between\ day\ (X)\ and\ day\ (X+n)} \tag{1}
$$

$$
Translocation\ factor\ (TF)\ of\ a\ mineral = \frac{Concentration\ of\ the\ mineral\ in\ the\ shoot}{Concentration\ of\ the\ mineral\ in\ the\ root} \tag{2}
$$

2.6. Statistical Analyses

The data were subjected to two-way analysis of variance (ANOVA) followed by Tukey's HSD test. The analysis was performed with the Jump software, version 16 (SAS 2016, Cary, NC, USA). The two factors analyzed in the two-way ANOVA are sampling time (T) and genotype (G).

3. Results and Discussion

The present study examined the dynamics of mineral nutrients uptake and plant function of cannabis plants throughout their cultivation cycle, and achieved a temporal resolution and better understanding of the plant's nutritional demands, and in relation to its physiological activity. The results facilitated the development of temporal trends of mineral uptake and deposition rates for the improvement of precision fertilization, minimizing agricultural inputs, reduced environmental pollution, and obtaining higher yields.

3.1. Plant Development: Biomass Accumulation and Visual Appearance

Modern cannabis cultivars have a diverse genetic background, and consequently demonstrate considerable morphological and chemical diversity that may impact the plant's growth and yield potential [52,53] as well as the requirements for mineral nutrients. A morphological and physiological variability was also reflected in the three cultivars investigated in the present study, which demonstrated a considerable variability in key parameters such as plant height and the exposure time to short photoperiod required for maturation (Figure 1). RM is the tallest variety and has a longer maturation period than DQ, which is shorter (Figure 1). ANP plants are of intermediate height and maturation period (the ANP plants appear high in Figure 2 as they began the reproductive phase taller than the other two cultivars). Nevertheless, the biomass accumulation patterns of the three cultivars are very similar, as the shoot biomass increased steadily over time during both the vegetative and the reproductive phases (Figure 2A), in accord with the steady increase in stem and root biomass (Figure 3B,C), and the increase in leaves biomass during the vegetative growth phase (Figure 3A). In all cultivars, leaf biomass accumulation became more moderate and even decreased with the progression of the reproductive growth phase, likely due to allocation of resources toward development of inflorescences (Figure 3A). This is supported by the result that the decrease in leaves biomass in the second half of the reproductive growth phase paralleled a substantial increase in inflorescences biomass in all cultivars (Figure 3D). These opposing trends suggest that inflorescences development is the cause of the decrease in leaves biomass production and of leaf senescence, as is also common for other plant species under reproductive development [54,55]. In addition, the root:shoot ratio substantially decreased with the transition from vegetative to reproductive growth and further decreased as the inflorescences evolved (Figure 2B), demonstrating that growth of inflorescences and stems was not in line with root growth, a phenomenon well known to occur in plants [56–58]. As mineral nutrients are required for the production of new tissue, i.e., to support new growth, ample mineral supply to the developing inflorescences is required to facilitate the intensive reproductive development. Such supply can be achieved via root uptake during the reproductive stage, or by *in planta* remobilization from storage pools in other plant organs, as will be further discussed.

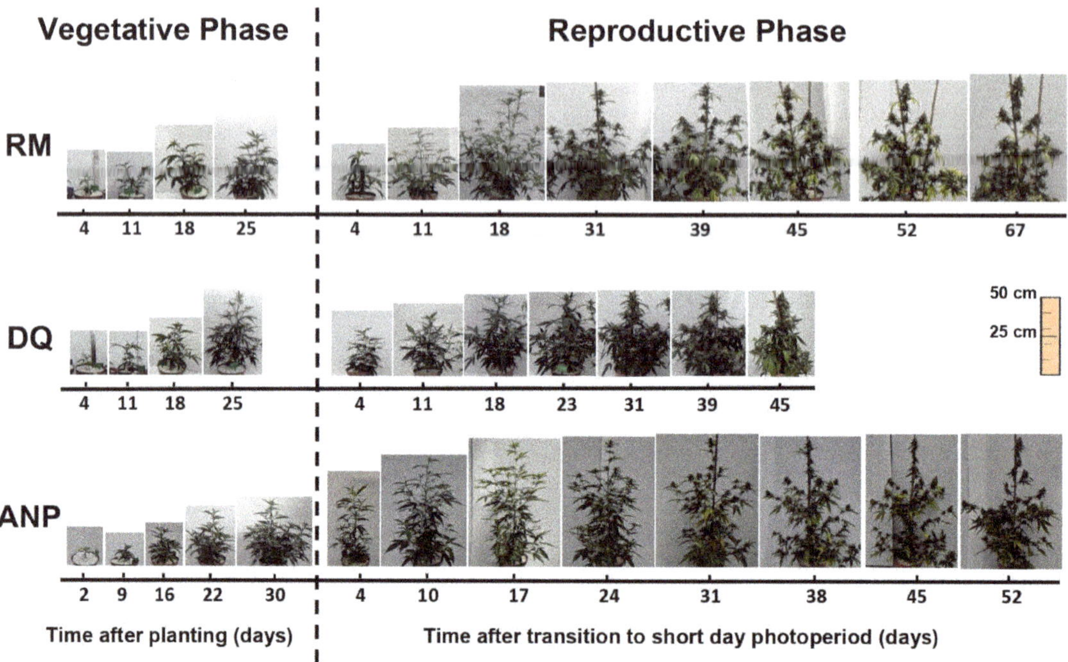

Figure 1. Visual appearance of three medical cannabis cultivars Royal Medic (RM—**top** row), Desert Queen (DQ—**middle** row), and Annapurna (ANP—**bottom** row) during vegetative and reproductive development.

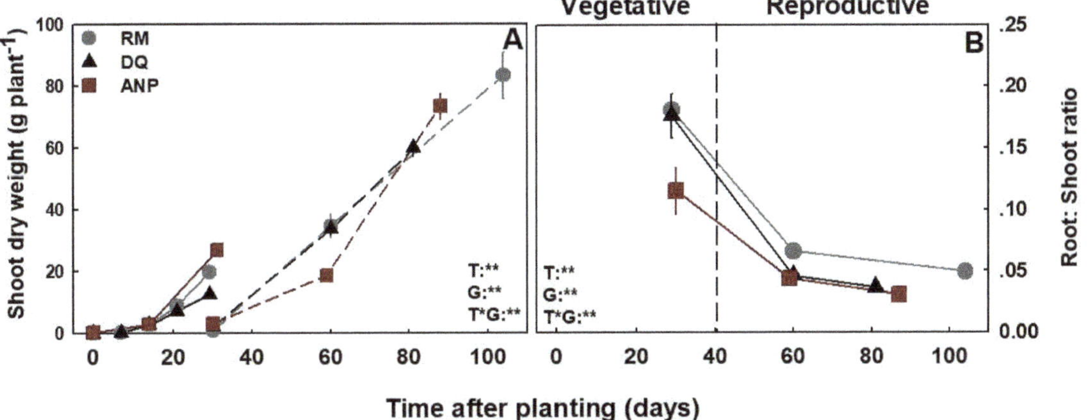

Figure 2. Biomass accumulation in three medical cannabis cultivars (RM, DQ, ANP) during vegetative and reproductive development: Plant dry weight (**A**) and root:shoot ratio (**B**). In (**A**), solid lines represent vegetative growth (long photoperiod); scattered lines represent reproductive growth (short photoperiod). In (**B**), the dashed line marks the transition to the reproductive growth phases. Presented data are averages $\pm$ SE (n = 5). Results of two-way ANOVA indicated as ** $p < 0.05$, F-test; NS, not significant $p > 0.05$, F-test. In the ANOVA results, T*G represents the interaction between time and genotype. p values are presented at the Supplementary Materials, File S1.

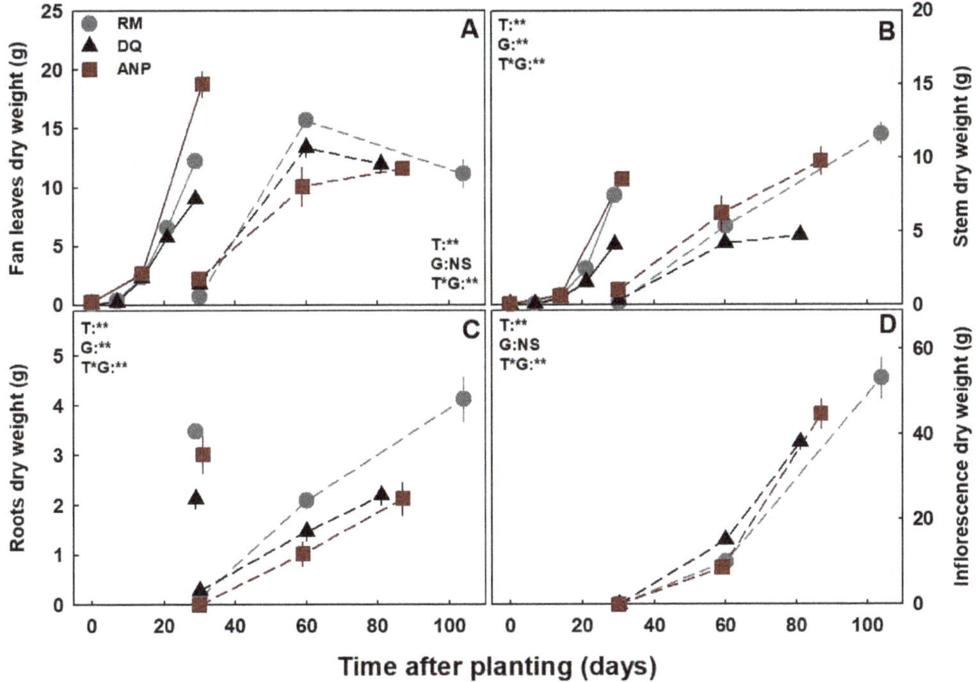

Figure 3. Biomass accumulation in organs of three medical cannabis cultivars (RM, DQ, ANP) during vegetative and reproductive development: Dry weights of fan leaves (**A**), stem (**B**), roots (**C**), and inflorescences (**D**). Solid lines represent vegetative growth (long photoperiod); dashed lines represent

reproductive growth (short photoperiod). Presented data are averages ± SE (n = 5). Root biomass was measured once at the vegetative phase. Results of two-way ANOVA indicated as **, $p < 0.05$, F-test; NS, not significant $p > 0.05$. In the ANOVA results, T*G represents the interaction between time and genotype. p values are presented at the Supplementary Materials, File S1.

3.2. Gas Exchange, Water Relations, and Photosynthetic Pigments

The three cultivars tested differ in physiological function and the way it changes over time during plant development (Figure 4): (i) RM generally performed best under vegetative growth, and its physiological function declined under reproductive growth and with the progression of the reproductive stage; (ii) ANP as well performed best at the vegetative growth phase, but its function declined in the middle of the reproductive growth phase and increased again before harvest; and (iii) DQ's performance was generally lowest under vegetative growth, increased at the reproductive growth phase, and was highest before harvest. These data indicate that trends of physiological function vary between cannabis cultivars and throughout plant development, as was also reported for oil palm [59], corn [60], and olives [61]. The differences in physiological activity between cultivars may result from variations in the duration of exposure to short-photoperiod required for maturation. There were no significant differences between cultivars in activity levels at the vegetative growth phase since all plants (and inspected leaves) were of the same age (Figures 4 and 5). At the reproductive growth phase, although DQ demonstrated higher gas exchange activity than ANP, as was reflected by higher rates of photosynthesis, transpiration, and stomatal conductance, they both demonstrated an increase in these parameters with the progression of the reproductive stage (Figure 4A–C). As RM has a longer reproductive development phase than ANP and DQ, it was analyzed later than the other cultivars and was, therefore, older during the last measurement. Hence, it is not surprising that an older plant (and leaf) demonstrated lower physiological function, as is already well documented for a range of plant species [22,62–65].

However, some of the physiological parameters demonstrated a uniform trend across cultivars: Intercellular CO_2 concentration was lowest in the middle of the reproductive growth phase (Figure 4D); relative water content (RWC) and water-use efficiency (WUEi) were highest in the middle of the reproductive phase (Figure 4E); membrane leakage was lowest at the vegetative growth phase and increased over time (Figure 4H); and photosynthetic pigment contents were highest at the vegetative growth phase and decreased over time (Figure 5). The temporal changes in the physiological function of the plants over time may reflect developmental trends characteristic of annual plants [66]. Toward the end of the reproductive development, it is inevitable that source-sink relationships and resource partitioning in the plant will change, and the translocation of nutrients and carbohydrates from vegetative to reproductive organs will increase, to support inflorescence and seed development. This mechanism, which has already been demonstrated in other plant species [67–69], correlates with the substantial increase in inflorescence production (Figure 3D) and the cessation of leaf production (Figure 3A) in the second half of the reproductive phase. Taken together, these data suggest that the leaves' physiological function decreases by the end of the reproductive growth phase on account of divergence of resources to the reproductive inflorescences, as can be seen for the RM cultivar (Figures 3 and 4). As DQ and ANP were harvested after a shorter duration of reproductive development than DQ, they had less time to translocate resources from leaves to inflorescences. Thus, they still demonstrate high performance before harvest (Figure 4). We suggest that had they been grown for a longer duration, their gas exchange activity would have likely decreased similarly to RM (Figure 4), and their mineral uptake would have changed accordingly.

The second reason for the relatively high physiological performance of the plants at the vegetative growth phase compared to the reproductive growth phase is the difference in light spectrum and light intensity. During the vegetative growth phase, the plants received lighting from Metal Halide bulbs, which supply a relatively enriched blue light spectrum with an intensity of 380 $\mu mol \cdot m^{-2} \cdot s^{-1}$. During the reproductive phase, the plants received lighting from High-Pressure Sodium bulbs, which supply a spectrum enriched in the red range with an intensity of 860 $\mu mol \cdot m^{-2} \cdot s^{-1}$. As the light intensity and the red:blue ratio increased with the transition to reproductive growth, the photosynthetic and gas exchange mechanisms could have reduced. This is supported by our finding that concentrations of photosynthetic pigments were generally highest under the vegetative growth conditions (Figure 5), potentially to increase light capture per unit leaf area since light intensity was lower at that phase. It is important to note that the relative decrease in plant function under reproductive growth conditions might have resulted from the high light intensity, which could have caused photo-inhibition damages and induced oxidative stress that damaged cell homeostasis [70,71]. Indeed, our results show that membrane stability was lower and electrolyte leakage from the cell membrane was higher under the reproductive growth conditions (Figure 4H), as was already shown to occur in other plant species [72,73].

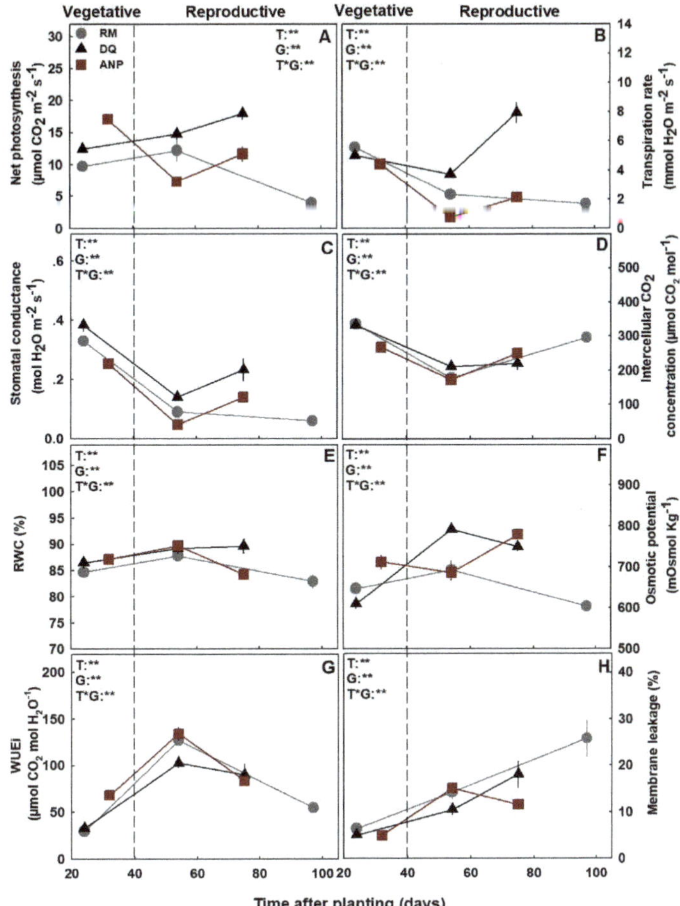

Figure 4. Gas exchange activity and physiological characteristics of three medical cannabis cultivars (RM, DQ, ANP) during vegetative and reproductive development: Photosynthesis (**A**), transpiration (**B**), stomatal conductance (**C**), intercellular CO_2 concentration (**D**), relative water content (RWC) (**E**), osmotic potential (**F**), intrinsic water-use efficiency (WUEi) (**G**), and membrane leakage (**H**). Dashed line marks the transition to the reproductive growth phases. Presented data are averages $\pm$ SE (n = 5). Results of two-way ANOVA indicated as ** $p < 0.05$, *F*-test; NS, not significant $p > 0.05$, *F*-test. T*G represents the interaction between time and genotype. *p* values are presented at the Supplementary Materials, File S1.

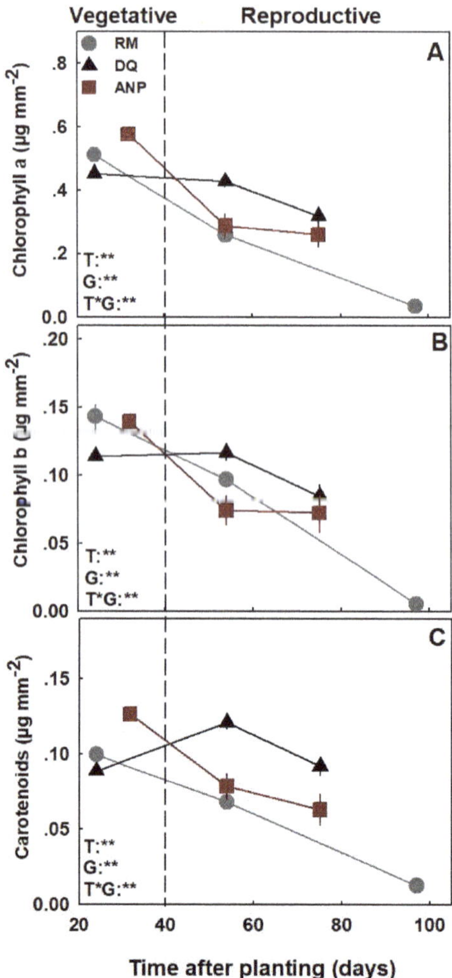

Figure 5. Changes in concentrations of photosynthetic pigments throughout the development of cannabis plants, in three cultivars of medical cannabis (RM, DQ, ANP), during vegetative and reproductive development: Chlorophyll a (**A**), chlorophyll b (**B**), and carotenoids (**C**). The dashed line marks the transition to the reproductive growth phases. Presented data are averages $\pm$ SE (n = 5). Results of two-way ANOVA indicated as ** $p < 0.05$, *F*-test; NS, not significant $p > 0.05$, *F*-test. T*G represents the interaction between time and genotype. The p value for G (genotype) was 0.0123 in subfigure B, and <0.001 for all other variables in all subfigures.

3.3. Nutrient Uptake and Deposition

Nutrient uptake and distribution in the plant body are selective and essential metabolic processes performed by all plants throughout their development [32,74,75]. As the need for mineral uptake into plants lies in their necessity for vital growth and metabolic activity of plant cells, it is not surprising that mineral uptake is a selective process and is affected by the plant's need for nutrients [32,33,76]. In this study, two parameters related to nutrient accumulation are reported: **nutrient uptake**, which presents the total amount of individual minerals present in the plant at a specific time (Figure 6); and **nutrient deposition rate**, which presents the rate of accumulation of individual minerals into the plant over a defined period of time, e.g., mg per day (Figure 7).

As arises from Figure 6, the amount of each of the examined nutrients, excluding Ca, increased gradually over time in the plant during both growth phases. As the data match the trend obtained for shoot biomass production (Figure 2A), it is concluded that for most minerals, the plant's biomass production is the governing factor for their uptake into the plant. Also, the uptake of most nutrients into the plant, including N, P, Fe, Mn, and Zn, was relatively similar between cultivars, as is apparent from the comparable amounts in the plant (Figure 6). A specific difference was obtained for Ca uptake at the second stage of the reproductive growth phase (Figure 6E): for RM and DQ, Ca uptake was very limited, as is also apparent from the zero Ca deposition at that stage, while for ANP, the deposition rate of Ca only mildly decreased at that stage, as Ca uptake continued (Figures 6E and 7E). Another significant difference is the contradicting trends of Na and Cu accumulation during the vegetative growth phase, as RM and DQ accumulated more Cu and less Na, and ANP presented an opposing trend (Figure 6I,J). In addition, ANP accumulated more K than the other cultivars during the vegetative growth phase (Figure 6C). The differences between cultivars can be explained by the differences in the fertilization regime that RM and DQ received compared to ANP (Table S1); RM and DQ were supplied and thus accumulated more Ca and Cu and less K and Na than ANP (Figure 6C,E,I,J). This is supported by recent results for medical cannabis by our group and others that demonstrated that when the supply of a specific nutrient is elevated, its accumulation tends to increase, and vice versa [3,6,26,12,13,77]. Furthermore, as the uptake curves of RM and DQ are highly similar for most nutrients (including N, P, K, Mg, Cu, and Na; Figure 6A–C,F,I,J), although their physiological function and time to maturation differ, we conclude that the fertilization regime governs the plant nutrient uptake, in addition to biomass accumulation.

Since mineral uptake (the total amount of minerals in the plant) is derived from the plant size and biomass, it should be addressed that bigger plants will show higher mineral accumulation, and thus growth practices and plant architectural manipulations may affect mineral uptake. Therefore, in order to understand the in-plant changes in mineral uptake over time, nutrient deposition rates need to be examined. The rates of mineral deposition into the cannabis plants were indeed highly affected by plant age, and were nutrient-specific (Figure 7). At the vegetative growth phase, the deposition rate of most minerals increased gradually over time (Figure 7), reflecting the increase in plant biomass (Figure 2A). The cultivars RM and DQ demonstrated unexpected trends of deposition rates for K, Mg, and Zn, as the rates decreased at the end of the vegetative growth phase (Figure 7C,F,H). At the reproductive growth phase, the deposition rates for N, P, K, Zn, and Na generally increased with time, whereas the deposition rates of Ca and Mg decreased, and Fe and Cu deposition was steady (Figure 7).

The differences found in the rates of mineral deposition between time points likely reflect different plant demands during the growing season. The differences between minerals, point to selective absorption following plant demand. This observation correlates with the known ability of plants to regulate specific ion uptake following plant requirements [32,78,79]. However, it should not be overlooked that plants may overconsume some minerals, and cannabis has already been shown to take up more K than is required for optimal function, without affecting plant performance [26]. Despite the differences in physiological performance and maturation periods, the deposition rates of RM and DQ were similar for most nutrients (Figure 7). ANP, which received a different fertilization regime, demonstrated different uptake curves than RM and DQ (Figure 6), and consequently, its deposition rates for the minerals differed from RM and DQ (Figure 7). These results suggest again that the differences between cultivars are not an outcome of genetic differences, but arise from the variability of environmental factors (fertilization regimes) to which the plants were subjected. Resolving this issue will require evaluation of trends of changes in nutrient deposition rates into genetically different cultivars, under a range of fertilization conditions.

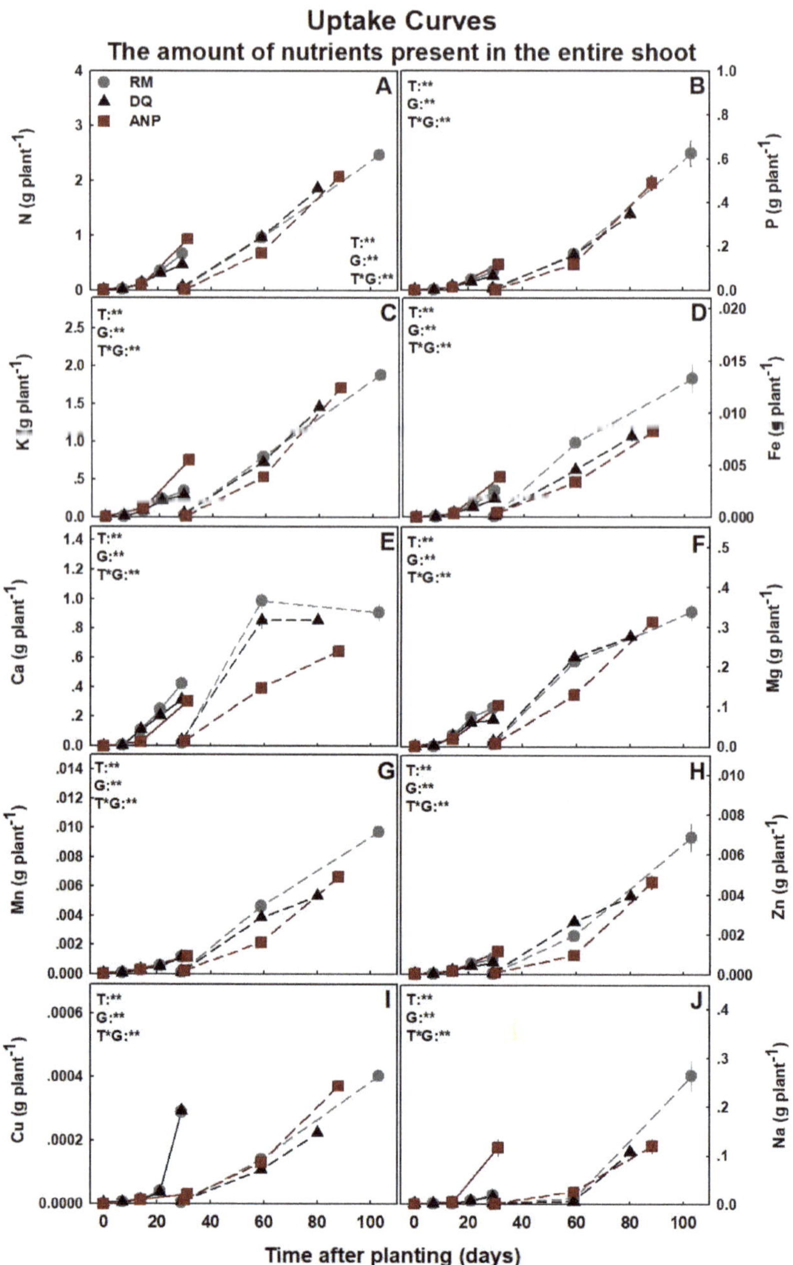

Figure 6. Uptake curves. The amount of nutrients present in the shoot of cannabis plants during vegetative and reproductive development, for three medical cannabis cultivars (RM, DQ, ANP): N (**A**), P (**B**), K (**C**), Fe (**D**), Ca (**E**), Mg (**F**), Mn (**G**), Zn (**H**), Cu (**I**), and Na (**J**). Solid lines—vegetative growth (long photoperiod); scattered lines—reproductive growth (short photoperiod). Presented data are averages ± SE (n = 5). Results of two-way ANOVA indicated as ** $p < 0.05$, F-test; NS, not significant $p > 0.05$, F-test. T*G represents the interaction between time and genotype. p values are presented at the Supplementary Materials, File S1.

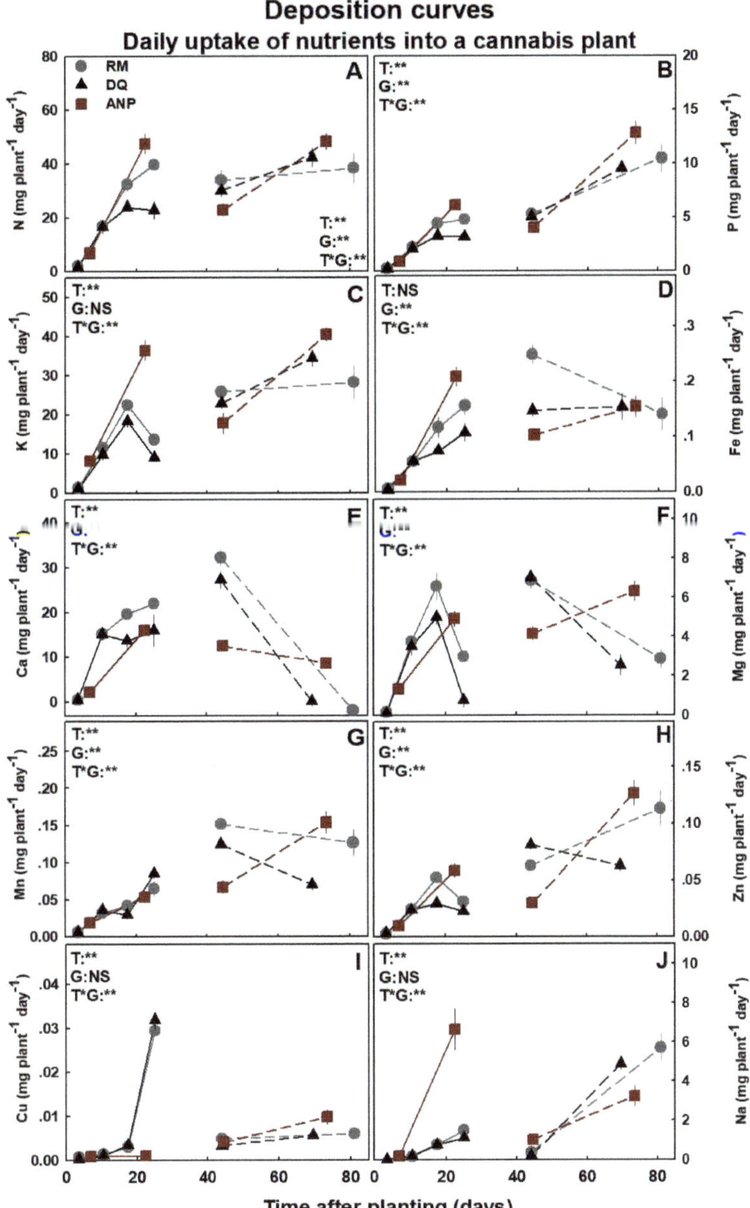

Figure 7. Deposition Rate Curves. Deposition rates of minerals into cannabis plants, for three medical cannabis cultivars (RM, DQ, ANP). Presented data are the daily amounts of minerals taken up by a plant, throughout the vegetative and reproductive development: N (**A**), P (**B**), K (**C**), Fe (**D**), Ca (**E**), Mg (**F**), Mn (**G**), Zn (**H**), Cu (**I**), and Na (**J**). Solid lines represent vegetative growth, and scattered lines—reproductive growth. Presented data are averages $\pm$ SE (n = 5). Results of two-way ANOVA indicated as ** $p < 0.05$, F-test; NS, not significant $p > 0.05$. In the ANOVA results, T*G represents the interaction between time and genotype. p values are presented at the Supplementary Materials, File S1.

3.4. Nutrient Translocation and Root:Shoot Ratio

For nutrients to arrive at the leaves and inflorescences, they must first be absorbed into the root and translocated to the shoot. Deficient supply of nutrients to shoot organs can thereby result from limited root uptake as well as restricted root-to-shoot translocation or remobilization of nutrients in the shoot [80–82]. Therefore, this study examined the translocation of individual nutrients from root to the shoot in cannabis plants using a calculated translocation factor (Figure 8). A clear trend arising from the analysis is that the translocation of most nutrients, including P, K, Mg, Mn, Zn, and Na, was higher during reproductive than vegetative growth (Figure 8B,C,F–H,J). Specifically, the translocation of P and Mg was about three times higher under reproductive than vegetative growth in all cultivars, demonstrating a substantial increase in translocation to the shoot (Figure 8B,F). Iron, Ca, and Cu translocation was generally higher under vegetative growth, compared to reproductive growth (Figure 8D,E,I), while N translocation did not substantially change between growth phases (Figure 8A).

Another important trend is that root–shoot translocation of all micronutrients was small, as their TF were all <1 (Figure 8). Specifically, Fe, Mn, Zn, Cu, and Na accumulated to higher concentrations in the root in all cultivars (Figure 8D,G–J). Moreover, the TF of Fe and Na was ~0.2, reflecting that the concentration in the root was five times higher than in the shoot (Figure 8D,J). On the contrary, the TF of most macronutrients, including N, K, Ca, and Mg, was $\geq$1, reflecting that they generally tended to accumulate in the shoot and not in the root (Figure 8A,C,F–G). These findings concerning the high accumulation of micronutrients in the roots, and higher translocation of macronutrients to the shoot are not surprising, as similar trends were identified in numerous plant species [34,36,37,83,84]. Furthermore, these results align with results published by our group in previous studies, showing that cannabis plants tend to accumulate micronutrients such as Fe, Cu, and Zn in the root > shoot, and to accumulate macronutrients such as N, K, Ca, and Mg in the shoot > root [3,5,6,26,43].

The changes in mineral translocation between growth phases reflect on the physiological performance, and provide indications for the requirements and roles of individual nutrients in the plant. The root:shoot ratio was dramatically smaller under reproductive growth, revealing that the increase in shoot development was higher than of the root during reproductive growth (Figure 2B). As the translocation of the majority of nutrients to the shoot increased in parallel to the decrease in root:shoot ratio (Figures 2B and 8), we suggest that this escalation was required to support the increase in shoot development, and specifically the substantial increase in inflorescences biomass (Figure 3D). The argument that an increase in the shoot's demand for nutrients promotes the translocation of essential nutrients to the shoot in general, and to the reproductive organs in particular, is supported by similar trends identified for other plant species [41,85]. Interestingly, the decrease in Fe, Ca, and Cu with the transition to reproductive growth and the decrease in root:shoot ratio (Figures 2B and 8D,E,I) may imply that these micronutrients do not play a significant role in increasing shoot (and inflorescence) development over root formation. Alternatively, it may reflect that their concentration is more affected by other parameters, such as water movement [86] and availability of chelates [87], which are frequently correlated with Ca and Fe translocation, respectively. As N concentration was generally stable and did not vary between growth phases (Figure 8A), we conclude that its demand and deposition are more stable, and it is less involved in the physiological transition of the decrease in root:shoot ratio and inflorescence formation. This is supported by findings of our latest studies, which revealed that the demand of cannabis for N, and therefore N supply, is similar for the vegetative and the reproductive growth phases [5,42].

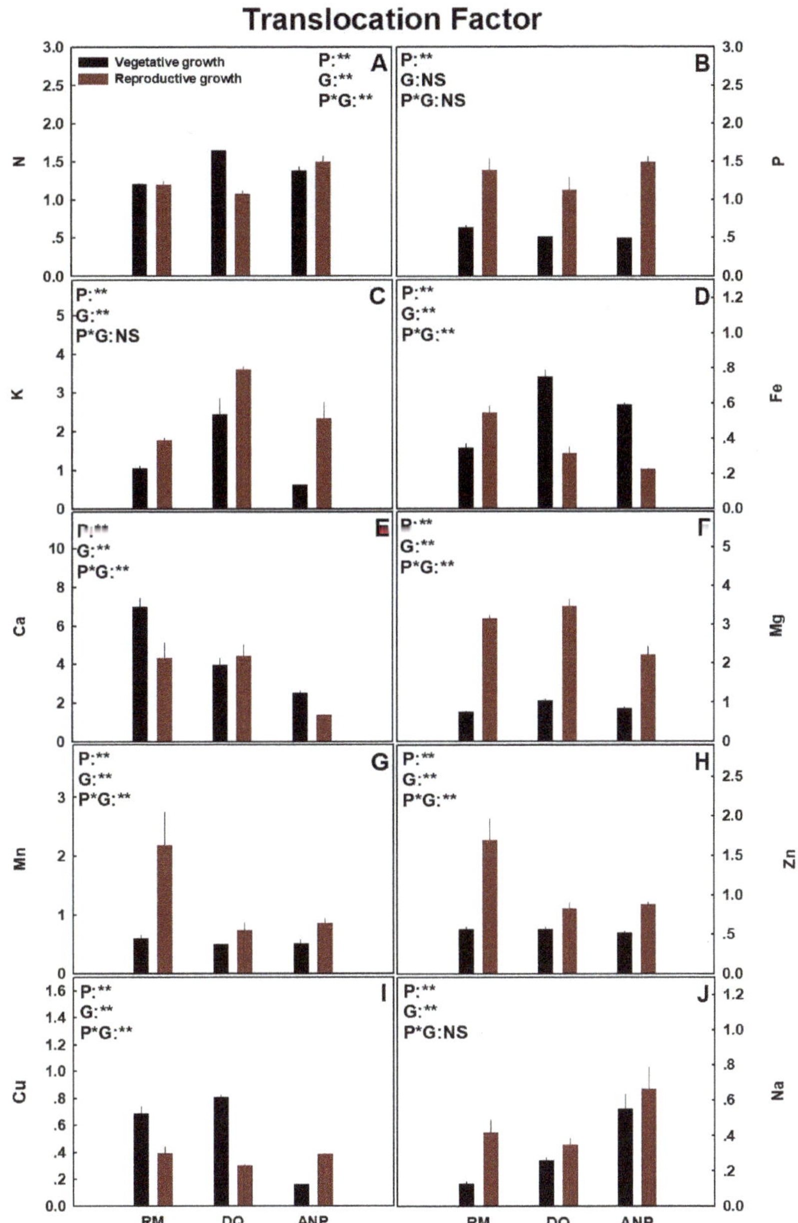

Figure 8. Translocation factor (TF) of three medical cannabis cultivars (RM, DQ, ANP) during vegetative and reproductive development: N (**A**), P (**B**), K (**C**), Fe (**D**), Ca (**E**), Mg (**F**), Mn (**G**), Zn (**H**), Cu (**I**), and Na (**J**). Presented data are averages ± SE (n = 5). Results of two-way ANOVA indicated as ** $p < 0.05$, *F*-test; NS, not significant $p > 0.05$. P*G represents the interaction between growth phase (P) and genotype (G). *p* values are presented at the Supplementary Materials, File S1.

3.5. Agronomic Considerations

The results of this study that demonstrate substantial changes in accumulation and deposition rates of minerals into cannabis plants over time have agronomic implications. First, the deposition rate of most nutrients increases over time while Ca deposition decreases at the second stage of the reproductive growth phase, revealing that the fertilization regime needs to be adjusted accordingly. Second, as nutrient supply was one of the main factors affecting uptake of nutrients by the plants, an optimal and uniform nutritional regime between cultivation batches is required for standardization and optimization of plant growth and secondary metabolism. Third, as the accumulation of most nutrients increases with the increase in biomass production, it is concluded that bigger plants require larger amounts of nutrients for their overall development regardless of genotype. Furthermore, the demand of the cannabis plant for nutrients continues up to plant maturation and does not stop at the second stage of the reproductive growth phase, raising the question whether the "flushing" practice performed by cannabis cultivators (i.e., irrigating with water without nutrients at the last 7–10 days prior to harvest) is necessary, beneficial, or harmful. This issue is currently under investigation in our laboratory. Finally, while using the nutritional requirements shown in this study, possible effects of cultivation conditions including growth media, climatic conditions, plant architecture manipulations, and plant phenotypical traits that may differ between agricultural systems, need to be considered.

4. Conclusions

The present study examined trends of uptake, deposition, and translocation of mineral nutrients during the vegetative and reproductive development of three medical cannabis cultivars. The results, which include also the analyses of temporal trends of physiological activity, demonstrate that (i) The uptake of most nutrients increases gradually during plant development. (ii) The mineral deposition rate is nutrient-specific and highly sensitive to the plant's nutritional regime. (iii) Root-to-shoot translocation is nutrient-specific as well, but most nutrients demonstrated higher translocation during reproductive growth as the inflorescence biomass rose and root:shoot ratio decreased. (iv) The length of the maturation period of a cultivar, and plant age, were identified as key factors for the observed differences in physiological activity between cultivars. The results provide a first step toward understanding the plant's temporal physiological activity and nutritional requirements over the crop cultivation cycle.

Supplementary Materials: The following supporting information can be downloaded at: https://www.mdpi.com/article/10.3390/agronomy13122865/s1, File S1: *p* values for the results presented in the figures; Table S1. Mineral composition of the irrigation solutions during the vegetative and the reproductive growth phases.

Author Contributions: N.B., conceptualization, funding acquisition, methodology, recourses, supervision, writing; A.S., formal analysis, data curation, writing. All authors have read and agreed to the published version of the manuscript.

Funding: This study was funded by the Chief Scientist Fund of the Ministry of Agriculture in Israel, grant no. 20-03-0018.

Data Availability Statement: The data is contained within the manuscript.

Acknowledgments: We thank Yael Sade, Nadav Danziger, Sivan Shiponi, Geki Shoef, Ayana Neta, and Dalit Morad for technical assistance, and Shiran Cohen for assistance with N and P analysis. We thank Rami Levy, Neri Barak, and Adriana Kamma from Canndoc Ltd., the largest certified medical cannabis commercial cultivator in Israel, and Gerry Kolin from Teva Adir Ltd., for cooperation and for the supply of the plant material for the study.

Conflicts of Interest: The authors declare no conflict of interest.

References

1. Saloner, A.; Bernstein, N. Effect of potassium (K) supply on cannabinoids, terpenoids and plant function in medical cannabis. *Agronomy* **2022**, *12*, 1242. [CrossRef]
2. Bevan, L.; Jones, M.; Zheng, Y. Optimisation of nitrogen, phosphorus, and potassium for soilless production of *Cannabis sativa* in the flowering stage using response surface analysis. *Front. Plant Sci.* **2021**, *12*, 764103. [CrossRef]
3. Shiponi, S.; Bernstein, N. The highs and lows of P supply in medical cannabis: Effects on cannabinoids, the ionome, and morpho-physiology. *Front. Plant Sci.* **2021**, *12*, 657323. [CrossRef]
4. Saloner, A.; Bernstein, N. Nitrogen source matters: High NH_4/NO_3 ratio reduces cannabinoids, terpenoids, and yield in medical cannabis. *Front. Plant Sci.* **2022**, *13*, 830224. [CrossRef]
5. Saloner, A.; Bernstein, N. Nitrogen supply affects cannabinoid and terpenoid profile in medical cannabis (*Cannabis sativa* L.). *Ind. Crops Prod.* **2021**, *167*, 113516. [CrossRef]
6. Morad, D.; Bernstein, N. Response of medical cannabis to magnesium (Mg) supply at the vegetative growth phase. *Plants* **2023**, *12*, 2676. [CrossRef] [PubMed]
7. Danziger, N.; Bernstein, N. Light matters: Effect of light spectra on cannabinoid profile and plant development of medical cannabis (*Cannabis sativa* L.). *Ind. Crops Prod.* **2021**, *164*, 113351. [CrossRef]
8. Eichhorn Bilodeau, S.; Wu, B.S.; Rufyikiri, A.S.; MacPherson, S.; Lefsrud, M. An update on plant photobiology and implications for cannabis production. *Front. Plant Sci.* **2019**, *10*, 296. [CrossRef] [PubMed]
9. Rodriguez-Morrison, V.; Llewellyn, D.; Zheng, Y. Cannabis inflorescence yield and cannabinoid concentration are not increased with exposure to short-wavelength ultraviolet-B radiation. *Front. Plant Sci.* **2021**, *12*, 725078. [CrossRef]
10. Rodriguez-Morrison, V.; Llewellyn, D.; Zheng, Y. Cannabis yield, potency, and leaf photosynthesis respond differently to increasing light levels in an indoor environment. *Front. Plant Sci.* **2021**, *12*, 456. [CrossRef] [PubMed]
11. Westmoreland, M.; Kusuma, P.; Bugbee, B. Cannabis lighting: Decreasing blue photon fraction increases yield but efficacy is more important for cost effective production of cannabinoids. *PLoS ONE* **2021**, *16*, e0248988. [CrossRef]
12. Yep, B.; Gale, N.V.; Zheng, Y. Aquaponic and hydroponic solutions modulate NaCl-induced stress in drug-type *Cannabis sativa* L. *Front. Plant Sci.* **2020**, *11*, 1169. [CrossRef] [PubMed]
13. Yep, B.; Gale, N.V.; Zheng, Y. Comparing hydroponic and aquaponic rootzones on the growth of two drug-type *Cannabis sativa* L. cultivars during the flowering stage. *Ind. Crops Prod.* **2020**, *157*, 112881. [CrossRef]
14. Danziger, N.; Bernstein, N. Too dense or not too dense: Higher planting density reduces cannabinoid uniformity but increases yield/area in drug-type medical cannabis. *Front. Plant Sci.* **2022**, *13*, 713481. [CrossRef]
15. Danziger, N.; Bernstein, N. Plant architecture manipulation increases cannabinoid standardization in 'drug-type' medical cannabis. *Ind. Crops Prod.* **2021**, *167*, 113528. [CrossRef]
16. Danziger, N.; Bernstein, N. Shape matters: Plant architecture affects chemical uniformity in large-size medical cannabis plants. *Plants* **2021**, *10*, 1834. [CrossRef]
17. Clarke, R.C.; Merlin, M.D. Cannabis domestication, breeding history, present-day genetic diversity, and future prospects. *CRC. Crit. Rev. Plant Sci.* **2017**, *35*, 293–327. [CrossRef]
18. Moher, M.; Jones, M.; Zheng, Y. Photoperiodic response of in vitro *Cannabis sativa* plants. *HortScience* **2021**, *56*, 108–113. [CrossRef]
19. Weller, J.L.; Kendrick, R.E. Photomorphogenesis and photoperiodism in plants. In *Photobiology: The Science of Life and Light*, 2nd ed.; Björn, L.O., Ed.; Springer: New York, NY, USA, 2008; pp. 417–463, ISBN 9780387726540.
20. Vitale, L.; Arena, C.; Carillo, P.; di Tommasi, P.; Mesolella, B.; Nacca, F.; de Santo, A.V.; Fuggi, A.; Magliulo, V. Gas exchange and leaf metabolism of irrigated maize at different growth stages. *Plant Biosyst. Int. J. Deal. Asp. Plant Biol.* **2011**, *145*, 485–494. [CrossRef]
21. Reekie, E.G.; Bazzaz, F.A. Reproductive effort in plants. 3. Effect of reproduction on vegetative activity. *Am. Nat.* **1987**, *129*, 907–919. [CrossRef]
22. Bielczynski, L.W.; Łącki, M.K.; Hoefnagels, I.; Gambin, A.; Croce, R. Leaf and plant age affects photosynthetic performance and photoprotective capacity. *Plant Physiol.* **2017**, *175*, 1634–1648. [CrossRef] [PubMed]
23. Murray, G.A. The relationship of light quality, duration, and intensity to vegetative and reproductive growth in alfalfa (*Medicago sativa* L.). PhD Thesis, The University of Arizona, Tucson, AZ, USA, 1967.
24. Nadalini, S.; Zucchi, P.; Andreotti, C. Effects of blue and red LED lights on soilless cultivated strawberry growth performances and fruit quality. *Eur. J. Hortic. Sci.* **2017**, *82*, 12–20. [CrossRef]
25. Schwarz, D.; Thompson, A.J.; Kläring, H.P. Guidelines to use tomato in experiments with a controlled environment. *Front. Plant Sci.* **2014**, *5*, 625. [CrossRef] [PubMed]
26. Saloner, A.; Sacks, M.M.; Bernstein, N. Response of medical cannabis (*Cannabis sativa* L.) genotypes to K supply under long photoperiod. *Front. Plant Sci.* **2019**, *10*, 1369. [CrossRef]
27. Muntendam, R.; Happyana, N.; Erkelens, T.; Bruining, F.; Kayser, O. Time dependent metabolomics and transcriptional analysis of cannabinoid biosynthesis in *Cannabis sativa* var. Bedrobinol and Bediol grown under standardized condition and with genetic homogeneity. *Online Int. J. Med. Plant Res* **2012**, *1*, 31–40. [CrossRef]
28. Apicella, P.V.; Sands, L.B.; Ma, Y.; Berkowitz, G.A. Delineating genetic regulation of cannabinoid biosynthesis during female flower development in *Cannabis sativa*. *Plant Direct* **2022**, *6*, e412. [CrossRef] [PubMed]

29. Aizpurua-Olaizola, O.; Soydaner, U.; Öztürk, E.; Schibano, D.; Simsir, Y.; Navarro, P.; Etxebarria, N.; Usobiaga, A. Evolution of the cannabinoid and terpene content during the growth of *Cannabis sativa* plants from different chemotypes. *J. Nat. Prod.* **2016**, *79*, 324–331. [CrossRef]

30. Bauerle, W.L.; McCullough, C.; Iversen, M.; Hazlett, M. Leaf age and position effects on quantum yield and photosynthetic capacity in hemp crowns. *Plants* **2020**, *9*, 271. [CrossRef] [PubMed]

31. Smith, P.F. Mineral analysis of plant tissues. *Annu. Rev. Plant Physiol.* **1962**, *13*, 81–108. [CrossRef]

32. White, P.J. Ion uptake mechanisms of individual cells and roots: Short-distance transport. In *Marschner's Mineral Nutrition of Higher Plants*, 3rd ed.; Marchner, P., Ed.; Academic Press: London, UK, 2012; pp. 7–47, ISBN 9780123849052.

33. Wang, H.; Inukai, Y.; Yamauchi, A. Root development and nutrient uptake. *CRC Crit. Rev. Plant Sci.* **2006**, *25*, 279–301. [CrossRef]

34. Nestby, R.; Lieten, F.; Pivot, D.; Raynal Lacroix, C.; Tagliavini, M. Influence of mineral nutrients on strawberry fruit quality and their accumulation in plant organs: A review. *Int. J. fruit Sci.* **2008**, *5*, 139–156. [CrossRef]

35. Garcia, C.B.; Grusak, M.A. Mineral accumulation in vegetative and reproductive tissues during seed development in *Medicago truncatula*. *Front. Plant Sci.* **2015**, *6*, 622. [CrossRef] [PubMed]

36. Hawkesford, M.; Horst, W.; Kichey, T.; Lambers, H.; Schjoerring, J.; Skrumsager Møller, I.; White, P. Functions of macronutrients. In *Marschner's Mineral Nutrition of Higher Plants*; Marschner, P., Ed.; Academic Press: London, UK, 2012; pp. 135–190, ISBN 9780123849052.

37. Vittori Antisari, L.; Carbone, S.; Gatti, A.; Vianello, G.; Nannipieri, P. Uptake and translocation of metals and nutrients in tomato grown in soil polluted with metal oxide (CeO_2, Fe_3O_4, SnO_2, TiO_2) or metallic (Ag, Co, Ni) engineered nanoparticles. *Environ. Sci. Pollut. Res.* **2015**, *22*, 1841–1853. [CrossRef] [PubMed]

38. Wu, C.Y.; Lu, L.L.; Yang, X.E.; Feng, Y.; Wei, Y.Y.; Hao, H.L.; Stoffella, P.J.; He, Z.L. Uptake, translocation, and remobilization of zinc absorbed at different growth stages by rice genotypes of different Zn densities. *J. Agric. Food Chem.* **2010**, *58*, 6767–6773. [CrossRef] [PubMed]

39. Akhter, M.F.; Macfie, S. Species-specific relationship between transpiration and cadmium translocation in lettuce, barley and radish. *J. Plant Stud.* **2012**, *1*, 2. [CrossRef]

40. Clarkson, D.T.; Hanson, J.B. The mineral nutrition of higher plants. *Annu. Rev. Plant Physiol.* **1980**, *31*, 239–298. [CrossRef]

41. Engels, C.; Marschner, H. Root to shoot translocation of macronutrients in relation to shoot demand in maize (*Zea mays* L.) grown at different root zone temperatures. *Z. Pflanzenernähr. Bodenkd.* **1992**, *155*, 121–128. [CrossRef]

42. Saloner, A.; Bernstein, N. Response of medical cannabis (*Cannabis sativa* L.) to nitrogen supply under long photoperiod. *Front. Plant Sci.* **2020**, *11*, 1517. [CrossRef]

43. Shiponi, S.; Bernstein, N. Response of medical cannabis (*Cannabis sativa* L.) genotypes to P supply under long photoperiod: Functional phenotyping and the ionome. *Ind. Crops Prod.* **2021**, *161*, 113154. [CrossRef]

44. Hall, J.; Bhattarai, S.P.; Midmore, D.J. Review of flowering control in industrial hemp. *J. Nat. Fibers* **2012**, *9*, 23–36. [CrossRef]

45. Clarke, R. Botany of the genus Cannabis. In *Advances in Hemp Research*; Haworth Press: Binghamton, NY, USA, 1999; pp. 1–19.

46. Bernstein, N.; Gorelick, J.; Zerahia, R.; Koch, S. Impact of N, P, K, and humic acid supplementation on the chemical profile of medical cannabis (*Cannabis sativa* L.). *Front. Plant Sci.* **2019**, *10*, 736. [CrossRef]

47. Angelini, L.; Tavarini, S.; Cestone, B.; Beni, C. Variation in mineral composition in three different plant organs of five fibre hemp (*Cannabis sativa* L.) cultivars. *Agrochimica* **2014**, *58*, 1–18.

48. Small, E. *Cannabis: A Complete Guide*; CRC Press: Boca Raton, FL, USA, 2016; ISBN 9781315367583.

49. Bernstein, N.; Ioffe, M.; Bruner, M.; Nishri, Y.; Luria, G.; Dori, I.; Matan, E.; Philosoph-Hadas, S.; Umiel, N.; Hagiladi, A. Effects of supplied nitrogen form and quantity on growth and postharvest quality of *Ranunculus asiaticus* flowers. *HortScience* **2005**, *40*, 1879–1886. [CrossRef]

50. Lichtenthaler, H.K.; Wellburn, A.R. Determinations of total carotenoids and chlorophylls a and b of leaf extracts in different solvents. *Biochem. Soc. Trans.* **1983**, *11*, 591–592. [CrossRef]

51. Bernstein, N.; Shoresh, M.; Xu, Y.; Huang, B. Involvement of the plant antioxidative response in the differential growth sensitivity to salinity of leaves vs roots during cell development. *Free Radic. Biol. Med.* **2010**, *49*, 1161–1171. [CrossRef]

52. Chouvy, P.A. Cannabis cultivation in the world: Heritages, trends and challenges. In *EchoGéo*; OpenEdition Press: Marseille, France, 2019. [CrossRef]

53. Small, E. Botanical Classification and Nomenclatural Issues. In *Cannabis: A Complete Guide*; CRC Press: Boca Raton, FL, USA, 2016; pp. 477–504.

54. Munné-Bosch, S.; Alegre, L. Die and let live: Leaf senescence contributes to plant survival under drought stress. *Funct. Plant Biol.* **2004**, *31*, 203–216. [CrossRef]

55. Wingler, A.; Purdy, S.; MacLean, J.A.; Pourtau, N. The role of sugars in integrating environmental signals during the regulation of leaf senescence. *J. Exp. Bot.* **2006**, *57*, 391–399. [CrossRef]

56. Wolstenholme, B.N. Root, shoot or fruit. *South Afr. Avocado Grow. Assoc. Yearb.* **1981**, *4*, 27–29.

57. Bonifas, K.D.; Walters, D.T.; Cassman, K.G.; Lindquist, J.L. Nitrogen supply affects root:shoot ratio in corn and velvetleaf (*Abutilon theophrasti*). *Weed Sci.* **2005**, *53*, 670–675. [CrossRef]

58. Iwasa, Y.; Roughgarden, J. Shoot/root balance of plants: Optimal growth of a system with many vegetative organs. *Theor. Popul. Biol.* **1984**, *25*, 78–105. [CrossRef]

59. Tezara, W.; Domínguez, T.S.T.; Loyaga, D.W.; Ortiz, R.N.; Chila, V.H.R.; Ortega, M.J.B. Photosynthetic activity of oil palm (*Elaeis guineensis*) and interspecific hybrid genotypes (*Elaeis oleifera* × *Elaeis guineensis*), and response of hybrids to water deficit. *Sci. Hortic.* **2021**, *287*, 110263. [CrossRef]
60. Haldimann, P. Low growth temperature-induced changes to pigment composition and photosynthesis in *Zea mays* genotypes differing in chilling sensitivity. *Plant. Cell Environ.* **1998**, *21*, 200–208. [CrossRef]
61. Bacelar, E.A.; Moutinho-Pereira, J.M.; Gonçalves, B.C.; Lopes, J.I.; Correia, C.M. Physiological responses of different olive genotypes to drought conditions. *Acta Physiol. Plant.* **2009**, *31*, 611–621. [CrossRef]
62. Bond, B.J. Age-related changes in photosynthesis of woody plants. *Trends Plant Sci.* **2000**, *5*, 349–353. [CrossRef]
63. Niinemets, Ü. Stomatal conductance alone does not explain the decline in foliar photosynthetic rates with increasing tree age and size in *Picea abies* and Pinus sylvestris. *Tree Physiol.* **2002**, *22*, 515–535. [CrossRef]
64. Kennedy, R.A.; Johnson, D. Changes in photosynthetic characteristics during leaf development in apple. *Photosynth. Res.* **1981**, *2*, 213–223. [CrossRef]
65. Munné-Bosch, S.; Alegre, L. Plant aging increases oxidative stress in chloroplasts. *Planta* **2002**, *214*, 608–615. [CrossRef]
66. Bonini, S.A.; Premoli, M.; Tambaro, S.; Kumar, A.; Maccarinelli, G.; Memo, M.; Mastinu, A. *Cannabis sativa*: A comprehensive ethnopharmacological review of a medicinal plant with a long history. *J. Ethnopharmacol.* **2018**, *227*, 300–315. [CrossRef]
67. Sklensky, D.E.; Davies, P.J. Resource partitioning to male and female flowers of *Spinacia oleracea* L. in relation to whole-plant monocarpic senescence. *J. Exp. Bot.* **2011**, *62*, 4323–4336. [CrossRef]
68. Fan, K.; Zhang, Q.; Liu, M.; Ma, L.; Shi, Y.; Ruan, J. Metabolomic and transcriptional analyses reveal the mechanism of C, N allocation from source leaf to flower in tea plant (*Camellia sinensis* L.). *J. Plant Physiol.* **2019**, *232*, 200–208. [CrossRef]
69. Clifford, P.E.; Neo, H.H.; Hew, C.S. Regulation of assimilate partitioning in flowering plants of the monopodial orchid *Aranda* Noorah Alsagoff. *New Phytol.* **1995**, *130*, 381–389. [CrossRef]
70. Poulson, M.E.; Thai, T. Effect of high light intensity on photoinhibition, oxyradicals and artemisinin content in *Artemisia annua* L. *Photosynthetica* **2015**, *53*, 403–409. [CrossRef]
71. Allen, R.D.; Webb, R.P.; Schake, S.A. Use of transgenic plants to study antioxidant defenses. *Free Radic. Biol. Med.* **1997**, *23*, 473–479. [CrossRef] [PubMed]
72. Faisal, M.; Anis, M. Changes in photosynthetic activity, pigment composition, electrolyte leakage, lipid peroxidation, and antioxidant enzymes during ex vitro establishment of micropropagated *Rauvolfia tetraphylla* plantlets. *Plant Cell. Tissue Organ Cult.* **2009**, *99*, 125–132. [CrossRef]
73. Cakmak, I.; Kurz, H.; Marschner, H. Short-term effects of boron, germanium and high light intensity on membrane permeability in boron deficient leaves of sunflower. *Physiol. Plant.* **1995**, *95*, 11–18. [CrossRef]
74. Maathuis, F.J. Physiological functions of mineral macronutrients. *Curr. Opin. Plant Biol.* **2009**, *12*, 250–258. [CrossRef]
75. Mitra, G. Essential plant nutrients and recent concepts about their uptake. In *Essential Plant Nutrients*; Naeem, M., Ansari, A., Gill, S., Eds.; Springer: Cham, Switzerland, 2017; pp. 3–36, ISBN 9783319588414.
76. Bassirirad, H. Kinetics of nutrient uptake by roots: Responses to global change. *New Phytol.* **2000**, *147*, 155–169. [CrossRef]
77. Cockson, P.; Schroeder-Moreno, M.; Veazie, P.; Barajas, G.; Logan, D.; Davis, M.; Whipker, B.E. Impact of phosphorus on Cannabis sativa reproduction, cannabinoids, and terpenes. *Appl. Sci.* **2020**, *10*, 7875. [CrossRef]
78. Pardo, J.M.; Quintero, F.J. Plants and sodium ions: Keeping company with the enemy. *Genome Biol.* **2002**, *3*, 1–4. [CrossRef]
79. Monaci, F.; Leidi, E.O.; Mingorance, M.D.; Valdés, B.; Oliva, S.R.; Bargagli, R. Selective uptake of major and trace elements in *Erica andevalensis*, an endemic species to extreme habitats in the Iberian Pyrite Belt. *J. Environ. Sci.* **2011**, *23*, 444–452. [CrossRef]
80. Qin, S.; Liu, H.; Nie, Z.; Rengel, Z.; Gao, W.; Li, C.; Zhao, P. Toxicity of cadmium and its competition with mineral nutrients for uptake by plants: A review. *Pedosphere* **2020**, *30*, 168–180. [CrossRef]
81. Drechsler, N.; Zheng, Y.; Bohner, A.; Nobmann, B.; von Wirén, N.; Kunze, R.; Rausch, C. Nitrate-dependent control of shoot K homeostasis by the nitrate transporter1/peptide transporter family member NPF7.3/NRT1.5 and the Stelar K+ outward rectifier SKOR in arabidopsis. *Plant Physiol.* **2015**, *169*, 2832–2847. [CrossRef] [PubMed]
82. White, P.J. Long-distance Transport in the Xylem and Phloem. In *Marschner's Mineral Nutrition of Higher Plants*, 3rd ed.; Marchner, P., Ed.; Elsevier Ltd.: Amsterdam, The Netherlands, 2012; pp. 49–70, ISBN 9780123849052.
83. Broadley, M.; Brown, P.; Cakmak, I.; Rengel, Z.; Zhao, F. Function of nutrients: Micronutrients. In *Marschner's Mineral Nutrition of Higher Plants*; Marschner, P., Ed.; Academic Press: Cambridge, MA, USA, 2012; pp. 191–248, ISBN 9780123849052.
84. Rahman, H.; Sabreen, S.; Alam, S.; Kawai, S. Effects of nickel on growth and composition of metal micronutrients in barley plants grown in nutrient solution. *J. Plant Nutr.* **2007**, *28*, 393–404. [CrossRef]
85. Siebrecht, S.; Herdel, K.; Schurr, U.; Tischner, R. Nutrient translocation in the xylem of poplar—Diurnal variations and spatial distribution along the shoot axis. *Planta* **2003**, *217*, 783–793. [CrossRef]
86. Mclaughlin, S.B.; Wimmer, R. Calcium physiology and terrestrialecosystem processes. *New Phytol.* **1999**, *142*, 373–417. [CrossRef]
87. Kobayashi, T.; Nishizawa, N.K. Iron uptake, translocation, and regulation in higher plants. *Annu. Rev. Plant Biol.* **2012**, *63*, 131–152. [CrossRef] [PubMed]

agronomy

Article

Synergistic Effects of Subsoil Calcium in Conjunction with Nitrogen on the Root Growth and Yields of Maize and Soybeans in a Tropical Cropping System

Murilo De Souza, Jéssica Pigatto de Queiroz Barcelos and Ciro A. Rosolem *

Department of Crop Science, College of Agricultural Sciences, São Paulo State University, Botucatu 18610-034, Brazil; jessica.pqb@gmail.com (J.P.d.Q.B.)
* Correspondence: ciro.rosolem@unesp.br

Abstract: A large part of Brazilian maize is double-cropped after soybeans, when water shortages are very frequent. A larger root system can mitigate drought stress and enable better nitrogen (N) use. Alleviating acidity and applying gypsum can increase root growth and N-use efficiency in maize, which has a more aggressive root system than soybeans. However, it is not known how these factors interact in integrated cropping systems, or how soybeans respond to them. Soybean and maize root growth and grain yields as affected by soil Ca enrichment using lime and gypsum, along with the N rates applied to maize intercropped with Guinea grass (*Megathyrsus maximus*), were assessed in a medium-term field experiment. Liming resulted in greater root growth for both crops; however, when lime was used in conjunction with gypsum, root growth was further enhanced. The total maize grain yield was 35% higher compare to the control when gypsum was used in conjunction with lime; however, subsoil Ca enrichment increased the total soybean grain yield by 8% compared to the control. Nitrogen fertilization increased the total maize grain yield by 36%, with a more expressive increase when applying 160 kg ha^{-1} or more, and despite a positive effect on soybean grain yields in the long term, this response seems not to be a direct effect of the N applied to the maize. Both subsoil Ca enrichment and N application to maize increase root growth and the total yield of the system.

Keywords: gypsum; lime; Guinea grass; intercropping systems; acid soils

Citation: De Souza, M.; Barcelos, J.P.d.Q.; Rosolem, C.A. Synergistic Effects of Subsoil Calcium in Conjunction with Nitrogen on the Root Growth and Yields of Maize and Soybeans in a Tropical Cropping System. *Agronomy* **2023**, *13*, 1547. https://doi.org/10.3390/agronomy13061547

Academic Editors: Christos Noulas, Shahram Torabian and Ruijun Qin

Received: 26 April 2023
Revised: 25 May 2023
Accepted: 29 May 2023
Published: 2 June 2023

1. Introduction

Soil acidity affects approximately 30% of the world's potential food production area [1]. Acidic tropical soils are usually calcium (Ca)-deficient and show aluminum (Al) toxicity, which inhibits root growth and decreases agricultural production [2,3]. Surface liming has been effective in alleviating topsoil acidity; however, alleviating subsoil acidity by using lime alone is challenging due to its low solubility [4]. Agricultural gypsum—hydrated calcium sulfate (CaSO$_4$2H$_2$O)—has been used in conjunction with lime in acidic soils as an alternative to increase calcium (Ca) contents and alleviate Al toxicity in the subsoil. Some no-till studies have shown lime's effects in the subsoil even when applied on the soil surface, but this takes time [5,6], whereas the alleviation of subsoil acidity is faster with gypsum. Due to its higher mobility, gypsum application increases Ca^{2+} and SO$_4$$^{2-}$ in the soil solution, facilitating leaching of these elements in the soil profile [7], and reduces aluminum activity and toxicity in the subsoil, favoring deep root growth [8]. A better root system results in higher soil exploration [9,10] and plays a crucial role in water acquisition and NUE [11–13] by avoiding N leaching [3]. In no-till systems, when lime is applied without incorporation, the pH is steeply increased close to the soil surface. At pH levels above 5.5, nitrification is enhanced, and the applied N is converted to nitrate, which can leach and take Ca^{2+} with it, significantly improving the effect of lime [3]. Moreover, it has been shown that N fertilization improves maize root branching and growth [14], which can further increase root growth in the soil profile.

Double-cropping of maize with forage grasses after soybeans has proven to be a sound and economical agricultural practice. However, as maize is grown after soybeans, and in many years this period is characterized by rainfall scarcity in tropical climate regions, there is a risk of decreased yields due to water stress. Therefore, improving subsoil conditions to establish a deep root system is paramount for the success of this cropping system [15,16]. Accordingly, greater crop responses to gypsum have been reported in water-deficient growing seasons [17,18]. Furthermore, a deeper root system results in increased N uptake by maize, higher N cycling within the system, and less N loss by leaching [3]. Forage grasses with vigorous root systems have been grown as cover crops or in association/consortium with maize in integrated systems in subtropical and tropical regions, and Guinea grass has been shown to be better than *Urochloa*, especially under N fertilization [19]. However, there is no reliable recommendation of N fertilization for maize/forage systems, especially when cropped after soybeans, since it is assumed that some of the atmospheric N fixed by soybeans could be available for maize. Therefore, considering the lower maize yield potential in this system, the higher N cycling, and the soybean contribution, the optimal N rate would be lower than when maize is grown as a lone crop.

Despite the general belief that mineral N application is unnecessary for inoculated soybeans, since the nutrient can be fully supplied by biological N fixation or by the soil [20], Salvagiotti et al. [21] reported that high-yield soybeans require large amounts of N to sustain their aboveground biomass and seeds with high protein content. These authors speculated that supplying N without decreasing the nodule activity could increase the soybean yield, and promising options include applying N before sowing or at depths below the nodulation zone—that is, increasing the availability of N in the soil profile. However, this has been not demonstrated so far.

Therefore, the hypothesis is that subsoil Ca enrichment with lime and gypsum, in a production system with maize double-cropped with Guinea grass after soybeans, will increase root growth in the soil profile, resulting in higher yields of maize and soybeans, while better root growth of the grasses along with N fertilization and higher dry matter production can also improve soybeans' root growth and grain production. Although the effect of Ca in the subsoil increasing root growth is not new, there is a gap in knowledge about its legacy effect for the next crop, and its interaction with N in integrated cropping systems with maize cropped after soybeans has not yet been addressed. The objective of this study was to evaluate the effects of enriching the subsoil with Ca using lime alone or in conjunction with gypsum and N fertilizer on root growth and grain yields in a no-till cropping system.

2. Materials and Methods

2.1. Site Description

The experiment was carried out in Botucatu, São Paulo State, Brazil, from 2016 to 2020, in a clayey, kaolinitic, thermic Typic Haplorthox [22] located at 48°25′38.84″ W, 22°49′50.90″ S, 790 m above sea level. The climate is Cwa, i.e., tropical with dry winters and warm, rainy summers. Precipitation and temperatures were recorded during the experiment at a meteorological station located 300 m from the experimental area (Figure 1).

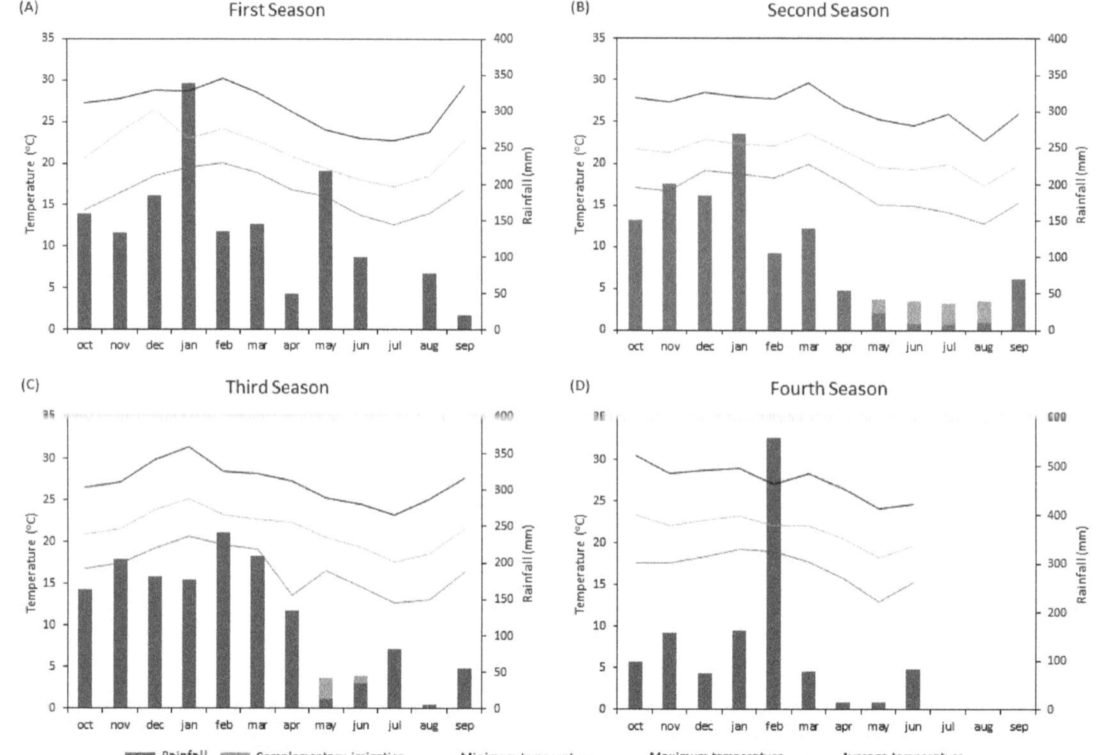

Figure 1. Rainfall and minimum, average, and maximum temperatures in the first (**A**), second (**B**), third (**C**), and fourth (**D**) growing seasons: 2016/2017, 2017/2018, 2018/2019, and 2019/2020, respectively. Botucatu, SP.

Before the experiment, the area was under fallow, with a mix of grasses and some broad leaves. In September 2016, the soil was sampled for chemical characterization analyses [23], and the results are shown in Table 1.

Table 1. Soil chemical characteristics at four depths in the experimental area before initiating the experiment, August 2016.

Depth	pH [#]	OM [†]	P	S	K	Ca	Mg	Al	H + Al	CEC [‡]	BS [§]
m	$CaCl_2$	g dm^{-3}	mg dm^{-3}			-------------- mmol$_c$ dm^{-3} --------------					%
0.00–0.10	4.4	18	13	5	2.1	12	12	9	52	78	32
0.10–0.20	4.3	14	10	5	1.1	9	8	9	48	66	27
0.20–0.40	4.1	11	6	22	0.5	4	4	12	55	63	13
0.40–0.60	4.0	11	6	21	0.4	3	4	20	90	97	8

[#] Soil pH measured in calcium chloride solution. [†] Organic matter. [‡] Cation-exchange capacity. [§] Base saturation.

2.2. Experimental Design and Treatments

The treatments were lime, lime + gypsum, and control, with 0, 80, 160, and 240 kg ha^{-1} of N applied to the maize, arranged in a 3 × 4 factorial scheme in completely randomized blocks with four replications. The lime ($CaCO_3$) rate was calculated to raise the soil's base saturation to 70%, and the gypsum ($CaSO_4 2H_2O$) rate was calculated using the average soil clay content from 0 to 0.4 m multiplied by six, as recommended by Duarte et al. (2022).

The rates corresponding to 2.92 Mg ha^{-1} lime and 2.0 Mg ha^{-1} gypsum were applied on the soil surface. Gypsum was applied to the respective plots immediately after the application of lime. In September 2017, lime and gypsum were reapplied at the same rates. The four rates of N were applied annually to maize intercropped with Guinea grass (*Megathyrsus maximus*).

2.3. Experiment Management

The plots consisted of 10 soybean or maize rows that were 10 m long and spaced 0.45 m apart from one another. The spontaneous vegetation was desiccated using glyphosate (2.30 kg^{-1} a.i.). The lime + gypsum treatment was applied in September 2016 and October 2017. Soybean, cv. TMG 7062 IPRO (Tropical Improvement and Genetics), was planted each year in November, over the desiccated residues of the previous spontaneous species or maize/Guinea grass, which was desiccated two weeks before soybean planting each year. The seeds were inoculated with *Bradyrhizobium japonicum*. Phosphorus (P) and potassium (K) were applied in the seed furrow at 26 kg ha^{-1} and 50 kg ha^{-1}, respectively, each year, as triple superphosphate [Ca(H$_2$PO$_4$)$_2$] and potassium chloride (KCl). Soybean was harvested 125, 145, 136, and 128 days after emergence in the growing seasons 2017, 2018, 2019, and 2020, respectively, and the grain yield was adjusted to 13% moisture.

Maize (Dow AgroSciences Hybrid 2B 587 RRBTPW) was planted after the soybean harvest each year, with a population of 55,000 plants ha^{-1}, intercropped with Guinea grass using 10 kg ha^{-1} of pure live seeds. The forage seeds were mixed with the phosphate fertilizer and applied at a depth of 0.08 m. Each plot received 35 kg ha^{-1} P as triple superphosphate. Potassium chloride was used to supply K at 82 kg ha^{-1} at sowing, plus 41 kg ha^{-1} at V4 (i.e., plants with four fully developed leaves). Ammonium sulfate [(NH$_4$)$_2$SO$_4$] was applied to supply N at 30 kg ha^{-1} at sowing, completed with 50, 130, and 210 kg ha^{-1} side-dressed 0.1 m from the plant line to the respective treatments at stage V4. For a yield of 6–8 Mg ha^{-1}, 120 kg ha^{-1} N would be recommended [24]. The rates used in this experiment ranged from low to very high, because we wanted to know the effects on the subsequent soybean crop. Maize was harvested at, 155, 122, and 132 days after plant emergence in the growing seasons 2017, 2018, 2019 and 2019, respectively. The grain moisture was corrected to 13%.

2.4. Root Sampling and Dry Matter Determination

Root samples were collected at the depths 0 to 0.1, 0.1 to 0.2, 0.2 to 0.4, and 0.4 to 0.6 m, using a steel probe with a 0.075 m internal diameter. Five root subsamples were randomly taken per plot for soybeans and maize, in the planting row and between rows. For soybeans, sampling was performed at R2—the full flowering stage [25]—on 24 January 2017, 8 January 2018, 15 January 2019, and 13 January 2020. For maize intercropped with Guinea grass, the samples were taken on 22 June 2017, 29 July 2018, and 7 July 2019. The roots were carefully separated from the soil and other residues by washing them under a flow of swirling water over a 0.5 mm mesh sieve. Then, the roots were immersed in 30% ethyl alcohol solution, placed in plastic pots, and stored under refrigeration at 2 °C. Afterward, the roots were scanned [26] using an optical scanner (Scanjet 4C/T, HP) at 300 dpi resolution and analyzed with WinRHIZO version 3.8-b (Regent Instrument Inc., Quebec, QC, Canada). The samples were dried in a forced-air oven (Fanen, model 32 E, Brazil) at 60 °C for 48 h to assess the roots' dry matter.

2.5. Soil Sampling and Chemical Analysis

Soil samples were collected 12, 24, and 36 months after the first lime application, in September 2017, 2018, and 2019, respectively, up to the depth of 0.6 m. Four subsamples were randomly collected and combined into a composite sample, and exchangeable Ca was extracted with ion-exchange resin [23].

2.6. Statistical Analysis

Data from each year and each soil layer were analyzed separately. After testing for normality and homoscedasticity, the root length density, dry matter, and grain yield data were subjected to ANOVA. Blocks were considered as random effects, and for the first soybean crop, one-way ANOVA was used. For the remaining years, a factorial ANOVA was used based on a completely randomized block design with two factors (corrective and nitrogen rates). When the ANOVA result was significant, the modified *t*-test (Fisher's protected least significant difference (LSD) at $p \leq 0.05$) was used to separate the means. SAS software, version 9.4, was used. Pearson's correlation coefficients between maize and soybean root length densities were determined ($p \leq 0.05$).

3. Results

Soil Ca^{2+} was increased by liming in the second, third, and fourth growing seasons up to the depth of 0.60 m (Figure 2A–C). However, when gypsum was used in conjunction with lime, the Ca^{2+} concentrations were higher in the soil profile. Nitrogen application further increased the percolation of Ca^{2+} through the soil profile, with rates over 160 kg ha^{-1}, but in the fourth growing season there was generally no significant effect of N fertilization (Figure 2F).

There were no significant interactions of lime application with N rates for root length density (RLD), root dry matter (RDM), soil Ca^{2+} concentrations, or grain yield for any of the crops and in any of the growing seasons. In the upper soil layers, RLD was generally higher—both in the soybean plant rows and between rows—with lime or lime + gypsum compared to the control (Figure 3), except in the first growing season (2016/2017) in the 0.10 to 0.20 m layer (Figure 3(A1)). In the third growing season, the RLD was higher when lime was used in conjunction with gypsum compared with isolated lime—both in the soybean plant rows and between rows—in the 0.00 to 0.10 m layer (Figure 3(E1,F1)). In the fourth growing season (2019/2020), a higher RLD was also observed between soybean rows in the 0.10 to 0.20 m layer when lime was applied in conjunction with gypsum, with values higher than the other treatments (Figure 3(H1)). In the subsoil (0.40–0.60 m layer), the application of lime in conjunction with gypsum increased the RLD compared with the control, except for between soybean rows in the first growing season (Figure 3(B1)). In the third and fourth growing seasons, there was no difference between lime and lime + gypsum in the soybean plant rows (Figure 3(E1,G1)). However, in the fourth growing season, the RLD between soybean rows was higher when lime was used in conjunction with gypsum (Figure 3(H1)). The soybean RDM was higher in almost the entire soil profile after alleviating acidity (Figure 3). However, in the 0.00 to 0.10 m layer, there was no difference in the plant rows in the second growing season (2017–2018) or between rows in the second, third, and fourth growing seasons. Furthermore, lime + gypsum resulted in higher RDM only in the third growing season at the 0.10 to 0.20 m layer.

The use of lime in conjunction with gypsum resulted in higher RDM of soybeans in the subsoil compared with the control (Figure 3), regardless of the sampling location. Comparing lime + gypsum with lime, there was an effect on RDM only in the plant rows in the 0.20 to 0.40 m layer in the second and third growing seasons (Figure 3(C2,E2)), and in the 0.40 to 0.60 m layer in the second and fourth growing seasons (Figure 3(C2,G2)). However, between rows, the RDM was higher in the 0.20 to 0.40 m layer in the first, third, and fourth growing seasons (Figure 3(B2,F2,H2)), and in the 0.40 to 0.60 m layer in the second, third, and fourth growing seasons (Figure 3(D2,F2,H2)). The differences found in the results for the plant rows and between rows probably occurred due to the water deficit during the experiment (Figure 1).

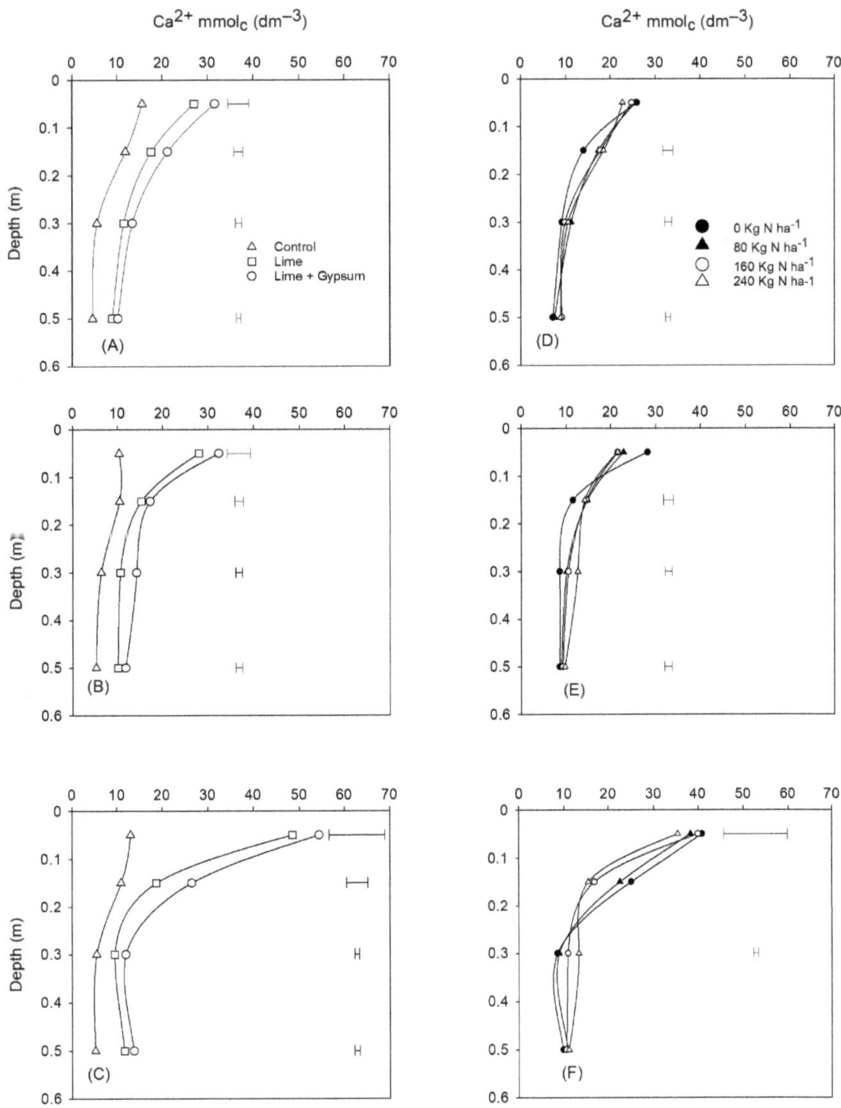

Figure 2. Calcium (Ca) concentration in the soil as affected by lime and lime + gypsum in the growing seasons 2017 (**A**), 2018 (**B**), and 2019 (**C**), and N rates in the growing seasons 2017 (**D**), 2018 (**E**), and 2019 (**F**).

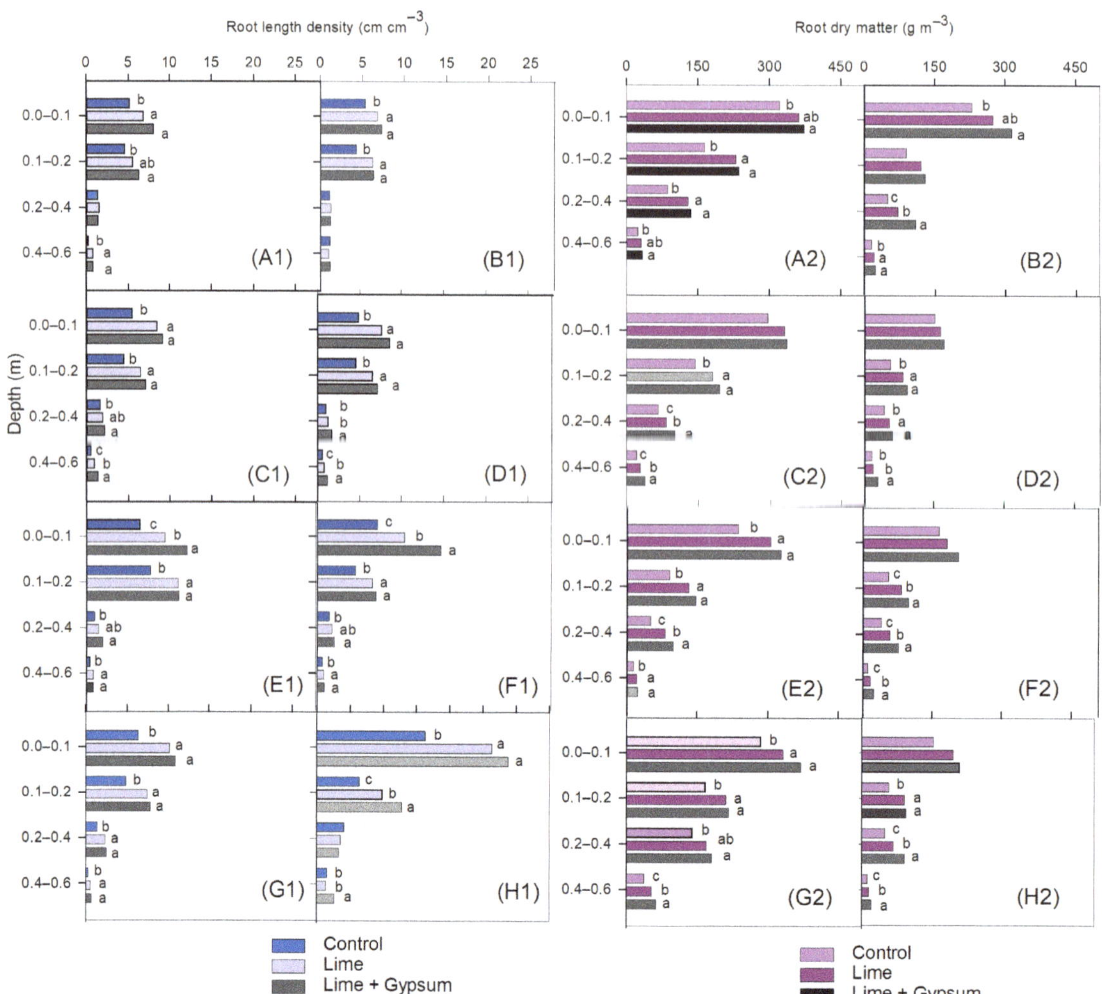

Figure 3. Soybean root length density within rows—growing seasons 2016/2017 (**A1**), 2017/2018 (**C1**), 2018/2019 (**E1**), and 2019/2020 (**G1**)—and between rows—growing seasons 2016/2017 (**B1**), 2017/2018 (**D1**), 2018/2019 (**F1**), and 2019/2020 (**H1**)—and soybean root dry matter in rows—growing seasons 2016/2017 (**A2**), 2017/2018 (**C2**), 2018/2019 (**E2**), and 2019/2020 (**G2**)—and between rows—growing seasons 2016/2017 (**B2**), 2017/2018 (**D2**), 2018/2019 (**F2**), and 2019/2020 (**H2**)—as affected by lime and gypsum application. Different letters indicate means that are statistically different according to the Turkey HSD test ($p \leq 0.05$).

When N was applied to maize at 160 and 240 kg ha^{-1} N, the soybean RLD was higher in the plant rows at the 0.00 to 0.10 m soil layer and in the subsoil (0.40–0.60 m) in the second soybean growing season (Figure 4(A1)). Between the soybean rows, the RLD was higher with 240 kg ha^{-1} N up to 0.20 m, and in the subsoil the highest soybean RLD was observed in the treatments receiving 160 and 240 kg ha^{-1} N (Figure 4(B1)). In the third growing season, a higher RLD was observed in the surface layer and the 0.40 to 0.60 m subsoil layer under the plant rows when soybeans were grown after N-fertilized maize, regardless of the applied rate (Figure 4(C1)). However, between the soybean rows (Figure 4(D1)), the RLD was higher in the treatments receiving N at 160 and 240 kg ha^{-1}, except in the 0.00 to 0.10 m layer, where the increase was observed only with 240 kg ha^{-1} compared with the control. In the fourth growing season, the soybean RLD was increased in the plant rows by N fertilization up to 160 kg ha^{-1}, only in the 0.20 to 0.40 m layer (Figure 4(E1)). The soybean RDM was higher in the plant rows with the use of higher N rates compared with the control in the uppermost soil layer (Figure 4(A2)). In the subsoil (0.20–0.40 layer), the RDM was higher in the treatments with N in the second (Figure 4(A2)) and third (Figure 4(C2)) growing seasons compared with the control. In the 0.40 to 0.60 m layer, both in the plant rows and between rows, N rates of 160 and 240 Kg ha^{-1} resulted in higher RDM compared with the treatment without N in the second, third, and fourth growing seasons (Figure 4(A2–C2,E2,F2)), except between rows in the third growing season (Figure 4(D2)).

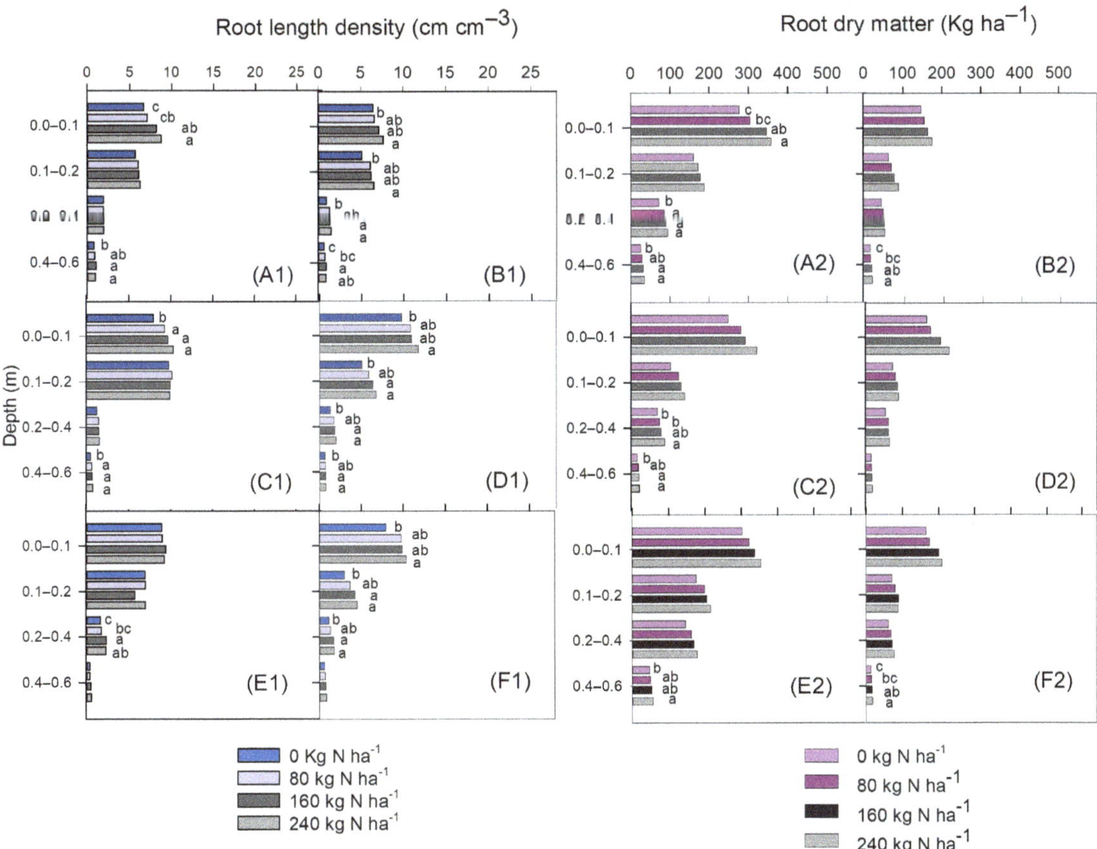

Figure 4. Soybean root length density within plant rows—growing seasons 2017/2018 (**A1**), 2018/2019 (**C1**), and 2019/2020 (**E1**)—and between rows—growing seasons 2017/2018 (**B1**), 2018/2019 (**D1**), and 2019/2020 (**F1**)—and soybean root dry matter in rows—growing seasons 2017/2018 (**A2**), 2018/2019 (**C2**), and 2019/2020 (**E2**)—and between rows—growing seasons 2017/2018 (**B2**), 2018/2019 (**D2**), and 2019/2020 (**F2**)—as affected by N rates. Different letters indicate means that are statistically different according to the Tukey HSD test ($p \leq 0.05$).

The RLD in the maize/Guinea grass intercropping was increased by lime (whether used in conjunction with gypsum or not) in all growing seasons, both in the plant rows and between rows, with a few exceptions (Figure 5). In the second and third growing seasons, the RLD was higher from 0.20 to 0.60 m (Figure 5(C1–F1)) when lime was used in conjunction with gypsum. In general, it was higher with the use of gypsum in the subsoil than when lime was applied alone. However, generally, there was no difference in the RDM when lime was applied with or without gypsum (Figure 5(A2–F2)). The response of RLD to N in the maize/Guinea grass consortium was significant up to 240 kg ha^{-1} in almost all soil layers (Figure 6). However, N application did not increase the RDM in the plant rows (Figure 6), except for the 0.40 to 0.60 m layer in the first growing season (Figure 6(A2)) and the 0.20 to 0.40 m layer in the second growing season (Figure 6(C2)). Between rows, an increase in RDM was observed only in the second growing season in the 0.40 to 0.60 m layer for the highest rate (Figure 6(D2)), and in the third growing season in the 0.00 to 0.10 m layer for all N rates, compared with the control (Figure 6(F2)).

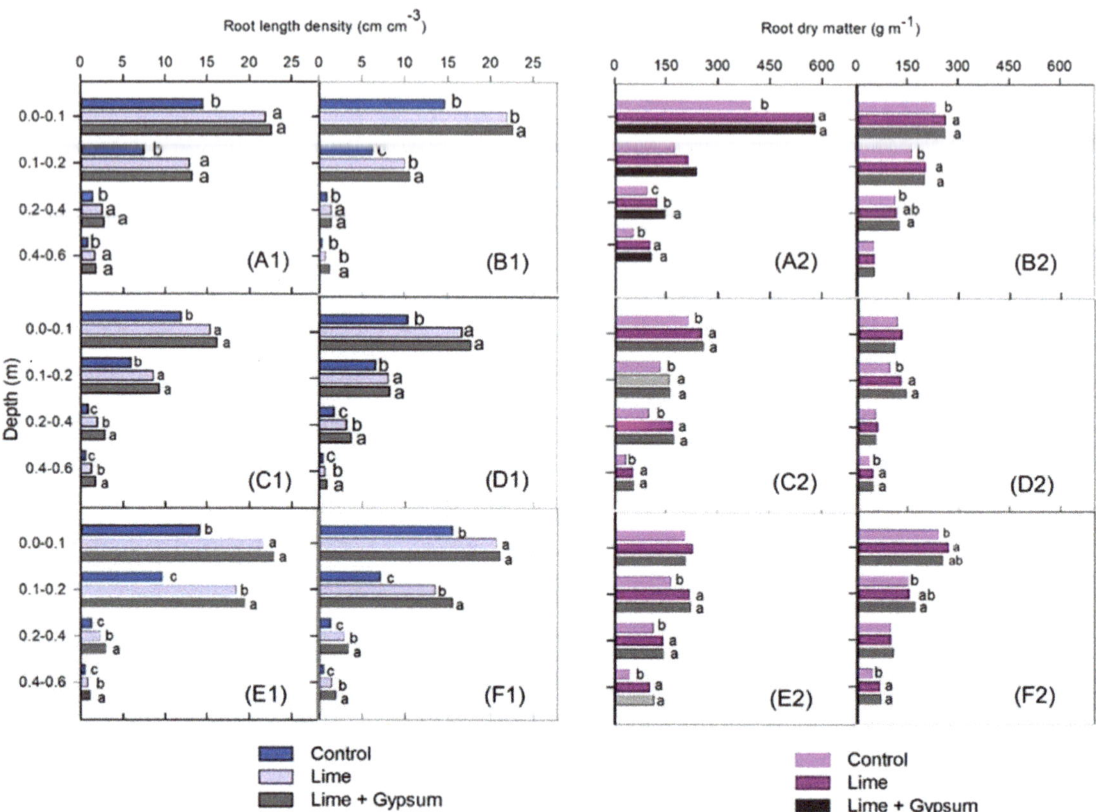

Figure 5. Root length density of maize intercropped with Guinea grass within plant rows—growing seasons 2017 (**A1**), 2018 (**C1**), and 2019 (**E1**)—and between rows—growing seasons 2017 (**B1**), 2018 (**D1**), and 2019 (**F1**)—and root dry matter in rows—growing seasons 2017 (**A2**), 2018 (**C2**), and 2019 (**E2**)—and between rows—growing seasons 2017 (**B2**), 2018 (**D2**), and 2019 (**F2**)—as affected by lime and gypsum application. Different letters indicate means that are statistically different according to the Tukey HSD test ($p \leq 0.05$).

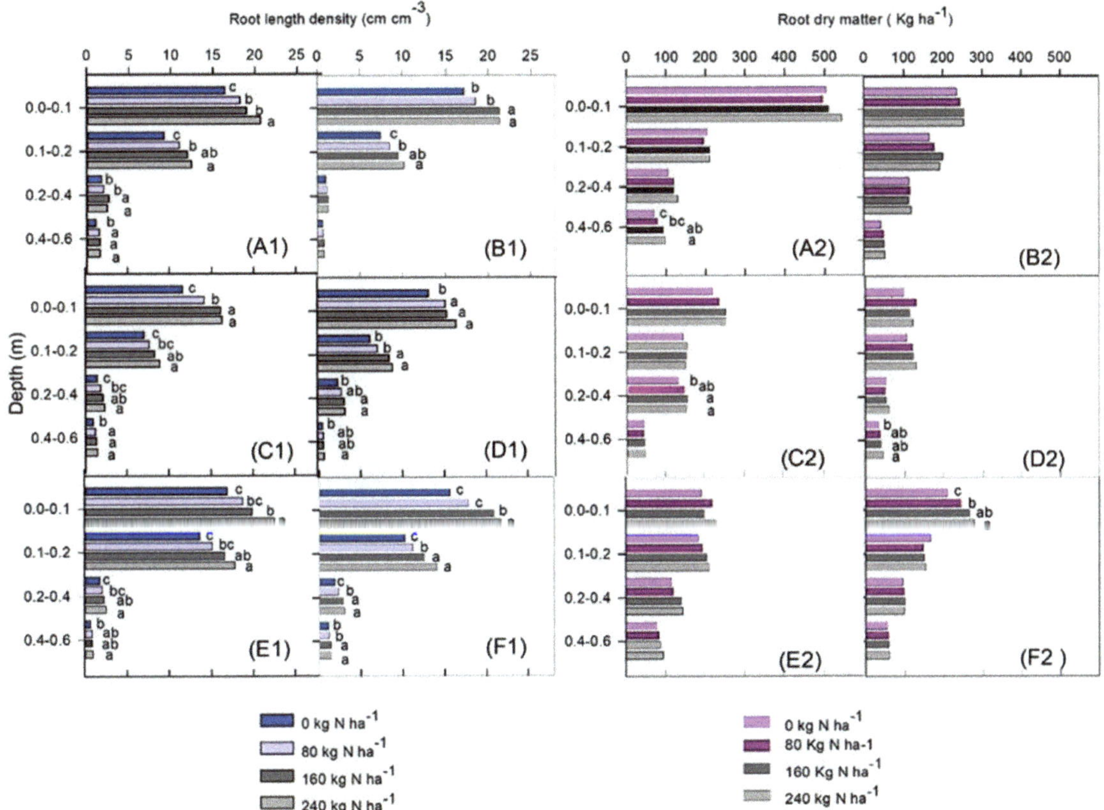

Figure 6. Root length density of maize intercropped with Guinea grass within plant rows—growing seasons 2017 (**A1**), 2018 (**C1**), and 2019 (**E1**)—and between rows—growing seasons 2017 (**B1**), 2018 (**D1**), and 2019 (**F1**)—and root dry matter in rows—growing seasons 2017 (**A2**), 2018 (**C2**), and 2019 (**E2**)—and between rows—growing seasons 2017 (**B2**), 2018 (**D2**), and 2019 (**F2**)—as affected by N rates. Different letters indicate means that are statistically different according to the Tukey HSD test ($p \leq 0.05$).

It is interesting to observe that the soybean RLD was correlated with the RLD of the intercrop in all seasons and soil depths, with a few exceptions (Table 2).

Table 2. Correlation coefficients of root length density between maize intercropped with Guinea grass and soybeans (plant rows and between rows) in growing seasons 2017/2018, 2018/2019, and 2019/2020.

Depth (m)	2017/2018		2018/2019		2019/2020	
	Row	Inter-Row	Row	Inter-Row	Row	Inter-Row
0–0.10	0.641 **	0.588 **	0.733 **	0.686 **	0.493 **	0.597 **
0.10–0.20	0.481 **	0.630 **	0.319 *	0.544 **	0.407 **	0.661 **
0.20–0.40	0.292 *	0.269 ns	0.433 **	0.302 *	0.565 **	−0.066 ns
0.40–0.60	0.699 **	0.732 **	0.735 **	0.448 **	0.387 **	0.462 **

* Significant $p < 0.05$. ** Significant $p < 0.01$. ns means not significant.

When the effects of lime and gypsum were compared year by year, there was no difference in soybean grain yield up to the third growing season. However, in the fourth growing season, lime increased the grain yield by 10.8% compared with the control, and no further increase was observed when used in conjunction with gypsum. However, when looking at the accumulated soybean grain yield, the response to lime was also 5.8% higher than in the control (Figure 7A).

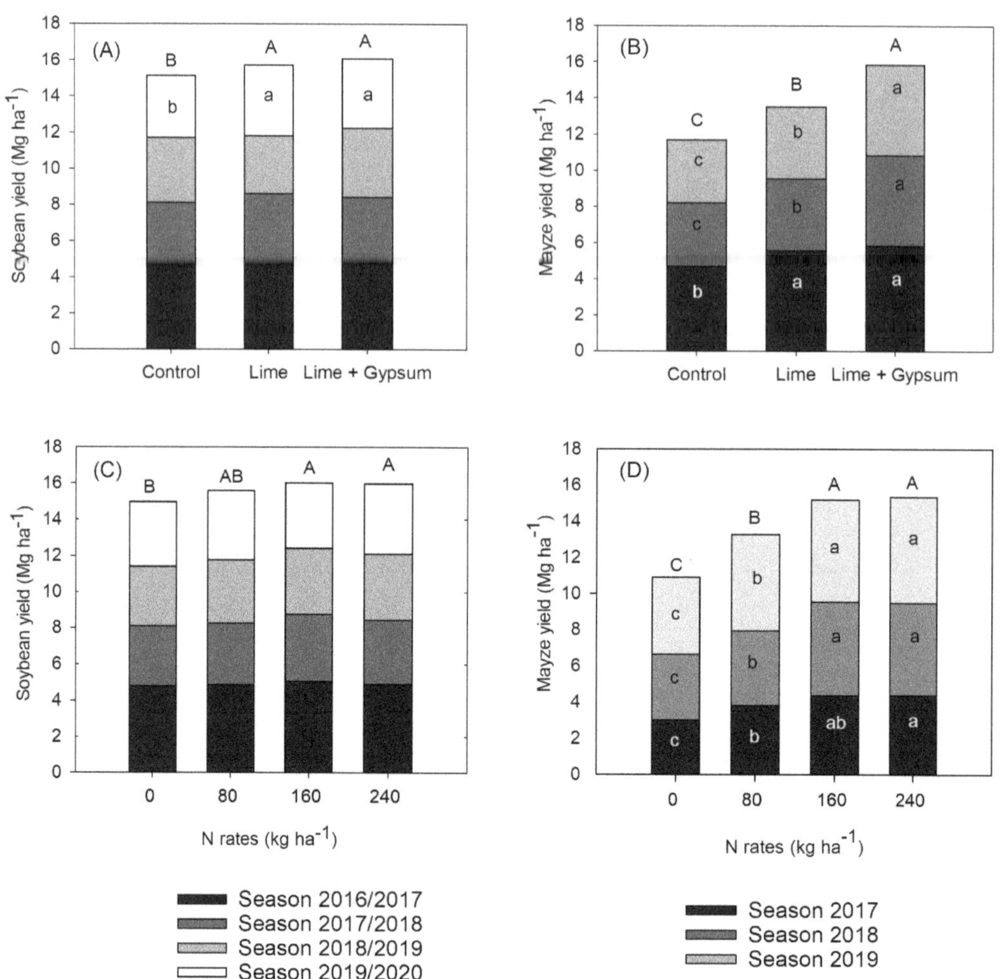

Figure 7. Soybean and maize yields as affected by lime and lime + gypsum (**A** and **B**, respectively) and by nitrogen rates (**C** and **D**, respectively). Means followed by a common letter are not significantly different between amendments or N rates (LSD, $p < 0.05$).

Again, there was no effect of N rates on soybean grain yield when analyzed year by year. The N concentrations in soybean leaves were not affected by the treatments and averaged 52.5 mg kg^{-1} over four years. However, the accumulated grain yield was higher with 160 and 240 kg ha^{-1} applied to the consortium compared with the treatment without N (Figure 7C). It is important to note that in the first soybean crop, the effect of the N dose was not considered, since nitrogen was applied only to the maize crop.

Maize was more responsive than soybeans to liming (Figure 7B). In 2017, the application of lime and lime + gypsum resulted in an average increase of 17.8% compared with the control. However, in 2018, the combination of gypsum with lime resulted in higher maize grain yields: 20.5% and 29.8% higher compared to lime alone and the control, respectively. In 2019, again, the addition of gypsum increased the grain yield by 7.2% compared with lime alone and 20.2% compared with the control, and these increases were reflected in the accumulated grain yields (Figure 7B). However, there was no significant effect of lime or gypsum on maize leaf N concentrations.

Maize grown in association with Guinea grass as a relay crop after soybeans responded to N fertilization up to 160 kg ha^{-1} (Figure 7B), and the leaf N content was increased from 28.7 mg kg^{-1} to 34.2 mg kg^1 on average. Although the N rate of 80 kg ha^{-1} increased the yield by 2988 kg ha^{-1} compared with the control, when 160 and 240 kg ha^{-1} were applied there was an average increase of 5104 kg ha^{-1} in the accumulated grain yields.

4. Discussion

The enrichment of the soil profile with Ca^{2+} after the application of gypsum was expected; however, this has been seldom observed when lime is applied alone [4,27]. Gypsum is considered to be an important alternative to improve root system distribution in the soil profile [3], and its association with lime is efficient in improving the Ca_2^+ content in the soil.

As we hypothesized, Ca^{2+} leaching to the subsoil was improved by N fertilization (Figure 2D–F). This result can be explained, since when lime is applied to the soil surface there is a sharp pH increase close to the surface, and the N from the fertilizer is transformed into nitrate. According to Rosolem et al. [28], nitrate is mobile in the soil profile and tends to follow the water infiltration flow, which is facilitated under no-till conditions due to both less evaporation and better soil profile structuring. Pearson et al. [29] reported that nitrogen fertilization increased the movement of Ca^{2+} along the soil profile when acidity amendments were applied. The authors attributed this effect to the formation of soluble salts such as $Ca(NO_3)_2$, subject to leaching by the downward movement of water. Additionally, the movement of small particles of lime [4] may have played a role in increasing Ca throughout the soil profile.

Root growth was positively influenced by soil acidity alleviation (Figure 3). Calcium is essential for root elongation [30], so root development is impaired in acidic soils or under low concentrations of Ca^{2+}, especially in the subsoil [3]. There are reports of regular growth of soybean roots with 10 mmolc dm^{-3} Ca^{2+} [31], and the response is expected to be low when the soil concentrations are over 12 mmolc dm^{-3} [32], but the response was significant with higher Ca^{2+} contents in the deeper soil layers in this experiment. This could have been due to the weather conditions, in response to dry spells. In the present study, although lime was found to be efficient in increasing the soybean root system, greater development was observed in the subsoil when it was used in conjunction with gypsum (Table 1), probably due to the higher Ca^{2+} content at this depth (Figure 2). Despite a report of greater soybean root growth when the soil Ca^{2+} content was increased after the application of gypsum, the effect of Ca on root growth after soil acidity correction was inconsistent for soybeans.

The effect of soil and subsoil acidity alleviation on root growth in the intercropped maize/Guinea grass was more evident than that observed in soybeans after liming, with or without gypsum. This suggests that maize is more responsive to soil acidity correction. Variations in responses between different plant species occur because of genetic factors linked to the plants' efficiency in acquiring soil Ca^{2+}, since this only occurs in new, non-suberized parts of the roots, and the availability of Ca^{2+} in situ is paramount for constant absorption by very young roots. Therefore, the better root distribution in the subsoil observed in the present study could be related to the Ca^{2+} concentrations, since the application of gypsum in conjunction with N fertilization was efficient in increasing its contents in the subsoil compared to those observed with the application of lime alone. On average, the Ca^{2+} content in the subsoil was increased in this experiment (Figure 2), reaching up to

20.0 mmolc dm^{-3} (0.20–0.60 m layer). However, there is evidence of maize root responses to Ca^{2+} in the order of 15 mmolc dm^{-3} [33]. In addition, an effect on root growth was observed throughout the soil profile after N supply (Figures 4 and 6). Nitrate (NO$_3{}^-$) is known to play a signal-regulating role in many physiological processes, including root growth, and an interconnection of the concentrations of NO$_3{}^-$ and Ca^{2+} [34,35] with auxin is possible [36]. These processes are controlled by several concentrations of gene transcription, which are regulated by NO$_3{}^-$ [37]. Studies have shown that the growth of the root system is regulated by NO$_3{}^-$ and auxin signaling pathways [38,39].

It has been demonstrated [9] that intercropping systems with forage grasses are efficient in the use of N, avoiding N leaching, and it has been reported that Guinea grass is highly demanding with respect to N [40]. In addition, deep root growth may also have been favored by the presence of Guinea grass, which has an aggressive root system and was alive up to the desiccation before the next soybean crop. Soybeans' root growth is improved as a result of the previous root growth of cover crops [41], probably because of the biopores present in the soil profile. Under no-till conditions, without soil disturbance, continuous channels are formed by decomposing roots, which serve as paths that favor the root growth of subsequent crops in the soil profile [42,43]. This is supported by the positive correlation between soybean root growth and the root length density of the intercrop (Table 2). As *Urochloa* species have a vigorous, abundant, and deep root system, these plants can explore a large volume of soil and take up greater amounts of nutrients available in soil regions that are far from the roots of the consortium's grain-producing crop, which are usually more superficial and sparser [3]. In addition, its aggressive root system improves the soil's physical condition by increasing its pore continuity [9], which may result in greater soil microporosity, and the root length density is improved at higher soil microporosity [44].

Grain Yield

Soybean yields were higher when soil acidity was corrected, with significant differences for the last season and the accumulated yields. Several studies have shown no soybean yield responses to the superficial application of lime and/or gypsum when there was no water shortage [45,46]; thus, the lack of response was attributed to the adequate rainfall conditions during crop development. In addition, greater organic matter and nutrient accumulation on the soil surface under NT decreases Al toxicity through the formation of Al–organic complexes [47]. This may explain the lack of soybean response in the first three harvests of this study, since there was no water deficit in this period (Figure 1). However, the 2018/2019 season was marked by low rainfall (total precipitation of 493 mm) and, eventually, the soybean yield was lower in the control treatment (3.4 Mg ha^{-1}) than after the application of lime (3.9 Mg ha^{-1}) and lime + gypsum (3.8 Mg ha^{-1}), corroborating previous observations. The increase in grain yield may have been associated with the increased concentrations of Ca^{2+} in the soil [17]. The authors observed responses in soybean grain yield when rainfall conditions were unfavorable, applying a similar rate of gypsum to that used in the present study. Thus, the increase in soil Ca^{2+} concentration (Figure 6) with the application of lime and gypsum improved root development (Figure 3), which lessened the effects of water shortage on plant growth and yield. This study demonstrates that soil acidity alleviation is efficient in increasing soybean yields throughout the seasons, justifying the need for correction even when year-to-year comparisons do not show significant differences.

The application of N rates of 160 and 240 kg ha^{-1} to maize intercropped with Guinea grass positively affected the accumulated soybean grain yields (Figure 7). Biological nitrogen fixation (BNF) can supply the N requirements of soybeans through the use of adapted rhizobia strains selected for tropical conditions [20]. However, there are studies reporting that the decrease in BNF generally observed after flowering may restrict N's availability to soybeans, which would not be able to acquire adequate amounts of the nutrient for high yields. Salvagiotti et al. [21] analyzed several field studies in different regions after a comprehensive literature review and observed that high-yield soybeans require large amounts of N to support their aboveground biomass and protein-rich seeds. According to the authors, the supply of N without decreasing the nodule activity could increase crop yield and, as promising options, they suggested supplying N before sowing or at depths below the nodulation zone—that is, increasing N's availability in the soil profile. Despite the efficiency of N supply by BNF for soybeans in tropical soils [20], there was a response to the N applied to the previous maize. Crop and cover crop residues accumulated on the soil surface constitute an important nutrient reserve, whose availability can be fast and intense [3] or slow and gradual, depending on precipitation, temperature, soil microbial activity, and the quality and quantity of plant residue [48]. Additionally, Guinea grass has a high N-cycling capacity in the system; therefore, the amount of N recycled by Guinea grass certainly increased the total N availability as the system progressed. However, in this study, the N concentration in soybean leaves was not affected by the application of N to maize. Therefore, the effect of maize fertilization on soybean yields was not a direct effect of the nutrient but, rather, a consequence of a general improvement of the system.

In the first season of maize intercropped with Guinea grass, the grain yield was increased by liming, with no further increase when gypsum was applied (Figure 7). Favorable rainfall conditions in this season (Figure 1) can explain this result, as a greater crop response to gypsum has been reported [17] when there was a water deficit. In the second and third seasons of maize intercropped with Guinea grass, lime + gypsum resulted in a higher yield when compared with lime alone. In addition, the accumulated maize grain yield was higher when gypsum was used in conjunction with lime (Figure 7). These results confirm that the use of gypsum is an important tool to increase maize yields when cropped after soybeans in a period in which there is usually a decrease in soil water availability that results in plant drought stress. Caires et al. [46] also observed significant increases in maize yields due to the higher Ca availability in deep soil, and the authors related these results to a better distribution of the crop's root system. A long-term experiment with maize concluded that the combined application of gypsum and lime resulted in a 17% increase in yields [4], highlighting the importance of this tool in maximizing the grain yield of this species under water shortage during crop development. Penariol et al. [49] showed that maize's grain yield can be compromised if there is a water deficit during flowering—a phase that determines the number of ovules to be fertilized and, consequently, grain production. This explains the lower yields observed in the 2018 season compared with 2017 and 2019.

In general, an increase in maize grain yield was observed when N was added after soil acidity alleviation compared with the control treatment, even when a low rate (80 kg ha^{-1}) was applied. These results suggest a response of maize grown as a relay crop to lower N doses, which can be explained by the high N residues deposited on the soil surface by soybeans, plus the N recycled by the grass. There is a fast decomposition of soybean residues due to the low C/N ratio, which benefits the next crop. The recommendations for N fertilization in an intercropped system are not yet clear. The best yields were obtained in this study with the application of 160 kg ha^{-1} N, which is comparable with the findings of Souza and Soratto [50], who also observed an increase in maize grain yield grown as a relay crop when 120 kg ha^{-1} was applied. These results suggest that intercropping maize with Guinea grass is an alternative to avoid N losses [39], due to its potential for deep soil exploration.

5. Conclusions

The results of this study provide important information on the effects of soil acidity alleviation and N supply on soybean and maize root growth in a no-till system. Soil acidity correction and N supply result in better distribution of the soybean and maize root systems in the soil profile, increasing soil exploration, which facilitates water extraction in periods of scarcity and nutrient absorption in deeper layers of the soil, ultimately resulting in higher yields.

The combination of lime with gypsum is an important alternative to increase Ca^{2+} concentrations in the soil profile and improve the distribution of the crop root systems, and N fertilization helps in improving this system, not only in resulting higher maize yields but also improving soybean yields.

These results show the need for and the benefits of N application to maize on the next soybean crop, and they should be considered in recommending soil acidity correction and N fertilization in integrated cropping systems. Future research on this topic should focus on the recovery of the N applied to maize by the next crop.

Author Contributions: C.A.R.—experiment planning, funding, writing; M.D.S.—field and lab work, first draft; J.P.d.Q.B.—field and lab work, writing. All authors have read and agreed to the published version of the manuscript.

Funding: This work was supported by FAPESP, São Paulo Research Foundation, grant 2017/22134-0.

Data Availability Statement: Data will be made available upon reasonable request.

Conflicts of Interest: The authors declare no conflict of interest.

References

1. Jones, D.L.; Ryan, P.R. Aluminum Toxicity. In *Encyclopedia of Applied Plant Sciences*; Elsevier: Amsterdam, The Netherlands, 2017. [CrossRef]
2. Ritchey, D.K.; Feldhake, C.M.; Clark, R.B.; de Sousa, D.M.G. Improved Water and Nutrient Uptake from Subsurface Layers of Gypsum-Amended Soils. In *Agricultural Utilization of Urban and Industrial By-Products*; Karlen, D., Wright, R., Kemper, W., Eds.; ACS Special Publication: Washington, DC, USA, 1995. [CrossRef]
3. Rosolem, C.A.; Ritz, K.; Cantarella, H.; Galdos, M.V.; Hawkesford, M.J.; Whalley, W.R.; Mooney, S.J. Enhanced Plant Rooting and Crop System Management for Improved N Use Efficiency. *Adv. Agron.* **2017**, *146*, 205–239. [CrossRef]
4. Caires, E.F.; Garbuio, F.J.; Alleoni, L.R.F.; Cambri, M.A. Calagem superficial e cobertura de aveia preta antecedendo os cultivos de milho e soja em sistema plantio direto. *Rev. Bras. Ciência Solo* **2006**, *30*, 87–98. [CrossRef]
5. Caires, E.F.; Joris, H.A.W.; Churka, S. Long-term effects of lime and gypsum additions on no-till corn and soybean yield and soil chemical properties in southern Brazil. *Soil Use Manag.* **2011**, *27*, 45–53. [CrossRef]
6. Costa, A.; Rosolem, C.A. Liming in the transition to no-till under a wheat–soybean rotation. *Soil Tillage Res.* **2007**, *97*, 207–217. [CrossRef]
7. Raij, B.V. Reações de Gesso em Solos Ácidos. In *Seminário Sobre o Uso do Gesso na Agricultura*; Ibrafos: Uberaba, Brazil, 1992; pp. 105–120.
8. Pivetta, L.A.; Castoldi, G.; Pivetta, L.G.; Maia, S.C.M.; Rosolem, C.A. Gypsum application, soil fertility and cotton root growth. *Bragantia* **2019**, *78*, 264–273. [CrossRef]
9. Galdos, M.V.; Brown, E.; Rosolem, C.A.; Pires, L.F.; Hallett, P.D.; Mooney, S.J. Brachiaria species influence nitrate transport in soil by modifying soil structure with their root system. *Sci. Rep.* **2020**, *10*, 5072. [CrossRef] [PubMed]
10. Maeght, J.-L.; Rewald, B.; Pierret, A. How to study deep roots—And why it matters. *Front. Plant Sci.* **2013**, *4*, 299. [CrossRef]
11. Jing, J.; Rui, Y.; Zhang, F.; Rengel, Z.; Shen, J. Localized application of phosphorus and ammonium improves growth of maize seedlings by stimulating root proliferation and rhizosphere acidification. *Field Crops Res.* **2010**, *119*, 355–364. [CrossRef]
12. Shu, L.; Shen, J.; Rengel, Z.; Tang, C.; Zhang, F. Cluster Root Formation by *Lupinus Albus* is Modified by Stratified Application of Phosphorus in a Split-Root System. *J. Plant Nutr.* **2007**, *30*, 271–288. [CrossRef]
13. Williamson, L.C.; Ribrioux, S.P.C.P.; Fitter, A.H.; Leyser, H.M.O. Phosphate Availability Regulates Root System Architecture in Arabidopsis. *Plant Physiol.* **2001**, *126*, 87–882. [CrossRef]
14. Souza, E.A.; Ferreira-Eloy, N.R.; Grassmann, C.S.; Rosolem, C.A.; White, P.J. Ammonium improves corn phosphorus acquisition through changes in the rhizosphere processes and root morphology. *Pedosphere* **2019**, *29*, 534–539. [CrossRef]
15. Farina, M.P.W.; Channon, P.; Thibaud, G.R. A Comparison of Strategies for Ameliorating Subsoil Acidity I. Long-Term Growth Effects. *Soil Sci. Soc. Am. J.* **2000**, *64*, 646–651. [CrossRef]
16. Sumner, M.E. Amelioration of Subsoil Acidity with Minimum Disturbance. In *Advances in Soil Science: Subsoil Management Techniques*; Jayawardane, N.S., Stewart, B.A., Eds.; Lewis Publishers: Boca Raton, FL, USA, 1995; pp. 147–185.
17. Nora, D.D.; Amado, T.J.; Nicoloso, R.D.; Gruhn, E.M. Modern High-Yielding Maize, Wheat and Soybean Cultivars in Response to Gypsum and Lime Application on No-Till Oxisol. *Rev. Bras. Ciência Solo* **2017**, *41*, 1–21. [CrossRef]
18. Tiecher, T.; Pias, O.H.; Bayer, C.; Martins, A.P.; Denardin, L.G.; Anghinoni, I. Crop Response to Gypsum Application to Subtropical Soils Under No-Till in Brazil: A Systematic Review. *Rev. Bras. Ciência Solo* **2018**, *42*, 1–17. [CrossRef]
19. Rocha, K.F.; de Souza, M.; Almeida, D.S.; Chadwick, D.R.; Jones, D.L.; Mooney, S.J.; Rosolem, C.A. Cover crops affect the partial nitrogen balance in a maize-forage cropping system. *Geoderma* **2020**, *360*, 114000. [CrossRef]

20. Hungria, M.; Franchini, J.C.; Campo, R.J.; Crispino, C.C.; Moraes, J.Z.; Sibaldelli, R.N.R.; Mendes, I.C.; Arihara, J. Nitrogen nutrition of soybean in Brazil: Contributions of biological N_2 fixation and N fertilizer to grain yield. *Can. J. Plant Sci.* **2006**, *86*, 927–939. [CrossRef]
21. Salvagiotti, F.; Cassman, K.G.; Specht, J.E.; Walters, D.T.; Weiss, A.; Dobermann, A. Nitrogen uptake, fixation and response to fertilizer N in soybeans: A review. *Field Crops Res.* **2008**, *108*, 1–13. [CrossRef]
22. USDA. *Keys to Soil Taxonomy*, 12th ed.; Soil Survey Staff: Washington, DC, USA, 2014.
23. Van Raij, B.; Quaggio, J.A.; Da Silva, N.M. Extraction of phosphorus, potassium, calcium and magnesium from soils by an ion-exchange resin procedure. *Commun. Soil Sci. Plant Anal.* **1986**, *17*, 544–566. [CrossRef]
24. Duarte, A.P.; Cantarella, H.; Quaggio, J.A. Milho (Zea Mays). In *Boletim 100: Recomendações de adubação e Calagem para o Estado de São Paulo*; Cantarella, H., Quaggio, J.A., Mattos, D., Jr., Boaretto, R.M., Van Raij, B., Eds.; Instituto Agronomico (IAC): Campinas, SP, Brazil, 2022; pp. 199–205.
25. Fehr, W.R.; Caviness, C.E.; Burmood, D.T.; Pennington, J.S. Stage of Development Descriptions for Soybeans, *Glycine Max* (L.) Merrill [1]. *Crop Sci.* **1971**, *11*, 929–931. [CrossRef]
26. Tennant, D. A Test of a Modified Line Intersect Method of Estimating Root Length. *J. Ecol.* **1975**, *63*, 995–1001. [CrossRef]
27. Rosolem, C.A.; Pace, L.; Crusciol, C.A.C. Nitrogen management in maize cover crop rotations. *Plant Soil* **2004**, *264*, 261–271. [CrossRef]
28. Rosolem, C.A.; Foloni, J.S.; Oliveira, R.H. Dinâmica do nitrogênio no solo em razão da calagem e adubação nitrogenada, com palha na superfície. *Pesqui. Agropecuária Bras.* **2003**, *38*, 301–309. [CrossRef]
29. Pearson, R.W.; Abruna, F.; Vicente-Chandler, J. Effect of lime and nitrogen applications on downward movement of calcium and magnesium in two humid tropical soils of puerto rico. *Soil Sci.* **1962**, *93*, 77–82. [CrossRef]
30. Hanson, J.B. The functions of calcium in plant nutrition. *Adv. Plant Nutr.* **1984**, 149–208.
31. Ritchey, K.D.; Silva, J.R.; Costa, U.F. Calcium deficiency in clayey B horizons of savanna oxisols. *Soil Sci.* **1980**, *133*, 378–382. [CrossRef]
32. Rosolem, C.A.; Bicudo, S.J.; Marubayashi, O.M. Soybean Yield and Root Growth as Affected by Lime Rate and Quality. In *Plant-Soil Interactions at Low pH: Principles and Management*; Date, R.A.A., Grundon, N.J., Rayment, G.E., Probert, M.E., Eds.; Kluwer Academic Publishers: Dordrecht, The Netherlands, 1995; pp. 543–548.
33. Rosolem, C.A.; Vale, L.S.R.; Grassi Filho, H.; Moraes, N.H. Sistema radicular e nutrição do milho em função da calagem e da compactação do solo. *Rev. Bras. Ciência Solo* **1994**, *18*, 491–497.
34. Chen, Y.H.; Kao, C.H. Calcium is involved in nitric oxide- and auxin-induced lateral root formation in rice. *Protoplasma* **2012**, *249*, 1085–1091. [CrossRef] [PubMed]
35. Hasenstein, K.-H.; Evans, M.L. Calcium Dependence of Rapid Auxin Action in Maize Roots. *Plant Physiol.* **1986**, *81*, 439–443. [CrossRef]
36. Sun, C.-H.; Yu, J.-Q.; Hu, D.-G. Nitrate: A Crucial Signal during Lateral Roots Development. *Front. Plant Sci.* **2017**, *8*, 485. [CrossRef]
37. Muday, G.K.; Haworth, P. Tomato root growth, gravitropism, and lateral development: Correlation with auxin transport. *Plant Physiol. Biochem. PPB* **1994**, *32*, 193–203.
38. Vidal, E.A.; Gutiérrez, R.A. A systems view of nitrogen nutrient and metabolite responses in Arabidopsis. *Curr. Opin. Plant Biol.* **2008**, *11*, 521–529. [CrossRef] [PubMed]
39. Guo, F.; Wang, R.; Crawford, N.M. The Arabidopsis dual-affinity nitrate transporter gene AtNRT1.1 (CHL1) is regulated by auxin in both shoots and roots. *J. Exp. Bot.* **2002**, *53*, 835–844. [CrossRef] [PubMed]
40. Walch-Liu, P.; Forde, B.G. Nitrate signalling mediated by the NRT1.1 nitrate transporter antagonises l-glutamate-induced changes in root architecture. *Plant J.* **2008**, *54*, 820–828. [CrossRef] [PubMed]
41. Rocha, K.F.; Mariano, E.; Grassmann, C.S.; Trivelin, P.C.O.; Rosolem, C.A. Fate of 15N fertilizer applied to maize in rotation with tropical forage grasses. *Field Crops Res.* **2019**, *238*, 35–44. [CrossRef]
42. Calonego, J.C.; Rosolem, C.A. Soybean root growth and yield in rotation with cover crops under chiseling and no-till. *Eur. J. Agron.* **2010**, *33*, 242–249. [CrossRef]
43. Williams, S.M.; Weil, R.R. Crop Cover Root Channels May Alleviate Soil Compaction Effects on Soybean Crop. *Soil Sci. Soc. Am. J.* **2004**, *68*, 148–153. [CrossRef]
44. Hernandez-Ramirez, G.; Lawrence-Smith, E.J.; Sinton, S.M.; Tabley, F.; Schwen, A.; Beare, M.H.; Brown, H.E. Root responses to alterations in microporosity and penetrability in a silt loam soil. *Soil Sci. Soc. Am. J.* **2014**, *78*, 1392–1403. [CrossRef]
45. Oliveira, E.L.; Pavan, M.A. Control of soil acidity in no-tillage system for soybean production. *Soil Tillage Res.* **1996**, *38*, 47–57. [CrossRef]
46. Caires, E.F.; Alleoni, L.R.F.; Cambri, M.A.; Barth, G. Surface Application of Lime for Crop Grain Production Under a No-Till System. *Agron. J.* **2005**, *97*, 791–798. [CrossRef]
47. Franchini, J.C.; Miyazawa, M.; Pavan, M.A.; Malavolta, E. Dinâmica de íons em solo ácido lixiviado com extratos de resíduos de adubos verdes e soluções puras de ácidos orgânicos. *Pesqui. Agropecuária Bras.* **1999**, *34*, 2267–2276. [CrossRef]
48. Pariz, C.M.; Andreotti, M.; Buzetti, S.; Bergamaschine, A.F.; Ulian, N.D.; Furlan, L.C.; Meirelles, P.R.; Cavasano, F.A. Straw decomposition of nitrogen-fertilized grasses intercropped with irrigated maize in an integrated crop-livestock system. *Rev. Bras. Ciência Solo* **2011**, *35*, 1–17. [CrossRef]

49.   Penariol, F.G.; Fornasieri Filho, D.; Coicev, L.; Bordin, L.; Farinelli, R. Comportamento de Cultivares de Milho Semeadas em Diferentes Espaçamentos entre Linhas e Densidades Populacionais, na Safrinha. *Rev. Bras. Milho E Sorgo* **2003**, *2*, 52–60. [CrossRef]

50.   Souza, E.F.C.; Soratto, R.P. Efeito de Fontes e Doses de Nitrogênio em Cobertura, no Milho Safrinha, em Plantio Direto. *Rev. Bras. Milho E Sorgo* **2006**, *5*, 395–405. [CrossRef]

Article

Relationship between Phosphorus and Nitrogen Concentrations of Flax

Yaping Xie [1,2], Lingling Li [2,3], Limin Wang [1], Jianping Zhang [1,*], Zhao Dang [1], Wenjuan Li [1], Yanni Qi [1], Wei Zhao [1], Kongjun Dong [1], Xingrong Wang [1], Yanjun Zhang [1], Xiucun Zeng [4], Yangchen Zhou [2], Xingzhen Wang [1], Linrong Shi [5] and Gang Wu [6]

1. Crop Research Institute, Gansu Academy of Agricultural Sciences (GASS), Lanzhou 730070, China
2. College of Agronomy, Gansu Agricultural University (GSAU), Lanzhou 730070, China
3. State Key Laboratory of Aridland Crop Science, Gansu Agricultural University (GSAU), Lanzhou 730070, China
4. Key Laboratory of Hexi Corridor Resources Utilization of Gansu, Hexi University, Zhangye 734000, China
5. College of Mechanical and Electrical Engineering, Gansu Agricultural University (GSAU), Lanzhou 730070, China
6. Zhangye Water-Saving Agricultural Experimental Station, Gansu Academy of Agricultural Sciences (GAAS), Zhangye 734000, China
* Correspondence: zhangjpzw3@gsagr.ac.cn; Tel.: +86-093-1761-1081

Abstract: Tools quantifying phosphorus (P) status in plants help to achieve efficient management and to optimize crop yield. The objectives of this study were to establish the relationship between P and nitrogen (N) concentrations of flax (*Linum usitatissimum* L.) during the growth season to determine the critical P concentration for diagnosing P deficiency. Field experiments were arranged as split plots based on a randomized complete block design. Phosphorus levels (0, 40, 80, 120, and 160 kgP$_2$O$_5$ ha^{-1}) were assigned to the main plots, and cultivars (Dingya 22, Lunxuan 2, Longyaza 1, Zhangya 2, and Longya 14) were allocated to the subplots. Shoot biomass (SB) and P and N concentrations were determined at 47, 65, 74, 98, and 115 days after emergence. Shoot biomass increased, while P and N concentrations and the N:P ratio declined with time in each year. The P concentration in respect of N concentration was described using a liner relationship (P = 0.05, N + 1.68, R^2 = 0.76, p < 0.01) under non-limiting P conditions, in which the concentrations are expressed in g kg^{-1} dry matter (DM). The N:P ratio was fitted to a second-order polynomial equation (N:P = 11.56 × SB$^{-0.1}$, R^2 = 0.71, p = 0.03), based on the SB of flax. This research first developed a predictive model for critical P concentration in flax, as a function of N concentration in shoots of flax. The critical P concentration can be used as a promising alternative tool to quantify the degree of P deficiency of flax during the current growing season.

Keywords: nitrogen; phosphorus; shoot biomass; N:P ratio

Citation: Xie, Y.; Li, L.; Wang, L.; Zhang, J.; Dang, Z.; Li, W.; Qi, Y.; Zhao, W.; Dong, K.; Wang, X.; et al. Relationship between Phosphorus and Nitrogen Concentrations of Flax. *Agronomy* **2023**, *13*, 856. https://doi.org/10.3390/agronomy13030856

Academic Editors: Shahram Torabian, Ruijun Qin and Christos Noulas

Received: 29 January 2023
Revised: 9 March 2023
Accepted: 13 March 2023
Published: 15 March 2023

1. Introduction

Rock phosphate reserves are finite, non-renewable, and rapidly shrinking due to use in phosphorus (P) fertilizers [1]. Further, with the growing human population, the oversupply of P fertilizers in agriculture to maximize crop yield has resulted in a series of environmental, ecological, and human health issues [2,3]. Therefore, diagnosing P nutrient status and optimizing P fertilizer management have become important topics in agriculture production.

Flax (*Linum usitatissimum* L.) is an important oil crop and is used as an industrial material [4,5]. Many studies have shown that improving the productivity of flax to meet growing demands is worth investigating as flaxseed has functional nutritional ingredients for human health [4,5], flax oil is used in biodiesel production [6–9], and flax shives are used as a biosorbent and biochar after processing of fiber [10,11]. Thereafter, precise management of P fertilization of flax has become a core area of research [12].

A method to diagnose P nutrition, based on the relationship between P and N concentrations during growth, was proposed earlier on the basis of the relationship between P and N concentrations during growth [13]. A number of researchers have documented that N and P concentrations decrease with increasing plant biomass during crop growth [14–16]. With the dilution of N and P in plant biomass, P concentrations decrease when N is limiting [17]. The positive relationship between P and N concentrations was proposed by Kamprath [18] and reflects the dilution of both elements with increasing shoot biomass as well as indicating the effect of crop N status on P absorption in plants. The relationship between shoot N and P concentrations has also been reported in wheat [19–21], maize [22], rapeseed [23], and forage grasses [15,24]. The coupling of P and N is carried out through different mechanisms, such as N availability in accelerating P cycling [25], the availability of P on N cycling [26,27] and the control of biological N fixation [28]. Furthermore, P regulates N uptake and translocation, and vice versa [14,17,29]. Given the close balance and synergy of N and P in crops [14,30], it is crucial to assess the level of N deficiency and estimate the critical P concentration. At the same time, due to the decrease in N and P concentrations with increasing shoot biomass, use of the N:P ratio was also proposed for diagnostic purposes [10]. Use of the N:P ratio to detect the nature of nutrient limitations was also proposed by Koerselman and Meuleman [31] on natural ecosystems and by Sinclair et al. [32] on cut white clover/ryegrass swards. Moreover, Güsewell [14] and Greenwood et al. [33] reported that the N to P ratio decreases as plants grow larger. This could be because the relative decline in P concentration is lower compared to that in N concentration [24,33]. The necessity of a multi-element integrated approach to nutrition in crops is highlighted by the interaction between N and P in crops [16]. However, this relationship between P and N concentrations in shoots of flax under different P fertilization levels and the application of this the relationship in evaluating the critical P concentration have not been extensively studied in flax.

The objectives of this study were to elucidate the relationship between P and N concentrations of flax using data from experiments with five P rates, and with various flax cultivars grown under non-limiting P conditions. Specifically, we wanted to determine the critical P concentration using the relationship for shoot growth, which could be used to diagnose and quantify P deficiency in flax.

2. Materials and Methods

2.1. Site Description, Experimental Design and Treatments

Field experiments 1 and 2 were conducted at Dingxi Academy of Agricultural Science (34.26° N, 103.52° E, and altitude 2060 m) in 2017 and 2018 in Gansu, China; experiments 3 and 4 were carried out at Yongdeng (36°02′ N, 103°40′ E, altitude 2149 m) in 2018 and 2019, in Gansu, China. The two sites have a continental climate. The soil type is Arenosols [19]. Wheat was the previous crop for the four experiments.

Monthly mean temperatures over the growing season, from March to August, ranged from −4 to 26 °C at Dingxi and from −5 to 26 °C at Yongdeng. The lowest temperature was recorded in March and the highest value in July for the four-year sites. The monthly mean temperatures each year was close to the long-term average (30 yr). In brief, total precipitation over the growing season in March to August was from 264 to 259 mm at Dingxi and from 275 to 262 mm at Yongdeng.

The experiments were arranged as split plots based on a randomized complete block design with three replicates, with a plot size of $5.0 \text{ m} \times 4.0 \text{ m}$. Five P rates (0, 40, 80, 120, and 160 kg P_2O_5 ha^{-1}) were assigned to the main plots, with five cultivars (Dingxi: Lunxuan 2 and Dingya 22; and Yongdeng: Longyaza 1, Zhangya 2, and Longya 14, respectively) allocated to the subplots. Urea, calcium superphosphate, and potassium sulfate were incorporated into the top 30 cm of soil prior to sowing. The K rate was 52.5 kg K_2O ha^{-1}, and the N rate was 80 kg N ha^{-1} to flax. All of the P and K was used as the base fertilizer for flax while 75% of N was applied as the base fertilizer before sowing, and 25% as topdressing at the budding stage. The crop was only irrigated once, prior to flowering, and each plot received 40 mm of irrigation with pipes of 13 cm in diameter. A water meter installed at the discharging end of the pipes measured and recorded the amount of irrigation. Other crop management procedures followed along local agricultural practices to ensure maximum potential productivity.

2.2. Preplant Soil Sampling and Analysis

The soil samples were collected from a depth of 0–30 cm before the application of P fertilizer and were air dried. Soil pH was determined using the solution of 10 g soil: 10 mL water [34]. The analysis of available P was determined using the Colorimetric Molybdenum-Blue method according to Olsen et al. [35]. The micro-Kjeldahl method was used to quantify total N concentration [36]. Available K was measured by flame emission spectroscopy [36]. In brief, the basic information of soil in Dingxi of 2017 plots contains an organic matter of 10.2 g kg^{-1}, alkali-hydrolyzable N of 48.9 mg kg^{-1}, available P of 11.7 mg kg^{-1} and available K of 122.5 mg kg^{-1} and pH of 7.9. The soil in Dingxi of 2018 contains an organic matter of 11.0 g kg^{-1}, alkali-hydrolyzable N of 50.6 mg kg^{-1}, available P of 12.6 mg kg^{-1} and available K of 135.4 mg kg^{-1} and pH of 8.1. Additionally, the soil at Yongdeng was described as follows: an organic matter of 9.8 and 7.6 g kg^{-1}, alkali-hydrolyzable N of 53.9 and 48.2 mg kg^{-1}, available P of 8.0 and 8.7 mg kg^{-1} and available K of 178.3 and 141.6 mg kg^{-1}, pH of 7.5 and 8.2 in 2018 and 2019, respectively. When P concentration is below 10 mg kg^{-1}, it is considered low; the optimum soil P concentration is considered to be above 20 mg kg^{-1} [37].

2.3. Plant Sampling and Analysis

Each year growth period, 30 plants per plot were gathered to measure shoot biomass (SB) (namely, shoot dry matter), the P concentration and N concentration in shoot at 47, 65, 74, 98, and 115 days after emergence (DAE). At each year-site sampling date, 30 plants was randomly selected from the two central rows of a plot then separated above ground parts and roots. For chemical analysis, all above ground parts were rinsed with deionized water, then samples were oven-dried at 105 °C for half an hour, and then at 80 °C until they reached a constant weight and shoot biomass was weighed. The dry matter (DM) of shoot was ground to pass a 1 mm sieve for measuring P and N concentrations. The P and N concentrations were determined by the H_2SO_4-H_2O_2 digestion method, then P and N concentrations were quantified using the Colorimetric Molybdenum-Blue method [35] and the micro-Kjeldahl method [36], respectively.

Seed yield was measured in each plot by harvesting manually with a sickle.

2.4. Data Analysis

All data were subjected to analysis of variance (ANOVA) to compare differences between all studied parameters caused by the variation in P levels and cultivar, using the SPSS (version 19, Inc., Chicago, IL, USA) at a probability level of 5%. The differences among the treatments were calculated using the least significant difference (LSD) test at the 95% confidence level. Different P levels and cultivars were investigated as fixed impacts when present in all experiments. The relationship between P and N concentrations under nonlimiting conditions was described by linear regressions of SPSS by a combined analysis. A non-P-limiting treatment was defined as one in which P application did not lead to an increase in shoot biomass; however, there was a significant increment in shoot P concentration (SPC). The critical P concentration, which is defined as the lowest P concentration required to obtained highest shoot growth [12].

The P nutrition index (PNI) was determined by dividing the P concentration in shoot by the critical P concentration, similar to an approach previously used on flax [12]. The relative shoot biomass (RSB) and relative seed yield (RY) were the rates of shoot biomass and seed yield gained for a given P level to their respective peak values observed at a specific year [12]. The coefficients of determination (R^2) were calculated using SPSS 20.0.

3. Results

3.1. Shoot Biomass at Different P Levels

Phosphorus fertilization significantly improved the SB of flax in all years-sites excluded at 47 DAE (Tables 1 and 2). Moreover, the SB of flax were not significant difference among P_{80}, P_{120}, and P_{160} treatments. Moreover, the SB increased gradually, as plants grew from 47 to 115 DAE. Over the two years, the average SB of Luanxuan 2 ranged in between 0.82 to 7.18 t ha^{-1} and Dingya 22 ranged within 0.92 to 7.22 t ha^{-1}, while in case of Longyaza 1 ranged in between 0.82 and 7.26 t ha^{-1}, Zhangya 2 ranged from 0.87 to 7.21 t ha^{-1}, and Longya 14 varied from 0.91 to 7.38 t ha^{-1}, respectively (data not shown).

Table 1. Shoot biomass (t ha^{-1}) at Dingxi in 2017 and 2018 with two cultivars flax and five phosphorus rates.

Year	Treatment	DAE 47	DAE 65	DAE 74	DAE 98	DAE 115
	Cultivar					
2017	Lunxuan 2	1.22	2.03	3.56	4.68 b	6.19
	Dingya 22	1.30	2.46	3.62	4.94 a	6.33
2018	Lunxuan 2	1.13	2.00 b	3.50	4.66	6.15
	Dingya 22	1.23	2.31 a	3.59	4.74	6.21
	P rate					
	P_0	0.89	1.30 c	2.33 c	3.41 c	4.51 c
	P_{40}	1.20	1.82 b	3.11 b	4.43 b	5.51 b
2017	P_{80}	1.34	2.57 ab	4.21 a	5.40 a	6.94 a
	P_{120}	1.43	2.72 a	4.08 a	5.42 a	7.09 a
	P_{160}	1.47	2.83 a	4.25 a	5.40 a	7.26 a
	P_0	0.85	1.30 c	2.39 c	3.21	4.49 c
	P_{40}	1.17	1.70 b	2.96 b	3.96	5.47 b
2018	P_{80}	1.28	2.52 a	4.15 a	5.44	6.79 a
	P_{120}	1.31	2.61 a	4.12 a	5.46	7.01 a
	P_{160}	1.29	2.65 a	4.12 a	5.44	7.14 a
	Source of variance (SOV)					
	C	0.3142	0.5276	0.5441	0.0032	0.2183
2017	P	0.5051	0.0042	0.0313	0.0113	0.0372
	C × P	0.9011	0.0192	0.7465	0.7698	0.3521
	C	0.8326	0.0120	0.4798	0.5266	0.7145
2018	P	0.1425	<0.0001	0.0244	0.0158	0.0116
	C × P	0.3764	0.0197	0.5218	0.2671	0.6899

C, cultivar. P, phosphorus. DAE, days after emergence. Means (n = 3) with a different letter in each column are significantly different at the 5% probability level according to the least significant difference test. P_0, P_{40}, P_{80}, P_{120}, and P_{160} represent 0, 40, 80, 120, and 160 kg P_2O_5 ha^{-1}, respectively.

Table 2. Shoot biomass (t ha^{-1}) at Yongdeng in 2018 and 2019 with three cultivars flax and five phosphorus rates.

Year	Treatment	DAE 47	DAE 65	DAE 74	DAE 98	DAE 115
	Cultivar					
2018	Longyaza 1	1.22	2.46	3.55	4.68	6.48
	Zhangya 2	1.14	2.46	3.38	4.68	6.35
	Longya 14	1.27	2.60	3.57	4.63	6.55
2019	Longyaza 1	1.18	2.44	3.51	4.48 c	6.38
	Zhangya 2	1.22	2.51	3.49	4.66 b	6.38
	Longya 14	1.17	2.49	3.74	4.83 a	6.41
	P rate					
	P_0	0.87	1.48 c	2.38 c	3.08 c	4.87 c
	P_{40}	1.09	1.97 b	2.75 b	4.06 b	5.78 b
2018	P_{80}	1.35	2.95 ab	4.09 a	5.34 a	7.10 ab
	P_{120}	1.35	3.05 a	4.12 a	5.41 a	7.24 a
	P_{160}	1.40	3.07 a	4.16 a	5.44 a	7.31 a
	P_0	0.85	1.41 c	2.43 b	3.35 c	4.79 c
	P_{40}	1.08	2.02 b	2.95 b	4.01 b	5.62 b
2019	P_{80}	1.29	2.93 ab	4.17 a	5.29 a	7.12 a
	P_{120}	1.34	3.01 a	4.12 a	5.35 a	7.17 a
	P_{160}	1.39	3.03 a	4.23 a	5.28 a	7.26 a
	Source of variance (SOV)					
	C	0.3481	0.5664	0.0602	0.0722	0.1233
2018	P	0.527	<0.0001	0.0247	0.0196	0.0416
	C × P	0.644	0.4152	0.0809	0.1009	0.0208
	C	0.241	0.0815	0.0941	0.0247	0.2258
2019	P	0.089	<0.0001	0.0125	<0.0001	0.0200
	C × P	0.529	0.0864	0.2145	0.0992	0.7431

C, cultivar. P, phosphorus. DAE, days after emergence. Means (n = 3) with different letters in each column are significantly different at the 5% probability level according to the least significant difference test. P_0, P_{40}, P_{80}, P_{120}, and P_{160} represent 0, 40, 80, 120, and 160 kg P_2O_5 ha^{-1}, respectively.

The SB was affected by cultivar, the SB of Dingya 22 was greater than that of Lunxuan 2 at 98 DAE of 2017 and at 65 DAE of 2018 in Dingxi (Table 1), and Longya 14 had highest SB, moreover, there were significant difference among three cultivars at 98 DAE in Yongdeng of 2019 (Table 2). In addition, the interaction between cultivar and P influenced the SB. The relationship between SB and P rate over two cultivars at 65 DAE in Dingxi of 2018 can be described through the linear functions: $SB = 0.398P_{rate} + 1.054$ ($R^2 = 0.90$, $p < 0.01$) in Dingxi of 2017 and $SB = 0.360P_{rate} + 1.075$ ($R^2 = 0.88$, $p < 0.01$). Additionally, the SB was affected by the interaction between cultivar and P rate at 115 DAE in Yongdeng of 2018, the relationship between SB and P rate over three cultivars can be described using the linear functions: $SB = 0.595P_{rate} + 4.615$ ($R^2 = 0.86$, $p < 0.01$).

3.2. Shoot P Concentration at Different P Levels

With the exception of sampling date at 47 DAE in each site-year and at 74 DAE in Dingxi of 2017, P fertilizer significantly influenced shoot P concentration of flax (Tables 3 and 4). Nevertheless, there were no differences in shoot P concentration between P_{120} and P_{160} treatments in Dingxi and no differences in shoot P concentration among P_{80}, P_{120}, and P_{160} treatments in Yongdeng. In general, shoot P concentration increased with P levels, increasing at the same sampling date and the same cultivar. Averaged over the P_{40}, P_{80}, P_{120}, and P_{160} treatments, the fertilized flax increased the P concentration in the shoot by 18, 18, 17, and 17% in Dingxi of 2017 and 2018, in Yongdeng of 2018 and 2019, respectively, compared with the zero P control. Furthermore, P concentration decreased with time from 47 DAE to 115 DAE. Across sampling dates and site years, P concentration ranged from 1.77 to 5.36 g kg^{-1} DM (data not shown).

Table 3. Shoot P concentration (g kg^{-1} DM) at Dingxi in 2017 and 2018 with two cultivars flax and five phosphorus rates.

Year	Treatment	DAE 47	DAE 65	DAE 74	DAE 98	DAE 115
	Cultivar					
2017	Lunxuan 2	5.11	3.47	2.58 b	2.49	2.14
	Dingya 22	5.06	3.45	2.89 a	2.29	2.06
2018	Lunxuan 2	5.02	3.31 b	3.01	2.70	2.40
	Dingya 22	5.07	3.52 a	3.14	2.59	2.16
	P rate					
2017	P_0	4.79	3.01 b	2.31	1.88 c	1.78 b
	P_{40}	4.97	3.15 b	2.58	2.03 bc	1.86 b
	P_{80}	5.18	3.46 a	2.68	2.39 b	1.98 ab
	P_{120}	5.20	3.78 a	2.89	2.75 a	2.35 a
	P_{160}	5.31	3.92 a	3.22	2.91 a	2.55 a
2018	P_0	4.80	3.08 b	2.58 c	2.11 c	1.86 b
	P_{40}	4.98	3.22 b	2.83 b	2.35 bc	1.98 b
	P_{80}	5.03	3.41 ab	3.03 b	2.67 b	2.25 ab
	P_{120}	5.17	3.49 a	3.43 a	2.98 a	2.59 a
	P_{160}	5.25	3.88 a	3.51 a	3.13 a	2.73 a
	Source of variance (SOV)					
2017	C	0.5277	0.06431	0.0128	0.3440	0.0812
	P	0.0976	0.0129	0.0906	<0.0001	0.0215
	C × P	0.1254	0.4685	0.0342	0.7411	0.4125
2018	C	0.0708	0.0366	0.7164	0.4962	0.3588
	P	0.0924	0.0218	0.0324	<0.0001	0.0400
	C × P	0.1457	0.2586	0.3457	0.2568	0.6215

P, phosphorus. DM, dry matter. C, cultivar. DAE, days after emergence. Means (n = 3) with different letters in each column are significantly different at the 5% probability level according to the least significant difference test. P_0, P_{40}, P_{80}, P_{120}, and P_{160} represent 0, 40, 80, 120, and 160 kg P_2O_5 ha^{-1}, respectively.

Table 4. Shoot P concentration (g kg^{-1} DM) at Yongdeng in 2018 and 2019 with three cultivars flax and five phosphorus rates.

Year	Treatment	DAE 47	DAE 65	DAE 74	DAE 98	DAE 115
	Cultivar					
	Longyaza 1	4.84	3.68 a	3.10	2.54	2.16
2018	Zhangya 2	4.92	3.39 b	3.18	2.71	2.14
	Longya 14	4.71	3.66 a	3.29	2.78	2.37
	Longyaza 1	4.83	3.51	3.21 a	2.74	2.33
2019	Zhangya 2	4.79	3.76	2.69 b	2.52	2.35
	Longya 14	4.86	3.50	3.19 a	2.69	2.51
	P rate					
	P_0	4.44	3.14 c	2.79 c	2.24 c	1.90 b
	P_{40}	4.60	3.41 b	2.96 b	2.37 bc	2.14 b
2018	P_{80}	4.85	3.55 ab	3.18 a	2.60 ab	2.18 ab
	P_{120}	5.08	3.72 a	3.35 a	3.02 a	2.40 a
	P_{160}	5.13	4.05 a	3.65 a	3.13 a	2.51 a
	P_0	4.53	3.37 b	2.73 c	2.16 b	1.94 b
	P_{40}	4.70	3.42 b	2.91 b	2.36 b	2.20 b
2019	P_{80}	4.73	3.51 ab	2.96 ab	2.70 ab	2.45 a
	P_{120}	5.04	3.70 a	3.13 ab	2.93 a	2.62 a
	P_{160}	5.12	3.95 a	3.42 a	3.11 a	2.76 a
	Source of variance (SOV)					
	C	0.4188	0.0259	0.0912	0.2438	0.6257
2018	P	0.0954	0.0329	<0.0001	0.0241	0.0115
	C $\times$ P	0.0325	0.2549	0.4752	0.3892	0.2567
	C	0.1230	0.4258	0.0344	0.0615	0.1281
2019	P	0.4785	0.0329	<0.0001	0.0274	0.0315
	C $\times$ P	0.6352	0.0174	0.7548	0.6351	0.5322

P, phosphorus. DM, dry matter. C, cultivar. DAE, days after emergence. Means (n = 3) with different letters in each column are significantly different at the 5% probability level according to the least significant difference test. P_0, P_{40}, P_{80}, P_{120}, and P_{160} represent 0, 40, 80, 120, and 160 kg P_2O_5 ha^{-1}, respectively.

The shoot P concentration was affected by cultivar, the P concentration of Dingya 22 was greater than that of Lunxuan 2 at 74 DAE of 2017 and at 65 DAE of 2018 in Dingxi (Table 3); and Zhangya 2 was observed the lowest P concentration at 65 DAE of 2018 and at 74 DAE of 2019 in Yongdeng (Table 4). Furthermore, cultivar and P interaction significantly affected the P concentration was at 74 DAE in Dingxi 2017, at 47 and 115 DAE of 2018 and at 65 DAE of 2019 in Yongdeng. The relationship between P concentration and P rate at 74 DAE over two cultivars can be described through the linear functions: $P_{concentration} = 0.213P_{rate} + 2.097$ ($R^2 = 0.79$, $p < 0.01$) in Dingxi of 2017, and the relationship between P concentration and P rate over three cultivars can be described through the linear functions: $P_{concentration} = 0.187P_{rate} + 4.261$ ($R^2 = 0.88$, $p < 0.01$), $P_{concentration} = 0.206P_{rate} + 1.775$ ($R^2 = 0.89$, $p < 0.01$), and $P_{concentration} = 0.142P_{rate} + 3.164$ ($R^2 = 0.83$, $p < 0.01$), at 47 and 115 DAE in 2018 and at 65 DAE of 2019 in Yongdeng, respectively.

3.3. Shoot N Concentration at Different P Levels

Shoot N concentration was significantly affected by P fertilization apart from at 47 DAE in each site-year and at 65 DAE in Yongdeng of 2019 (Tables 5 and 6). Averaged over the P_{40}, P_{80}, P_{120}, and P_{160} treatments, the fertilized flax increased the N concentration in shoot. Similar to shoot P concentration, shoot N concentration also decreased with plant growth from 47 to 115 DAE. In the study, over sampling dates all years, N concentration varied between 16.77 and 54.78 g kg^{-1} DM (Tables 5 and 6).

Table 5. Shoot N concentration (g kg^{-1} DM) at Dingxi in 2017 and 2018 with two cultivars flax and five phosphorus rates.

Year	Treatment	DAE 47	DAE 65	DAE 74	DAE 98	DAE 115
	Cultivar					
2017	Lunxuan 2	53.83	35.71	25.40 b	21.73	18.22
	Dingya 22	53.42	34.65	27.59 a	21.83	18.51
2018	Lunxuan 2	54.93	36.51	30.43	23.39	19.97
	Dingya 22	53.31	35.76	31.03	24.83	20.16
	P rate					
	P_0	53.40	32.93 b	24.61 c	19.10 c	16.77 b
	P_{40}	53.55	33.83 b	25.93 b	20.30 b	17.66 b
2017	P_{80}	53.68	34.82 ab	26.21 ab	22.60 ab	17.88 ab
	P_{120}	53.77	36.99 a	26.90 a	23.18 a	19.48 a
	P_{160}	53.74	37.36 a	28.83 a	23.73 a	20.06 a
	P_0	52.96	33.93 c	27.50 c	21.92 c	18.86 b
	P_{40}	54.16	34.81 b	29.90 b	23.40 b	18.99 b
2018	P_{80}	54.30	36.44 ab	30.55 ab	24.60 a	20.44 a
	P_{120}	54.43	37.50 a	32.69 a	25.34 a	20.50 a
	P_{160}	54.78	38.00 a	33.02 a	25.31 a	21.53 a
	Source of variance (SOV)					
	C	0.0942	0.7415	0.0195	0.4578	0.6942
2017	P	0.2043	0.0125	<0.0001	0.0005	0.0200
	C × P	0.4108	0.4256	0.3289	0.3452	0.0981
	C	0.0922	0.6500	0.0672	0.1280	0.3211
2018	P	0.1288	0.0248	0.0324	0.0288	0.0109
	C × P	0.3145	0.0904	0.2584	0.0127	0.0992

N, nitrogen. DM, dry matter. P, phosphorus. C, cultivar. DAE, days after emergence. Means (n = 3) with different letters in each column are significantly different at the 5% probability level according to the least significant difference test. P_0, P_{40}, P_{80}, P_{120}, and P_{160} represent 0, 40, 80, 120, and 160 kg P_2O_5 ha^{-1}, respectively.

Table 6. Shoot N concentration (g kg^{-1} DM) at Yongdeng in 2018 and 2019 with three cultivars flax and five phosphorus rates.

Year	Treatment	DAE 47	DAE 65	DAE 74	DAE 98	DAE 115
	Cultivar					
	Longyaza 1	54.66	39.97	32.99	24.84	20.15
2018	Zhangya 2	52.36	35.80	32.71	26.89	20.96
	Longya 14	54.88	38.32	33.23	26.76	21.90
	Longyaza 1	53.74	39.16	33.94 a	26.22	21.87
2019	Zhangya 2	51.97	35.74	24.64 c	23.11	20.26
	Longya 14	54.10	37.48	29.62 b	23.81	22.47
	P rate					
	P_0	54.16	37.18 b	31.43 b	24.38 b	19.79 b
	P_{40}	53.86	37.53 b	32.15 b	24.73 b	20.27 b
2018	P_{80}	53.59	38.34 ab	33.15 a	25.69 ab	20.82 ab
	P_{120}	54.32	38.41 a	33.74 a	27.81 a	21.42 a
	P_{160}	53.91	39.89 a	34.42 a	28.20 a	22.03 a
	P_0	52.84	36.60	28.30 b	21.73 c	18.95 c
	P_{40}	52.95	36.74	28.51 b	22.38 c	20.02 b
2019	P_{80}	52.97	37.25	29.04 ab	24.97 b	21.94 b
	P_{120}	53.76	38.03	29.38 ab	25.83 a	23.06 a
	P_{160}	53.82	38.69	31.78 a	26.99 a	23.67 a
	Source of variance (SOV)					
	C	0.0815	0.2708	0.0815	0.0679	0.4352
2018	P	0.2431	0.0142	0.0403	0.0279	0.0183
	C × P	0.1688	0.4215	0.0855	0.2144	0.3216
	C	0.0621	0.2789	<0.0001	0.0740	0.1259
2019	P	0.3219	0.1528	0.0224	0.0165	0.0411
	C × P	0.1452	0.0578	0.0298	0.0911	0.2016

N, nitrogen. DM, dry matter. P, phosphorus. C, cultivar. DAE, days after emergence. Means (n = 3) with different letters in each column are significantly different at the 5% probability level according to the least significant difference test. P_0, P_{40}, P_{80}, P_{120}, and P_{160} represent 0, 40, 80, 120, and 160 kg P_2O_5 ha^{-1}, respectively.

The shoot N concentration was affected by cultivar at 74 DAE in Dingxi of 2017 (Table 5). In Yongdeng, N concentration was affected by cultivar at 74 DAE of 2019, Longyaza 1 exhibited the maximum value, followed by Longya 14 and Zhangya 2 (Table 6). Moreover, the interaction between cultivar and P rate impacted the N concentration at 98

DAE in Dingxi of 2018 and at 74 DAE in Yongdeng of 2019 (Tables 5 and 6). The relationship between N concentration and P rate over two cultivars at 98 DAE of 2018 can be described through the linear functions: $N_{concentration} = 0.871P_{rate} + 21.498$ ($R^2 = 0.71$, $p < 0.01$), and the relationship between N concentration and P rate over three cultivars at 74 DAE of 2019 can be described by the linear functions: $N_{concentration} = 0.782P_{rate} + 27.053$ ($R^2 = 0.76$, $p < 0.01$).

3.4. Phosphorus and N Concentration Relationships in Shoot

This study was to elucidate the relationship between P and N concentrations of flax throughout the growing period, namely from 47 to 115 DAE under non-limiting P conditions. Hence, the data from the experiments conducted in Dingxi of 2017 and 2018 under non-limiting P conditions were pooled with data obtained under non-limiting P conditions in Yongdeng of 2018 and 2019. In the current, shoot N concentration increased with increasing P concentration at four sites-years. Obviously, the relationship between N and P concentrations in shoot under non-limiting P conditions can be described through the linear function: $P = 0.05N + 1.68$ ($R^2 = 0.82$, $p < 0.01$) (Figure 1A), in which both concentrations are expressed in g kg^{-1} DM. This relationship approximates the critical P concentration under non-limiting P conditions that is, the minimum P concentration needed when attained the highest shoot growth.

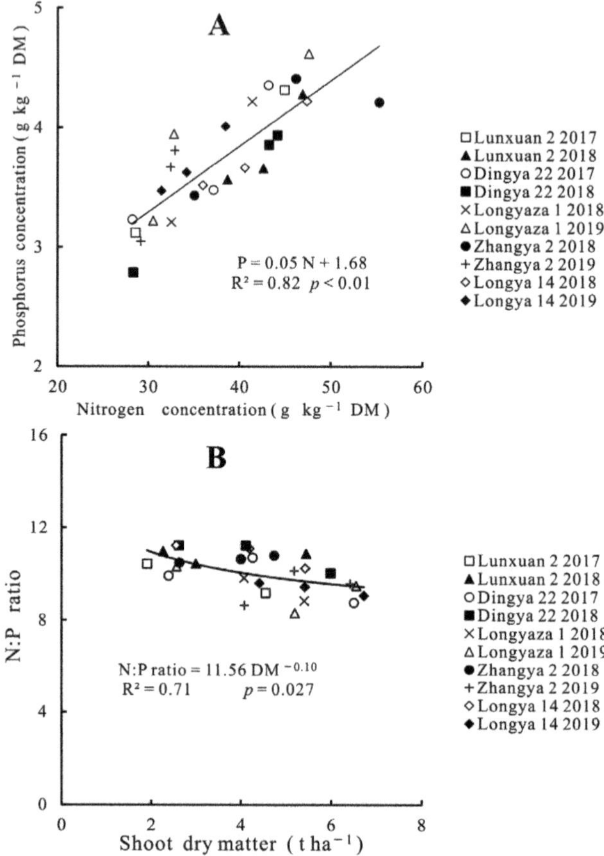

Figure 1. Relationship between P and N concentrations in shoot (**A**) and between the ration N concentration to P concentration in shoot of flax under non-limiting P conditions (**B**). P, phosphorus; N, nitrogen; SB, shoot biomass.

3.5. Implications for P Diagnostic in Flax

The rate of N:P has also been suggested to diagnose purposes. In the present study, the N:P ratio declined with increasing biomass ($R^2 = 0.71$, $p = 0.027$) (Figure 1B), similar to the changes trend of N and P concentrations. The N:P ratio was 11.56, corresponding to 1 t ha^{-1} DM, within the range of 10:20 (mass basis) reported to be optimal by Güsewell [14]. Moreover, the dilution coefficient of the N:P curve was 0.10.

To diagnose P nutrition status, the PNI and relative seed yield (RY) was applied for all sampling dates. The values of PNI < 1, P deficiency, while values > 1, P excess, and PNI was 1, P optimal. Our results showed that the relationship between the PNI and RY was well fitted using a second-order polynomial equation (RY = -1.58 PNI^2 + 3.38 PNI $-$ 0.82) ($R^2 = 0.88$, $p < 0.001$) (Figure 2). Value PNI = 1, the RY was near 1.0, while PNI > 1 or PNI < 1, the RY reduced on the basis of those relationships. Those indicate that inadequate and excessive of P application both lower the RY, while the optimal P rate leads to the maximum RY.

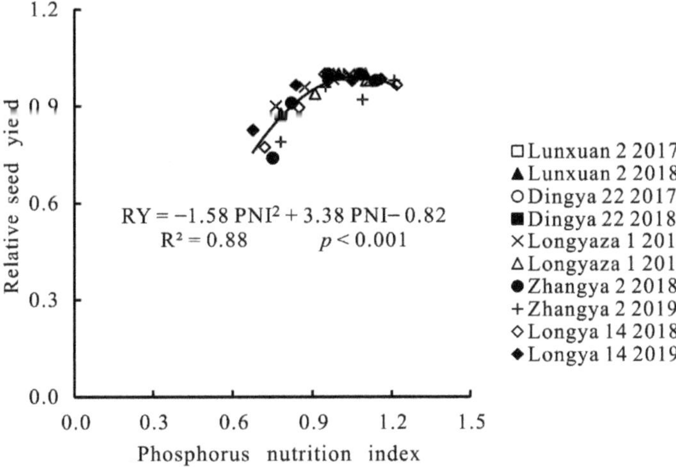

Figure 2. Relationship between relative seed yield (RY) and phosphorus nutrition index (PNI) with five flax cultivars. The PNI data were averaged over three years and two sites.

4. Discussion

This study first suggests the notion of the critical P concentration through the relationship between P and N concentrations in shoot of flax to quantify the degree of P deficiency during the current growing season. This diagnostic model based on this relationship between P and N concentrations can be used to guide agricultural field production.

4.1. Shoot Dry Matter, Phosphorus and Nitrogen Concentrations

The values of shoot biomass on flax in the current study are lower than those reported by Flénet et al. [38] in a study conducted in northern France with four levels of fertilizer N. The growing season in Gansu, China, has cold and dry conditions. However, the water holding capacities of soils are great due to the cool and humid climate of northern France. Hence, there is little risk of water deficit. Irrigation was triggered when the soil water content of the 0–30 cm layer was below half of the soil water availability. Lower biomass in our study (ranged from 1 to 7 t ha^{-1}) compared to those conducted in northern France (1 to 10 t ha^{-1}) could be correlated with differences in water availability, cultivars, and fertilizer type.

Our results observed that P concentration in shoot varied from 1.77–5.36 g kg^{-1} DM from 47 to 115 DAE, namely seedling to maturity, which a wider range than those in C3 crops reported by Bélanger et al. [20] on wheat (1.6–4.6 g kg^{-1} DM) from vegetative to late heading stages of development in Canada and by Cadot et al. [39] of rapeseed (5.38–6.52 g kg^{-1} DM) between inflorescence emergence and ripening and of wheat (3.84–4.53 g kg^{-1} DM) between tillering and joint stage in Switzerland. This probably due to the following: (i) sampling dates were at different stages of growth and development among them, which might have caused the different the capability of uptake P; (ii) the difference may be correlated to species, of which organ weight ratios and P concentration of organs were significant different; and (iii) the difference could be correlated to environments, especial soils properties and soil microbial phosphorus. Further research is required to do for exploring the cause.

Shoot N concentration in flax ranged between 16.12 g kg^{-1} DM and 55.70 g kg^{-1} DM in our study, which a wider range than that reported on linseed in northern France [38], and others C3 crops, such as winter wheat in Canada (14.4–43.4 g kg^{-1} DM), in Finland (17.3–49.8 g kg^{-1} DM), and in China (17.3–32.6 g kg^{-1} DM) [20]. The difference among crops was probably explained with diversity in P and N absorption and utilization among different species with various organ weight ratios and N concentration of organs, and significant differences existing in N concentrations under the environment's conditions.

In addition, the effect of cultivar and the interaction between cultivar and P doses on SB, N, and P concentrations were few, with a sampling date from 47 to 113 DAE. However, this study intended to elucidate the relationship between P and N concentrations of flax throughout the growing period under within a range of P levels, hence, we concentrated on analysis of the effect of P doses on SB, N, and P concentrations.

4.2. Diagnosis of Phosphorus Nutrition Status

Our results exhibited that the N and P concentrations in shoot of flax existed a significant positive correlation. This confirms the powerful inter-dependence between N and P in crops [40], as observed in previous studies on grasses [13,24], wheat leaves [19,22], grassland swards [15], and canola [20]. Additionally, in the present study, the value of N:P ratio is in the extent of reported previously in terrestrial plants (10–20) [14] and oilseed crops (1.5–20) [40]. These strongly supported the results of the present study.

In 2008, Greenwood et al. [33] established the unifying N:P dilution curve for several crops, in which assumed non-limiting nutrient availability. The N:P dilution curve in our study was higher than those developed by Greenwood et al. [33]. This is possibly due to differences in N and P requirements between flax and the others crop species. The N:P ratios in shoot of flax (11.56) are close to the value of 11.83 for growth-related tissues, identifying that the interpretation of the N:P ratio for diagnosis of N and P sufficiency must be closely connected to the biomass [15,33,37].

According to the opinion of Güsewell [14] there is a 'critical N:P ratio' below which growth is limited only by N and above which growth is limited only by P. In the current study, the N:P ratios in shoot were in the range of 10–20, which validates that N and P were in sufficiency for the data set used to develop the critical P concentration curve.

Our results clearly showed that there were significant quadratic relationships between PNI and relative seed yield. The quadratic relationships correlated to seed yield increased with the P fertilizer up to 120 kg P_2O_5 ha^{-1} and then declined with further P application rate. The positive effect of appropriate P application on seed yield may be a consequence of improving photosynthesis [41] and increasing photosynthesis efficiency [42]. Value of relative seed yield was 1, PNI was 1, and the seed yield reached the peak, P optimal; PNI > 1 or PNI < 1, P excess or deficiency, those production decreased. Obviously, P nutrition status can be estimated by sampling biomass of shoot and this would represent a fast and cost effective option. Therefore, our results showed that the relationship between the P and N concentrations provided tools to evaluate the critical P concentration, in turn, to assess P status of flax during growing season.

5. Conclusions

Effective management of P fertilizer is critical to crop production and environment. In the current study, the N and P concentrations as well as N:P ratio of shoot all declined with increasing shoot biomass. The positive relationship between P and N concentrations under non-limiting P conditions for shoot growth identified in the strong stoichiometry between P and N in flax. Moreover, studies showed that the relationship between P and N concentrations and the N:P ratio in shoot are potential indices of P nutrition sufficiency, however, considering in relation to shoot biomass amount. When PNI was 1, the maximum seed yield obtained. Therefore, the current study provides diagnosis tool by the relationship between P and N concentrations to estimate the critical P concentration for quantifying the degree of P deficiency in flax production. This tool can be used to adjust P fertilization in the following growing seasons for the species-specific condition at the pH approximately to 8 of soil.

Author Contributions: Conceptualization, Writing—original draft, review and editing, Y.X. and J.Z.; Methodology, W.L.; Software, L.W., X.Z. and X.W. (Xingzhen Wang); Validation, Investigation, K.D., L.L., Z.D., Y.Q., X.W. (Xingrong Wang), Y.Z. (Yanjun Zhang), Y.Z. (Yangchen Zhou), L.S., G.W. and W.Z. All authors have read and agreed to the published version of the manuscript.

Funding: This research was funded by the Key Research and Development Projects of Gansu Academy of Agricultural Sciences (2021GAAS020), the National Natural Science Programs of China (31660368), the Major Special Projects of Gansu Province (21ZD4NA022-02), Gansu Intellectual Property Program (21ZSCQ026), and China Agriculture Research System of MOF and MARA (CARS-17-GW-04).

Data Availability Statement: The datasets used and/or analyzed during the current study are available from the corresponding author upon reasonable request.

Acknowledgments: We gratefully thank the journal's editors and two anonymous reviewers for offering some excellent, constructive suggestions for further improvement our manuscript. We also appreciate the support and help from Junyi Niu.

Conflicts of Interest: The authors declare no conflict of interest.

References

1.   Elser, J.; Bennett, E. Phosphorus cycle: A broken biogeochemical cycle. *Nature* **2011**, *478*, 29–31. [CrossRef]
2.   MacDonald, G.K.; Bennett, E.M.; Potter, P.A.; Ramankutty, N. Agronomic phosphorus imbalances across the world's croplands. *Proc. Natl. Acad. Sci. USA* **2011**, *108*, 3086–3091. [CrossRef]
3.   Li, B.; Boiarkina, I.; Yu, W.; Ming, H.; Munir, T.; Qian, G.; Young, B.R. Phosphorous recovery through struvite crystallization: Challenges for future design. *Sci. Total Environ.* **2019**, *648*, 1244–1256. [CrossRef] [PubMed]
4.   Mueed, A.; Shibli, S.; Jahangir, M.; Jabbar, S.; Deng, Z.Y. A comprehensive review of flaxseed (*Linum usitatissimum* L.): Health-affecting compounds, mechanism of toxicity, detoxification, anticancer and potential risk. *Crit. Rev. Food Sci. Nutr.* **2022**, 1–24. [CrossRef]
5.   Parikh, M.; Maddaford, T.G.; Austria, J.A.; Aliani, M.; Netticadan, T.; Pierce, G.N. Dietary flaxseed as a strategy for improving human health. *Nutrients* **2019**, *11*, 1171. [CrossRef] [PubMed]
6.   Dixit, S.; Kanakraj, S.; Rehman, A. Linseed oil as a potential resource for bio-diesel: A review. *Renew. Sust. Energy Rev.* **2012**, *16*, 4415–4421. [CrossRef]
7.   Bacenetti, J.; Restuccia, A.; Schillaci, G.; Failla, S. Biodiesel production from unconventional oilseed crops (*Linum usitatissimum* L. and *Camelina sativa* L.) in Mediterranean conditions: Environmental sustainability assessment. *Renew. Energy* **2017**, *112*, 444–456. [CrossRef]
8.   Ahmad, T.; Danish, M.; Kale, P.; Geremew, B.; Adeloju, S.B.; Nizami, M.; Ayoub, M. Optimization of process variables for biodiesel production by transesterification of flaxseed oil and produced biodiesel characterizations. *Renew. Energy* **2019**, *139*, 1272–1280. [CrossRef]
9.   Danish, M.; Kale, P.; Ahmad, T.; Ayoub, M.; Geremew, B.; Adeloju, S. Conversion of flaxseed oil into biodiesel using KOH catalyst: Optimization and characterization dataset. *Data Brief* **2020**, *29*, 105225. [CrossRef]
10.   Dey, P.; Mahapatra, B.S.; Juyal, V.K.; Pramanick, B.; Negi, M.S.; Paul, J.; Singh, S.P. Flax processing waste -A low-cost, potential biosorbent for treatment of heavy metal, dye and organic matter contaminated industrial wastewater. *Ind. Crop. Prod.* **2021**, *174*, 114195. [CrossRef]

11. Khiari, B.; Amel, I.; Azzaz, A.; Jellali, S.; Limousy, L.; Jeguirim, M. Thermal conversion of flax shives through slow pyrolysis process: In-depth biochar characterization and future potential use. *Biomass Convers. Bior.* **2021**, *11*, 325–337. [CrossRef]
12. Xie, Y.P.; Li, Y.; Wang, L.M.; Nizamani, M.M.; Lv, Z.C.; Dang, Z.; Li, W.J.; Qi, Y.N.; Zhao, W.; Zhang, J.P.; et al. Determination of Critical Phosphorus Dilution Curve Based on Capsule Dry Matter for Flax in Northwest China. *Agronomy* **2022**, *12*, 2819. [CrossRef]
13. Bélanger, G.; Richards, J.E. Relationships between P and N concentrations in timothy. *Can. J. Plant Sci.* **1999**, *79*, 65–70. [CrossRef]
14. Güsewell, S. N:P ratios in terrestrial plants: Variation and functional significance. *New Phytol.* **2004**, *164*, 243–266. [CrossRef]
15. Bélanger, G.; Ziadi, N.; Lajeunesse, J.; Jouany, C.; Virkajarvi, P.; Sinaj, S.; Nyiraneza, J. Shoot growth and phosphorus-nitrogen relationship of grassland swards in response to mineral phosphorus fertilization. *Field Crop. Res.* **2017**, *204*, 31–41. [CrossRef]
16. Lemaire, G.; Sinclair, T.; Sadras, V.; Bélanger, G. Allometric approach to crop nutrition and implications for crops diagnosis and phenotyping. A review. *Agron. Sust. Dev.* **2019**, *39*, 27. [CrossRef]
17. Duru, M.; Ducrocq, H. A nitrogen and phosphorus herbage index as a tool for assessing the effect of N and P supply on the dry matter yield of permanent pastures. *Nutr. Cycl. Agroecosyt.* **1997**, *47*, 56–69. [CrossRef]
18. Kamprath, E.J. Enhanced Phosphorus Status of Maize Resulting from Nitrogen Fertilization of High Phosphorus Soils. *Soil Sci. Soc. Am. J.* **1987**, *51*, 1522–1526. [CrossRef]
19. Ziadi, N.; Belanger, G.; Cambouris, A.N.; Tremblay, N.; Nolin, M.C.; Claessens, A. Relationship between phosphorus and nitrogen concentrations in spring wheat. *Agron. J.* **2008**, *100*, 80–86. [CrossRef]
20. Bélanger, G.; Ziadi, N.; Pageau, D.; Grant, C.; Högnasbacka, M.; Virkajärvi, P.; Hu, Z.; Lu, J.; Lafond, J.; Nyiraneza, J. A model of critical phosphorus concentration in the shoot biomass of wheat. *Agron. J.* **2015**, *107*, 963–970. [CrossRef]
21. Fontana, M.; Bélanger, G.; Hirte, J.; Ziadi, N.; Elfouki, S.; Bragazza, L.; Liebisch, F.; Sinaj, S. Critical plant phosphorus for winter wheat assessed from long-term field experiments. *Eur. J. Agron.* **2021**, *126*, 126263. [CrossRef]
22. Ziadi, N.; Bélanger, G.; Cambouris, A.N.; Tremblay, N.; Nolin, M.C.; Claessens, A. Relationship between P and N concentrations in corn. *Agron. J.* **2007**, *99*, 833–841. [CrossRef]
23. Bélanger, G.; Ziadi, N.; Pageau, D.; Grant, C.; Lafond, J.; Nyiraneza, J. Shoot growth, phosphorus–nitrogen relationships, and yield of canola in response to mineral phosphorus fertilization. *Agron. J.* **2015**, *107*, 1458–1464. [CrossRef]
24. Bélanger, G.; Ziadi, N. Phosphorus and Nitrogen Relationships during Spring Growth of an Aging Timothy Sward. *Agron. J.* **2008**, *100*, 1757–1762. [CrossRef]
25. Marklein, A.R.; Houlton, B.Z. Nitrogen inputs accelerate phosphorus cycling rates across a wide variety of terrestrial ecosystem. *New Phytol.* **2012**, *193*, 696–704. [CrossRef] [PubMed]
26. Rufty, T.W., Jr.; MacKown, C.T.; Israel, D.W. Phosphorus stress effects on assimilation of nitrate. *Plant Physiol.* **1990**, *94*, 328–333. Available online: https://www.jstor.org/stable/4273089 (accessed on 24 May 2016). [CrossRef]
27. Rufty, T.W., Jr.; Israel, D.W.; Volk, R.J.; Qiu, J.; Sa, T. Phosphate regulation of nitrate assimilation in soybean. *J. Exp. Bot.* **1993**, *44*, 879–891. [CrossRef]
28. Crews, T.E. Phosphorus regulation of nitrogen fixation in a traditional Mexican agroecosystems. *Biogeochem* **1993**, *21*, 141–166. [CrossRef]
29. Güsewell, S.; Bollens, V.; Ryser, P.; Klötzli, F. Contrasting effects of nitrogen, phosphorus and water regime on first year and second year growth of 16 wetland plant species. *Funct. Ecol.* **2003**, *11*, 754–765. [CrossRef]
30. Briat, J.F.; Gojon, A.; Plassard, C.; Rouached, H. Reappraisal of the central of soil nutrient availability in nutrient management in light of recent advances in plant nutrition at crop and molecular levels. *Eur. J. Agron.* **2020**, *116*, 126069. [CrossRef]
31. Koerselman, W.; Meuleman, A.F.M. The vegetation N:P ratio: A new tool to detect the nature of nutrient limitation. *J. Appl. Ecol.* **1996**, *33*, 1441–1450. [CrossRef]
32. Sinclair, A.G.; Morrison, J.D.; Smith, L.C.; Dodds, K.G. Effects and interactions of phosphorus and sulphur on a mown white clover/ryegrass sward. *New Zeal. J. Agr. Res.* **1997**, *39*, 421–433. [CrossRef]
33. Greenwood, D.J.; Karpinets, T.V.; Zhang, K.; Bosh-Serra, A.; Boldrini, A.; Karawulova, L. A unifying concept for the dependence of whole-crop N:P ratio on biomass: Theory and experiment. *Ann. Bot.* **2008**, *102*, 967–977. [CrossRef]
34. Hendershot, W.H.; Lalande, H.; Duquette, M. Soil reaction and exchangeable acidity. In *Soil Sampling and Methods of Analysis*, 2nd ed.; Carter, M.R., Ed.; Taylor and Francis: Boca Raton, FL, USA, 2008; pp. 173–178.
35. Olsen, S.R.; Cole, C.V.; Watanabe, F.B.; Dean, L.A. *Estimation of Soil Available Phosphorus in Soils by Extraction with Sodium Bicarbonate*; US Department of Agriculture, Circular: Washington, DC, USA, 1954; p. 939.
36. Lithourgidis, A.S.; Matsi, T.; Barbayiannis, N.; Dordas, C.A. Effect of liquid cattle manure on corn yield, composition, and soil properties. *Agron. J.* **2007**, *99*, 1041–1047. [CrossRef]
37. Ozturk, L.; Eker, S.; Torun, B.; Cakmak, I. Variation in phosphorus efficiency among 73 bread and durum wheat genotypes grown in a phosphorus-deficient calcareous soil. *Plant Soil* **2005**, *269*, 69–80. [CrossRef]
38. Flénet, F.; Gúerif, M.; Boiffin, J.; Dorvillez, D.; Champolivier, L. The critical N dilution curve for linseed (*Linum usitatissimum* L.) is different from other C3 species. *Eur. J. Agron.* **2006**, *24*, 367–373. [CrossRef]
39. Cadot, S.; Bélanger, G.; Ziadi, N.; Morel, C.; Sinaj, S. Critical plant and soil phosphorus for wheat, maize, and rapeseed after 44 years of P fertilization. *Nutr. Cycl. Agroecosyst.* **2018**, *112*, 417–433. [CrossRef]
40. Sadras, V.O. The N: P stoichiometry of cereal, grain legume and oilseed crops. *Field Crop. Res.* **2006**, *95*, 13–29. [CrossRef]

41. Taliman, N.A.; Dong, Q.; Echigo, K.; Raboy, V.; Saneoka, H. Effect of Phosphorus Fertilization on the Growth, Photosynthesis, Nitrogen Fixation, Mineral Accumulation, Seed Yield, and Seed Quality of a Soybean Low-Phytate Line. *Plants* **2019**, *8*, 119. [CrossRef]

42. Li, P.; Yu, J.; Feng, N.; Weng, J.; Rehman, A.; Huang, J.; Tu, S.; Niu, Q.L. Physiological and Transcriptomic Analyses Uncover the Reason for the Inhibition of Photosynthesis by Phosphate Deficiency in *Cucumis melo* L. *Int. J. Mol. Sci.* **2022**, *23*, 12073. [CrossRef]

Article

Ammonia Volatilization and Marandu Grass Production in Response to Enhanced-Efficiency Nitrogen Fertilizers

Juliana Bonfim Cassimiro [1,*], Clayton Luís Baravelli de Oliveira [2], Ariele da Silva Boni [1], Natália de Lima Donato [1], Guilherme Constantino Meirelles [1], Juliana Françoso da Silva [1], Igor Virgilio Ribeiro [1] and Reges Heinrichs [1]

[1] Department of Crop Science, São Paulo State University—UNESP, Rodovia SP 294, km 651, Dracena 17900-000, SP, Brazil
[2] Soil and Fertilizers Laboratory, Universidade do Oeste Paulista—UNOESTE, Rodovia Raposo Tavares, Presidente Prudente 19067-175, SP, Brazil
* Correspondence: bonfimjuliana70@gmail.com

Abstract: The objective of this study was to evaluate dry matter (DM) production of *Urochloa brizantha* cv. Marandu and ammonia volatilization in response to rates and sources of enhanced-efficiency N fertilizers. The experiment was took place in a pasture area, two growing seasons. A randomized block design with four replications was used, in a $4 \times 2 + 1$ factorial arrangement, consisting of four N sources (Urea—Ur_{Conv}; Ammonium nitrate—AN; Urea + NBPT—Ur_{NBPT}; Urea + Duromide—$Ur_{Duromide}$) and two nitrogen rates (100 and 200 kg ha^{-1} year), plus a treatment without nitrogen fertilization (control). At both N rates, ammonia volatilization from $Ur_{Conv100/200}$ was greatest. Ammonia volatilization was less after Ur_{NBPT} and $Ur_{Duromide}$ application, with values similar to AN. Ammonia losses from $Ur_{Duromide}$ tend to be lower than from Ur_{NBPT}. The N use efficiency in dry matter production of Marandu was influenced by the N sources and rates. At both N rates, the efficiency of $Ur_{Duromide}$ and Ur_{NBPT} was greater than that of Ur_{Conv}. With regard to total DM and leaf percentage in response to N rates, DM production increased after 200 kg N ha^{-1} rates in response to all sources, in both years. The $Ur_{Duromide}$ reduce N losses by volatilization compared to Ur_{NBPT} and Ur_{conv}, and resulted in greater total DM production and relative leaf production of Marandu, in comparison to Ur_{NBPT}, AN and Ur_{conv}.

Keywords: dry matter production; Duromide; pasture; urease inhibitor; *Urochloa brizantha*

check for updates

Citation: Cassimiro, J.B.; de Oliveira, C.L.B.; Boni, A.d.S.; Donato, N.d.L.; Meirelles, G.C.; da Silva, J.F.; Ribeiro, I.V.; Heinrichs, R. Ammonia Volatilization and Marandu Grass Production in Response to Enhanced-Efficiency Nitrogen Fertilizers. *Agronomy* **2023**, *13*, 837. https://doi.org/10.3390/agronomy13030837

Academic Editors: Christos Noulas, Shahram Torabian and Ruijun Qin

Received: 18 January 2023
Revised: 10 February 2023
Accepted: 13 February 2023
Published: 13 March 2023

1. Introduction

Urochloa brizantha cv. Marandu is native to tropical Africa and adapts well to soils of medium fertility, obtaining high yields in fertile soils [1]. With a short cycle and perennial, it grows in clump form, its stems have a dense hairiness, and the plant has good digestibility and palatability. When grown in medium to high fertility soils, the plants exceed 1.5 m in height [2]. The predominance of the genus Urochloa in Brazil is more common in sown pastures, with about 50 million hectares of arable land, mainly because it is robust, and is associated with a high productive potential, great nutritional quality and widely adaptable in various edaphoclimatic environments [3–5].

The forages used for pasture, extremely relevant for livestock production, are predominantly grasses, which are very responsive to nitrogen fertilization. This nutrient (N) is applied in large amounts to pasture, so sound agronomic practices and modern technology are needed in the sector [6].

Nitrogen fertilization can improve yield and crude protein content of forages [7,8]. Urea is the most frequently used nitrogen source, mainly due to its low cost. However, it is highly susceptible to losses by ammonia volatilization, caused by changes in soil moisture, temperature and pH, wind speed, soil organic C and N, and the applied urea rate [9].

Certain products have increase urea efficiency, e.g., urea treated with N–(n-butyl) thiophosphoric triamide (NBPT), marketed since 1996 in the United States and, more recently, in Brazil. This urease inhibitor is currently the only commercially available option for agriculture and is sold in more than 70 countries [10,11]. The main benefit of stabilizing urea by adding substances for this purpose is the reduction of volatilization, which: (a) extends the period (number of days) of chemical stability between nitrogen fertilization and soil incorporation by rainwater or irrigation, reducing N losses by volatilization; (b) reduces the N volatilization losses caused by urea hydrolysis on the soil surface; (c) increases N uptake, fertilization efficiency and crop yield and quality [10].

The retardant NBPT is a conventional inhibitor that loses its efficiency under acidic soil pH conditions and temperatures above 30 °C. A novel urease inhibitor called Duromide is the active ingredient of a new urease inhibitor generation [12]. The new molecule has the same chemical function as conventional NBPT, having the same mode of action to block the urease active site by binding to it. On the other hand, the rest of its chemical structure differs from NBPT, making it more stable. The greater stability of this new active principle raises expectations of allowing more durable storage and applications in wider ranges of soil pH and temperature. Based on these characteristics, enhanced urease inhibition and accordingly, reduced N volatilization losses are expected [12]. In view of the above, the objective was to evaluate ammonia volatilization and dry matter production of *Urochloa brizantha* cv. Marandu in response to rates and sources of enhanced-efficiency N fertilizers.

2. Material and Methods

2.1. Experimental Area

The experiment was installed in an experimental field of the Faculty of Agricultural and Technological Sciences of UNESP, Campus de Dracena (21°27′ S; 51°36′ W), with a tropical climate, classified as Aw by Köppen [13], mean annual rainfall of approximately 1300 mm, mean annual air temperature of 24 °C and mean maximum of 31 °C and minimum of 19 °C. An area in the process of pasture formation of *Urochloa brizantha* cv. Marandu was evaluated over two growing seasons (2018/2019 and 2019/2020) (Figure 1).

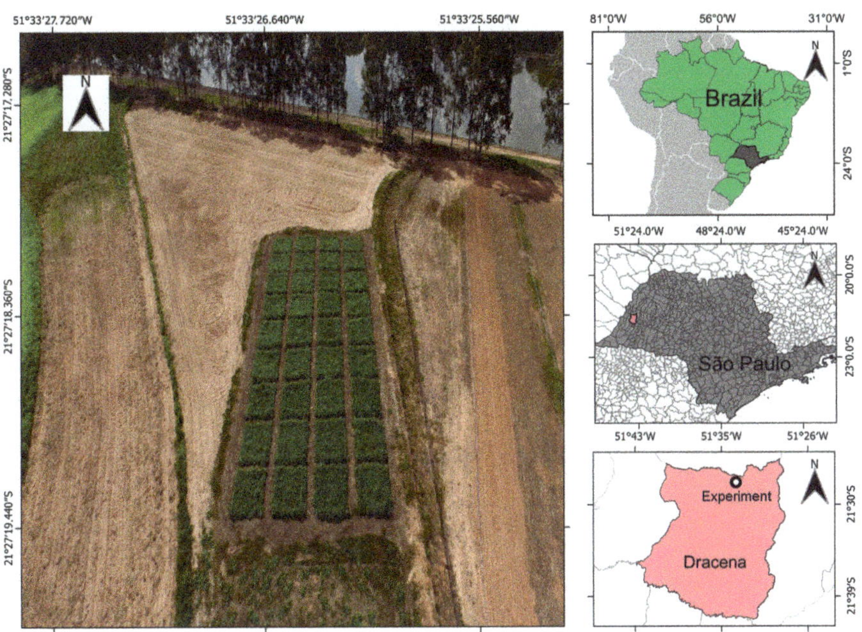

Figure 1. Location of experimental area, Dracena-SP, Brazil.

The soil of the experimental area was classified as Argissolo Vermelho Amarelo distrófico soil with sandy texture [14], corresponding to a dystrophic Ultisol [15]. Soil chemical and particle-size analyses (0.00–0.20 m layer), showed the following results: 13 g dm^{-3} organic matter; pH (CaCl$_2$) 4.5; 3 mg dm^{-3} P (Resin); 5.0 mmol$_c$ dm^{-3} Ca^{2+}; 3.0 mmol$_c$ dm^{-3} Mg^{2+}; 1.4 mmol$_c$ dm^{-3} K$^+$; 24.4 mmol$_c$ dm^{-3} cation exchange capacity; 120 g kg^{-1} clay; 30 g kg^{-1} silt; and 850 g kg^{-1} sand.

2.2. Experimental Design

The experiment was arranged in randomized blocks with four replicates, in a $4 \times 2 + 1$ factorial design, consisting of four N sources: conventional urea (Ur$_{Conv100}$ and Ur$_{Conv200}$), ammonium nitrate (AN$_{100}$ and AN$_{200}$), urea treated with NBPT (Ur$_{NBPT100}$ and Ur$_{NBPT200}$), urea treated with Duromide (Ur$_{Duromide100}$ and Ur$_{Duromide200}$) and two nitrogen rates (100 and 200 kg ha^{-1} year), plus a treatment without nitrogen fertilization (control), resulting in a total of 36 plots. The annual rates of 100 and 200 kg N ha^{-1} were split in four applications of 25 and 50 kg ha^{-1}, respectively, broadcast on the soil surface. The first fertilization was applied 30 days after sowing and the others subsequently after each cut. In the second year, the first rate was applied in the beginning of the rainy season, in October 2019, and the others after the next three cuts, resulting in a total of four applications.

2.3. Soil Management, Sowing and Cultural Treatments

In August 2018, dolomitic limestone was incorporated to a soil depth of 0.20 m, to raise base saturation to 60%. Three months after liming, immediately before sowing, 80 kg P$_2$O$_5$ ha^{-1} as single superphosphate and 30 kg K$_2$O ha^{-1} as potassium chloride were broadcast [16].

Marandu grass was sown in rows (December 2018), spaced 25 cm apart, at a density of 10 kg ha^{-1} of pure, healthy seeds. The plot size was 4×4 m and spacing between plots and blocks was 1 m.

In January 2019, the first nitrogen fertilization was carried out in the treatments, at the respective rates. In each growing season, four cuts were made in the rainy and one in the dry season, resulting in a total of five forage cuts per growing season. In the second year, the same phosphorus and potassium rates as in the first were applied, together with the first N fertilizer application.

2.4. Ammonia Volatilization

Volatilization cylinders similar to those described by [17–20] were used (Figure 2). In the first year of evaluation, ammonia volatilization was evaluated by analyzing the polyethylene foam strips soaked in phosphoric acid, which were collected from the cylinders and exchanged on day 2, 5, 9, 14, 20 and 26 after fertilization with each of the four N rates. In the second year, these foams were collected and exchanged twice (tow cycles), on day 1, 2, 5, 9, 14, 20 and 26 after the 2nd and 4th application of N rates, to assess ammonia volatilization.

The polyethylene cylinders were fitted on top of round PVC bases (diameter 9 cm, height 10 cm) (Figure 2). In each plot, six bases per cylinder were installed (one per evaluation). Since the contact between rain and fertilizer was impeded within the cylinders, they were shifted to a subsequent base at each foam exchange. In this way, in the following period, NH$_3$ losses from the fertilizer treatments were evaluated under exposure to the same conditions (rain, temperature, wind, etc.) as in the rest of the experimental field.

At the moment of fertilizer broadcasting in the total area of the plots, the volatilization cylinders were sealed and the respective relative amount of fertilizer of each treatment was individually weighed and applied to the area within the bases underneath the cylinders. In each cylinder, the amount of fertilizers applied corresponded to rates of 100 and 200 kg N ha^{-1}, divided into four applications of 25 and 50 kg ha^{-1}, respectively, per growing cycle.

To determine volatilization, the retained ammonia was extracted from the foam strips by washing four times with 10 mL deionized water and measuring this solution in a 100 mL volumetric flask, Thereafter, an aliquot of 20 mL was distilled and the volatilized NH_3 was determined by subsequent titration (H_2SO_4 0.0025 mol L^{-1}) [21].

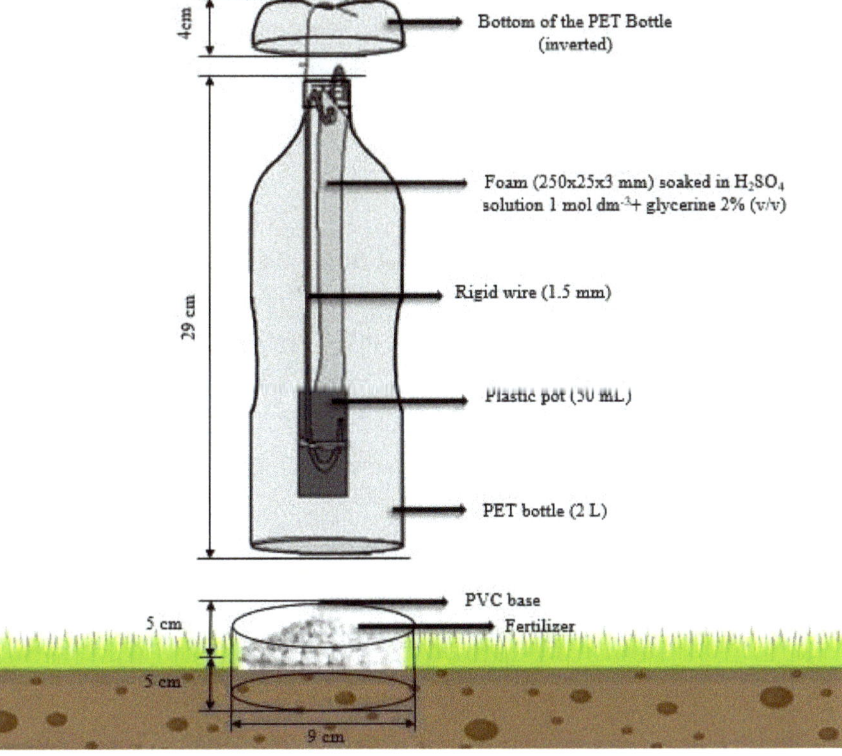

Figure 2. Semi-open static NH_3 collection cylinder.

2.5. Grass Production

Four cuts were taken, at intervals regulated by the mean plant height of 28 cm (i.e., 95% light interception) in the best treatment [22]. In the rainy season, the cutting height was reached within intervals of approximately 30 days.

Leaf fresh matter was measured in one sample of 0.5 m^2 (1 × 0.5 m) per experimental unit, taken with a rectangular iron sampler. The sampler was placed randomly at representative points of each plot and the forage within the rectangle was cut at 15 cm above the ground and immediately weighed to determine leaf fresh matter. Then, a forage subsample was taken, immediately weighed and dried to constant weight in a forced air circulation oven at 65 °C, to determine dry matter production [23].

2.6. Morphological Composition of Forage

To determine the relative participation of each morphological component, a subsample of leaf fresh matter was removed, separated in leaf blades and pseudostem (stems + sheaths) and separately dried to constant weight in a forced-air circulation oven, at 65 °C [23].

2.7. Data Analysis

Data were evaluated for error normality and homogeneity and the results subjected to analysis of variance and mean comparison by the Tukey test at 5% significance ($p < 0.05$).

Pearson's correlation (Minitab statistical software version 21.2.0, State College, PA, USA) was calculated to investigate the relationship between total dry matter, leaf percentage, leaf dry matter, N loss reduction, N use efficiency and percent N loss. All statistical analyses were performed using the Statistical Analysis System software (SAS OnDemand for Academics 2022). Graphs were plotted using Sigmaplot® version 14.5 (Systat Software, Inc., San Jose, CA, USA, www.sigmaplot.com, accessed on 12 December 2022). The model was selected according to the Akaike Information Criterion (CIA) [24] by choosing the models with the least CIA. After selecting the model, the data were subjected to non-linear regression, using the logistic model represented by Equation (1), as described by [25]. This model is traditionally used to estimate cumulative ammonia volatilization [26–28].

$$\hat{Y} = \frac{\alpha}{1 + exp^{[-(t-\beta)/\gamma]}} \tag{1}$$

where $\hat{Y}$ is the amount of N volatilized in form of NH_3-N (kg ha^{-1}) at time t; α is the maximum cumulative volatilization; β the moment when 50% of the losses had occurred, corresponding to the inflection point of the curve (day of maximum daily NH_3-N loss); t the time (days); and γ a parameter of the equation used to calculate the maximum daily loss (MDL) of NH_3-N, as shown in Equation (2).

$$MDL = \frac{\alpha}{4\gamma} \tag{2}$$

To evaluate the reduction in ammonia loss compared to urea, as shown in Equation (3).

$$NH_3 - N = 100 - \frac{N\ loss\ by\ fertilizer \times 100}{N\ loss\ by\ urea} \tag{3}$$

3. Results

3.1. NH_3-N Volatilization Losses

Figure 3a,c,e,g and Figure 4a,c show ammonia volatilization over a 26-day period. The climatic conditions of each evaluation period are shown in Figure 3b,d,f,h and Figure 4b,d. In the four evaluations of both annual N rates, ammonia volatilization from Ur_{Conv} was observed to be highest. In turn, Ur_{NBPT} and $Ur_{Duromide}$ proved more efficient in reducing ammonia volatilization compared to conventional urea, reaching values close to AN.

In the first experiment (2019) (Figure 3a), two days after the first split applications of $Ur_{Conv100}$ and $Ur_{Conv200}$, respectively, 2.82 and 7.45 kg ha^{-1} of the applied N was lost, with losses peaking on the 5th day after fertilization, with 5.41 and 13.93 kg N ha^{-1}, respectively. Subsequently, the N losses from $Ur_{NBPT200}$ reached approximately 6.52 kg N ha^{-1}, 3.76 kg N ha^{-1} from $Ur_{Duromide200}$ and approximately 0.32 kg N ha^{-1} from AN_{200}. In the first experiment, NH_3 losses were less after application of fractional rates of 25 kg N ha^{-1} (Table 1). Compared with urea, NH_3 losses from AN, Ur_{NBPT} and $Ur_{Duromide}$ for the split rates of 25 kg N ha^{-1} were reduced by 94.4%, 35.1% and 52.4%, respectively, and by 97.7%, 54.2% and 74.7% in response to rates of 50 kg N ha^{-1} (Table 1).

The second evaluation of the first year (Figure 3c) showed that 3.94 and 7.83 kg N ha^{-1} was lost from $Ur_{Conv100/200}$ on the second day of data collection, while the sources $Ur_{NBPT100/200}$ and $Ur_{Duromide100/200}$ lost approximately 1 kg N ha^{-1}. During this period, it should be mentioned that rain (8 mm) fell on the first day after fertilization and another rainfall (7 mm) occurred on the second day, after data collection (Figure 3d). The reduction in ammonia loss in this cycle was greater than in the first, reaching 96.2% and 98% at rates of 25 and 50 kg N ha^{-1} compared to AN. This performance was better than that of $Ur_{NBPT100/200}$, with respective reductions of about 73% and 81.4% and for $Ur_{Duromide100/200}$, with respective reductions of about 81.9% and 87%, exceeding those in response to Ur_{NBPT} (Table 1).

After the third fertilization (Figure 3e,f), the first rainfall occurred only nine days after application, and during this rain-free period, volatilization was practically nonexistent. As of the onset of rainfall (37.6 mm), intense volatilization occurred in response to $Ur_{Conv100/200}$, peaking at 14 days after fertilization, with respective losses of 4.04 and 10.43 kg N ha^{-1} at the end of the cycle. Losses were least from $Ur_{Duromide100}$ (0.42 kg N ha^{-1}), and $Ur_{Duromide100}$ reached a 90.8% reduction in NH$_3$ losses (Table 1). The efficiency of urease inhibitor with urea in reducing ammonia volatilization was confirmed. Thus, N volatilization losses from unprotected urea proved to be higher than from urea with urease inhibitor, in other words, volatilization losses from UrDuroimide$_{100/200}$ were 90.8% and 88.4% less than from $Ur_{Conv100/200}$ (Table 1).

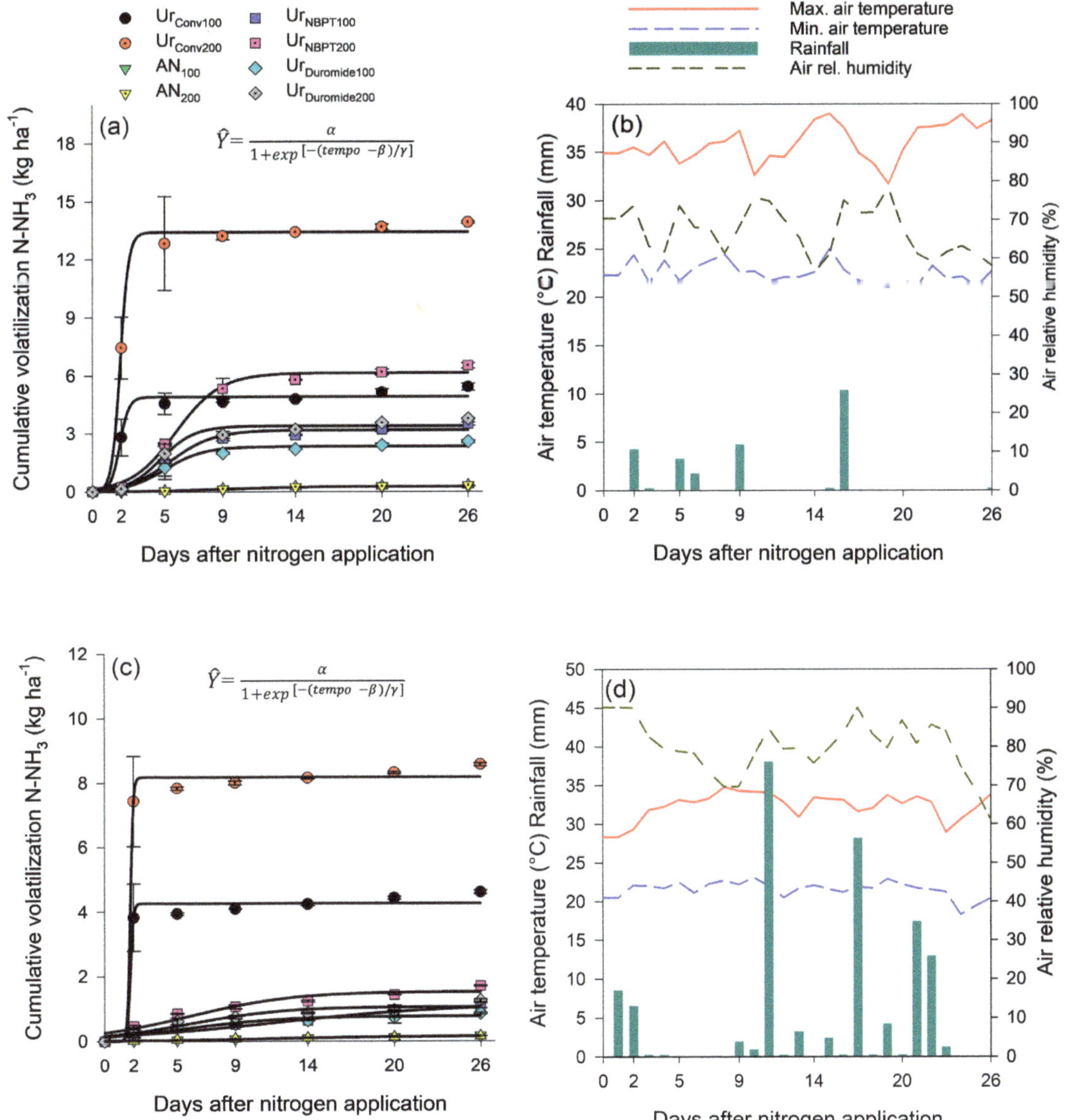

Figure 3. *Cont.*

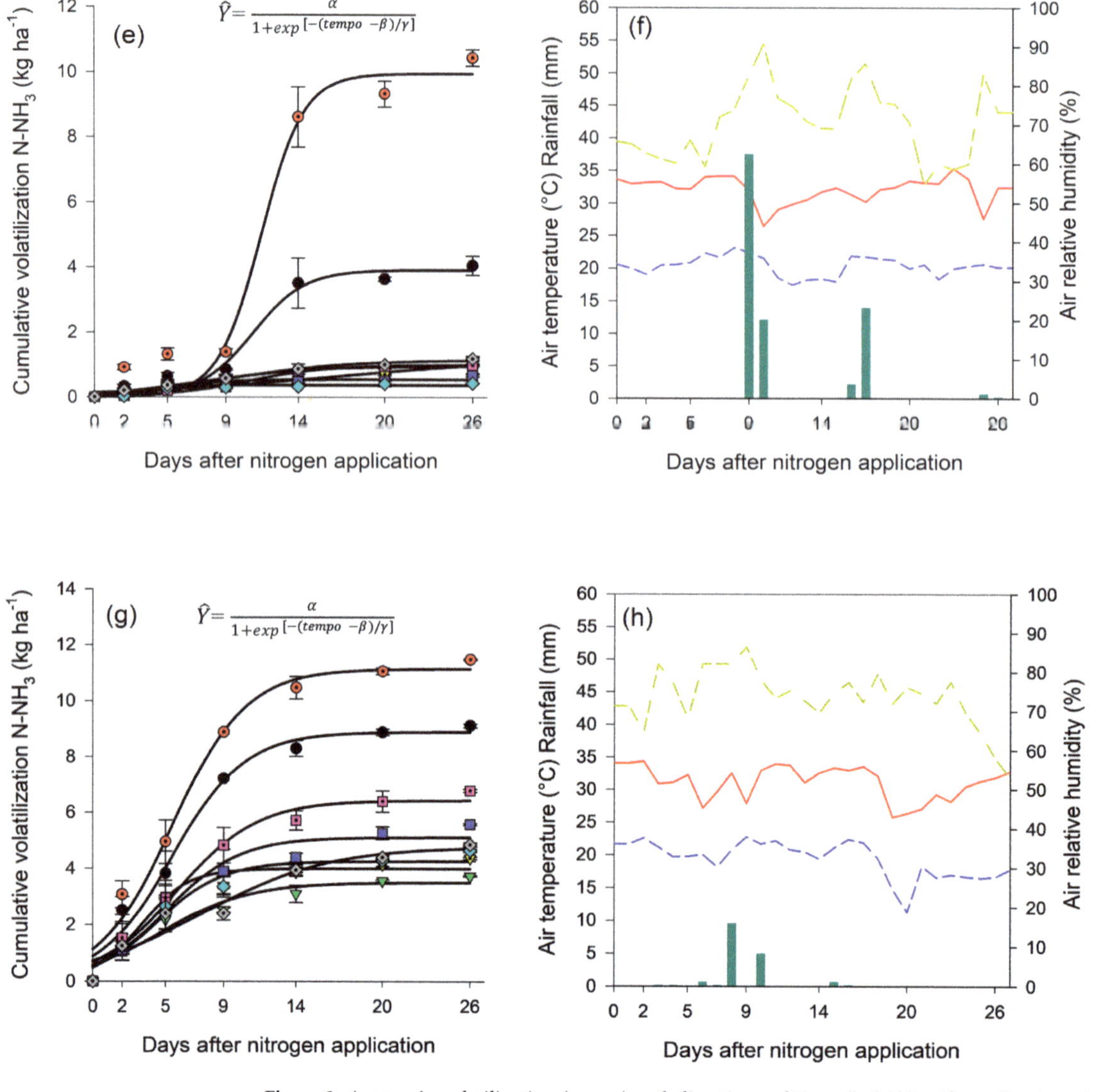

Figure 3. Ammonia volatilization (**a,c,e,g**) and climatic conditions (**b,d,f,h**) with application of 25 and 50 kg N ha^{-1}, referring to $\frac{1}{4}$ of the total rate of 100 kg ha^{-1} and 200 kg N ha^{-1}, respectively. (**a**) 1st application (**c**) 2nd application (**e**) 3rd application (**g**) 4th application. Growing season 2018/2019.

The rainfall (9.7 mm) in the first days after fertilization in the fourth evaluation intensified ammonia losses from Ur$_{\text{Conv100}}$, to about 11.47 kg N ha^{-1} (32% of applied N) (Figure 3g,h). The losses from Ur$_{\text{Duromide100/200}}$ were slightly less than from Ur$_{\text{NBPT100/200}}$ at both rates, with values very close to AN$_{100/200}$.

Figure 4 shows the values of ammonia volatilization after two forage cuts. In both evaluations and at both rates, ammonia volatilization from Ur$_{\text{Conv}}$ was higher. In turn, Ur$_{\text{NBPT100/200}}$ and Ur$_{\text{Duromide100/200}}$ proved to be efficient in reducing ammonia volatilization, reaching values similar to AN$_{100/200}$.

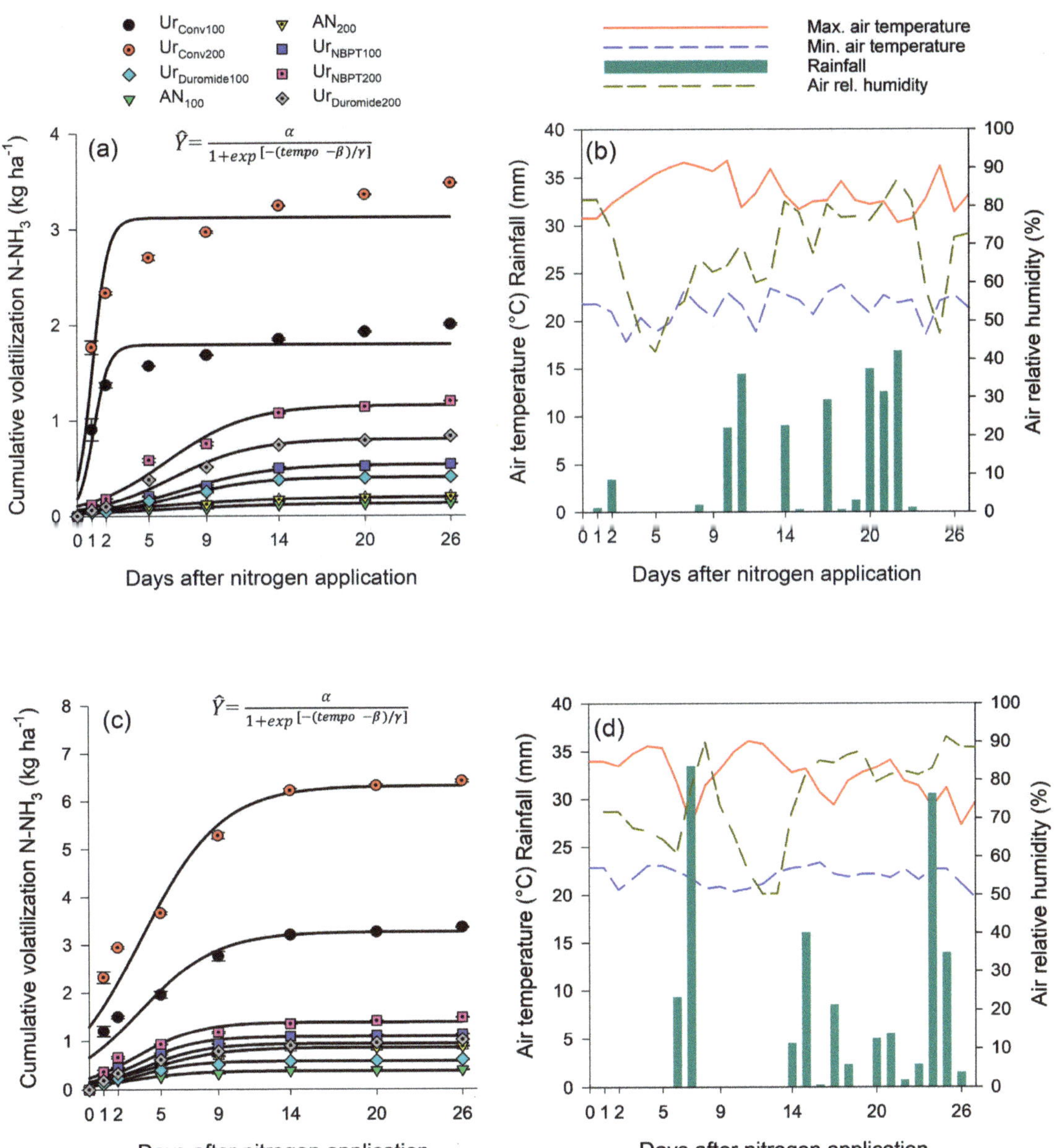

Figure 4. Ammonia volatilization (**a,c**) and climatic conditions (**b,d**) after application of 25 and 50 kg N ha^{-1}, corresponding to $\frac{1}{4}$ of the total rate of 100 and 200 kg N ha^{-1}, respectively. (**a**) 2nd application (**c**) 4th application. Growing season 2019/2020.

At the second fertilizer rate application, the first rainfall occurred one day after fertilization, but was insufficient to solubilize the fertilizers and incorporate them into the soil, so Ur$_{Conv100}$ reached 7.8% volatilization (2.00 kg N ha^{-1}) and Ur$_{Conv200}$ 6.8% (3.47 kg N ha^{-1}) (Figure 3a). Compared to conventional urea, the responses to AN in the second growing season (2019/2020) in the second cycle provided reductions of 92.7% and 93.9%, respectively, in response to N rates of 100 and 200 kg N ha^{-1}. For sources with urease inhibitors, losses from $Ur_{Duromide100/200}$ were reduced by 77.6% and 74.3 and from $Ur_{NBPT\ 100/200}$ by 70.3% and 63.1%, in response to the same respective rates (Table 2).

Table 1. Parameters of the nonlinear (logistic) model fitted to the cumulative NH_3-N losses for N rates of 100 and 200 kg ha^{-1}, split into four applications, and reduction of NH_3-N losses in comparison with urea. Growing season 2018/2019.

| Treatments | Growing Season 2018/2019 Cycle | Parameters | | | | MDL kg ha^{-1} day^{-1} NH_3-N | Reduction of NH_3-N Losses in Comparison with Urea (%) |
		α kg NH_3-N ha^{-1}	Γ	β Day	R^2		
$Ur_{Conv100}$	1	4.9	0.31	1.9	0.96	3.86	-
	2	4.26	0.1	1.77	0.97	10.65	-
	3	3.9	1.81	10.79	0.96	0.53	-
	4	8.88	2.49	5.48	0.97	0.89	-
AN_{100}	1	0.27	3.1	8.83	0.99	0.02	94.4
	2	0.16	5.23	11.87	0.95	0.00	96.2
	3	0.95	3.69	7.39	0.93	0.06	75.6
	4	3.5	2.99	4.77	0.89	0.29	60.6
$Ur_{NBPT100}$	1	3.18	1.48	5.38	0.97	0.53	35.1
	2	1.15	6.25	11.74	0.89	0.04	73.0
	3	0.53	2.01	4.76	0.85	0.06	86.4
	4	5.1	2.48	5.16	0.92	0.51	42.6
$Ur_{Duromide100}$	1	2.33	1.32	5.02	0.96	0.44	52.4
	2	0.77	3.43	5.47	0.88	0.05	81.9
	3	0.36	0.93	4.13	0.89	0.09	90.8
	4	4.25	2.26	4.47	0.93	0.47	52.1
$Ur_{Conv200}$	1	13.42	0.29	1.93	0.99	11.41	-
	2	8.17	0.09	1.77	0.99	22.69	-
	3	9.93	1.48	11.4	0.96	1.67	-
	4	11.13	2.52	5.5	0.97	1.10	-
AN_{200}	1	0.31	2.99	9.62	0.99	0.02	97.7
	2	0.16	5.23	11.9	0.95	0.00	98.0
	3	1.28	8.08	16.92	0.89	0.03	87.1
	4	3.99	1.66	3.3	0.92	0.60	64.2
$Ur_{NBPT200}$	1	6.14	1.54	5.72	0.98	0.99	54.2
	2	1.52	3.6	5.7	0.91	0.10	81.4
	3	0.98	2.8	10.5	0.98	0.08	90.1
	4	6.42	2.74	5.83	0.96	0.58	42.3
$Ur_{Duromide200}$	1	3.39	1.14	4.7	0.96	0.74	74.7
	2	1.06	3.3	5.71	0.84	0.08	87.0
	3	1.15	4.28	9.26	0.96	0.06	88.4
	4	4.77	4.37	7.47	0.9	0.27	42.9

MDL: maximum daily NH_3-N loss.

After the fourth fertilization, the first rainfall occurred only five days after fertilization, showing that during this rainless period, volatilization was very low. At the onset of rains, volatilization became intense in the treatments with $Ur_{Conv200}$, reaching losses of approximately 13% in 14 days after fertilization, while losses from $Ur_{Duromide100/200}$, $Ur_{NBPT100/200}$ and $AN_{100/200}$ were less throughout the entire period. In this experiment, at rates of 25 and 50 kg N ha^{-1} as AN, volatilization reduction was smaller than in the previous cycle, with 88.6% and 86.5%, respectively, while volatilization from $Ur_{NBPT100/200}$ was reduced by 57.6% and 82.9% and from $Ur_{Duromide100/200}$ by 82.5% and 85.1%, respectively (Table 2). In summary, of the total of twelve volatilization comparisons, in eleven of them $Ur_{Duromide}$ resulted in less losses than Ur_{NBPT}.

Table 2. Parameters of the nonlinear (logistic) model fitted to the cumulative NH₃-N losses for N rates of 100 and 200 kg ha⁻¹, split into four applications, and reduction of NH₃-N losses in relation to urea. Growing season 2019/2020.

Treatments	Growing Season 2019/2020 Cycle	Parameters α kg NH₃-N ha⁻¹	γ	β Day	R^2	MDL kg ha⁻¹ day⁻¹ NH₃-N	Reduction of NH₃-N Losses in Comparison with Urea (%)
Ur$_{Conv100}$	2	1.79	0.53	1.15	0.93	0.23	-
	4	3.26	2.65	3.6	0.9	2.15	-
An$_{100}$	2	0.13	4.37	5.34	0.83	0.14	92.74
	4	0.37	2.15	2.83	0.88	0.19	88.65
Ur$_{NBPT100}$	2	0.53	3.03	7.23	0.98	0.40	70.39
	4	1.38	2.13	3.34	0.91	0.73	57.67
Ur$_{Duromide100}$	2	0.4	2.74	6.99	0.98	0.27	77.65
	4	0.57	0.01	1.53	0.96	0.00	82.52
Ur$_{Conv200}$	2	3.12	0.54	1.07	0.89	0.42	-
	4	6.31	2.76	3.71	0.89	4.35	-
An$_{200}$	2	0.19	4.56	4.86	0.79	0.21	93.91
	4	0.85	2.53	4.2	0.94	0.53	86.53
Ur$_{NBPT200}$	2	1.15	2.79	6.29	0.96	0.80	63.14
	4	1.08	1.90	3.3	0.93	0.52	62.00
Ur$_{Duromide200}$	2	0.8	2.75	6.6	0.96	0.55	74.36
	4	0.94	2.1	3.92	0.95	0.48	85.10

MDL: maximum daily NH₃-N loss.

3.2. Nitrogen Use Efficiency for Dry Matter Production

Nitrogen use efficiency in Marandu dry matter production in the first growing season ($p < 0.001$) was influenced by N sources and rates. The N use efficiency of Ur$_{Duromide200}$ (50.6 kg dry matter/kg N) was greater than that of Ur$_{Conv100}$ (5.1 kg dry matter/kg N). In turn, Ur$_{Duromide}$ 100 (37.1 kg dry matter/kg N) and Ur$_{NBPT100/200}$ (30 and 34.8 kg dry matter/kg N, respectively) also proved more efficient than Ur$_{Conv100/200}$ (kg dry matter/kg N). The efficiency of dry matter production in response to fertilization with AN$_{100/200}$ was close to that observed for Ur$_{Conv100/200}$ (Figure 5).

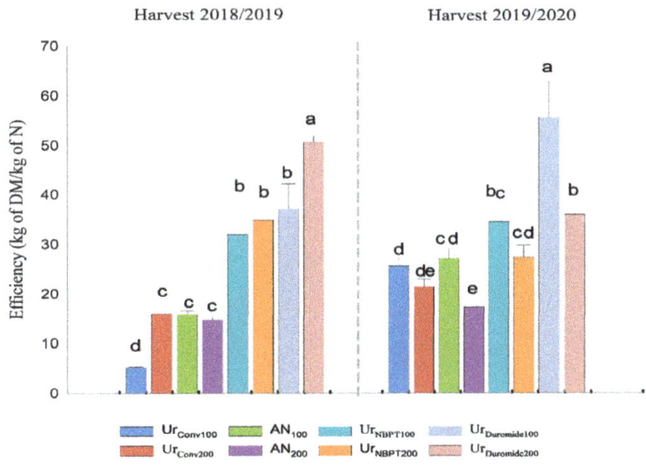

Figure 5. Dry matter production efficiency of *Urochloa brizantha* cv. Marandu fertilized with nitrogen rates and sources for pasture formation and maintenance. Growing seasons 2018/2019 and 2019/2020. Means followed by distinct vertical letters were significantly different according to Tukey's test ($p < 0.05$).

In the second year of evaluations, the forage used more N from $Ur_{Duromide100/200}$ than from $Ur_{Conv100/200}$ and $AN_{100/200}$ ($p < 0.001$). The efficiency of $Ur_{Duromide\ 100}$ was higher (55.5 kg dry matter/kg N) than that of the sources without urease inhibitor. In the case of AN_{200}, the production efficiency ratio was less (17.3 5 kg dry matter/kg N) (Figure 5).

The total dry matter and leaf production in response to N rates differed significantly between treatments, and the rate of 200 kg N ha^{-1} differed for all sources, in both years of evaluation. In the first year, $Ur_{Duromide200}$ stood out from the other sources, with increases of 73.6% dry matter production and 20.4% leaf production in relation to $Ur_{Conv100/200}$. On the other hand, compared to $Ur_{Conv100/200}$, $Ur_{Duromide\ 100}$ provided an increase of 47.6% dry matter and 33% leaf production (Figure 6a,c).

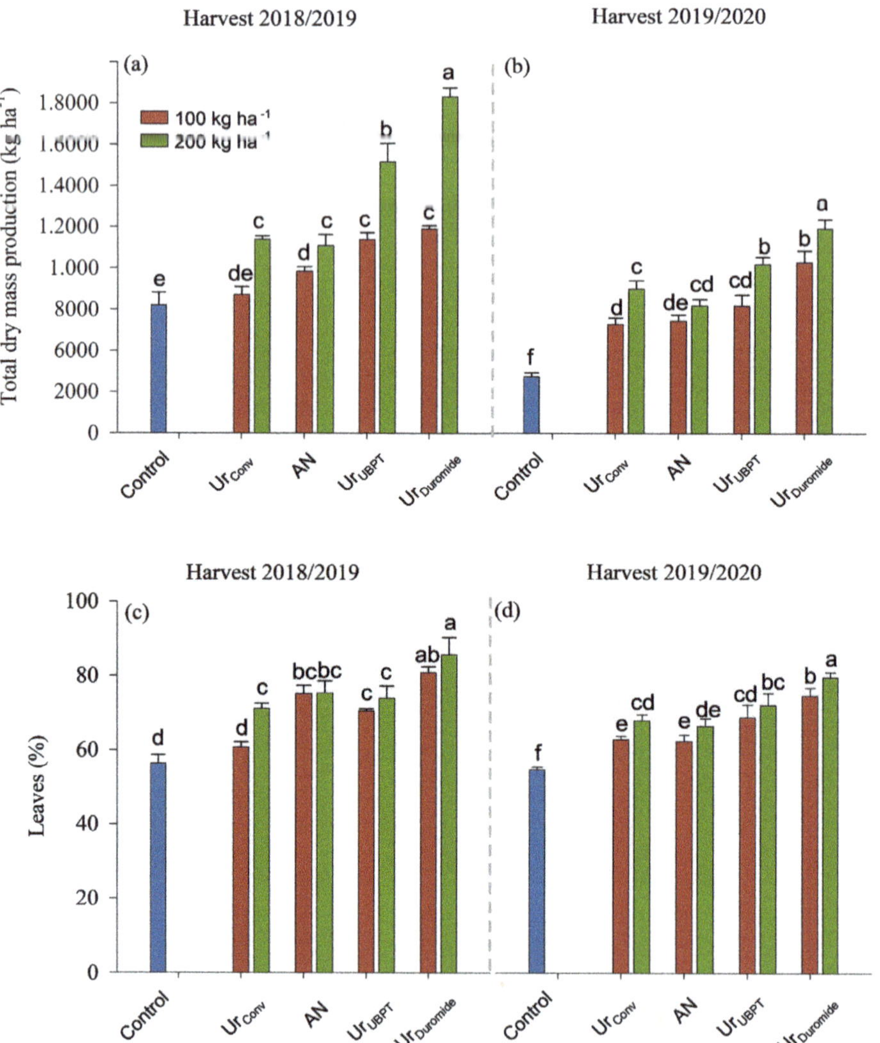

Figure 6. Total dry matter production and relative leaf production of *Urochloa brizantha* cv. Marandu, in two growing seasons, fertilized with nitrogen sources and rates. Means followed by distinct vertical letters were significantly different according to Tukey's test ($p < 0.05$).

In the second year, dry matter and leaf production were relatively less than in the previous year (2018/2019), although dry matter production in response to $Ur_{Duromide100/200}$ still exceeded that of $Ur_{Conv100/200}$, with about 56.3% dry matter, i.e., more than in the previous year at 100 kg N ha^{-1}. The increase at 200 kg N ha^{-1} was 41.3% in comparison with Ur_{Conv} (Figure 6b,d)

In the second year, nitrogen fertilization induced an increase in leaf percentage over the control, mainly in response to $Ur_{Duromide200}$, resulting in an increase of 17.1% in leaf production (Figure 6d).

3.3. Correlation

Pearson's correlation analysis between pasture production parameters and nitrogen rates detected a positive correlation between dry matter production and leaf production, leaf dry matter, reduction of N losses and N use efficiency. Total dry matter and N loss percentage were negatively correlated, i.e., dry matter increases as N losses decrease. This result was also observed for leaf percentage and N use efficiency, which increased with decreasing N losses (Table 3).

Table 3. Pearson's correlation of total dry matter production and leaf production with parameters of N dynamics in a *Urochloa brizantha* cv. Marandu pasture, in two growing seasons and in response to two annual N rates, applied in four split rates.

Parameters	2018/2019		2019/2020	
	100 (kg ha^{-1})	200 (kg ha^{-1})	100 (kg ha^{-1})	200 (kg ha^{-1})
Total dry matter/% leaves	0.87	0.88	0.98	0.99
Total dry matter/leaf dry matter	0.97	0.99	0.99	0.99
Total dry matter/N loss reduction	0.73	0.38	0.43	0.24
Total dry matter/N use efficiency	1.00	1.00	0.97	0.99
Total dry matter/% N loss	−0.75	−0.35	−0.44	−0.26
% leaves/N loss reduction	0.85	0.54	0.40	0.31
% leaves/N use efficiency	0.77	0.83	0.97	0.99
% leaves/% N loss	−0.86	−0.52	−0.40	−0.33
N use efficiency/% N loss	−0.73	−0.35	−0.44	−0.26

4. Discussion

In general, ammonia losses were high in the Ur_{Conv} treatment (Figures 3 and 4), which contributed to reduce N availability, reflecting in lower dry matter production than of $Ur_{Duromide100/200}$ and $Ur_{NBPT100/200}$ (Figure 6). The high ammonia losses from Ur_{Conv} can be explained by the greater saturation of the sites of urease enzyme action, due to the higher ammonium availability in soil fertilized with untreated urea, as stated by [29].

In both years, NH$_3$-N losses occurred as of the 2nd day after fertilizer application to the soil, except for the third cycle, with the onset of volatilization on the 10th day (Figure 3e,f). The amplitude of ammonia losses is very variable after urea application to the soil surface, and depends on the rates, sources and prevailing climatic conditions during the evaluation period. The hydrolysis rate of Ur_{Conv} by the enzyme urease was higher in the first 2–3 days after fertilization, according to soil temperature, moisture and fertilizer volume applied per area [29].

The intense volatilization from the third fertilization plot of the first year may also have been caused by the collector, where $Ur_{Conv100/200}$ was possibly prevented from being incorporated in the area of volatilization measurement (inside the collector) by rainwater. The rain that fell around the collector probably caused surface wetting of the soil, favoring urea solubilization, but not enough for soil incorporation. With regard to the sources with urease inhibitor and ammonium nitrate, the values were much lower during the whole period, demonstrating the inhibition efficiency under adverse conditions (Figure 3e,f). According to Afshar et al. [30], the variation in the amount of volatilized NH$_3$ was significant, which may be a result of the measurement method and environmental conditions.

Indeed, the total emissions were possibly not fully captured because measurements were not made daily.

In all evaluations, losses ceased after the volatilization peak, which can be explained by the rainfall volume, which was insufficient for fertilizer aggregation in the soil, resulting in a stabilization of the losses after a period with higher soil water concentrations. This may have occurred due to the rapid hydrolysis of urea by urease, since losses were greater in the first seven days immediately after soil fertilization [30]. Similar results were reported for coffee, where the authors observed that up to 35% N can be lost from Ur_{Conv} [31].

The results of this study confirmed that the amount of volatilized N was lower from sources with urease inhibitors and ammonium nitrate. This finding is in agreement with Otto et al. [19], who observed that fertilization with ammonium nitrate resulted in N volatilization losses of less than 1%. However, under favorable climatic conditions, the N fertilizers commonly used in agriculture, as is the case with Ur_{Conv} and AN, can be used as fertilizers for pasture growth. However, fertilizers containing ammonia or urea promote soil acidification in the production system, especially when used at high rates [32].

Losses from $Ur_{Duromide}$ were a little lower than from Ur_{NBPT} at both annual rates and very close to AN, and also promoted a slightly delayed onset of N losses. A delay in the initial loss of N from Ur_{NBPT} and $Ur_{Duromide}$ was confirmed by Cassim et al. [33], increasing the chances that urea is incorporated into the soil by rain, for example, which would reduce NH_3 volatilization. However, it is worth remembering that the ammonia loss measurement method in this study was not able to detect losses by denitrification and leaching, which occur more easily from AN.

The reduction in ammonia loss from fertilizer with Ur_{NBPT} was greater than that found by Souza [34]. Comparing the peak of fertilizer volatilization, the NH_3 loss from Ur_{NBPT} was reduced by 77% in relation to Ur_{Conv} (fourth day after fertilizer application). Other studies also concluded that the urease inhibitor reduces NH_3 losses by 54% in comparison with Ur_{Conv} [29]. Sources with urease-inhibiting technologies ($Ur_{Duromide}$ and Ur_{NBPT}) were more efficient than Ur_{Conv}, with reductions of over 50% compared to Ur_{Conv}. According to Cassim et al. [33], $Ur_{Duromide}$ and Ur_{NBPT} were extremely efficient in reducing NH_3 volatilization (reductions of 35–54%) at 45 and 90 kg N ha^{-1}. In comparison with Ur_{Conv}, the new Duromide stabilizing technology reduced NH_3 losses by up to 33%, compared to NBPT alone.

The dry matter production efficiency of *Urochloa brizantha* cv. Marandu, in the first growing season with $Ur_{Duromide200}$ was more efficient than that of $Ur_{Conv100}$, however in the second year, the efficiency of $Ur_{Duromide200}$ as well as AN_{200} decreased with increasing nitrogen rates. According to Rowlings et al. [35], a reduction in fertilization efficiency is common with increasing N rates. Nitrogen rates and sources positively influence dry matter and leaf production of Marandu grass, and should be taken into account to achieve improved forage quantity and quality [36,37]. A study on N fertilization of Mombasa grass also showed an increase in dry matter accumulation, in addition to increasing nutrient accumulation and improving pasture maintenance [38]. It is worth emphasizing that not only volatilization was reduced by the new $Ur_{Duromide}$ technology, but that leaf percentage, dry matter production and N use efficiency were also increased.

5. Conclusions

The urease inhibitors Ur_{NBPT} and $Ur_{Duromide}$ reduced N losses by ammonia volatilization, to values similar to those observed for ammonium nitrate. Ammonia losses from $Ur_{Duromide}$ tend to be lower than from Ur_{NBPT}. In the separate evaluation of each cut, in the 2018/2019 and 2019/2020 growing seasons, volatilization losses from $Ur_{Conv100/200}$ were greater than from AN, Ur_{NBPT} and $Ur_{Duromide}$.

The total dry matter and the leaves dry matter production of Marandu grass in the first and second year, were highest in response to $Ur_{Duromide100/200}$ in comparison with the other sources (Ur_{Conv}, AN and Ur_{NBPT}).

It is concluded that compared to conventional urea, Ur_{NBPT} and $Ur_{Duromide}$ reduce N losses by NH_3 volatilization. The total dry matter and leaf yield of Marandu grass in the first and second year were higher in response to $Ur_{Duromide100/200}$ compared to the other sources (Ur_{Conv}, AN and Ur_{NBPT}).

Thus, the combination of duromide and NBPT is a very promising technology for reducing losses by ammonia volatilization and greater utilization of nitrogen from the fertilizer by the crop and higher productivity.

Author Contributions: Conceptualization, J.B.C. and R.H.; methodology J.B.C., C.L.B.d.O., A.d.S.B., N.d.L.D., G.C.M., I.V.R. and R.H.; software and validation, J.B.C., R.H. and C.L.B.d.O.; Resource, R.H.; formal analysis, J.B.C., C.L.B.d.O., A.d.S.B., N.d.L.D., G.C.M. and R.H.; investigation, J.B.C., C.L.B.d.O.; data curation, J.B.C., C.L.B.d.O.; writing—original draft preparation, J.B.C. and C.L.B.d.O.; writing—review and editing, J.B.C., C.L.B.d.O., A.d.S.B., J.F.d.S. and R.H.; visualization, J.B.C. and R.H.; supervision, R.H. All authors have read and agreed to the published version of the manuscript.

Funding: This research received no external funding.

Data Availability Statement: The data presented in this study are available on request from the corresponding author.

Conflicts of Interest: The authors declare no conflict of interest.

References

1. Cezário, A.S.; Ribeiro, K.G.; Santos, S.A.; de Campos Valadares Filho, S.; Pereira, O.G. Silages of Brachiaria Brizantha Cv. Marandu Harvested at Two Regrowth Ages: Microbial Inoculant Responses in Silage Fermentation, Ruminant Digestion and Beef Cattle Performance. *Anim. Feed. Sci. Technol.* **2015**, *208*, 33–43. [CrossRef]
2. Nunes, S.G.; Boock, A.; Penteado, M.D.O.; Gomes, D.T. *Brachiaria Brizantha Cv. Marandu*; Brazilian Agricultural Research Corporation (Empresa Brasileira de Pesquisa Agropecuária): Campo Grande, Brazil, 1984.
3. Macedo, M.C.M.; Araujo, A.R. *de Sistemas de Produção em Integração: Alternativa para Recuperação de Pastagens Degradadas*; Empresa Brasileira de Pesquisa Agropecuária—Embrapa: Brasília, Brasil, 2019.
4. Bezerra, J.D.D.V.; Neto, J.V.E.; Alves, D.J.D.S.; Neta, I.E.B.; Neto, L.C.G.; Santos, R.D.S.; Difante, G.D.S. Características produtivas, morfogênicas e estruturais de cultivares de Brachiaria brizantha cultivadas em dois tipos de solo. *Res. Soc. Dev.* **2020**, *9*, e129972947. [CrossRef]
5. Leite, R.D.C.; dos Santos, J.G.D.; Silva, E.L.; Alves, C.R.C.R.; Hungria, M.; Leite, R.D.C.; dos Santos, A.C. Productivity Increase, Reduction of Nitrogen Fertiliser Use and Drought-Stress Mitigation by Inoculation of Marandu Grass (Urochloa Brizantha) with Azospirillum Brasilense. *Crop Pasture Sci.* **2019**, *70*, 61. [CrossRef]
6. Francisco, E.A.B.; Bonfim-Silva, E.M.; Teixeira, R.A. Aumento da produtividade de carne via adubação de pastagens. *Inf. Agron.* **2017**, *257*, 6–12.
7. Delevatti, L.M.; Cardoso, A.S.; Barbero, R.P.; Leite, R.G.; Romanzini, E.P.; Ruggieri, A.C.; Reis, R.A. Effect of Nitrogen Application Rate on Yield, Forage Quality, and Animal Performance in a Tropical Pasture. *Sci. Rep.* **2019**, *9*, 7596. [CrossRef] [PubMed]
8. Termonen, M.; Korhonen, P.; Kykkänen, S.; Kärkönen, A.; Toivakka, M.; Kauppila, R.; Virkajärvi, P. Effects of Nitrogen Application Rate on Productivity, Nutritive Value and Winter Tolerance of Timothy and Meadow Fescue Cultivars. *Grass Forage Sci.* **2020**, *75*, 111–126. [CrossRef]
9. Faria, L.D.A.; Karp, F.H.S.; Machado, M.C.; Abdalla, A.L. Ammonia Volatilization Losses from Urea Coated with Copper, Boron, and Selenium. *Semin. Ciênc. Agrár.* **2020**, *41*, 1415–1420. [CrossRef]
10. Guelfi, D. Fertilizantes Nitrogenados Estabilizados, de Liberação Lenta Ou Controlada. *Inf. Agron.* **2017**, *157*, 1–14.
11. Timilsena, Y.P.; Adhikari, R.; Casey, P.; Muster, T.; Gill, H.; Adhikari, B. Enhanced Efficiency Fertilisers: A Review of Formulation and Nutrient Release Patterns. *J. Sci. Food Agric.* **2015**, *95*, 1131–1142. [CrossRef]
12. Koch The Duromide Advantage. Available online: https://kochagronomicservices.com/solutions/agricultural-nitrogen-efficiency/duromide/ (accessed on 3 February 2022).
13. Alvares, C.A.; Stape, J.L.; Sentelhas, P.C.; Gonçalves, J.L.D.M.; Sparovek, G. Köppen's Climate Classification Map for Brazil. *Meteorol. Z.* **2013**, *22*, 711–728. [CrossRef]
14. Santos, H.G.D.; Jacomine, P.K.T.; Anjos, L.H.C.D.; Oliveira, V.A.D.; Lumbreras, J.F.; Coelho, M.R.; Almeida, J.A.D.; Araujo Filho, J.C.D.; Oliveira, J.B.D.; Cunha, T.J.F. *Sistema Brasileiro de Classificação de Solos*; Embrapa: Brasília, Brazil, 2018; ISBN 978-85-7035-817-2.
15. Soil Survey Dvision. *Keys to Soil Taxonomy*, 12th ed.; United States Department of Agriculture, Natural Resources Conservation Service: Washinton, DC, USA, 2014.
16. Cantarella, H.; Quaggio, J.A.; Mattos, D., Jr.; Boaretto, R.M.; van Raij, B. *Boletim 100: Recomendações de Adubação e Calagem Para o Estado de São Paulo*, 1st ed.; Instituto Agronômico de Campinas: Campinas, Brazil, 2022.

17. Araújo, E.D.S.; Marsola, T.; Miyazawa, M.; Soares, L.H.D.B.; Urquiaga, S.; Boddey, R.M.; Alves, B.J.R. Calibração de câmara semiaberta estática para quantificação de amônia volatilizada do solo. *Pesqui. Agropecu. Bras.* **2009**, *44*, 769–776. [CrossRef]
18. Nascimento, C.A.C.D.; Vitti, G.C.; Faria, L.D.A.; Luz, P.H.C.; Mendes, F.L. Ammonia Volatilization from Coated Urea Forms. *Rev. Bras. Ciênc. Solo* **2013**, *37*, 1057–1063. [CrossRef]
19. Otto, R.; Zavaschi, E.; Souza Netto, G.M.D.; Machado, B.D.A.; Mira, A.B.D. Ammonia Volatilization from Nitrogen Fertilizers Applied to Sugarcane Straw. *Rev. Ciênc. Agron.* **2017**, *48*, 413–418. [CrossRef]
20. Meirelles, G.C.; Heinrichs, R.; Lira, M.; Ribeiro Virgílio, I.; Felipe Melo dos Santos, L.; Bonfim Cassimiro, J.; Luis Ruffo, M.; Viega Soares Filho, C.; Moreira, A. Ammonia Volatilization and Pasture Yield of Urochloa Decumbens Fertilized with Nitrogen Sources. *Arch. Agron. Soil Sci.* **2022**, 1–9. [CrossRef]
21. Araujo, R.S.; Hungira, M.D.C. *Manual de Métodos Empregados em Estudos de Microbiologia Agrícola*; Embrapa: Brasília, Brazil, 1994; pp. 149–177.
22. Corrêa, L.A. *Simpósio de Forragicultura e Pastagens*; Temas em Evidência—Lavras: UFLA: Lavras, Brazil, 2000; pp. 149–177.
23. Silva, D.J.; Queiroz, A.C. *Análises de Alimentos—Métodos Químicos e Biológicos*, 3rd ed.; UFV: Viçosa, Brazil, 2002; ISBN 85-7269-105-7.
24. Akaike, H. A New Look at the Statistical Model Identification. *IEEE Trans. Autom. Control* **1974**, *19*, 716–723. [CrossRef]
25. Seber, G.A.; Wild, C.J. *Nonlinear Regression*; John Wiley & Sons: Hoboken, NJ, USA, 2003; Volume 62, p. 1238.
26. Silva, T.R.D.; Cazetta, J.O.; Carlin, S.D.; Telles, B.R. Drought-induced alterations in the uptake of nitrogen, phosphorus and potassium, and the relation with drought tolerance in sugar cane Drought-induced alterations in the uptake of nitrogen, phosphorus and potassium, and the relation with drought tolerance in sugar cane. *Ciênc. Agrotecnol.* **2017**, *41*, 117–127. [CrossRef]
27. Cantarella, H.; Otto, R.; Soares, J.R.; Silva, A.G. de B. Agronomic Efficiency of NBPT as a Urease Inhibitor: A Review. *J. Adv. Res.* **2018**, *13*, 19–27. [CrossRef]
28. Minato, E.A.; Besen, M.R.; Cassim, B.M.A.R.; Mazzi, F.L.; Inoue, T.T.; Batista, M.A. Modeling of Nitrogen Losses Through Ammonia Volatilization in Second-Season Corn. *Commun. Soil Sci. Plant Anal.* **2019**, *50*, 2733–2741. [CrossRef]
29. Soares, J.R.; Cantarella, H.; Menegale, M.L.D.C. Ammonia Volatilization Losses from Surface-Applied Urea with Urease and Nitrification Inhibitors. *Soil Biol. Biochem.* **2012**, *52*, 82–89. [CrossRef]
30. Afshar, R.; Lin, R.; Mohammed, Y.A.; Chen, C. Agronomic Effects of Urease and Nitrification Inhibitors on Ammonia Volatilization and Nitrogen Utilization in a Dryland Farming System: Field and Laboratory Investigation. *J. Clean. Prod.* **2018**, *172*, 4130–4139. [CrossRef]
31. Dominghetti, A.W.; Guelfi, D.R.; Guimarães, R.J.; Caputo, A.L.C.; Spehar, C.R.; Faquin, V. Nitrogen Loss by Volatilization of Nitrogen Fertilizers Applied to Coffee Orchard. *Ciênc. Agrotecnol.* **2016**, *40*, 173–183. [CrossRef]
32. Caires, E.F.; Milla, R. Adubação nitrogenada em cobertura para o cultivo de milho com alto potencial produtivo em sistema de plantio direto de longa duração. *Bragantia* **2015**, *75*, 87–95.
33. Cassim, B.M.A.R.; Kachinski, W.D.; Besen, M.R.; Coneglian, C.F.; Macon, C.R.; Paschoeto, G.F.; Inoue, T.T.; Batista, M.A. Duromide Increase NBPT Efficiency in Reducing Ammonia Volatilization Loss from Urea. *Rev. Bras. Ciênc. Solo* **2021**, *45*, e0210017. [CrossRef]
34. Souza, J.R.D.; Lemos, L.B.; Leal, F.T.; Magalhães, R.S.; Ribeiro, B.N.; Cabral, W.F.; Gissi, L.D. Volatilization of Ammonia from Conventional Sources of Nitrogen and Compacted Urea under Controlled Conditions. *Rev. Bras. Ciênc. Agrár.* **2020**, *15*, 1–6. [CrossRef]
35. Rowlings, D.W.; Scheer, C.; Liu, S.; Grace, P.R. Annual Nitrogen Dynamics and Urea Fertilizer Recoveries from a Dairy Pasture Using 15N; Effect of Nitrification Inhibitor DMPP and Reduced Application Rates. *Agric. Ecosyst. Environ.* **2016**, *216*, 216–225. [CrossRef]
36. Cassimiro, J.B.; Rochetti, A.C.A.; Heinrichs, R.; Castillo, E.O.F. Volatilização da amônia e avaliação do capim-marandu sob doses e fontes de fertilizantes nitrogenados. *RSD* **2020**, *9*, e526985823. [CrossRef]
37. Cecagno, D.; Costa, S.E.V.G.D.A.; Anghinoni, I.; Brambilla, D.M.; Nabinger, C. Acidificação do solo sob fertilização nitrogenada de longo prazo em campo nativo com introdução de azevém. *Rev. Ciênc. Agrovet.* **2019**, *18*, 263–267. [CrossRef]
38. Galindo, F.S.; Buzetti, S.; Filho, M.C.M.T.; Dupas, E.; Ludkiewicz, M.G.Z. Acúmulo de matéria seca e nutrientes no capim-mombaça em função do manejo da adubação nitrogenada. *Rev. Agric. Neotrop.* **2018**, *5*, 1–9. [CrossRef]

Article

Response of Multi-Stressed *Olea europaea* Trees to the Adjustment of Soil pH by Acidifying Agents: Impacts on Nutrient Uptake and Productivity

Hamada R. Beheiry [1], Ahmed A. M. Awad [2],* and Hamdy A. Z. Hussein [1]

[1] Horticulture Department, Faculty of Agriculture, Fayoum University, Fayoum 63514, Egypt
[2] Soil and Natural Resources, Faculty of Agriculture and Natural Resources, Aswan University, Aswan 81528, Egypt
* Correspondence: ahmed.abdelaziz@agr.aswu.edu.eg; Tel.: +20-10-0042-1124

Abstract: Soil pH is the most important factor in evaluating plant nutritional status due to its close association with nutrient availability. In the 2018 and 2019 seasons, two field experiments were conducted to evaluate the performance of olive trees (*Olea europaea*, Picual cv.) grown in sandy clay loam soil under multi-abiotic stresses with the application of three different acidifying agents (AAs), acetic (AC), citric (CA), and sulfuric (SA) acid, at two doses (25 and 50 cm^3; AC_1 and AC_2, CA_1 and CA_2, and SA_1 and SA_2, respectively), as compared with a control treatment. This study was established according to a randomized complete block design. In general, our results showed that all the AAs applied surpassed the control treatment with respect to all the studied parameters except for the leaf iron content. Furthermore, the trees treated with CA yielded the best results in terms of the leaf nitrogen, calcium, and magnesium contents; the physiological and growth parameters (except for the performance index); the total fruit weight, flesh weight, and flesh dry matter; the fruit diameter; the oil content; and the total olive yield. Furthermore, the maximum leaf potassium, manganese, zinc, and copper contents were obtained in the trees growing in soil injected with AC. The correlation coefficient fluctuated between positive and negative among the studied characteristics.

Keywords: soil reaction; olive trees; acetic; citric and sulfuric acids; leaf nutrient content; growth and physiological parameters; productivity and its attributes

 check for updates

Citation: Beheiry, H.R.; Awad, A.A.M.; Hussein, H.A.Z. Response of Multi-Stressed *Olea europaea* Trees to the Adjustment of Soil pH by Acidifying Agents: Impacts on Nutrient Uptake and Productivity. *Agronomy* **2023**, *13*, 539. https://doi.org/10.3390/agronomy13020539

Academic Editors: Christos Noulas, Shahram Torabian and Ruijun Qin

Received: 31 December 2022
Revised: 9 February 2023
Accepted: 10 February 2023
Published: 14 February 2023

1. Introduction

Soil pH is an influential factor in the adsorption/absorption and availability of nutrients in the soil owing to the close relationship between them. In addition, most of the chemical, fertility, and biological properties of soil are strongly associated with soil pH; thus, in turn, it also affects plant growth and development [1,2]. Chemically, soil pH can be defined as the negative logarithm of the active hydrogen (H^+) or hydroxyl ion concentration (OH^-) or, simply, pH = − log [H^+]; pOH = − log [OH^-] [3,4]. A scale ranging from 0 to 14 is used to describe the acidity and alkalinity of soil. pH values of less than 7 refer to acidic conditions, while those above 7 indicate an alkaline environment; however, pH values at 7 are considered neutral [3,5]. Soils in arid and semi-arid regions are commonly alkaline with a high pH [6] as a result of water scarcity, in addition to low precipitation and potential evapotranspiration [7,8], as indicated by the negative correlations between soil pH and temperature and between soil pH and precipitation. Recently, some studies have reported that the soil pH in Egypt varies from neutral to strongly alkaline as an inherent characteristic of the soil, resulting from the nature of the parent material [9] along with the prevailing climatic conditions [10–12]. The results obtained from the studies of [13–15] indicate that the low availability of some nutrients, especially phosphorus (P) and other micronutrients (except molybdenum), is strongly related to an increase in soil pH. The decrease in micronutrient availability in alkaline soil could be explained by the fact that these cations become strongly bonded with soil organic matter (SOM) and soil colloids as the pH approaches 8.

The available information regarding the strong association between soil pH and nutrient availability suggests that the ability of plants to uptake nutrients through their root hairs and root tips is affected by soil pH. Although this issue is worthy of attention, since alkaline soils cover more than one-fourth of Earth's surface and the maximum availability of most nutrients occurs at a pH less than 6 [16], studies related to soil alkalinity modifications are still required. Adjustments in soil pH are often caused by the application of either acidic fertilizers or synthetic chemicals; however, these changes are resisted by the soil's buffering capacity (SBC). Accordingly, the availability of nutrients depends indirectly on the SBC [17,18]. Furthermore, soil pH is not only affected by climatic conditions but also by soil type; calcareous soils are particularly alkaline owing to their high $CaCO_3$ content [19]. In addition, salinity and sodicity are considered as major factors in increasing soil pH. Globally, the total area of salt-affected soils is about 935,000,000 ha [20]; about 560,000,000 ha of this area is characterized by saline–sodic soils [21,22]. Moreover, several factors, such as soil texture, soil mineralogical compounds, and SOM, as well CEC, have an appreciable influence on soil pH [23].

Since the 1960s, many efforts have been made to bring soils to a desired pH [24]. Accordingly, several acidifying materials, either chemical or organic, have been applied to alkaline soil in an attempt to obtain a more optimal pH. For example, the application of organic manures has a significant impact on soil pH [25] due to their role in producing acidic organics and enhancing the cation exchange capacity (CEC), which consequently increases the acid saturation percentage [26]. In addition, the application of some acidifying agents, such as gypsum ($CaSO_4 \cdot 2H_2O$), has been shown to result in a significant adjustment in the pH at a lower cost [27]. Sulfuric acid (SA) and polyacrylamide are also considered as vital chemical treatments [28,29] for lowering soil pH. In addition, other acidifying agents have been applied, including nitric acid (HNO_3), hydrochloric acid (HCl), or the salts of trivalent metal ions, including aluminum (Al) and iron (Fe), as well as materials containing ammonium (NH_4^+) ions [30]. Recently, the influential role of organic acids (OAs) as acidifying agents for lowering soil pH has been reported in some studies. In addition to their effect on soil pH, OAs generally play a pivotal role in enhancing the solubility of micronutrients through chelation and complexation, which in turn improves their uptake by plants [31]. Acetic (AA) and citric acid (CA), both classified as OAs, are natural substances that exhibit a low toxicity to microorganisms; their chemical formulas are CH_3COOH and $C_3H_5O(COOH)_3$, respectively. In their study on three different textured soils (sandy clay loam, clay loam and silt loam) using four chemicals, including aluminum sulfate $Al_2(SO_4)_3$, hydrogen peroxide (H_2O_2), hydrochloric acid (HCl), and sulfuric acid H_2SO_4, [16] reported that $Al_2(SO_4)_3$ had the greatest effect in terms of lowering pH either alone or in combination with other chemicals. Similarly, H_2O_2, HCl, and H_2SO_4 are helpful in the conditioning of soil pH.

In this study, to understand the potential influence of the application of acidifying agents on the availability of nutrients and the physiological aspects of plants, olive (*Olea europaea* L.) trees grown under multi-stress conditions were selected because of their socioeconomic importance and the fact that the majority of olive trees are cultivated in newly reclaimed soils [32,33]. According to the FAO, 2022 [34], the total area cultivated with olive trees around the world is nearly 11 million ha, more than 90% of which is found in the Mediterranean basin countries, such as Spain, Italy, Greece, Turkey, and Tunisia [35]. In Egypt, olives rank fourth after citrus, mango, and table grape cultivation [36,37]. Egypt is considered as a good competitor in global markets as it produces 13% of the total global yield [34]. In the last thirty years, Egypt has achieved unprecedented progress in terms of the total cultivated area and the total production of several varieties, including Picual, Kalamata, Teffahi, Waleken, etc. In 2017, the total cultivated area reached 101,326 ha, with the total production estimated at about 874,748 tons according to the Agricultural Affairs Sector.

For this purpose, two field experiments were conducted in the growing seasons of 2018 and 2019 on olive trees (*Olea europaea*, Picual cv.) growing under multi-abiotic stresses (pH = 7.95 vs. 7.87; $CaCO_3$ = 9.1 vs. 9.8%; and ECe 6.5 vs. 7.4 $dS \cdot m^{-1}$) in a sandy loam clay soil using two doses (25 and 50 cm^3) of three different acidifying agents (AAs): acetic acid (AC, AC_1, and AC_2), citric acid (CA, CA_1, and CA_2), and sulfuric acid (SA, SA_1, and SA_2) in an attempt to adjust the soil pH, which is the main problem with Egyptian soils, and evaluated their potential impacts on nutrient availability, which in turn influences the physiological and growth parameters, the yield of table and oil olives, and the fruit's physical attributes.

2. Materials and Methods

2.1. Field Experiment Site and Climatic Conditions

Two field experiments were performed in the Kawm Ushim region, situated between 29°32′39″ N latitude and 30°52′35″ E longitude and located on the Cairo-Fayoum Desert Road, Egypt, during the 2018 and 2019 seasons, to evaluate the potential performance of some acidifying agents in an attempt to decrease the soil pH within an optimal range for plant uptake. This investigation focused on olive (*Olea europaea* L.) trees grown in sandy loam clay textured soil in both seasons. The average weather data from January to December for both growing seasons are shown in Table 1.

Table 1. Monthly climate averages for Kawm Ushim region, Fayoum, Egypt, during the 2018 and 2019 seasons.

Month	ADT	ANT	ARH	AWS	AM-PEC-A	AP
	°C		(%)	(ms^{-1})	(mmd^{-1})	(mmd^{-1})
January	22.15	2.05	60.41	2.89	3.58	0.14
February	27.49	4.43	54.00	2.43	4.13	0.10
March	31.59	5.77	46.78	2.96	4.78	0.04
April	36.89	7.83	40.60	3.23	5.59	0.05
May	43.98	12.82	32.03	3.55	6.71	0.00
June	42.40	17.06	35.56	3.75	6.77	0.00
July	42.67	19.59	39.88	3.71	7.55	0.00
August	40.67	19.74	43.41	3.51	6.87	0.00
September	38.95	17.18	50.35	3.62	6.65	0.00
October	35.50	12.33	51.63	3.14	6.41	0.10
November	31.56	8.95	54.60	2.44	5.62	0.07
December	22.55	4.97	65.72	2.90	4.33	0.70

ADT °C = average day temperature, ANT °C = average night temperature, ARH = average relative humidity, AWS = average wind speed, AM-PEC-A= average measured pan evaporation class A, and AP = average precipitation.

2.2. Plant Material and Agricultural Practices

The tested Picual variety trees were about 15 years old and were propagated using leafy cuttings planted at a distance of about 5×8 m^2 between trees under a drip irrigation system. The trees were carefully selected to ensure that they were free from fungal and insect diseases.

All the horticulture practices were accomplished according to the recommendations of the Egyptian Ministry of Agriculture, including the amount and schedule for water irrigation and weed control (glyphosate 48% sprayed prophylactically with Kocide 54.8% copper hydroxide). Based on the technical bulletin NO_2, issued by the General Administration of Agriculture in 2016, the fertilization program of olive trees aged over 6 years included the application of ammonium sulfate [$(NH_4)_2SO_4$ N $\approx$ 20.6], granular calcium super phosphate [$Ca(H_2PO_4)_2$ P_2O_5 $\approx$ 15.5%], potassium sulfate [K_2SO_4 K_2O $\approx$ 48%], and magnesium sulfate [$MgSO_4 \cdot 7H_2O$ MgO $\approx$ 24%] as sources of nitrogen (N), phosphorus (P), potassium (K), and magnesium at a rate of 394, 500, 810, and 400 g per tree, respectively.

2.3. Treatments, Application Timing, and Experimental Design

Three acidifying agents—acetic acid (AC), CH_3COOH; citric acid (CA), $C_6H_8O_7$; and sulfuric acid (SA), H_2SO_4, 99.99%—were injected individually as soil applications. For each acidifying agent (AA), two doses were applied and compared with a control treatment (without AAs) ($AC_1 = 25$ and $AC_2 = 50$ cm^3; $CA_1 = 25$ and $CA_2 = 50$ cm^3; and $SA_1 = 25$ and $SA_2 = 50$ cm^3). These doses were selected depending on the depth of the olive trees' roots, which ranged from 100 to 120 cm; the soil pH in the study area; and the pH target to reach, as shown in Table 2. All the AAs applied were purchased from Sigma Aldrich, St. Louis, MO, USA. The acidifying agents were prepared via dilution in 20 L of irrigated water and applied to the soil around the tree trunk three times in the middle of March, May, and July during the 2018 and 2019 seasons. The main experimental plots were arranged according to the two doses of the three acidifying agents, in addition to the untreated trees. Thus, there were (3 acidifying agents × 2 doses) six treatments and one control treatment (a total of seven treatments), and each plot included 3 trees that were ordered according to the randomized complete block design (RCBD).

Table 2. The description of treatments applied, application method, and applying time in the field experiment.

Symbol	Treatment Description	Application Method	Applying Time
Con.	No acids were applied and treated with irrigated water	Three times in four plots before direct fertilizer application	All acids were applied three times before fertilizer application for 2 days. However, the fertilizers were added in the middle of March, May and July. The treatments were applied in around the rhizosphere area
AC_1	Twenty-five cubic centimeters of acetic acid was added to 20 L of irrigated water per tree		
AC_2	Fifty cubic centimeters of acetic acid was added to 20 L of irrigated water per tree		
CA_1	Twenty-five cubic centimeters of citric acid was added to 20 L of irrigated water per tree		
CA_2	Fifty cubic centimeters of citric acid was added to 20 L of irrigated water per tree		
SA_1	Twenty-five cubic centimeters of sulfuric acid was added to 20 L of irrigated water per tree		
SA_2	Fifty cubic centimeters of sulfuric acid was added to 20 L of irrigated water per tree		

2.4. Soil Sampling and Determination of Chemical and Physical Properties

Soil samples were taken from five consecutive depths (0–25, 25–50, 50–75, 75–100, and 100–120 cm) in the root system area before AA application (in February 2018 and 2019) and transferred to the Soil, Water, and Plant Analysis Laboratory (SWPA) at the Faculty of Agriculture, Fayoum University, for the characterization of the chemical and physical properties, as shown in Table 3. The following properties were analyzed: soil texture was determined using the hydrometer method [38]. Soil pH and soil electrical conductivity (ECe) were determined from the saturation soil paste and extract soil paste using a pH meter (Jenway, UK) and an EC meter (LF 191 Conduktometer, Germany), according to [39,40], respectively.

The calcium carbonate ($CaCO_3$%) content was determined using a Collin's calcimeter, as described in [39]. The soil organic matter (SOM) was measured according to [41], Walkely and Black's method.

Soluble cations, such as Na^+, K^+, Ca^{2+}, and Mg^{2+}, were extracted with ammonium acetate 1 M $NH_4CH_3CO_2$, while Na^+ and K^+ were determined using a flame photometer [42]. On the other hand, both Ca^{2+} and Mg^{2+} were measured via the EDTA titration method. Soluble anions, such as HCO_3^-, CO_3^{2-}, and Cl^-, were determined via the titration method described by [39]. The SO_4^{2-} ions were calculated as the difference between the total soluble cations and anions. In addition, the macronutrients nitrogen (N), phosphorus (P), and potassium (K) were determined using the methods described by [43,44].

Table 3. Some soil chemical and physical properties.

Soil Property	2020 Season	2021 Season
Particle size distribution (%)		
Sand	47.32	48.49
Silt	19.56	20.20
Clay	33.12	31.31
Soil texture	Sandy clay loam	Sandy clay loam
pH (in soil paste)	7.78	7.89
ECe (dS m^{-1})	6.4	7.2
Organic matter (%)	0.63	0.52
$CaCO_3$ (%)	8.8	9.2
Soluble ions (mmol L^{-1})		
CO_3^{2-}	—	—
HCO_3^-	2.8	3.7
Cl^-	53.4	55.3
SO_4^{2-}	19.3	21.1
Ca^{2+}	39.6	41.2
Mg^{2+}	7.8	8.4
Na^+	22.4	24.3
K^+	5.7	6.2
Macronutrients (mg kg^{-1})		
Total N	414	640
Available P (extractable with $NaHCO_3$ pH = 8.5)	4520	4830
Available K (extractable with NH_4OAC pH = 7.0)	1337	1415
DTPA Extractible micronutriments (mg kg^{-1})		
Fe	10.7	11.2
Mn	4.5	6.3
Zn	0.15	0.14
Cu	0.48	0.38

2.5. Physiological and Growth Parameters

One-year-old shoots were randomly collected from each side of the orchard trees to measure shoot length (ShL, cm), the average number of leaves per meter (NLf), and the leaf area (LA, cm^2) of the third and fourth leaves from the top of the new spring shoots, which were estimated using a digital planimeter device (Planx 7 Tamaya).

The relative chlorophyll content (SPAD reading) was determined using a SPAD-502 m device (Minolta, Osaka, Japan). The variable fluorescence by maximum fluorescence (fv/fm) and the photosynthetic performance index (PPI) were measured using a fluorimeter (Handy PEA, Hansatech Instruments LTd., Kings Lynn, UK) as described by [45,46].

2.6. Evaluation of Leaf Nutrient Content

Random leaf samples were taken from twenty shoots selected from each tree, transferred to the laboratory, washed with distilled water, oven-dried at 70 °C to constant weight (72 h), and crushed to determine the N, P, K, Na, Ca, and Mg, as described by A.O.A.C, 2005. In addition, the total contents of Fe, Mn, Zn, and Cu were determined via inductively coupled plasma-optical emission spectrometry (ICP-OES, Perkin-Elmer OPTIMA-2100 DV, Norwalk, CT, USA), according to the methods described by Baird et al. [47].

2.7. Statistical Analysis

All parameters studied were analyzed according to a randomized complete block design (RCBD) with four replications. An analysis of variance (ANOVA) was conducted using the GenStat software, 12th edition [48]. Mean values were calculated using Duncan's multiple range test. Correlation was determined by the calculation of Pearson's linear correlation coefficient (r).

3. Results

3.1. Effect of Acidifying Agents on Soil pH Values

Initially, the tested soil was characterized as moderately alkaline, similar to the native soil (7.78 vs. 7.89) but also possessing some undesirable characteristics, such as high $CaCO_3$ content (9.1 vs. 9.8%) and salinity (ECe; 6.5 vs. 7.4 dS m^{-1}), which effectively contributed to raising its alkalinity, as shown in Table 3. The data presented in Figure 1 show the soil pH values obtained as a result of the application of the acidifying agents (acetic acid (AC_1 = 25 and AC_2 = 50 cm^3), citric acid (CA_1 = 25 and CA_2 = 50 cm^3), and sulfuric acid (SA_1 = 25 and SA_2 = 50 cm^3)), in comparison with those of the untreated soil (Con.) tested in this investigation. The overall trend indicated that the acidifying agents appreciably decreased the soil pH values; however, this occurred at different rates depending on the type of acidifying agent used and the application dose.

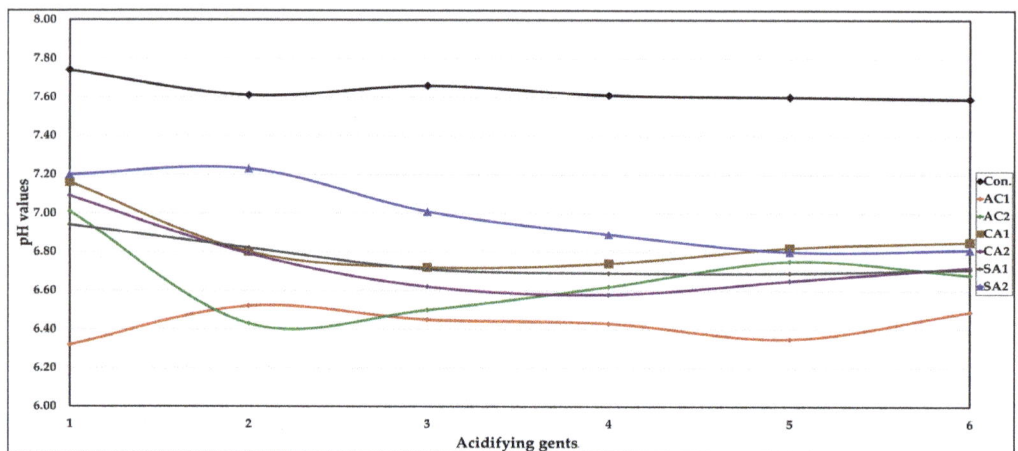

Figure 1. Effect of acidifying agents on the gradual change average in soil pH values during the 2018 and 2019 seasons.

It can be seen from the graphical representation of the data that the AC_1 treatment resulted in the lowest pH values (6.30 vs. 6.29) immediately after injection (i.e., on the first day), when compared to the other acidifying agents. Furthermore, the soil pH values oscillated, increasing and decreasing, such that there was no stable tendency observed. The lowest pH values detected were 6.30 vs. 6.29 for AC_1; 6.43 vs. 6.47 for AC_2; 6.72 vs. 6.73 for CA_1; 6.58 vs. 6.60 for CA_2; 6.69 vs. 6.70 for SA_1; and 6.80 vs. 6.82 for SA_2 for the 2018 and 2019 growing seasons, respectively.

In addition, decreasing percentages of the highest and lowest values were also observed: 3.37 vs. 3.82 for AC_1; 8.27 vs. 7.97 for AC_2; 6.15 vs. 6.14 for CA_1; 7.19 vs. 7.04 for CA_2; 13.57 vs. 15.51 for SA_1; and 5.56 vs. 4.75 for SA_2 in both seasons, respectively.

The results obtained from the statistical analysis revealed highly significant differences in soil pH with all the acidifying agents applied and a non-significant impact with the control treatment, where the changes in pH were very limited.

3.2. Leaf Macro and Micronutrient Content

The results shown in Table 4 indicate that all the applied acidifying agents appreciably improved the leaf macronutrient content. However, the highest leaf nitrogen content (LNC), with values of 2.50 vs. 2.32% in both seasons, and leaf sodium (LNaC), calcium (LCaC), and magnesium (LMgC) contents, with values of 0.65, 1.38, and 0.43%, respectively, in the first season, were only obtained with the application of CA_2.

Table 4. Effect of acidifying agents as a soil application on the leaf macronutrient content of olive trees (Picual cv.) grown in a sandy loam clay soil under multi-abiotic stresses (pH = 7.78 vs. 7.89; $CaCO_3$ = 8.8 vs. 9.2%; and ECe = 6.4 vs. 7.2 dS m^{-1}).

Treatment	LNC	LPC	LKC	LNaC	LCaC	LMgC
	(%, in DM of Leaves)					
2018 season						
Con.	1.43c ± 0.01	0.26c ± 0.01	0.59b ± 0.09	0.42cd ± 0.01	1.12c ± 0.01	0.27bc ± 0.01
AC_1	0.90d ± 0.01	0.32b ± 0.01	0.60b ± 0.01	0.50bc ± 0.03	1.13c ± 0.03	0.34ab ± 0.02
AC_2	0.71d ± 0.01	0.32b ± 0.01	0.86a ± 0.02	0.61a ± 0.07	1.21bc ± 0.06	0.28bc ± 0.04
CA_1	1.07cd ± 0.01	0.34ab ± 0.02	0.70ab ± 0.04	0.51b ± 0.05	1.29ab ± 0.05	0.18c ± 0.05
CA_2	2.50a ± 0.01	0.32b ± 0.01	0.69ab ± 0.07	0.65a ± 0.08	1.38a ± 0.01	0.43a ± 0.06
SA_1	0.89d ± 0.01	0.38a ± 0.02	0.67b ± 0.05	0.39d ± 0.03	1.13c ± 0.05	0.38ab ± 0.03
SA_2	1.97b ± 0.01	0.35ab ± 0.02	0.52b ± 0.03	0.43cd ± 0.01	1.20bc ± 0.03	0.28bc ± 0.04
2019 season						
Con.	2.14a ± 0.01	0.25c ± 0.01	0.69b ± 0.03	0.39a ± 0.01	1.16e ± 0.01	0.25b ± 0.01
AC_1	2.14a ± 0.01	0.43b ± 0.01	0.62c ± 0.01	0.36a ± 0.01	1.27d ± 0.08	0.43a ± 0.03
AC_2	1.61b ± 0.03	0.40b ± 0.01	0.83a ± 0.01	0.31b ± 0.03	1.51a ± 0.05	0.24b ± 0.11
CA_1	1.07c ± 0.01	0.43ab ± 0.01	0.72b ± 0.02	0.25c ± 0.03	1.08f ± 0.03	0.47a ± 0.05
CA_2	2.32a ± 0.01	0.24c ± 0.02	0.73b ± 0.04	0.26c ± 0.01	1.36c ± 0.10	0.15bc ± 0.01
SA_1	1.25bc ± 0.02	0.27c ± 0.06	0.71b ± 0.03	0.27c ± 0.01	0.94g ± 0.03	0.38a ± 0.03
SA_2	1.43bc ± 0.01	0.46a ± 0.02	0.60c ± 0.03	0.38a ± 0.03	1.44b ± 0.03	0.25c ± 0.01

Mean values (±SE), different letters in each column indicate significance at $p \leq 0.05$. CA_1, CA_2, AC_1, AC_2, SA_1, and SA_2 represent citric acid, acetic acid, and sulfuric acid applied at 25 and 50 cm^3, respectively.

Furthermore, the plants treated with AC_2 had the highest values (0.86 vs. 0.83%) for leaf potassium content (LKC) in both seasons and leaf calcium content (LCaC; 1.51%) in the second season only. While the application of sulfuric acid, irrespective of the applied concentration, was found to be the best treatment for leaf phosphorus content (LPC); maximum leaf values were achieved in the plants treated with SA_1 and SA_2 during the growing season (0.38 vs. 0.46%, respectively). Dissimilar data were observed for LNaC, LCaC, and LMgC in the second season; however, the highest values were produced in the untreated plants (Con.) and those treated with AC_2 and CA_1, (0.39, 1.51 and 0.47%, respectively). Concerning the lowest values, similar data were obtained, where the minimum values for LPC in the 2018 and 2019 growing seasons (0.26 vs. 0.25%, respectively) and LCaC in the 2018 season (1.12%) were only achieved in the untreated plants.

Furthermore, the values of 0.52 vs. 0.60% for LKC were produced with the SA_2 soil treatment in the 2018 and 2019 growing seasons, respectively. Furthermore, the lowest values of LNC and LNaC were similar in each season. However, the SA_1 treatment in the 2018 season and the CA_1 treatment in the 2019 season had the least impact on both nutrients, with values of 0.89 vs. 0.39% and 1.07 vs. 0.25% in the first and second seasons, respectively. Moreover, the CA treatment, regardless of the concentration applied, had the weakest effect on LMgC, which was recorded as 0.18 vs. 0.15% for the 2018 and 2019 seasons, respectively. The results depicted in Table 4 show that the increment rates were 180.90 vs. 116.82 for LNC; 46.15 vs. 84.00 for LPC; 65.39 vs. 38.33 for LKC; 28.89 vs. 28.00 for LNaC; 23.21 vs. 60.64 for LCaC; and 138.89 vs. 213.33 for LMgC in the 2018 and 2019 growing seasons, respectively.

The analysis of variance indicated that all the treatments had a significant influence on LNC, LPC, and LNaC in both growing seasons and LKC, LCaC, and LMgC in the second season only at $p \leq 0.01$. However, the impact on the contents of the latter three nutrients was significant in the first season at $p \leq 0.05$.

As can be seen from Table 5, the use of the acidifying agents applied in this study significantly improved the content of the leaf micronutrients, except for the leaf iron content (LFeC); the highest values (389.67 vs. 449.67 mg kg^{-1}) were produced in the untreated plants in both seasons. Furthermore, the general trends indicated that CA$_1$ was the superior treatment for the leaf zinc content (LZnC) and the leaf copper content (LCuC) in both growing seasons, as well as the leaf manganese content (LMnC) in the first season only, with maximum values of 22.17 vs. 40.03 mg kg^{-1} for LZnC and 7.09 vs. 8.09 mg kg^{-1} for LCuC in the 2018 and 2019 seasons, respectively, and 22.17 mg kg^{-1} for LMnC in the 2018 season. The greatest value for LMnC (20.86 mg kg^{-1}) was observed in the second season with the AC$_1$ treatment.

Table 5. Effect of acidifying agents as a soil application on the leaf micronutrient content of olive trees (Picual cv.) grown in a sandy loam clay soil under multi-abiotic stresses (pH = 7.78 vs. 7.89; CaCO$_3$ = 8.8 vs. 9.2%; and ECe = 6.4 vs. 7.2 dS m^{-1}).

Treatment	LFeC	LMnC	LZnC	LCuC
	(mg·kg^{-1}, in DM of Leaves).			
	2018 season			
Con.	389.67a ± 9.72	17.03cd ± 0.12	16.00e ± 0.48	2.25d ± 0.14
AC$_1$	259.42c ± 10.16	22.17a ± 0.19	22.17a ± 0.19	7.09a ± 0.34
AC$_2$	167.00e ± 4.61	19.92b ± 0.53	20.92b ± 0.05	6.83a ± 0.19
CA$_1$	195.25de ± 4.19	19.25b ± 0.24	19.25c ± 0.24	5.25b ± 0.05
CA$_2$	228.50cd ± 2.02	17.75c ± 0.14	17.75d ± 0.14	4.42c ± 0.14
SA$_1$	320.92b ± 4.76	14.75e ± 0.43	14.75f ± 0.33	5.17b ± 0.10
SA$_2$	231.42cd ± 10.83	16.59d ± 0.24	16.59e ± 0.23	4.33c ± 0.29
	2019 season			
Con.	449.67a ± 7.60	18.00d ± 0.48	27.41d ± 0.42	3.25c ± 0.14
AC$_1$	300.42b ± 2.45	23.67a ± 0.10	40.03a ± 0.86	8.09a ± 0.24
AC$_2$	198.50d ± 7.53	20.00bc ± 0.58	37.50b ± 0.73	7.33a ± 0.10
CA$_1$	216.25cd ± 1.59	20.68b ± 0.09	33.37c ± 0.76	5.75b ± 0.33
CA$_2$	238.50c ± 2.60	19.37c ± 0.08	37.80ab ± 0.68	4.92b ± 0.43
SA$_1$	329.50b ± 8.37	16.75e ± 0.14	39.12ab ± 0.29	5.67b ± 0.38
SA$_2$	246.25c ± 1.88	18.18d ± 0.01	38.91ab ± 0.53	5.33b ± 0.30

Mean values (±SE), different letters in each column indicate significance at $p \leq 0.05$. CA$_1$, CA$_2$, AC$_1$, AC$_2$, SA$_1$, and SA$_2$ represent citric acid, acetic acid, and sulfuric acid applied at 25 and 50 cm^3, respectively.

The least impactful treatment was the control treatment, for which we recorded minimum values for LZnC (16 vs. 27.41 mg kg^{-1}) and LCuC (2.25 vs. 3.25 mg kg^{-1}) in both growing seasons, followed by the CA$_2$ and SA$_1$ treatments, with values of 167.00 vs. 198.50 mg kg^{-1} for LFeC and 14.75 vs. 16.75 mg kg^{-1} for LMnC in the 2018 and 2019 seasons, respectively. It can be seen from Table 5 that the increasing percentages of the highest and lowest values were 133.34 vs. 126.53 for LFeC; 50.31 vs. 23.46 for LMnC; 38.56 vs. 46.04 for LZnC; and 215.11 vs. 138.29 for LCuC in the 2018 and 2019 seasons, respectively. The results obtained from the statistical analysis showed highly significant differences among the studied micronutrients in both seasons.

3.3. Growth and Physiological Attributes

The data pertaining to the impact of the applied acidifying agents, irrespective of the doses, indicated that CA generally surpassed the other acids, as shown in (Figure 2A–C). In particular, the application of a high dose of CA (CA$_2$) had the most desirable effect on all studied growth attributes except the number of leaves per m^2 (NLf) in the 2019 season.

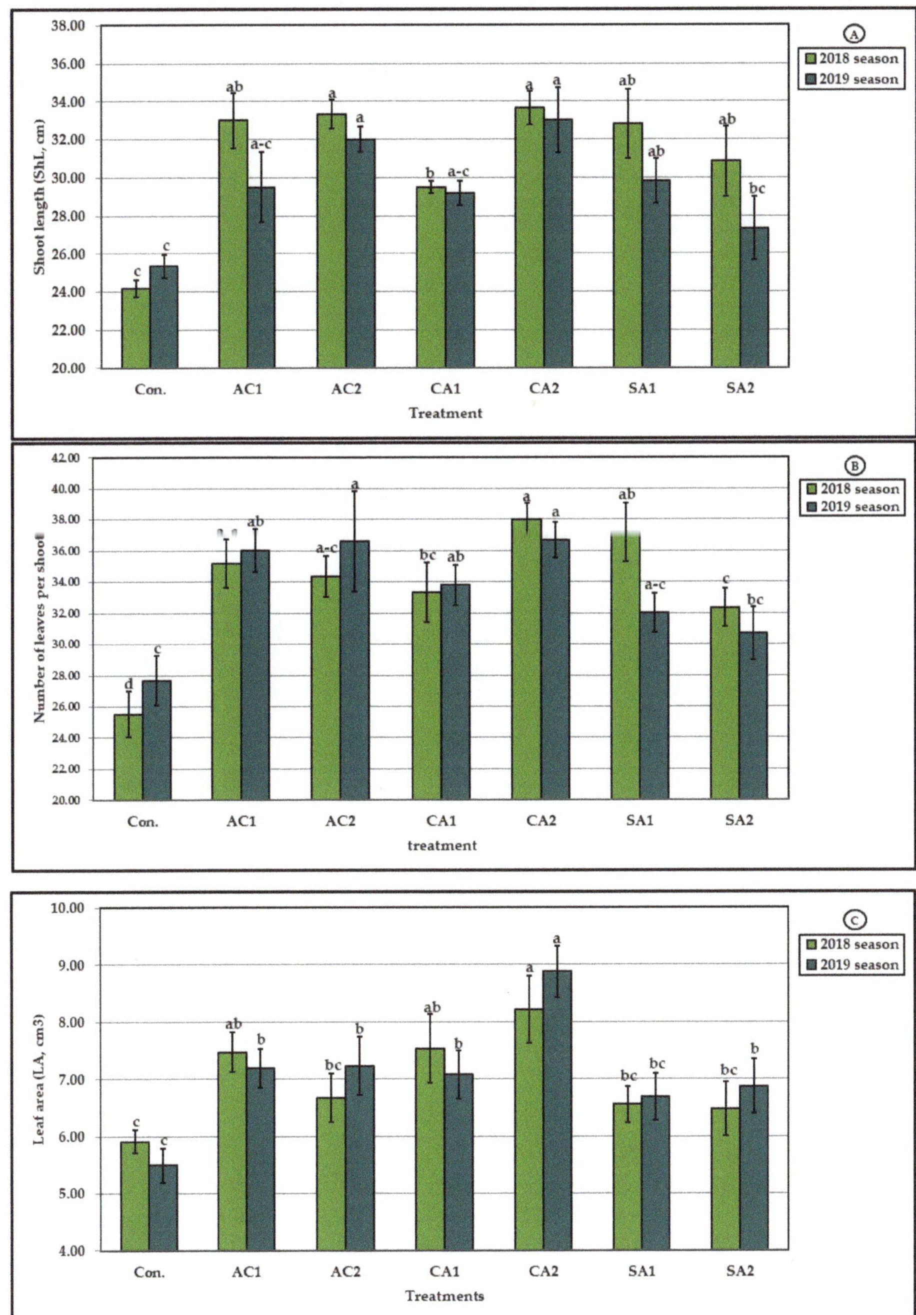

Figure 2. Effect of acidifying agents applied to (**A**) shoot length (ShL, cm); (**B**) number of leaves per shoot; and (**C**) leaf area (LA, cm^3) of olive trees picual cv. grown in a sandy loam clay soil under multi-abiotic stresses (pH = 7.78 vs. 7.89; CaCO$_3$ = 8.8 vs. 9.2%; and ECe = 6.4 vs. 7.2 dS m^{-1} and). Different letters in each column indicate significance at $p \leq 0.05$. CA$_1$, CA$_2$, AC$_1$, AC$_2$, SA$_1$, and SA$_2$ represent citric acid, acetic acid, and sulfuric acid applied at 25 and 50 cm^3, respectively.

The results (Figure 3A–C) showed that all applied acids markedly increased the studied physiological parameters, including SPAD chlorophyll, Fv/Fm, and the performance index (PI). The data obtained indicated that the AC_2 treatment was the superior treatment in both the 2018 and the 2019 seasons, with recorded values of 0.838 vs. 0.831 for fv/fm and 5.39 vs. 5.44 for PI and an increase in percentages of 3.08 vs. 2.72 and 42.21 vs. 39.85% when compared with the control treatment, which had the lowest values (0.813 vs. 0.809 for fv/fm and 4.17 vs. 3.89 for PI) in the 2018 and 2019 seasons, respectively. The plants treated with CA_2 had the highest values (82.45 vs. 82.52) for SPAD, while the lowest values (76.60 vs. 75.20) were produced in the untreated plants. The increase in percentages was 7.64 vs. 9.80% for the 2018 and 2019 seasons, respectively.

The results of the ANOVA indicated that all treatments had a significant impact on the SPAD reading and Fv/Fm in the first season at $p \leq 0.01$ and a significant influence on PI in the first season at $p \leq 0.05$; non-significant effects were observed for the SPAD reading, Fv/Fm, and PI in the second season.

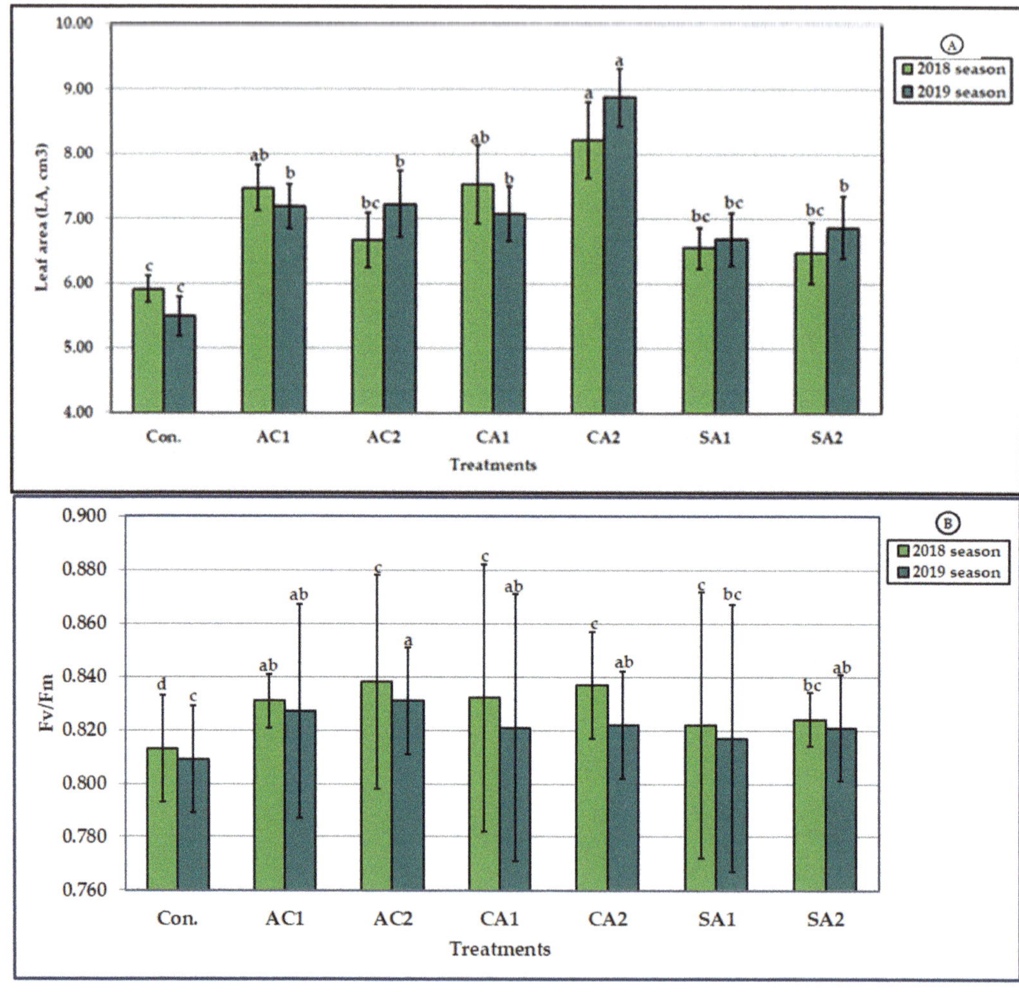

Figure 3. *Cont.*

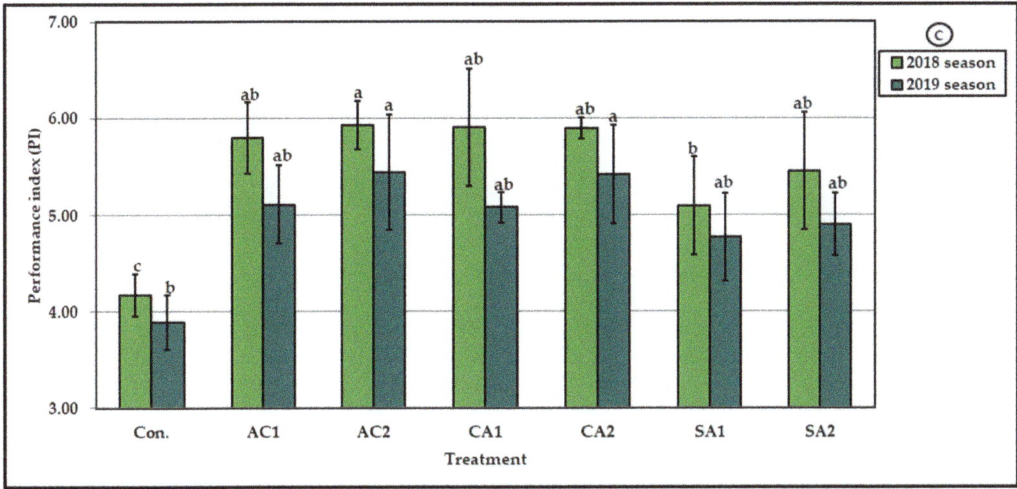

Figure 5. Impact of soil pH induced by some acidifying agents applied to (**A**) OPAD reading, (**B**) fv/fm, and (**C**) performance index (PI) of olive trees picual cv. grown in a sandy loam clay soil under multi-abiotic stresses (pH = 7.78 vs. 7.89; $CaCO_3$ = 8.8 vs. 9.2%; and ECe = 6.4 vs. 7.2 dS m^{-1}). Different letters in each column indicate significance at $p \leq 0.05$. CA_1, CA_2, AC_1, AC_2, SA_1, and SA_2 represent citric acid, acetic acid, and sulfuric acid applied at 25 and 50 cm^3, respectively.

3.4. Fruit Parameters, Olive Oil Yield, and Table Olive Yield

According to the data displayed in Table 6, the studied parameters improved appreciably as a result of the acid treatments. The general trend indicated that the application of AC and CA, irrespective of the dose, resulted in the highest values when compared with SA. However, the maximum values (4.53 vs. 4.52 g), (36.80 vs. 35.51%), and (30.25 vs. 40.75 kg tree^{-1}) for the total fruit weight (TFrW), olive oil content (OOC), and table olive yield (TOY), respectively, were observed in the plants treated with AC_2 in both seasons.

Furthermore, the maximum values for the flesh weight (FlW) (3.68 vs. 3.72 g) and the percentage flesh weight on the total fruit weight (FTW) (81.75 vs. 81.65%) in both growing seasons were found in the plants treated with AC_1 and CA_2, respectively. Dissimilar data were obtained for the flesh dry weight (FDrM), fruit length (FrL), and fruit diameter (FrD). The plants treated with AC_2 had the highest values for FlDrW (40.75%) and FrL (24.29 mm) in the first season. On the other hand, it was found that AC_1 was the best treatment for FDrM (38.73) and FrD (16.76) in the second season. The application of the CA_2 treatment gave the highest values for FrD in the 2018 season (17.86 mm) and FrL in the 2019 season (22.60 mm).

As presented in Table 6, the lowest values of all the studied parameters were recorded with the control treatment (Con.), except for FDrM and OOC in the second season (33.09 and 34.40, respectively). As shown in Table 6, the untreated plants had values of 3.29 vs. 3.41 for TFrW; 2.56 vs. 2.66 for FlW; 77.88 vs. 77.99 for FTW; 19.68 vs. 19.11 for FrL; 14.21 vs. 13.74 for FrD; and 23.00 vs. 31.50 for TOY. The increasing percentages of the maximum and minimum values were 37.69 vs. 25.75 for TFrW, 43.75 vs. 39.85 for FlW, 4.74 vs. 4.69 for FTW, 12.49 vs. 17.04 for FDrM, 23.43 vs. 18.26 for FrL, 25.69 vs. 21.98 for FrD, 10.48 vs. 9.60 for OOC, and 31.52 vs. 29.73 for TOY.

Table 6. Effect of acidifying agents as a soil application on fruit parameters, oil content, and total yield of olive trees (Picual cv.) grown in a sandy loam clay soil under multi-abiotic stresses (pH = 7.78 vs. 7.89; $CaCO_3$ = 8.8 vs. 9.2%; and ECe = 6.4 vs. 7.2 dS m^{-1}).

Treatment	TFrW	FhW	FhTW	FhDrM	FrL	FrD	OOC	TOY
	(g)		(%)		(mm)		(%)	(kg Tree^{-1})
	2018 season							
Con.	3.29c ± 0.10	2.56c ± 0.08	77.88d ± 0.39	36.23c ± 0.94	19.68d ± 0.53	14.21d ± 0.40	33.31b ± 0.43	23.00c ± 1.08
AC$_1$	3.92b ± 0.13	3.11b ± 0.13	79.40c ± 0.14	36.82c ± 0.57	21.76cd ± 0.29	16.26a–c ± 0.51	34.72b ± 0.63	24.50bc ± 1.66
AC$_2$	4.38a ± 0.19	3.58a ± 0.17	81.57ab ± 0.19	38.31b ± 0.61	23.99ab ± 0.73	17.86ab ± 0.54	33.97b ± 0.78	28.25ab ± 1.93
CA$_1$	4.51a ± 0.18	3.68a ± 0.12	81.46ab ± 0.61	37.80b ± 0.86	23.36a–c ± 0.91	17.28a–c ± 0.53	36.29a ± 0.55	26.75a–c ± 1.89
CA$_2$	4.53a ± 0.22	3.65a ± 0.14	80.81b ± 0.42	40.75a ± 0.95	24.29a ± 0.81	18.18a ± 0.82	36.80a ± 0.45	30.25a ± 1.89
SA$_1$	3.93b ± 0.23	3.10b ± 0.13	78.85c ± 0.13	38.53b ± 0.66	21.69cd ± 0.49	15.45cd ± 0.32	36.29a ± 0.56	24.75bc ± 1.31
SA$_2$	4.34a ± 0.19	3.56a ± 0.17	81.91a ± 0.19	38.04b ± 0.69	22.04bc ± 0.48	16.05b–d ± 0.46	36.36a ± 0.53	25.75a–c ± 1.08
	2019 season							
Con.	3.41f ± 0.24	2.66e ± 0.11	77.99g ± 0.22	35.15cd ± 0.32	19.11b ± 0.70	13.74b ± 0.64	32.75b ± 0.41	31.50c ± 0.96
AC$_1$	3.91e ± 0.22	3.10d ± 0.12	79.40e ± 0.21	33.09e ± 0.89	20.93ab ± 0.45	15.54a ± 0.14	32.40b ± 0.35	33.75c ± 1.25
AC$_2$	4.27c ± 0.20	3.49c ± 0.18	81.65a ± 0.22	36.24bc ± 0.78	22.60a ± 0.48	16.07a ± 0.14	34.63a ± 0.60	38.75ab ± 1.30
CA$_1$	4.51a ± 0.20	3.72a ± 0.13	81.46c ± 0.31	38.73a ± 0.73	22.45a ± 0.68	16.76a ± 0.61	35.50a ± 0.46	35.50bc ± 1.32
CA$_2$	4.52a ± 0.18	3.65a ± 0.17	80.91d ± 0.15	35.67c ± 0.81	21.95a ± 0.69	16.72a ± 0.67	35.51a ± 0.85	40.75a ± 1.25
SA$_1$	3.95d ± 0.22	3.11d ± 0.19	78.73f ± 0.22	34.35d ± 0.86	21.45ab ± 0.60	15.42ab ± 0.35	34.72a ± 0.41	32.75c ± 1.11
SA$_2$	4.40b ± 0.26	3.59b ± 0.13	81.56b ± 0.18	37.28b ± 0.60	21.61ab ± 0.57	15.96a ± 0.50	35.31a ± 0.39	34.00c ± 1.35

Mean values (±SE), different letters in each column indicate significance at $p \leq 0.05$. CA$_1$, CA$_2$, AC$_1$, AC$_2$, SA$_1$, and SA$_2$ represent citric acid, acetic acid, and sulfuric acid applied at 25 and 50 cm^3, respectively. TFrW = total fruit weight, FhW = flesh weight, FhTW = flesh weight/total fruit weight, FhDrM = flesh dry matter, FrL = fruit length, FrD = fruit diameter, OOC = olive oil content and TOY = total olive yield.

The results obtained from the statistical analysis indicated significant differences among some studied parameters at $p \leq 0.01$, including TFrW, FlW, FTW, FDrM, and OOC, in both growing seasons. Furthermore, significant effects were found for FrD and TOY at $p \leq 0.05$ in both seasons and FrL in the first season, while a non-significant influence was found for FrL in the second season.

3.5. The Heat Map of Correlation Coefficient and Stepwise Regression

The results of the correlation analysis between the leaf nutrient contents (N, P, K, Ca, Mg, and Na), the physiological and growth parameters (ShL, NLf, LA, and SPAD reading), and the yield and its attributes (TFrW, FrL, FrW, FrDrM, FiW, and TOY) are shown in (Figure 4A,B).

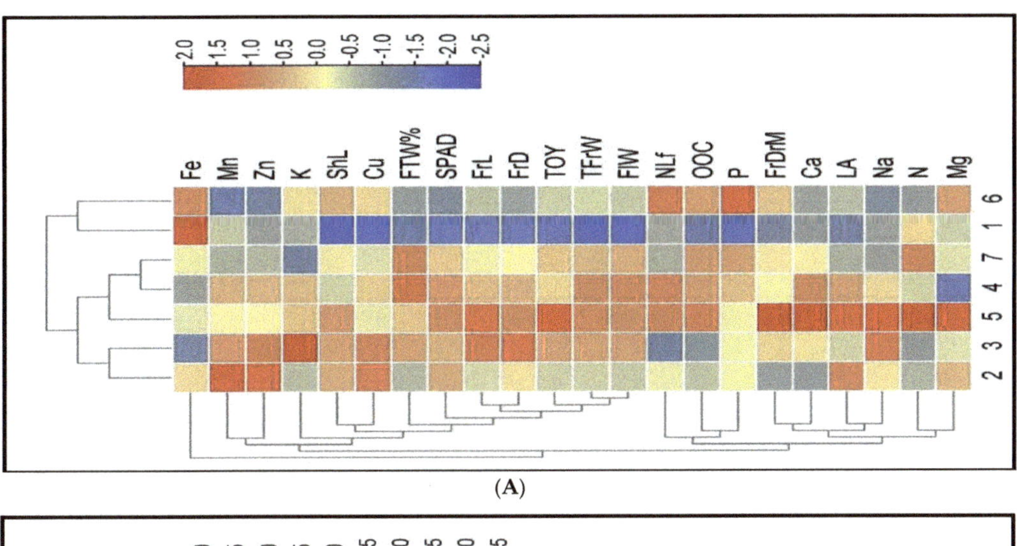

(**A**)

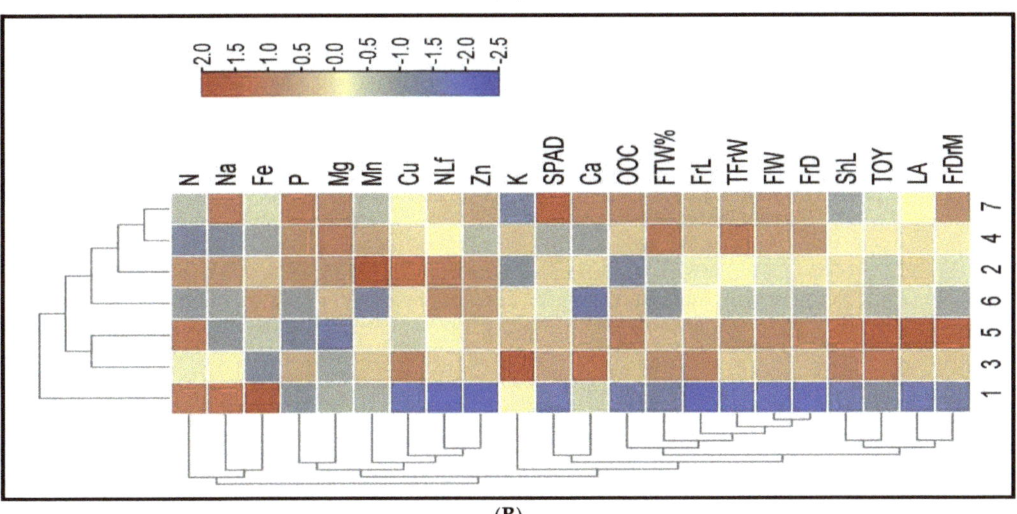

(**B**)

Figure 4. The heat map of Pearson's correlation coefficient matrix between leaf nutrient contents (N, P, K, Ca, Mg, and Na) with physiological parameters (ShL, NLf, LA, and SPAD reading) and yield and its attributes (TFrW, FiW%, FrL, FrW, FrDrM, and TOY) in the 2018 (**A**) and 2019 seasons (**B**).

The results obtained from the stepwise regression analysis, presented in Table 7, mention the relationship between the olive oil content (OOC,%) and the total olive yield (TOY, kg) as the dependent variables and the leaf macro- and micronutrients, the physiological and growth parameters, and the yield components as the independent variables of the olive plants (Picual cv.) grown under multi-abiotic stresses in the 2018 and 2019 seasons. Our results indicate that the variations in OOC are explained by the variations in the leaf nitrogen (LNU) and phosphorus uptake (LPU) in the first season and the leaf manganese uptake (LMnU) in the second season, while the differences in TOY were explained on the basis of the total fruit weight (TFrW) in the 2018 season and the shoot length (ShL) and number of leaves (NLf) in the 2019 season. In both seasons, most of the studied parameters contributed effectively to OOC and TOY. The adjusted R^2 values were (r= 0.671 vs. 0.708) for OOC and (r = 0.551 vs. 0.897) for TOY in the 2018 and 2019 seasons, respectively.

Table 7. Proportional contribution in predicting olive oil content (OOC) and total olive yield (TOY) using stepwise multiple linear regression for olive tree (Picual cv.) with three acidifying agents applied in two doses in comparison with untreated trees in 2018 and 2019 seasons.

Season	r	R^2	Adj. R^2	SEE	Significance	Fitted Equation
2018	0.671	0.451	0.390	1.368	***	OOC = 25.921 + 20.539LNU + 24.317LPU
	0.551	0.304	0.267	3.175	***	TOY = 7.298 + 4.410TFrW
2019	0.708	0.501	0.446	1.167	***	OOC = 29.023 + 3.083FiW − 0.257LMnU
	0.897	0.804	0.770	1.700	***	TOY = 2.956 + 0.772ShL + 1.015FrL − 0.118NLf

r = correlation coefficient, R^2 = coefficient of determination, Adj. R^2 = adjusted R^2, and SEE = standard error of estimates *** means differences at $p \leq 0.001$ probability levels.

4. Discussion

Soil pH is a very important property for plant growth and development due to its direct influence on nutrient availability. However, available information regarding this subject is still scarce and incomplete. In Egypt, excessive alkalinity is an inherent problem associated with many factors, the most important of which is the prevailing climate conditions, including high temperature and low precipitation, as presented in Table 1. Furthermore, the nature of the parent material and the predominance of basic cations, such as Ca^{2+}, Mg^{2+}, and Na^+, and their accumulation also have an effect on soil pH. In addition, unsuitable fertilization practices, such as applying alkaline fertilizers, can increase alkalinity, despite the high buffering capacity of the soil. The main method of our study relied on determining leaf nutrient contents as a measure of nutrient availability in the soil, which in turn affects physiological processes and improves the total yield and its attributes. The results presented in Figure 1 clearly show that the AC_1 treatment more quickly reduced the soil pH when compared to the other AAs. This significant decrease in soil pH was observed from the first day of injection. Although the pH values fluctuated between increase and decrease from the second to the sixth day, the increase was slight. These findings could be attributed to the ionization of the H+ ions of the carboxyl group (-COOH) being faster than that of the other acidifying agents; then, the decrease in pH would be observed. Regarding the control treatment, the results obtained suggested that the soil pH decreased, but in a non-significant manner. This slight decrease could be attributed to the decomposition of H_2O molecules into OH^- and H^+ ions in soil and the subsequent adsorption of H^+ ions on soil particles as a result of the ion substitution between H^+ and the other base ions, such as Ca^{2+}, Mg^{2+}, and Na^+. In both seasons, the general trend of our findings indicated that the acidifying agents could be ranked in descending order as follows: $AA_1 > AA_2 > CA_2 > CA_1 > SA_1 > SA_2$. The effect of the application of both AC and CA surpassed that of SA due to the fact that organic acids have a buffering influence, owing to their high ion exchange capacity [49].

The results shown in Table 4 reveal that CA, irrespective of the dose applied, was more influential in improving the leaf nitrogen (LNC) and magnesium content (LMgC) in both seasons, as well as the leaf calcium content (LCaC) in the first season, when compared with the other AAs. In other words, the increase in LCaC, LNaC, and LMgC may have been due to a disturbance of the ion balance by the acetate (-COOH$^-$) anion [50]. These results are in agreement with those of [51], who reported that the addition of increased levels of CA caused an increase in the nutrient content of wheatgrass plants, in addition to a significant decrease in soil pH. Furthermore, it was observed that the rate of decrease in soil pH increased with increasing molarity, although there was no strong evidence to support this assumption. Among the AAs applied, the SA treatment—whether applied at a low dose (as in the first season) or a high dose (as in the second season)—led to significant improvements in LPC. These enhancements are likely due to the pivotal role of SA in dissolving calcium, which is the element most closely related to P ions under alkaline soil conditions [52]. The reaction of H_2SO_4 with $CaCO_3$ can occur according to the first equation (open reaction) or the second equation (closed reaction) as follows:

$$H_2SO_4 + CaCO_3 \rightarrow Ca^{2+} + SO_4{}^{2-} + CO_2 + H_2O \tag{1}$$

$$H_2SO_4 + 2CaCO_3 \rightarrow 2Ca^{2+} + SO_4{}^{2-} + 2HCO_3{}^- \tag{2}$$

As can be seen from in Equation (2), the efficiency of the SA application in the amelioration of high soil pH may be due to the considerable amounts of hydrogen carbonate ($HCO_3{}^-$) ions formed in the closed reaction. In other words, SA, irrespective of the dose applied, introduces $SO_4{}^{2-}$ and replaces Ca^{2+} with Na^+ in the soil colloids [53,54]. In addition, CA has a high ionic ability to form complexes with soil nutrients [55].

Concerning the leaf micronutrient contents, as shown in Tables 4 and 5, the results obtained indicated that the AC treatment, regardless of the dose applied, had the greatest influence on the contents of most of the studied nutrients. It was also observed that the higher dose (AC_2) was associated with improved macronutrient measures, as was seen for LKC in both seasons and LCaC in the second season only. Conversely, the lower dose of CA (CA_1) was strongly related to increases in most of the studied micronutrients except for LFeC, including LMnC, LZnC, and LCuC, in both seasons. These findings could be explained by AC being too rapidly oxidized in the soil, which would cause a decrease in the soil pH values [56]. On the other hand, the decrease in LFeC could be due to competition for sorption sites. However, the accumulation of micronutrients depends not only on soil pH but also on other environmental factors [57]. Another explanation could be that AC is synthesized as a result of its small size and role in the tolerance to abiotic stresses [58]. These findings were confirmed in the studies by [59] on ryegrass plants, where AC production was shown to be enhanced for tolerance to salt stresses through hormone and antioxidant metabolism, improving K^+/Na^+ homeostasis. Furthermore, ref. [60] suggested that the improved effect of AC could perhaps be due to its large volume, which makes it difficult for it to enter the soil pores.

The results for the leaf nutrient contents were closely related to the improvement in the vegetative growth parameters (ShL, NLSh, and LA (Figure 2A–C) and physiological performance (SPAD reading, fm/fv, and PI (Figure 3A–C) of the olive trees. However, a significant increase in plant growth resulted from the application of a high dose of CA (CA_2), which played a significant role in decreasing soil pH and thus increasing the availability of some nutrients in the soil, subsequently enhancing LNC and LMgC. The obvious response to the application of N to olive trees was previously supported by the results of the studies by [61,62], in addition to its conspicuous roles in contributing to the formation of the chlorophyll molecule [63]. Furthermore, the vital role of Mg should be noted: it is a cofactor for many physiological processes that activate phosphorylation, and it is essential for amino acid and fat synthesis [64] and glutathione RNA polymerase, phosphatase, ATPase, and protein kinase activities [65,66]. In their previous studies, refs. [67–69] indicated the significant role of N in enhancing horticultural crops, in addition to its crucial role in the plant metabolism system, by increasing leaf area production and promoting photosynthetic processes. In addition, the influential role of Mg in carbohydrate formatting should also be mentioned [70]. In other words, the appreciable enhancements of all plant physiological and growth characteristics could perhaps be due to the improved influence of AC on micronutrient availability (except Fe), resulting from a significant decline in soil pH values, as presented in Table 5. In this context, the positive physiological impact of micronutrient availability was explained by several researchers [71–74]. Zn is known to be an essential component for many enzymes, such as carbonic anhydrase, and a principle component in tryptophan and indole acetic acid synthesis. Mn is an active component of the water-splitting system of photosystem II. Furthermore, Cu plays a crucial role in photosynthesis and the respiration and metabolism of N and carbon. Although most studies have indicated the positive effect of N on enhancing vegetative growth parameters, the findings of [75] were not in agreement with the majority; the authors reported that N had adverse impacts on root length. Furthermore, the application of AC increased IAA under abiotic stresses [59,76]. The only exception in our results is that SA provided the best results in terms of LPC. Although SA only had a slight effect on lowering soil pH when compared with the other AAs applied, it was the most influential for P uptake by plants. According to the studies by [77,78] on olive trees, P promotes new shoot growth and many physiological processes, such as cell division and the fixation of carbon from carbon dioxide during photosynthesis. These findings were in line with those of a study by [52] on sweet potato plants, which indicated that P played a crucial role in improving physiological and growth parameters. From our point of view, the minimum values for soil pH of 6.69 and 6.70 in the 2018 and 2019 seasons, respectively, as a result of the application of SA_1 and SA_2, prevented P fixation and reduced its linkage with the Ca^{2+} ions in the soil solution.

As a result of the positive influences of the application of AAs on the uptake of nutrients—either macro- or micronutrients, as reflected in the growth and physiological parameters—we observed that the total olive yield (TOY) and its attributes, in addition to the olive oil content (OOC), improved, as shown in Table 6. Our results evidently showed that the olive trees responded to the decrease in soil pH that resulted from the application of AAs as compared to the control treatment. The significant improvement in TOY and its attributes as well as the OOC of the Picual trees resulted from the positive effect of the AAs on the reduction in soil pH, which in turn improved the availability of the nutrients in the soil, enhancing their absorption, as shown in Tables 4 and 5. This trend was observed in our current investigation, where the control trees showed a significant decline in leaf nutrient contents, which, in turn, negatively affected TOY and its attributes.

5. Conclusions

Among the chemical properties of soil, pH is considered to be the most important due to its direct effect on nutrient solubility. Furthermore, simply controlling or changing soil pH is considered to be a very difficult task owing to the soil's buffering capacity, which resists any changes in pH. In addition, most nutrients have an optimal pH range between 6.0 and 6.5 with respect to their availability [79], which varies according to the element's behavior. In general, our results showed that all the studied AAs surpassed the control treatment in terms of reducing soil pH values. Furthermore, the daily soil pH measurements taken over the 6 days following application indicated that the AC treatment was faster in reducing soil pH values, while the CA treatment provided the most stable effect. Based on our work, the trees grown in soil injected with CA, followed by those treated with AC, gave the most desirable results when compared with the control treatment, which resulted in maximum leaf iron content, while the SA treatment provided maximum leaf phosphorus content in both seasons. In short, the results of our investigation can be summarized as the effectiveness of the treatments ranked and arranged in descending order, as follows: $AC_1 > AC_2 > CA_2 > CA_1 > SA_1 > SA_2$.

Author Contributions: Conceptualization, H.A.Z.H. and H.R.B.; data curation, A.A.M.A., H.A.Z.H. and H.R.B.; formal analysis, A.A.M.A. and H.A.Z.H.; investigation, H.A.Z.H., A.A.M.A. and H.R.B.; methodology, A.A.M.A. and H.A.Z.H.; resources, H.A.Z.H. and H.R.B.; software A.A.M.A. and H.R.B.; writing—original draft, A.A.M.A.; writing—review and editing, A.A.M.A. and H.A.Z.H. All authors have read and agreed to the published version of the manuscript.

Funding: This research received no external funding.

Data Availability Statement: The data presented in this study are available upon request from the corresponding author.

Conflicts of Interest: The authors declare no conflict of interest.

References

1. Ruan, J.; Ma, L.; Shi, Y.; Han, Y. The Impact of pH and calcium on the uptake of Fluoride by tea plants (*Camellia sinensis* L.). *Ann. Bot.* **2004**, *93*, 97–105. [CrossRef]
2. McCauley, A.; Jones, C.; Jacobsen, J. Soil pH and organic matter. *Nutr. Manag. Modul.* **2009**, *8*, 1–12.
3. Jackson, K.; Meetei, T.T. Influence of soil pH on nutrient availability: A review. *J. Emerg. Technol. Innov. Res.* **2018**, *5*, 707–713.
4. Zhang, Y.-Y.; Wu, W.; Liu, H. Factors affecting variations of soil pH in different horizons in hilly regions. *PLoS ONE* **2019**, *14*, e0218563. [CrossRef]
5. Schomberg, H.H.; Steiner, J.L. Estimating crop residue decomposition coefficients using substrate-induced respiration. *Soil Biol. Biochem.* **1997**, *29*, 1089–1097. [CrossRef]
6. Kheir, A.M.S.; Shabana, M.M.A.; Seleiman, M.F. Effect of Gypsum, Sulfuric Acid, Nano-Zeolite Application on Saline-Sodic Soil Properties and Wheat Productivity under Different Tillage Types. *J. Soil Sci. Agric. Eng. Mansoura Univ.* **2018**, *9*, 829–838. [CrossRef]
7. Chytrỳ, M.; Danihelka, J.; Ermakov, N.; Hájek, M.; Hájková, P.; Koči, M. Plant species richness in continental southern Siberia: Effects of pH and climate in the context of the species pool hypothesis. *Glob. Ecol. Biogeogr.* **2007**, *16*, 668–678. [CrossRef]
8. Ji, C.-J.; Yang, Y.-H.; Han, W.-X.; He, Y.-F.; Smith, J.; Smith, P. Climatic and edaphic controls on soil pH in alpine grasslands on the Tibetan Plateau, China: A quantitative analysis. *Pedosphere* **2014**, *24*, 39–44.
9. Reuter, H.I.; Lado, L.R.; Hengl, T.; Montanarella, L. Continental-scale digital soil mapping using European soil profile data: Soil pH. *Hambg. Beiträge Zur Phys. Geogr. Und Landsch.* **2008**, *19*, 91–102.
10. Mohamed, M.A.; Elgharably, G.A.; Rabie, M.H. Evaluation of Soil Fertility Status in Toshka, Egypt: Available Micronutrients. *World J. Agric. Sci.* **2019**, *15*, 1–6. [CrossRef]
11. Elwa, A.M.; Abou-Shady, A.M.; Sayed, A.; Showman, H. Impact of physical and chemical properties of soil on the growing plant in El Mounira-El Qattara New Valley. *Egypt. J. Appl. Sci.* **2021**, *36*, 148–175. [CrossRef]
12. Al-Soghir, M.M.A.; Mohamed, A.G.; El-Desoky, M.A.; Awad, A.A.M. Comprehensive assessment of soil chemical properties for land reclamation and cultivation purposes in the Toshka area, EGYPT. *Sustainability* **2022**, *14*, 15611. [CrossRef]
13. Song, K.; Xue, Y.; Zheng, X.; Lv, W.; Qiao, H.; Qin, Q.; Yang, J. Effects of the continuous use of organic manure and chemical fertilizer on soil inorganic phosphorus fractions in calcareous soil. *Sci. Rep.* **2017**, *7*, 1164. [CrossRef] [PubMed]
14. Rady, M.M.; El-Shewy, A.A.; Seif El-Yazal, M.A.; Abd El-Gawwad, I.F.M. Integrative application of soil P-solubilizing bacteria and foliar nano P improves *Phaseolus vulgaris* plant performance and antioxidative defense system components under calcareous soil conditions. *J. Soil Sci. Plant Nutr.* **2020**, *20*, 820–839. [CrossRef]
15. Awad, A.A.M.; Sweed, A.A.A.; Rady, M.M.; Majrashi, A.; Ali, E.F. Rebalance the nutritional status and the productivity of high CaCO$_3$-stressed sweet potato plants by foliar nourishment with zinc oxide nanoparticles and ascorbic acid. *Agronomy* **2021**, *11*, 1443. [CrossRef]
16. Fazal, T.; Ismail, B.; Shaheen, N.; Numan, A.; Adnan Khan, A. Effect of Different Acidifying Agents on Amendment in Buffering Capacity of Soil. *Res. Rev. J. Bot. Sci.* **2020**, *9*, 9–15.
17. Tanikawa, T.; Sobue, A.; Hirano, Y. Acidification processes in soils with different acid buffering capacity in *Cryptomeria japonica* and *Chamaecyparis obtusa* forests over two decades. *For. Ecol. Manag.* **2014**, *334*, 284–292. [CrossRef]
18. Latifah, O.; Ahmed, O.H.; Majid, N.M.A. Soil pH buffering capacity and nitrogen availability following compost application in a tropical acid soil. *Compos. Sci. Util.* **2018**, *26*, 1–15. [CrossRef]
19. Akay, A.; Şeker, C.; Negiş, H. Effect of enhanced elemental sulphur doses on pH value of a calcareous soil. *YYU J. Agric. Sci.* **2019**, *31*, 34–40.
20. Rengasamy, P. World salinization with emphasis on Australia. *J. Exp. Bot.* **2006**, *57*, 1017–1023. [CrossRef]
21. Zia, M.H.; Saifullah, M.; Sabir, A.; Ghafoor, A.G. Effectiveness of sulphuric acid and gypsum for the reclamation of a calcareous saline-sodic soil under four crop rotations. *J. Agron. Crop Sci.* **2007**, *193*, 262–269. [CrossRef]
22. Mahdy, A.M. Comparative effects of different soil amendments on amelioration of saline-sodic soils. *Soil Water Res.* **2011**, *4*, 205–216. [CrossRef]
23. Soaud, A.A.; Al darwish, F.H.; Saleh, M.E.; El-Tarabily, K.A.; Sofian-Azirun, M.; Rahman, M.M. Effects of elemental sulfur, phosphorus, micronutrients and *Paracoccus versutus* on nutrient availability of calcareous soils. *Australian J. Crop Sci.* **2011**, *5*, 554.

24. Shiwakoti, S.; Zheljazkov, V.D.; Gollany, H.T.; Xing, B.; Kleber, M. Micronutrient concentrations in soil and wheat decline by long-term tillage and winter wheat-pea rotation. *Agronomy* **2019**, *9*, 359. [CrossRef]

25. Awad, A.A.M.; Sweed, A.A.A. Influence of organic manures on soil characteristics and yield of Jerusalem artichoke. *Commun. Soil Sci. Plant Anal.* **2020**, *51*, 1101–1113. [CrossRef]

26. Garcia-Gill, J.C.; Ceppi, S.B.; Velasco, M.I.; Polo, A.; Senesi, N. Long-term effects of amendment with municipal solid waste compost on the elemental and acidic functional group composition and pH-buffer capacity of soil humic acids. *Geoderma* **2004**, *121*, 135–142. [CrossRef]

27. Hafez, E.H.; Abou El Hassan, W.H.; Gaafar, I.A.; Seleiman, M.F. Effect of gypsum application and irrigation intervals on clay saline-sodic soil characterization, rice water use efficiency, growth, and yield. *J. Agric. Sci.* **2015**, *7*, 208–219. [CrossRef]

28. Matosic, S.; Birkás, S.; Vukadinovic, S.; Kisic, I.; Bogunovic, I. Tillage, manure and gypsum use in reclamation of saline-sodic soils. *Agric. Conspec. Sci.* **2018**, *83*, 131–138.

29. Seleiman, M.F.; Kheir, A.M.S. Maize productivity, heavy metals uptake and their availability in contaminated clay and sandy alkaline soils as affected by inorganic and organic amendments. *Chemosphere* **2018**, *204*, 514–522. [CrossRef]

30. Islam, M.A.; Milham, P.J.; Dowling, P.M.; Jacobs, B.C.; Garden, D.L. Improved procedures for adjusting soil pH for pot experiments. *Commun. Soil Sci. Plant Anal.* **2004**, *35*, 25–37. [CrossRef]

31. Han, F.; Shan, X.Q.; Zhang, S.Z.; Wen, B.; Owens, G. Enhanced cadmium accumulation in maize roots-the impact of organic acids. *Plan Soil* **2006**, *289*, 355–368. [CrossRef]

32. Qureshi, F.F.K.; Khan, A.; Hassan, F.; Bibi, N. Effect of girdling and plant growth regulators on productivity in olive. *Pak. J. Agric. Res.* **2012**, *25*, 120–128.

33. El-Fouly, M.M.; El-Taweel, A.A.; Osman, I.M.S.; Saad El-Din, I.; Shaaban, S.H.A. Nutrient removal from different parts of Koroneiki olive trees grown in sandy soil as a base of fertilizer recommendation in Egypt. *Br. J. Appl. Sci. Technol.* **2014**, *4*, 1718–1728. [CrossRef]

34. FAO. Food and Agriculture Organization of the United Nations, Rome, Italy. 2022. Available online: http://www.fao.org/faostat/en/ (accessed on 9 February 2022).

35. Rallo, L.; Caruzo, T.; Diez, C.M.; Campisi, G. Olive growing in a time of change: From empiricism to genomics. In *The Olive Tree Genome*; Rugini, L.B., Muleo, R., Sebastiani, L., Eds.; Springer: Berlin/Heidelberg, Germany, 2016; pp. 55–64.

36. Shahin, M.F.M.; Genaidy, E.A.E.; Haggag, L.F. Impact of amino acids and humic acid as soil application on fruit quality and quantity of "Kalamata" olive trees. *Int. J. ChemTech Res.* **2015**, *8*, 75–84.

37. Mahmoud, T.M.; Emam, S.; Mohamed, A.; El-Sharony, T.F. Influence of foliar application withpotassium and magnesium on growth, yield and oil quality of "Koroneiki" olive trees. *Am. J. Food Technol.* **2017**, *12*, 209–220. [CrossRef]

38. Bouyoucos, C.J. Hydrometer method improved for making particle size analysis of soil. *Soil Sci. Soc. Proc.* **1981**, *26*, 446–465.

39. McLean, E.O. Soil pH and lime requirement. In *Methods of Soil Analysis. Part 2. Chemical and Microbiological Properties*; Page, A.L., Ed.; American Society of Agronomy: Madison, WI, USA, 1982; pp. 199–224.

40. Page, A.L.; Miller, R.H.; Keeney, D.R. *Method of Soil Analysis. Part 2 Chemical and Microbiological Methods*; American Society of Agronomy: Madison, WI, USA, 1982; pp. 225–246.

41. Walkey, A.; Black, C.A. An examination of the Degtjareff method for determining soil organic matter and a proposed modification of the chronic acidification method. *Soil Sci.* **1934**, *37*, 29–38. [CrossRef]

42. Jackson, M.L. *Soil Chemical Analysis*; Constable and Co. Ltd.: London, UK, 1962.

43. Olsen, S.R.; Cole, C.V.; Watanabe, F.S.; Dean, L.A. *Estimation of Available Phosphorus in Soils by Extraction with Bicarbonate*; Department of Agriculture: Washington, DC, USA, 1954.

44. Chapman, H.D. Cation-exchange capacity. In *Methods of Soil Analysis. Part II*; Dans, C., Black, A., Eds.; American Society of Agronomy Inc.: Madison, WI, USA, 1965; Chapter 57–58; pp. 891–903.

45. Maxwell, K.; Johnson, G.N. Chlorophyll fluorescence- a practical guide. *J. Exp. Bot.* **2000**, *51*, 659–668. [CrossRef]

46. Clark, A.J.; Landolt, W.; Bucher, J.B.; Strasser, R.J. Beech (*Fagus sylvatica*) response to ozone exposure assessed with a chlorophyll fluorescence performance index. *Environ. Pollut.* **2000**, *109*, 501–507. [CrossRef]

47. Baird, R.B.; Eaton, E.D.; Rice, E.W. *Standard Methods for the Examination of Water and Waste Water*, 23rd ed.; American Public Health Association: Washington, DC, USA, 2017.

48. Di Rienzo, J.A.; Casanoves, F.; Balzarini, M.G.; Gonzalez, L.; Tablada, M.; Robledo, C.W. InfoStat versión. Group InfoStat, FCA, Universidad Nacional de Córdoba, Argentina. 2011. Available online: http://www.infostat.com.ar (accessed on 29 September 2020).

49. Bohn, H.L.; McNeal, B.L.; O'Conner, G.A. *Soil Chemistry*, 2nd ed.; Wiley: New York, NY, USA, 1985.

50. Ogawa, D.; Suzuki1, Y.; Yokoo1, T.; Katoh, E.; Teruya, M.; Muramatsu, M.; Ma, J.F.; Yoshida, Y.; Isaji, S.; Ogo, Y.; et al. Acetic-acid-induced jasmonate signaling in root enhances drought avoidance in rice. *Sci. Rep.* **2021**, *11*, 6280. [CrossRef]

51. Tusei, C. The Effects of citric acid on pH and nutrient uptake in wheatgrass (*Triticum aestivum*). *IdeaFest Interdiscip. J. Creat. Work. Res. Cal Poly Humboldt* **2019**, *3*, 7.

52. Awad, A.A.M.; Ahmed, A.I.; Elazem, A.H.A.; Sweed, A.A.A. Mitigation of CaCO$_3$ Influence on *Ipomoea batatas* Plants Using *Bacillus megaterium* DSM 2894. *Agronomy* **2022**, *12*, 1571. [CrossRef]

53. Bello, S.K.; Alayafi, A.H.; AL-Solaimani, S.G.; Abo-Elyousr, K.A.M. Mitigating soil salinity stress with gypsum and bio-organic amendments: A review. *Agronomy* **2021**, *11*, 1735. [CrossRef]

54. Wang, X.L.; Guo, X.L.; Hou, X.G.; Zhao, W.; Xu, G.W.; Li, Z.Q. Effects of leaf zeatin and zeatin riboside induced by different clipping heights on the regrowth capacity of ryegrass. *Ecol. Res.* **2014**, *29*, 167–180. [CrossRef]
55. Schwab, A.P.; Zhu, D.S.; Banks, M.K. Influence of organic acids on the transport of heavy metals in soil. *Chemosphere* **2008**, *72*, 986–994. [CrossRef] [PubMed]
56. Saikia, J.; Sarma, R.K.; Dhandia, R.; Yadav, A.; Bharali, R.; Gupta, V.K.; Saikia, R. Alleviation of drought stress in pulse crops with ACC deaminase producing rhizobacteria isolated from acidic soil of Northeast India. *Sci. Rep.* **2018**, *8*, 3560. [CrossRef]
57. Qu, J.; Lou, C.Q.; Yuan, X.; Wang, X.H.; Cong, Q.; Wang, L. the effect of sodium hydrogen phosphate/ citric acid mixtures on phytoremediation by alfalfa & metals availability in soil. *J. Soil Sci. Plant Nutr.* **2011**, *11*, 85–95.
58. Kim, J.M.; Sasaki, T.; Ueda, M.; Sako, K.; Seki, M. Chromatin changes in response to drought, salinity, heat, and cold stresses in plants. *Front. Plant Sci.* **2015**, *6*, 114. [CrossRef]
59. Zhang, J.; Zhang, Q.; Xing, J.; Li, H.; Miao, J.; Xu, B. Acetic acid mitigated salt stress by alleviating ionic and oxidative damages and regulating hormone metabolism in perennial ryegrass (*Lolium perenne* L.). *Grass Res.* **2021**, *1*, 3. [CrossRef]
60. Shi, J.; Long, T.; Zheng, L.; Gao, S.; Wang, L. Neutralization of industrial alkali-contaminated soil by different agents: Effects and Environmental Impact. *Sustainability* **2022**, *14*, 5850. [CrossRef]
61. Erel, R.; Yermiyahu, U.; Van Opstal, J.; Ben-Gal, A.; Schwartz, A.; Dag, A. The importance of olive (*Olea europaea* L.) tree nutritional status on its productivity. *Sci. Hortic.* **2013**, *159*, 8–18. [CrossRef]
62. Haberman, A.; Dag, A.; Shtern, N.; Zipori, I.; Erel, R.; Ben-Gal, A.; Yermiyahu, U. Significance of proper nitrogenfertilizationforoliveproductivityinintensivecultivation. *Sci. Hortic.* **2019**, *246*, 710–717. [CrossRef]
63. El-Dissoky, R.A.; Al-Kamar, F.A.; Derar, R.M. Impact of Magnesium Fertilization on Yield and Nutrients Uptake by Maize Grown on two Different Soils. *Egypt. J. Soil Sci.* **2017**, *57*, 455–466. [CrossRef]
64. Mengel, K.; Kirkby, E.A. *Principles of Plant Nutrition*; International Potash Institute: Bern, Switzerland, 2001.
65. Zhao, H.; Zhou, Q.; Zhou, M.; Li, C.; Gong, X.; Liu, C.; Qu, C.; Wang, L.; Si, W.; Hong, F. Magnesium deficiency results in damage of nitrogen and carbon cross-talk of maize and improvement by cerium addition. *Biol. Trace Element Res.* **2012**, *148*, 102–109. [CrossRef] [PubMed]
66. Bloom, A.J.; Kameritsch, P. Relative association of Rubisco with manganese and magnesium as a regulatory mechanism in plants. *Physiol. Plant.* **2017**, *161*, 545–559. [CrossRef]
67. Fernández-Escobar, R. Use and abuse of nitrogen in olive fertilization. *Acta Hortic.* **2011**, *888*, 249–258. [CrossRef]
68. Kirkby, E. Introduction, definition and classification of nutrients. In *Marschner's Mineral Nutrition of Higher Plants*, 3rd ed.; Elsevier: Amsterdam, The Netherlands, 2012; pp. 3–5.
69. Hussein, H.A.Z.; Awad, A.A.M.; Beheiry, H.R. Improving nutrients uptake and productivity of stressed olive trees with mono-ammonium phosphate and urea phosphate application. *Agronomy* **2022**, *12*, 2390. [CrossRef]
70. Tränkner, M.; Tavakol, E.; Jákli, B. Functioning of potassium and magnesium in photosynthesis, photosynthate translocation and photoprotection. *Physiol. Plant.* **2018**, *163*, 414–431. [CrossRef]
71. Rate, A.W.; Sheikh-Abdullah, S.M. The geochemistry of calcareous forest soils in Sulaimani Governorate, Kurdistan Region, Iraq. *Geoderma* **2017**, *289*, 54. [CrossRef]
72. Ramírez-Pérez, L.; Morales-Díaz, A.; De Alba-Romenus, K.; González-Morales, S.; Benavides-Mendoza, A.; Juárez-Maldonado, A. Determination of micronutrient accumulation in greenhouse cucumber crop using a modeling approach. *Agronomy* **2017**, *7*, 79. [CrossRef]
73. Mam-Rasul, G.A. Zinc Sorption in Calcareous Soils of the Kurdistan Region of Iraq. *Soil Sci.* **2019**, *184*, 60. [CrossRef]
74. Salih, H.O. Effect of fertigation-applied sulfuric acid on phosphorus availability and some microelements for greenhouse cucumber. *Pol. J. Environ. Stud.* **2021**, *30*, 4901–4909. [CrossRef]
75. Kwon, S.-J.; Kim, H.-R.; Roy, S.K.; Kim, H.-J.; Boo, H.-O.; Woo, S.-H.; Kim, H.-H. Effects of nitrogen, phosphorus and potassium fertilizers on growth characteristics of two species of Bellflower (*Platycodon grandiflorum*). *J. Crop Sci. Biotechnol.* **2019**, *22*, 481–487. [CrossRef]
76. Shanthakumar, S.; Abinandan, S.; Venkateswarlu, K.; Subashchandrabose, S.R.; Megharaj, M. Algalization of acid soils with acid-tolerant strains: Improvement in pH, carbon content, exopolysaccharides, indole acetic acid and dehydrogenase activity. *Land Degrad. Dev.* **2021**, *32*, 3157–3166. [CrossRef]
77. Centeno, A.; Gómez del Campo, M. Response of mature olive trees with adequate leaf nutrient status to additional nitrogen, phosphorus, and potassium fertilization. *Acta Hortic.* **2011**, *888*, 277–280. [CrossRef]
78. Ferreiraa, I.Q.; Ângelo Rodriguesa, M.; Moutinho-Pereirab, J.M.; Correiab, C.M.; Arrobas, M. Olive tree response to applied phosphorus in field and pot experiments. *Sci. Hortic.* **2018**, *234*, 236–244. [CrossRef]
79. Ferrarezi, R.S.; Lin, X.; Neira, A.C.G.; Zambon, F.T.; Hu, H.; Wang, X.; Huang, J.-H.; Fan, G. Substrate pH influences the nutrient absorption and rhizosphere microbiome of Huanglongbing-affected graprfruit plants. *Front. Plant Sci.* **2022**, *13*, 856937. [CrossRef]

Article

Performance of Nitrogen Fertilization and Nitrification Inhibitors in the Irrigated Wheat Fields

Shahram Torabian [1,2], Salar Farhangi-Abriz [3], Ruijun Qin [1,*], Christos Noulas [4,*] and Guojie Wang [5]

[1] Hermiston Agricultural Research and Extension Center, Oregon State University, Hermiston, OR 97838, USA
[2] Agricultural Research Station, Virginia State University, Petersburg, VA 23806, USA
[3] Department of Plant Eco-Physiology, Faculty of Agriculture, University of Tabriz, Tabriz 5166616471, Iran
[4] Institute of Industrial and Forage Crops, Hellenic Agricultural Organization "Demeter", 41335 Larissa, Greece
[5] Department of Plant Science, The Pennsylvania State University, University Park, PA 16802, USA
* Correspondence: ruijun.qin@oregonstate.edu (R.Q.); noulaschristos@gmail.com (C.N.)

Abstract: Effective nitrogen (N) management practices are critical to sustain crop production and minimize nitrate (NO_3^-) leaching loss from irrigated fields in the Columbia Basin (U.S.), but studies on the applied practices are limited. Therefore, from 2014 to 2016, two separate field studies were conducted in sandy loam soils in the region to evaluate the performance of various N fertilizers in spring and winter wheat. The treatments consisted of two nitrification inhibitors (NIs) (Instinct® II and Agrotain® Ultra) in combination with two N fertilizers (urea and urea ammonium nitrate [UAN]) under two application methods (single vs. split-application) and two rates (100% vs. 85% of growers' standard). The results from these field trials demonstrated that N fertilizer treatments did not affect wheat grain yield (GY) and grain protein (GP). In the spring wheat trial, higher NH_4^+-N content but lower NO_3^--N content was observed in the UAN treatments (0–30 cm). However, the application of NIs had no considerable effect on soil N content. In the winter wheat trial, the split N application generally reduced NO_3^--N and total mineral nitrogen (TMN) content, especially at 30–60 cm, in comparison to a single application. The use of Instinct® II tended to reduce NO_3^--N and TMN contents, while Agrotain® Ultra was not effective in inhibiting nitrification. Our findings suggest that more studies on the effectiveness of NIs and N applications would enable growers to optimize N use efficiency and crop production in the region.

Keywords: nitrification inhibitors; nitrogen management; soil ammonium; soil nitrate; grain yield

Citation: Torabian, S.; Farhangi-Abriz, S.; Qin, R.; Noulas, C.; Wang, G. Performance of Nitrogen Fertilization and Nitrification Inhibitors in the Irrigated Wheat Fields. *Agronomy* **2023**, *13*, 366. https://doi.org/10.3390/agronomy13020366

Academic Editor: Wei Wu

Received: 23 December 2022
Revised: 21 January 2023
Accepted: 24 January 2023
Published: 27 January 2023

1. Introduction

Nitrogen (N) supply is highly relevant in wheat production, affecting yield and yield components. However, insufficient N in the rhizosphere is one of the most yield-limiting practices in intensive agricultural systems [1,2]. Balanced N management is key to sustainable wheat production and a cost-effective strategy to increase crop yields and improve long-term product quality [3]. While increased N fertilizer applications in intensive agriculture enhance yields, they increase the risk of N release into the environment (gaseous N loss, erosion, leaching) [4]. To maximize crop returns, farmers repeatedly apply N fertilizers in various forms (i.e., urea—$CO(NH_2)_2$, ammonium nitrate—NH_4NO_3, ammonium sulfate—$(NH_4)_2SO_4$, etc.), although these applications are not accompanied by proportional increases in N use efficiency (NUE) [5]. High NUE is paramount to reducing environmental pollution and guaranteeing acceptable yield while minimizing unnecessary fertilizer waste. Soil erosion, surface runoff, ammonia (NH_3) volatilization, nitrate (NO_3^-) leaching, and denitrification make N unavailable to plants. These processes are widely dependent on the cropping system, the form of fertilizer, and the method of application [6].

Ammonium (NH_4^+) and NO_3^- are the two N forms available for plants, and they have an important effect on crop growth and quality. More than 90% of soil N is in the organic form [7]. Physiologically, NH_4^+ uptake by plants from the soil is faster than NO_3^- [8].

Salsac et al. [9] found that assimilation of NH_4^+ requires 5 ATP mol^{-1} NH_4^+, while NO_3^- assimilation needs about 20 ATP mol^{-1} NO_3^-. Moreover, absorption of 1 mol NH_4^+ by plant roots consumes about 0.31 mol O_2, but 1 mol of NO_3^- absorption requires 1.5 mol O_2 [10]. Hence, NO_3^- absorption requires about five times more energy compared to NH_4^+ absorption. In addition, NH_4^+ can be directly used by plants to produce amino acids, but NO_3^- must be converted to NO_2^- and then to NH_4^+. Thus NO_3^- metabolism requires more energy than NH_4^+ metabolism.

Maintaining N in the NH_4^+ form in the soil would prevent its loss to nitrification and denitrification. In agricultural soils, NO_3^- originates from fertilizers, animal manure, atmospheric deposition, and nitrification of NH_4^+. During nitrification, NH_4^+ is converted by specific nitrifying microorganisms to NO_3^-, which is highly mobile and can leach after heavy rainfall or extensive irrigation management events [11]. NO_3^- readily moves with water from the root zone to deeper soil layers, depleting the plant-available N supply and causing environmental pollution [12]. The potential for nitrate-N to leach depends on soil type, N fertilizer source, and farm management strategies.

Management practices and the adoption of technologies such as controlled-release fertilizers and urease and nitrification inhibitors (NIs) can mitigate NO_3^- loss from the soil–plant system. Research suggests that the use of NIs is a promising approach to the reduction of nitrate leaching [13,14]. NIs diminish the transformation of NH_4^+ to NO_3^- in soil by reducing the activity of nitrifying microorganisms, with the benefit of decreasing NO_3^- leaching potential [15].

Common fertilizers usually contain N in one or more of the following forms: NO_3^-, NH_3, NH_4^+, or $CO(NH_2)_2$. Each form has specific properties determining its suitability for use. Most N fertilizers are used in the form of NH_4^+ or $CO(NH_2)_2$, which are easily converted to NO_3^- in the nitrification process. Previous studies have shown that adding NIs to NH_4^+ or NH_4-containing fertilizers decreases NO_3^--N formation, N leaching, and denitrification process, thus retaining N at the root zone, which is where the crops need it [16,17]. Lin and Hernandez-Ramirez [18] reported that NIs steadily increase concentrations of N in the NH_4^+ form. The stabilization of NH_4^+ by NIs allows for simplified fertilization strategies with reduced fertilizer applications [19,20]. This stabilization may increase crop yield while reducing negative environmental impacts. Bhatia et al. [21] showed that the application of urea with NIs *[S-benzylisothiouronium butanoate (SBT-butanoate)* and *S-benzylisothiouronium furoate (SBT-furoate)]* improves wheat yield. Liu et al. [22] revealed that NIs *(dicyandiamide* and DMPP) increase grain yield and NUE in a wheat–maize cropping system. Ma et al. [23] stated that the application of *dicyandiamide* and *chlorinated pyridine* as NIs increased wheat yield in conventional and no-till practices. In a recent study, Dawar et al. [24] showed that NIs preserve N in the rhizosphere and improve NUE in wheat. During the wet season in Montana, the application of Agrotain® Ultra (urease inhibitor, Koch Agronomic Services) with urea increased winter wheat yield, although no noticeable increase was found during the dry season [25].

There are some inconsistent reports about the efficiency of NIs. Dawar et al. [26] observed that urease inhibitor N-(n-butyl) *thiophosphoric triamide* (NBPT) reduced NH_4^+ concentrations for the first 5 days following fertilizer application. However, afterward, NH_4^+ concentration did not differ from the urea control. Chen et al. [27] observed that under moist (60% water-filled pore space) and mild conditions (15 °C), the effect of NI declined substantially after 14 days. Zaman and Blennerhassett [28] reported that spring applications of NIs did not significantly reduce NO_3^- leaching in a pasture system. They attributed this lack of response to a nine-month delay between the NIs application and the first leaching event. During this period, NIs may have been rendered ineffective by soil microorganisms. Following the application, bacteria gradually decompose NIs and they can be leached down the soil profile. NIs effectiveness is, therefore, governed by factors such as temperature, rainfall/drainage levels, soil organic matter, and pH [29–32]. It has been reported that the half-life of *dicyandiamide* (DCD) was 111–116 days at a soil temperature of 8 °C, while it became 18–25 days at a soil temperature of 20 °C [33]. The low effectiveness of DCD was attributed to late-season drainage occurring when soil DCD concentrations were likely low [34]. Suter et al. [35] reported that neither DMPP- nor NBPT-coated urea increased pasture yields. Cookson and Cornforth [36] did not find any increase in pasture dry matter from DCD use. Because Instinct® II (Cortiva Agriscience) did not induce positive effects on corn growth or yield, Sassman et al. [37] concluded that Instinct® II use with UAN solution at spring pre-plant would not be effective in enhancing fertilizer N availability to the crop, nor to increase corn production. Owing to the inconsistent efficiency of NIs under different climate conditions, further investigations are needed to optimize NUE from NI use.

Oregon's Columbia Basin is a main crop production region in the U.S. Soils are generally coarse-textured with low soil organic matter content and low water holding capacity. Therefore, there is a great potential for nitrate leaching, particularly in irrigated systems. Deteriorating groundwater quality has increased regulatory pressure to reduce nitrate leaching. Identification of optimal fertilization strategies could sustain or improve crop production while minimizing environmental hazards. However, little information is available on how nitrogen fertilizer sources, rates, and application methods impact crops and soils in the region. NIs might be effective tools to overcome nitrate-leaching issues in the Columbia Basin. Agrotain® Ultra urease inhibitor is marketed as an effective product for reducing N losses and improving crop NUE. Instinct® II is a nitrogen stabilizer containing nitrapyrin that delays the nitrification of ammoniacal and urea N fertilizers in soils by controlling the nitrification process. Thus, it can sustain or increase crop yield while reducing environmental issues. Both products are available in the region, but the information on their effectiveness is very limited. Therefore, comprehensive studies to evaluate the efficacy of these products in managing N could be of great importance in this region. For this purpose, we carried out two field trials with spring wheat and winter wheat from 2014 to 2016, using two kinds of NIs (Instinct® II and Agrotain® Ultra) on two N fertilizer sources, i.e., urea [$CO(NH_2)_2$] and urea ammonium nitrate UAN (liquid form; UAN32) with two N rates (85% vs. 100%) and two application methods (single application vs. split-application). We measured soil and plant parameters, including mineral soil N content (NO_3^{-}-N, NH_4^{+}-N and total mineral nitrogen-TMN), grain yield (GY), SPAD (leaf greenness), and grain protein (GP). These findings will provide growers with insights about N management strategies, improve NUE, and reduce environmental contamination.

2. Materials and Methods

2.1. The Experiments and Growing Conditions

Two field trials were conducted at the Oregon State University-Hermiston Agricultural Research and Extension Center, Hermiston, OR (Latitude: 45°50′43.9548″ N, Longitude: 119°17′33.5076″ W, elevation 140 m above sea level). The trial with spring wheat (*Triticum aestivum* L.) was conducted in 2014; the trial with winter wheat was conducted during the 2015−2016 growing seasons. The climate in the region is classified as *Csa* (= temperate, dry, and hot summer) by the Köppen-Geiger system [38]. In the 2014 growing season (March to July), cumulative precipitation was 60.0 mm, and the mean air temperature was 16.4 °C. March, with a mean temperature of 8.1 °C, was the coldest month, while July, with a mean temperature of 25.3 °C, was the warmest month (Figure 1). In the 2015–2016 growing season (October 2015 to July 2016), precipitation was 175.4 mm, and the mean air temperature was 11.2 °C. January was the coldest month, with a mean temperature of 2.0 °C, and July was the warmest month, with a mean temperature of 22.5 °C (Figure 1). Both trials were conducted on an Adkins fine sandy loam (Adkins coarse-loamy, mixed, superactive, *mesic Xeric Haplocalcid*). The basic soil properties for a soil depth of 0–30 cm were pH of 6.3, soil organic matter of 0.8%, soil available P of 39 mg kg^{-1}, soil available K of 330 mg kg^{-1}, and soil S of 49 mg kg^{-1}. Soil nitrogen (N) of 0–60 cm before the trials are compiled in Table 1.

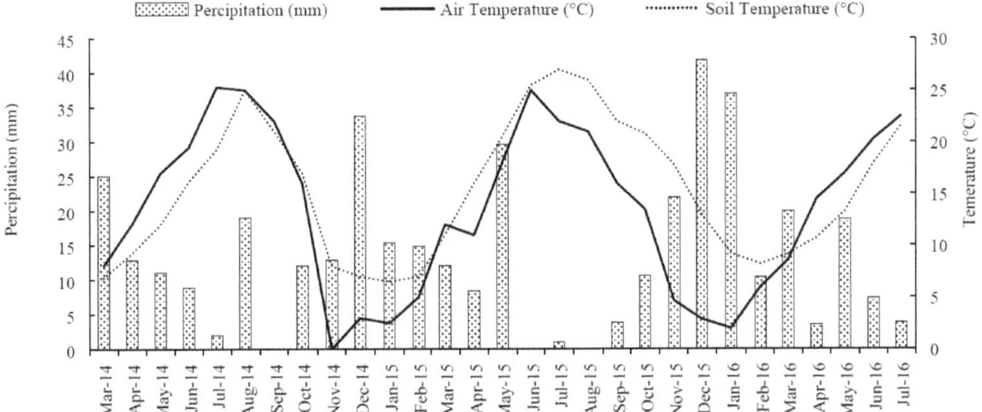

Figure 1. Average monthly air and soil temperatures (20 cm depth) and amount of precipitation during the two growing seasons (2014 for spring wheat and 2015–2016 for winter wheat).

Table 1. Soil nitrogen (mg kg^{-1} soil) before field trial establishment at Hermiston, OR, 2014–2016.

Growth Season	NH4$^+$-N	NO$_3$$^-$-N	NH4$^+$-N	NO$_3$$^-$-N
	(0–30 cm)		(30–60 cm)	
2014 (spring wheat)	10.3	34.0	10.8	16.0
2015–2016 (winter wheat)	7.1	17.1	4.1	9.2

In the spring wheat trial, the variety 'Westbred 528' was sown on March 10, 2014, at 135 kg ha^{-1} and harvested on July 23, 2014. In the winter wheat trial, the variety 'LCS Jet' was sown on 29 October 2015 at 135 kg ha^{-1} and harvested on 15 July 2016. Growers' standard pest and weed controls were applied throughout the growing season.

In both trials, each experimental plot was 9×9 m^2 in size containing 35 rows. The row-to-row distance was 0.26 m. The experiments were laid out as a randomized complete block design (RCBD), having eleven treatments with five replications in 2014 and four treatments with four replications in the 2015–2016 growing season.

The treatments in the 2014 trial included two sources of N, i.e., urea and UAN32, with two application rates (100% and 85%) in combination with two NIs as follows: (1) No-fertilizer Control (CK), (2) 85% Urea (85U), (3) 85% Urea + Instinct® II (85U + I), (4) 85% Urea + Agrotain® Ultra (85U + A), (5) 100% Urea (100U), (6) 100% Urea + Instinct® II (100U + I), (7) 85% UAN (85UAN), (8) 85% UAN + Instinct® II (85UAN + I), (9) 85% UAN + Agrotain® Ultra (85UAN + A), (10) 100% UAN (100UAN), and (11) 100% UAN + Instinct® II (100UAN + I). The treatments were applied three days after sowing. The N application rate of 100% U or 100% UAN was equivalent to 225 kg ha^{-1}. Both NIs were mixed with fertilizers before application. The mixing rates of Agrotain® Ultra to urea and UAN were 3.1 and 1.55 L ton^{-1}, respectively. The mixing rate of Instinct® II was according to its label rate of 1.1 kg ha^{-1}.

In the 2015–2016 trial, four treatments consisted of the following fertilizer-NIs combinations: (1) Single application of UAN + Instinct® II (100UAN + I), (2) Single application of UAN (100UAN), (3) Split application of UAN + Instinct® II (60% UAN in fall + Instinct® II and 40% UAN in spring; 60/40UAN + I), and (4) Split application of UAN (60% UAN in fall and 40% UAN in spring; 60/40UAN). For the treatments of 100 UAN + I and 100 UAN, the fertilizer was applied on October 27, while the Instinct® II was applied on October 28. For the treatments of 60/40UAN + I or 60/40UAN, 60% of UAN was applied on October 27, and the Instinct® II was applied on October 28, 2015, while 40% of UAN was applied on April 6 at the stem elongation stage. The N rate for the 100 UAN was 280 kg ha^{-1}. Fertilizers were applied by hand uniformly on the ground. The Instinct® II was only applied in fall with a rate of 1.1 kg ha^{-1} with the purpose of reducing nitrate loss through the wet winter.

2.2. Sampling and Measurements

2.2.1. Wheat Plant Parameters

The extended BBCH scale [39] was used to describe the phenological development of 50% of the plants in each treatment. During 2014, wheat parameters such as plant height (PH), leaf greenness (SPAD values), and total N content of flag leaf were measured at the flag leaf stage (BBCH stage 39). PH was recorded from 10 randomly selected plants from the inner plant rows in each plot. The SPAD meter readings were used as an indicator of leaf chlorophyll content per unit leaf area [40,41] and were determined on the blades, midway between the leaf edge and midrib [42] of fully expanded flag leaves using a SPAD-502 m (Minolta, Plainfield, IL, USA). Measurements were taken early in the morning and recorded as the mean of 10 randomly selected fully expanded leaves per plot. Total N content of the flag leaves was determined by the Kjeldhal method [43]. Moreover, at physiological maturity (BBCH stage 91 and 92), grain yield (GY) was assessed on a per plot basis and converted to tons per hectare (t ha^{-1}) after adjusting to 13% moisture content. Grain moisture (%) (GM) and grain protein content (%) (GP) were also measured. In the 2015–2016 trial, data collection was generally similar to the 2014 trial, with exceptions for PH and total N content of flag leaves.

2.2.2. Soil Nitrogen Content

In both trials, representative soil samples were collected to assess the contents of NH_4^+-N, NO_3^--N, and total mineral N (TMN) from 0–30 cm and 30–60 cm soil depths. Five well-distributed locations per plot were selected for soil sampling. The soils from the same depth were mixed uniformly into a composite sample and submitted for analysis. Soil sampling was conducted between irrigation events. The contents of NH_4^+-N and NO_3^--N were determined by potassium chloride extraction combined with cadmium reduction [44,45]. The TMN was calculated as the sum of the NH_4^+-N and NO_3^--N.

For the 2014 trial, soil N was measured at the 2nd, 4th, 6th, and 8th weeks after plant emergence (WAE). For the 2015–2016 trial, soil N was measured 3 weeks before the second split application and at the 4th and 8th week after the second split application.

2.3. Statistical Analyses

Data were subjected to analysis of variance (ANOVA) using the PROC GLM procedure of SAS (SAS version 9.4) for a randomized complete block design after checking for the normalcy of the variables with the Kolmogorov–Smirnov test. Means were compared using Fisher's least significant difference test (LSD) at $p \leq 0.05$. Data on NH_4^+-N, NO_3^--N, and TMN contents were analyzed with a split plot in time arrangement based on randomized complete block design because there were multiple measurements on the same experimental unit. Treatment and sampling were considered as main plot and subplot, respectively. It is noted that the presented table is a slice of the complete analysis. Figures were prepared in Excel version 2016 64-Bit Edition.

3. Results

3.1. First Experiment: Spring Wheat

Treatments significantly affected PH, leaf greenness (SPAD), and GY of spring wheat, and no effects were found for total leaf N content, GM, and GP (Table 2). Compared to control (no-fertilizer), PH, SPAD, and GY were 29.0%, 20.8%, and 31.0% higher when fertilizer/NIs combinations were applied. However, among all fertilization treatments, there was no significant difference in terms of GY. The tallest plants were recorded at 100U, followed by 85U + A and 85U + I, while the shortest plants were recorded at the treatments that included either 85U and 100U + I or any other UAN combination with NIs. The highest values of SPAD meter readings were recorded at 100UAN (Table 2).

Table 2. Analysis of variance (*p* values) of N fertilizer-nitrification inhibitor combinations on spring wheat plant height (PH), leaf greenness (SPAD), total N content of flag leaves measured at flag leaf stage, and grain yield (GY), grain moisture (GM) and protein content (GP) at physiological maturity in 2014.

Source of Variation	df	PH (cm)	SPAD	Total N of Flag Leaf mg kg^{-1}	GY (t ha^{-1})	GM (%)	GP (%)
Rep	4	0.61	0.28	0.01	0.08	0.25	0.78
Treatment	10	<0.01	<0.01	0.50	0.05	0.36	0.16
Control		47 ± 1.1 [d]	37 ± 1.3 [d]	3.5 ± 0.4	4.06 ± 0.45 [b]	5.5 ± 0.2	13.0 ± 0.6
85U		60 ± 0.4 [bc]	44 ± 0.4 [ab]	3.5 ± 0.2	5.30 ± 0.45 [a]	5.1 ± 0.1	14.2 ± 0.7
85U + I		62 ± 0.9 [ab]	46 ± 0.9 [ab]	4.1 ± 0.3	5.55 ± 0.41 [a]	5.3 ± 0.1	13.6 ± 0.3
85U + A		63 ± 1.2 [ab]	45 ± 1.0 [ab]	4.0 ± 0.2	4.91 ± 0.55 [a]	5.0 ± 0.1	14.5 ± 0.7
100U		64 ± 0.7 [a]	46 ± 0.7 [ab]	3.8 ± 0.1	5.45 ± 0.25 [a]	5.5 ± 0.1	14.8 ± 0.4
100U + I		60 ± 1.4 [bc]	43 ± 1.3 [bc]	3.9 ± 0.2	5.32 ± 0.30 [a]	5.3 ± 0.1	14.2 ± 0.5
85UAN		59 ± 0.8 [c]	41 ± 2.3 [c]	4.2 ± 0.4	5.45 ± 0.35 [a]	5.3 ± 0.2	13.0 ± 0.5
85UAN + I		59 ± 1.0 [c]	45 ± 0.6 [ab]	3.7 ± 0.2	5.48 ± 0.22 [a]	5.2 ± 0.1	14.5 ± 0.5
85UAN + A		60 ± 0.9 [bc]	45 ± 0.7 [ab]	3.9 ± 0.2	5.63 ± 0.23 [a]	5.2 ± 0.1	14.1 ± 0.5
100UAN		61 ± 0.5 [bc]	47 ± 1.1 [a]	3.6 ± 0.1	5.15 ± 0.25 [a]	5.1 ± 0.1	15.2 ± 0.5
100UAN + I		60 ± 1.2 [bc]	45 ± 0.2 [ab]	3.8 ± 0.3	4.95 ± 0.34 [a]	5.5 ± 0.1	14.5 ± 0.2

Means are averages of five replicates ±SE (standard error). Different letters within columns indicate means with significant differences according to least significant difference (LSD) at *p* < 0.05. No-fertilizer (Control), 85% Urea (85U), 85% Urea + Instinct® II (85U + I), 85% Urea + Agrotain® Ultra (85U + A), 100% Urea (100U), 100% Urea + Instinct® II (100U + I), 85% UAN (85UAN), 85% UAN + Instinct® II (85UAN + I), 85% UAN + Agrotain® Ultra (85UAN + A), 100% UAN (100UAN), 100% UAN + Instinct® II (100UAN + I).

The treatment effects, sampling time, and their interaction on NH_4^+-N, NO_3^--N, and TMN of soil are shown in Table 3. Compared to the control, all fertilizer treatments increased NH_4^+-N, NO_3^--N, and TMN contents in 0–30 cm significantly ($p < 0.01$). The use of UAN increased the average NH_4^+-N content of soil compared to urea. The highest soil NH_4^+-N content (14.6 mg N kg^{-1} soil) was found at 100UAN + I, followed by 100UAN, while the lowest one (9.7 mg N kg^{-1} soil) was observed in the treatment of 85U + I (Table 3).

The 100U treatment was associated with the highest soil NO_3^--N content, followed by 100U + I, 85U + A, and 85U, while the lowest NO_3^--N content (12.2 mg N kg^{-1} soil) was found in 85UAN + I, followed by 85UAN, 85U + I, and 85UAN + A. The 100U treatment was associated with the highest TMN at 0–30 cm. Except for the control (8.9 mg N kg^{-1} soil), the 85UAN + I treatment was associated with the lowest TMN (24.0 mg N kg^{-1} soil), followed by 85U + I, 85UAN, and 85UAN + A (Table 3). Urea application generally resulted in higher soil NO_3^--N and TMN than UAN, while the NH_4^+-N content was slightly higher following UAN applications. The effects of NIs on N forms found in soils were relatively limited when speciation was compared to the treatments without NIs. Between the two NIs, soil N contents tended to be higher with Agrotain® Ultra application.

Figure 2a–c show the NH_4^+-N, NO_3^--N, and TMN content in soil (0–30 cm depth). In all treatments, NH_4^+-N content was highest at 2 WAE; the highest value was shown in the treatment of 100UAN + I, while the lowest value was observed in the control, followed by 100U + I and 85U + I. In general, a sharp reduction was found from the 2nd WAE to the 4th WAE, and afterward, the reduction tended to be smoother (Figure 2a). At 8 WAE, no difference was found among the treatments.

Compared to control, the soil NO_3^--N content increased considerably by the second WAE. The NO_3^--N content increased continually up to the maximum at 4 WAE; the highest values were observed in the 85U and 85U + A treatments, while the lowest was observed in the 85UAN + I treatment. Afterward, NO_3^--N content decreased steadily. As a consequence, the lowest NO_3^--N content was obtained at 8 WAE (Figure 2b). On average, NO_3^--N content was less in treatments with UAN than with urea at all sampling events. The addition of Instinct[®] II tended to decrease NO_3^--N content. As expected, NO_3^--N content was lower in the treatments with lower N application rates. At the last sampling event (8 WAE), the highest and lowest NO_3^--N contents were found in the 100U and 85UAN + I, respectively.

Table 3. Analysis of variance of N fertilizer-nitrification inhibitor combinations and sampling time on soil NO_3^--N, NH_4^+-N, and total mineral N (TMN) at 0–30 cm and 30–60 cm in spring wheat, 2014.

Source of Variation	df	NH_4^+-N	NO_3^--N	TMN	NH_4^+-N	NO_3^--N	TMN
		(mg kg^{-1} soil)			**(mg kg^{-1} soil)**		
		0–30 cm			30–60 cm		
Rep	4	0.50	0.02	0.03	0.02	0.03	0.09
Treatment	10	<0.01	<0.01	<0.01	0.07	<0.01	0.09
Sampling	3	<0.01	<0.01	<0.01	<0.01	<0.01	<0.01
Sampling × Treatment	30	<0.01	<0.01	<0.01	0.25	0.29	0.19
Treatment							
Control		5.4 ± 1.2 [d]	3.4 ± 0.5 [d]	8.9 ± 1.5 [f]	5.3 ± 1.2	6.4 ± 0.8 [b]	11.7 ± 0.9
85U		12.1 ± 2.6 [b]	23 ± 3.8 [ab]	33.9 ± 4.4 [abc]	5.7 ± 1.4	8.5 ± 1.1 [ab]	14.3 ± 1.2
85U + I		9.7 ± 2.1 [c]	15.8 ± 2.3 [c]	24.9 ± 3.7 [de]	5.3 ± 1.4	9.3 ± 1.1 [a]	14.6 ± 1.3
85U + A		11.7 ± 2.7 [bc]	24.5 ± 3.4 [a]	36.4 ± 5 [ab]	5.3 ± 2.4	7.9 ± 1.3 [ab]	13.3 ± 3.3
100U		12.4 ± 2.7 [b]	26.7 ± 2.9 [a]	38.1 ± 4.4 [a]	5.1 ± 1.3	9.9 ± 0.9 [a]	15 ± 1.1
100U + I		10.6 ± 2.2 [bc]	26.1 ± 2.9 [a]	37.5 ± 4.2 [ab]	5.6 ± 1.3	8.2 ± 1 [ab]	14 ± 1.6
85UAN		10.9 ± 2.9 [bc]	14.1 ± 2.4 [c]	25.3 ± 4 [de]	5.7 ± 1.5	9.1 ± 0.8 [a]	14.8 ± 2.0
85UAN + I		11.7 ± 2.3 [bc]	12.2 ± 2 [c]	24.0 ± 3.6 [e]	6.8 ± 1.6	9.4 ± 0.8 [a]	16.3 ± 1.5
85UAN + A		11.7 ± 3.1 [bc]	16.4 ± 2.9 [c]	28.2 ± 4.2 [cde]	6.0 ± 1.4	8.3 ± 0.9 [ab]	14.3 ± 1.1
100UAN		12.4 ± 3 [ab]	17.7 ± 2.3 [bc]	30.1 ± 3.8 [bcd]	6.5 ± 1.8	10 ± 0.9 [a]	16.5 ± 1.7
100UAN + I		14.6 ± 3.4 [a]	16.8 ± 2.4 [c]	31.4 ± 5.2 [a-d]	6.5 ± 1.7	8.8 ± 0.7 [a]	15.4 ± 1.8
Sampling							
2 WAE		29.9 ± 1.1 [a]	17.4 ± 1.2 [b]	46.8 ± 1.9 [a]	16.7 ± 0.4 [a]	4.7 ± 0.2 [d]	21.6 ± 0.7 [a]
4 WAE		8.3 ± 0.5 [b]	32.1 ± 1.8 [a]	40.3 ± 2.1 [b]	3.5 ± 0.3 [b]	13 ± 0.3 [a]	16.6 ± 0.5 [b]
6 WAE		4.1 ± 0.4 [c]	15.5 ± 1.3 [b]	19.8 ± 1.6 [c]	1.8 ± 0.8 [c]	10.4 ± 0.5 [b]	12.2 ± 1.2 [c]
8 WAE		2.6 ± 0.1 [d]	6.56 ± 0.8 [c]	9.2 ± 0.8 [d]	1.3 ± 0.1 [c]	6.6 ± 0.4 [c]	7.97 ± 1.2 [d]

Means are averages of five replicates ± SE (standard error). Different letters within columns indicate means with significant differences according to least significant difference (LSD) at $p < 0.05$. No-fertilizer (Control), 85% Urea (85U), 85% Urea + Instinct[®] II (85U + I), 85% Urea + Agrotain[®] Ultra (85U + A), 100% Urea (100U), 100% Urea + Instinct[®] II (100U + I), 85% UAN (85UAN), 85% UAN + Instinct[®] II (85UAN + I), 85% UAN + Agrotain[®] Ultra (85UAN + A), 100% UAN (100UAN), 100% UAN + Instinct[®] II (100UAN + I). WAE (Weeks after Emergence).

The highest soil TMN content was found at 2 WAE, with the exception of 85U and 100U + I, which happened at 4 WAE (Figure 2c). Afterward, the TMN decreased gradually. At the final sampling event (8 WAE), the highest TMN content occurred in the 100U treatment, and the lowest were in the 85UAN and 85UAN + I treatments. A comparison of the two N sources indicated that TMN content was lower in the UAN treatments. The lower N rate naturally had the lower TMN content. The application of Instinct[®] II generally did not impact the TMN content or slightly reduced TMN content, while the application of Agrotain[®] Ultra tended to increase the TMN (Figure 2c).

In the 30–60 cm soil profile, fertilization treatments had a significant effect only on NO_3^--N content. The lowest NO_3^--N content (6.4 mg N kg^{-1} soil) was found in the control. Among the fertilization treatments, 85U, 85U + A, 100U + I, 85UAN + A had slightly lower NO_3^--N content. As with 0–30 cm, sampling time significantly affected mineral soil N content at the 30–60 cm soil depth. NH_4^+-N and TMN content in 30–60 cm decreased with sampling time; the lowest NH_4^+-N and TMN contents were observed at 8 WAE. Consistent with the pattern in 0–30 cm, the NO_3^--N content increased 176% at 4 WAF compared to 2 WAE and then gradually decreased (Table 3).

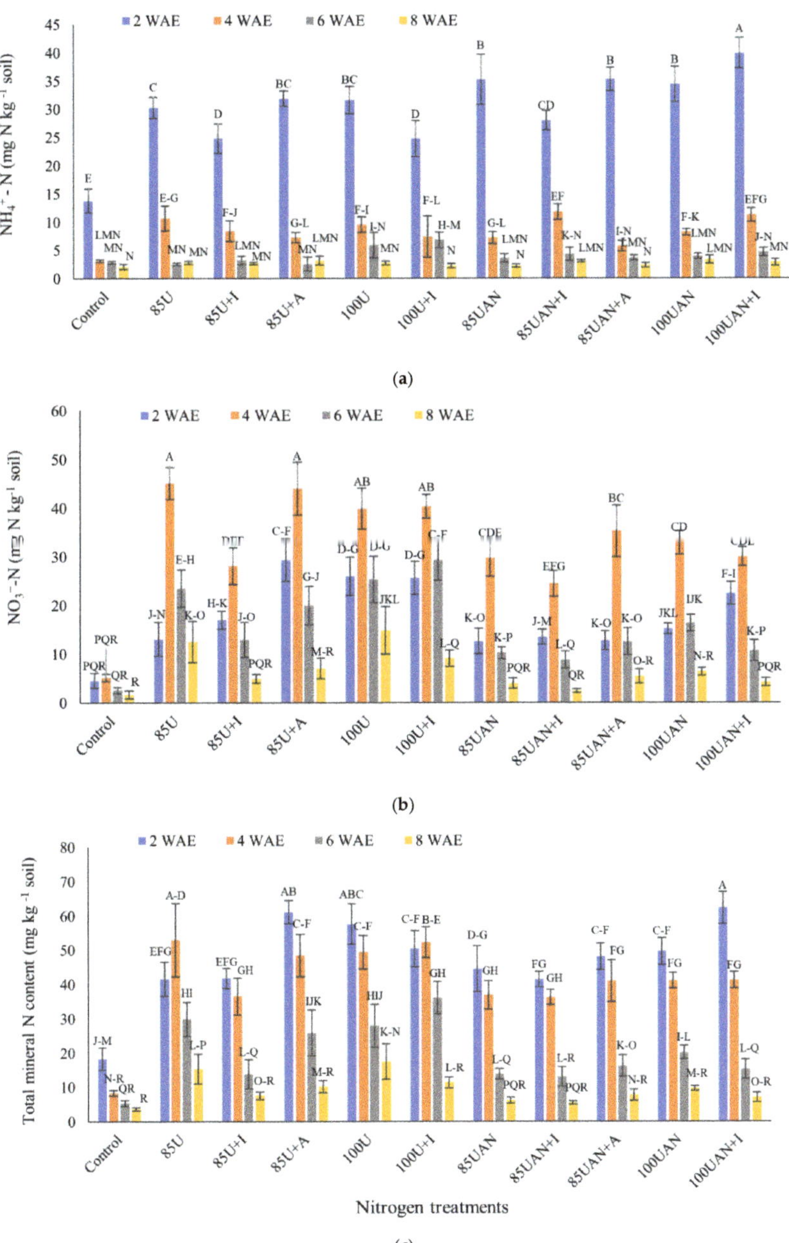

Figure 2. Changes of (**a**) NH$_4^+$-N, (**b**) NO$_3^-$-N, and (**c**) total mineral N content of soil during sampling times in spring wheat field at 0–30 cm. Data represent the averages of five replicates. Vertical bars indicate standard errors (±SE). Different letters indicate means with significant differences according to least significant difference (LSD) at $p < 0.05$. WAE (Weeks after Emergence). No-fertilizer (Control), 85% Urea (85U), 85% Urea + Instinct[®] II (85U + I), 85% Urea + Agrotain[®] Ultra (85U + A), 100% Urea (100U), 100% Urea + Instinct[®] II (100U + I), 85% UAN (85UAN), 85% UAN + Instinct[®] II (85UAN + I), 85% UAN + Agrotain[®] Ultra (85UAN + A), 100% UAN (100UAN), 100% UAN + Instinct[®] II (100UAN + I).

3.2. Second Experiment: Winter Wheat

Analysis of variance results showed that fertilization treatments, the combination of N fertilization and NIs, had no significant effects on flag leaf greenness (SPAD), GY, GM, and GP of winter wheat (Table 4). GY ranged from 8.51 t ha^{-1} (100UAN) to 9.01 t ha^{-1} (60/40UAN + I), GM ranged from 5.0% (100UAN +I) to 5.3% (100UAN), GP ranged from 9.4% (60/40UAN + I) to 10.0% [(100UAN + I) and (100UAN)], and leaf greenness ranged from 52 (100UAN + I) to 53 (Table 4).

Table 4. Analysis of variance of N fertilizer-nitrification inhibitor-application time combinations on leaf greenness (SPAD) at flag leaf stage and grain yield (GY), grain moisture (GM), and protein (GP) of winter wheat in 2015−2016.

Source of Variation	df	SPAD	GY (t ha^{-1})	GM (%)	GP (%)
Rep	3	0.35	0.015	0.78	0.19
Treatment	3	0.86	0.49	0.19	0.54
100UAN + I		52 ± 1.0	8.53 ± 0.51	5.0 ± 0.04	10.0 ± 0.2
100UAN		53 ± 0.9	8.51 ± 0.36	5.3 ± 0.09	10.0 ± 0.4
60/40UAN + I		53 ± 0.5	9.01 ± 0.22	5.1 ± 0.07	9.4 ± 0.2
60/40UAN		53 ± 0.8	8.77 ± 0.49	5.1 ± 0.10	9.8 ± 0.5

Means are averages of four replicates ± standard error (SE). Single application of UAN + Instinct® II (100UAN + I), single application of urea (100UAN), split application of UAN + Instinct® II (60% UAN in fall + Instinct® II and 40% UAN in spring; 60/40UAN + I), and split application of UAN (60% UAN in fall and 40% UAN in spring; 60/40UAN).

Fertilization treatments did not change NH_4^+-N, NO_3^--N, and TMN content in the 0–30 cm soil profile (Table 5). However, the soil N parameters differed significantly with sampling time ($p < 0.01$). Soil NH_4^+-N, NO_3^--N, and TMN contents decreased 17%, 69%, and 51%, respectively, at 4 weeks after the second split application (WAT), and 60%, 74%, and 69%, respectively, at 8 WAT compared to those found before the second split application (Table 5).

Table 5. Analysis of variance of treatment effects and sampling time on soil NO_3^--N, NH_4^+-N, and TMN at 0–30 cm and 30–60 cm in winter wheat in 2015−2016.

Source of Variation	df	NH_4^+-N	NO_3^--N	TMN	NH_4^+-N	NO_3^--N	TMN
		(mg kg^{-1} soil)			(mg kg^{-1} soil)		
		0–30 cm			30–60 cm		
Rep	3	0.92	0.60	0.51	0.81	0.16	0.17
Treatment	3	0.70	0.14	0.18	0.49	0.05	0.05
Sampling	2	<0.01	<0.01	<0.01	0.002	<0.01	<0.01
Sampling × Treatment	6	0.73	0.19	0.64	0.54	0.23	0.24
Treatment							
100UAN + I		3.6 ± 0.4	4.4 ± 1.1	8.1 ± 1.4	2 ± 0.3	6.8 ± 1.4 ab	8.9 ± 1.5 ab
100UAN		3.2 ± 0.3	5.2 ± 1.3	8.4 ± 1.5	1.7 ± 0.1	7.6 ± 2.5 a	10.1 ± 2.6 a
60/40UAN + I		3.2 ± 0.4	3.3 ± 0.5	6.5 ± 0.9	2 ± 0.2	4.3 ± 1 b	6.3 ± 1 b
60/40UAN		3.5 ± 0.4	4 ± 0.7	7.6 ± 1.1	1.8 ± 0.2	6.1 ± 1.6 ab	8 ± 1.7 ab
Sampling							
Before the second split application		4.6 ± 0.3 [a]	8.3 ± 0.9 [a]	13 ± 1 [a]	2.1 ± 0.2 [a]	12.9 ± 1.6 [a]	14.4 ± 1.6 [a]
4 WAT		3.8 ± 0.3 [a]	2.5 ± 0.1 [b]	6.3 ± 0.4 [b]	2.2 ± 0.2 [a]	3.8 ± 0.8 [b]	6.1 ± 0.9 [b]
8 WAT		1.8 ± 0.3 [b]	2.1 ± 0.1 [b]	4 ± 0.2 [c]	1.3 ± 0.1 [b]	2.8 ± 0.6 [b]	4.2 ± 0.6 [b]

Means are averages of four replicates ± SE (standard error). Different letters within columns indicate means with significant differences according to least significant difference (LSD) at $p < 0.05$. Single application of UAN + Instinct® II (100UAN + I), single application of urea (100UAN), split application of UAN + Instinct® II (60% UAN in fall + Instinct® II and 40% UAN in spring; 60/40UAN + I), and split application of UAN (60% UAN in fall and 40% UAN in spring; 60/40UAN). WAT (Weeks after the second split application).

Unlike 0–30 cm soil depth, the fertilization treatments in the 30–60 cm soil depth affected soil NO_3^--N and TMN content significantly. The highest levels of NO_3^--N and TMN were associated with 100UAN (7.6 mg kg^{-1} and 10.1 mg kg^{-1}, respectively) (Table 5). In contrast, the lowest NO_3^--N and TMN contents were associated with 60/40U + I (4.3 mg kg^{-1} and 6.3 mg kg^{-1}, respectively). On average, a split application of UAN reduced NO_3^--N and TMN contents by 27% and 24%, respectively, in comparison to a single UAN application.

Compared to treatments that did not include NIs, the addition of Instinct® II reduced NO_3^--N and TMN contents by 19% and 16%, respectively. Reduction trends were observed for NH_4^+-N, NO_3^--N, and TMN contents in the 30–60 cm soil profile. At 8 WAT, NH_4^+-N, NO_3^--N, and TMN levels were 38%, 78%, and 70% lower, respectively, than those before the second split application (Table 5).

4. Discussion

This study revealed that GY and GP of spring or winter wheat were not affected by the different N fertilizer treatments regardless of NIs application or N rates, although the PH and leaf greenness (SPAD) of spring wheat differed. Consistent with our results, some studies reported that the GY of barley [46], maize [46–48], and winter wheat [46,47] were not affected by NIs application. Other studies reported that NIs had a limited effect on biomass yield and crude protein of grass [11,19]. Several studies in pastures reported no significant effect of NIs on yields [50–53] nor in vegetable production [54,55]. However, a number of studies observed that the use of nitrification and urease inhibitors significantly increased wheat [24,56], maize [57–59], and vegetable yields [60]. A meta-analysis by Abalos et al. [61] indicated that, on average, the use of nitrification and urease inhibitors led to a 7.5% increase in yield, while effectiveness depended on environmental and management factors. Meng et al. [14] pointed out that yield improvement through NIs addition might be expected only when N is the limiting factor to plant growth. In our study, both the N application rates and the soil N data indicated sufficient N supply to the crops, and as a result, the response of N treatments on wheat yield and protein content was similar. Moreover, regardless of NIs use, treatments at a lower N rate did not affect the GY of spring wheat, which further confirmed that excessive N application occurred. Zhu [62] indicated that the N rate could be reduced by 7–24% without yield loss of rice or wheat.

Note that in the winter wheat trial, the numeric GY was higher, but the numeric GP was lower in the treatments with split-N fertilization, implying the advantages of split fertilization in improving NUE. Other studies showed that supplying a small portion of total N at planting coupled with multiple applications of the rest N according to crop N requirements can increase NUE and yield of rice, barley, wheat, potato, and maize [63–66].

In the trial with spring wheat, UAN treatments resulted in higher soil NH_4^+-N content and lower NO_3^--N content than urea treatments, which might be due to the fertilizer property, as UAN itself contain NH_4^+-N while urea may quickly convert to NO_3^--N. The higher NH_4^+-N contents in the soils indicate an increased risk of ammonia volatilization, another main pathway of N loss in agricultural systems [67,68]. Of the two N forms, NO_3^--N content was generally higher than NH_4^+-N, and TMN broadly followed the same pattern as NO_3^--N (Figure 2).

It was reported that NIs slow bacterial oxidation of ammonium (NH_4^+) to nitrite (NO_2^-) in soils by depressing the activity of ammonia mono-oxygenase released by *Nitrosomas* bacteria [15], extending the retention of NH_4^+-N in soil [69]. In addition, less NO_3^--N is produced, and NO_3^--N leaching potential is reduced as well. In our trial, we did not observe a significant effect of NIs on soil N contents. This may be related to the environmental conditions, as our studies were conducted under irrigation; frequent irrigation might have an impact on the properties of the NIs. Moreover, although NIs repeatedly have been shown to reduce N_2O and NO emissions from agricultural soils, their mitigation effect varies greatly, and the mechanism is still not well explored [70,71].

Similar to the surface soil depth of 0–30 cm, the N content in the 30–60 cm was not affected by different N treatments, suggesting that the NIs application did not affect the nitrate leaching potential. Over the monitoring period, the increase in soil NO_3^--N content from 2 WAE to 4 WAE may be sourced from nitrification. Afterward, it naturally decreased with time due to the plant uptake.

In the winter wheat field trial, the split application of N (60% UAN in the fall and 40% UAN in the spring) generally reduced NO_3^--N and TMN contents in comparison to a single application (100% UAN in the fall), suggesting a lower NO_3^--N leaching potential. When all N was applied at sowing, intensive and heavy precipitation plus irrigation could lead to NO_3^--N being leached more deeply into the soil [63].

Throughout the soil profile to 60 cm depth, in the spring wheat trial, soil NH_4^+-N, NO_3^--N, and TMN contents at 0–30 cm soil depth were higher than those recorded at 30–60 cm soil depth. However, in the winter wheat trial, the NH_4^+-N content decreased but NO_3^--N content increased with the soil depth because the former is relatively immobile while the latter is highly mobile. Differences in N distribution between trials might be due to the longer growing season for winter wheat.

Apart from the spring wheat trial, the application of Instinct[®] II reduced NO_3^--N and TMN contents, compared to the no-application of NI in the winter wheat trial, suggesting that Instinct[®] II reduced nitrification. Studies reported that the performance of NIs is significantly affected by the timing of application (growth stage), type of application (single or split), and rate of application [72–75]. Moreover, the variability of weather conditions, especially soil temperature, affects the effectiveness of NIs [28]. Because NIs degradation and nitrification increase with the increasing soil temperatures, the efficiency of NIs in winter wheat field were more pronounced, perhaps due to the decreasing soil temperature. Thus, the greater effectiveness of NIs in winter wheat could have been a result of overall reduced nitrification activity. Nair et al. [76] reported that the efficiency of NIs may be affected by soil conditions (texture, temperature, pH, and organic matter), through the activity of nitrifiers and denitrifiers, and through N distribution. The effectiveness of nitrapyrin at decreasing nitrification in soils depends on a number of interacting factors besides soil temperature [77]. Raza et al. [78] showed that nitrification was significantly affected by soil temperature and moisture levels. Soil temperature controls the persistence and performance of DMPP as a NI [79]. However, gross nitrification rates were reduced in the presence of nitrapyrin at both 20 °C and 40 °C soil temperatures [80]. Other studies have found that nitrapyrin can decrease nitrification at temperatures from 25 °C to 35 °C [27].

In general, there is no unwavering confirmation regarding the behavior of NIs in soil. It remains unclear how long NIs remain effective and exactly what factors can affect their efficiency. Therefore, there is a need for more studies to elucidate the influence factors on NIs efficiency. Such information would assist growers in using NIs correctly.

5. General Remark

The present findings were obtained from field trials with spring and winter wheat and indicated that the crops received sufficient N. Thus, a reduced N rate (e.g., 15% reduction) could result in similar yields. Between the two N sources, urea and urea ammonium nitrate-UAN, we observed that the application of UAN could significantly reduce soil NO_3^--N content in the 0–30 cm soil depth and may provide environmental benefits by reducing nitrate leaching potential and denitrification risk. Hence, the environmental advantages of UAN as an N source outweigh urea. Furthermore, splitting N applications could reduce soil NO_3^--N content compared to a single application. Application of Instinct[®] II with lower-rate urea and with UAN during cool temperatures seems to be a suitable strategy to reduce NO_3^--N leaching potential, while Agrotain[®] Ultra did not show any considerable effect. Our results demonstrated that selecting effective NIs, suitable N sources, reducing N rate, and splitting N fertilizers during the growing season can be regarded as practical strategies to reduce NO_3^--N leaching while not compromising crop yield. Although the findings from this were based on two crops, it should be noted that the data were only from a single-season observation for either crop. Ideally, it will be necessary to carry out trials based on multiple years and locations to make a solution conclusion on the effects of NIs and N management on potential yield benefits and the N dynamics of soils. In such trials, at least some treatments supplying suboptimal N should be included, as NIs might show their potential to significantly increase crop yields. Moreover, frequent field measurements on N contents should be conducted before and after fertilization as the N transformation occurs very rapidly.

Author Contributions: All authors contributed substantially to the work reported in this paper. Data analyzing, S.T.; writing—original draft preparation, S.T. and S.F.-A.; review and editing, R.Q., C.N. and G.W. All authors have read and agreed to the published version of the manuscript.

Funding: This research received no external funding.

Data Availability Statement: The data presented in this study are available upon request from the corresponding author.

Acknowledgments: We would like to express our appreciation for conducting field trials by Sandy DeBano, Dave Wooster, and Don Horneck from Oregon State University-Hermiston Agricultural Research and Extension Center, Hermiston, OR, 97838. Tim Weinke, students, and interns provided support in field measurements and data collection. Linda Brewer from Oregon State University provided thorough editorial comments for improving the written quality.

Conflicts of Interest: The authors declare no conflict of interest.

References

1. von Wirén, N.; Gazzarrini, S.; Frommer, W.B. Regulation of mineral nitrogen uptake in plants. *Plant Soil* **1997**, *196*, 191–199. [CrossRef]
2. Fageria, N.K.; Baligar, V.C. Enhancing nitrogen use efficiency in crop plants. *Adv. Agron.* **2005**, *8*, 97–185.
3. Xu, A.; Li, L.; Xie, J.; Wang, X.; Coulter, J.A.; Liu, C.; Wang, L. Effect of long-term nitrogen addition on wheat yield, nitrogen use efficiency, and residual soil nitrate in a semiarid area of the loess plateau of China. *Sustainability* **2020**, *12*, 1735. [CrossRef]
4. Ahmed, M.; Rauf, M.; Mukhtar, Z.; Saeed, N.A. Excessive use of nitrogenous fertilizers: An unawareness causing serious threats to environment and human health. *Environ. Sci. Pollut. Res.* **2017**, *24*, 26983–26987. [CrossRef] [PubMed]
5. Omara, P.; Aula, L.; Oyebiyi, F.; Raun, W.R. World cereal nitrogen use efficiency trends: Review and current knowledge. *Agrosyst. Geosci. Environ.* **2019**, *2*, 1–8. [CrossRef]
6. Cameron, K.C.; Di, H.J.; Moir, J.L. Nitrogen losses from the soil/plant system: A review. *Ann. Appl. Biol.* **2013**, *162*, 145–173. [CrossRef]
7. Farzadfar, S.; Knight, J.D.; Congreves, K.A. Soil organic nitrogen: An overlooked but potentially significant contribution to crop nutrition. *Plant Soil* **2021**, *462*, 7–23. [CrossRef]
8. Gaudin, R.; Dupuy, J. Ammoniacal nutrition of transplanted rice fertilized with large urea granules. *Agron. J.* **1999**, *91*, 33–36. [CrossRef]
9. Salsac, L.; Chaillou, S.; Morot-Gaudry, J.F.; Lesaint, C.H.; Jolivet, E. Nitrate and ammonium nutrition in plants. *Plant Physiol. Biochem.* **1987**, *25*, 805–812.
10. Bloom, A.J.; Meyerhoff, P.A.; Taylor, A.R.; Rost, T.L. Root development and absorption of ammonium and nitrate from the rhizosphere. *J. Plant Growth Regul.* **2002**, *21*, 416–431. [CrossRef]
11. Fageria, N.K. *The Use of Nutrients in Crop Plants*; CRC Press: Boca Raton, FL, USA, 2016.
12. Kishchenko, O.; Stepanenko, A.; Straub, T.; Zhou, Y.; Neuhäuser, B.; Borisjuk, N. Ammonium Uptake, Mediated by Ammonium Transporters, Mitigates Manganese Toxicity in Duckweed, Spirodela polyrhiza. *Plants* **2023**, *12*, 208. [CrossRef]
13. Barth, G.; von Tucher, S.; Schmidhalter, U.; Otto, R.; Motavalli, P.; Ferraz-Almeida, R.; Meinl Schmiedt Sattolo, T.; Cantarella, H.; Vitti, G.C. Performance of nitrification inhibitors with different nitrogen fertilizers and soil textures. *J. Plant Nutr. Soil Sci.* **2019**, *182*, 694–700. [CrossRef]
14. Meng, Y.; Wang, J.J.; Wei, Z.; Dodla, S.K.; Fultz, L.M.; Gaston, L.A.; Xiao, R.; Park, J.H.; Scaglia, G. Nitrification inhibitors reduce nitrogen losses and improve soil health in a subtropical pastureland. *Geoderma* **2021**, *388*, 114947. [CrossRef]
15. McCarty, G.W. Modes of action of nitrification inhibitors. *Biol. Fertil. Soils* **1999**, *29*, 1–9. [CrossRef]
16. Jiang, R.; Yang, J.; Drury, C.F.; Grant, B.B.; Smith, W.N.; He, W.; Reynolds, D.W.; He, P. Modelling the impacts of inhibitors and fertilizer placement on maize yield and ammonia, nitrous oxide and nitrate leaching losses in southwestern Ontario, Canada. *J. Clean. Prod.* **2023**, *384*, 135511. [CrossRef]
17. Sanz-Cobena, A.; Sánchez-Martín, L.; García-Torres, L.; Vallejo, A. Gaseous emissions of N_2O and NO and NO_3- leaching from urea applied with urease and nitrification inhibitors to a maize (*Zea mays*) crop. *Agric. Ecosyst. Environ.* **2012**, *149*, 64–73. [CrossRef]

18. Lin, S.; Hernandez-Ramirez, G. Nitrous oxide emissions from manured soils as a function of various nitrification inhibitor rates and soil moisture contents. *Sci. Total Environ.* **2020**, *738*, 139669. [CrossRef] [PubMed]

19. de Souza, T.L.; de Oliveira, D.P.; Santos, C.F.; Reis, T.H.P.; Cabral, J.P.C.; da Silva Resende, É.R.; Fernandes, T.J.; de Souza, T.R.; Builes, V.R.; Guelfi, D. Nitrogen fertilizer technologies: Opportunities to improve nutrient use efficiency towards sustainable coffee production systems. *Agric Ecosyst. Environ.* **2023**, *345*, 108317. [CrossRef]

20. Fettweis, U.; Mittelstaedt, W.; Schimansky, C.; Führ, F. Lysimeter experiments on the translocation of the carbon-14-labelled nitrification inhibitor 3,4-dimethylpyrazole phosphate (DMPP) in a gleyic cambisol. *Biol. Fertil. Soils* **2001**, *34*, 126–130. [CrossRef]

21. Bhatia, A.; Sasmal, S.; Jain, N.; Pathak, H.; Kumar, R.; Singh, A. Mitigating nitrous oxide emission from soil under conventional and no-tillage in wheat using nitrification inhibitors. *Agric. Ecosyst. Environ.* **2010**, *136*, 247–253. [CrossRef]

22. Liu, C.; Wang, K.; Zheng, X. Effects of nitrification inhibitors (DCD and DMPP) on nitrous oxide emission, crop yield and nitrogen uptake in a wheat–maize cropping system. *Biogeosciences* **2013**, *10*, 2427–2437. [CrossRef]

23. Ma, Y.; Sun, L.; Zhang, X.; Yang, B.; Wang, J.; Yin, B.; Yan, X.; Xiong, Z. Mitigation of nitrous oxide emissions from paddy soil under conventional and no-till practices using nitrification inhibitors during the winter wheat-growing season. *Biol. Fertil. Soils* **2013**, *49*, 627–635. [CrossRef]

24. Dawar, K.; Rahman, U.; Alam, S.S.; Tariq, M.; Khan, A.; Fahad, S.; Datta, R.; Danish, S.; Saud, S.; Noor, M. Nitrification Inhibitor and Plant Growth Regulators Improve Wheat Yield and Nitrogen Use Efficiency. *J. Plant Growth Regul.* **2022**, *41*, 1–11. [CrossRef]

25. Mohammed, Y.A.; Chen, C.; Jensen, T. Urease and nitrification inhibitors impact on winter wheat fertilizer timing, yield, and protein content. *Agron. J.* **2016**, *108*, 905–912. [CrossRef]

26. Dawar, K.; Zaman, M.; Rowarth, J.S.; Blennerhassett, J.; Turnbull, M.H. The impact of urease inhibitor on the bioavailability of nitrogen in urea and in comparison, with other nitrogen sources in ryegrass (*Lolium perenne* L.). *Crop Pasture Sci.* **2010**, *61*, 214–221. [CrossRef]

27. Chen, D.; Suter, H.C.; Islam, A.; Edis, R. Influence of nitrification inhibitors on nitrification and nitrous oxide (N_2O) emission from a clay loam soil fertilized with urea. *Soil Biol. Biochem.* **2010**, *42*, 660–664. [CrossRef]

28. Zaman, M.; Nguyen, M.L.; Šimek, M.; Nawaz, S.; Khan, M.J.; Babar, M.N.; Zaman, S. Emissions of Nitrous Oxide (N_2O) and Di-nitrogen (N_2) from the Agricultural Landscapes, Sources, Sinks, and Factors Affecting N_2O and N_2 Ratios. In *Greenhouse Gases—Emission, Measurement and Management*; Liu, G.X., Ed.; InTech.: London, UK, 2012; pp. 1–32.

29. Amberger, A. Research on dicyandiamide as a nitrification inhibitor and future outlook. *Commun. Soil Sci. Plant Anal.* **1989**, *20*, 1933–1955. [CrossRef]

30. Kelliher, F.M.; Clough, T.J.; Clark, H.; Rys, G.; Sedcole, J.R. The temperature dependence of dicyandiamide (DCD) degradation in soils: A data synthesis. *Soil Biol. Biochem.* **2008**, *40*, 1878–1882. [CrossRef]

31. Peixoto, L.; Petersen, S.O. Efficacy of three nitrification inhibitors to reduce nitrous oxide emissions from pig slurry and mineral fertilizers applied to spring barley and winter wheat in Denmark. *Geoderma Reg.* **2023**, *32*, e00597. [CrossRef]

32. Kim, D.G.; Palmada, T.; Berben, P.; Giltrap, D.; Saggar, S. Seasonal variations in the degradation of a nitrification inhibitor, dicyandiamide (DCD), in a Manawatu grazed pasture soil. In *Advanced Nutrient Management: Gains from the Past—Goals for the Future*; Currie, L.D., Christensen, C.L., Eds.; Massey University: Auckland, New Zealand, 2012; 7p.

33. Di, H.J.; Cameron, K.C. Treating grazed pasture soil with a nitrification inhibitor, eco-n™, to decrease nitrate leaching in a deep sandy soil under spray irrigation—A lysimeter study. *N. Z. J. Agric. Res.* **2004**, *47*, 351–361. [CrossRef]

34. Smith, L.C.; Orchiston, T.; Monaghan, R.M. The effectiveness of the nitrification inhibitor dicyandiamide (DCD) for mitigating nitrogen leaching losses from a winter grazed forage crop on a free draining soil in Northern Southland. *Proc. N. Z. Grassl. Assoc.* **2012**, *74*, 39–44.

35. Suter, H.; Lam, S.K.; Walker, C.; Chen, D. Nitrogen use efficiency for pasture production–impact of enhanced efficiency fertilisers and N rate. In Proceedings of the 17th Australian Society of Agronomy Conference, Hobart, Australia, 20–24 September 2015; pp. 20–24.

36. Cookson, W.R.; Cornforth, I.S. Dicyandiamide slows nitrification in dairy cattle urine patches: Effects on soil solution composition, soil pH and pasture yield. *Soil Biol. Biochem.* **2002**, *34*, 1461–1465. [CrossRef]

37. Sassman, A.M.; Barker, D.W.; Sawyer, J.E. Corn Response to Urea–Ammonium Nitrate Solution Treated with Encapsulated Nitrapyrin. *Agron. J.* **2018**, *110*, 1058–1067. [CrossRef]

38. Peel, M.C.; Finlayson, B.L.; McMahon, T.A. Updated world map of the Köppen-Geiger climate classification. *Hydrol. Earth Syst. Sci.* **2007**, *11*, 1633–1644. [CrossRef]

39. Meier, U. *Growth Stages of Mono- and Dicotyledonous Plants. BBCH Monograph*, 2nd ed.; Blackwell Science: Berlin, Germany, 2001; p. 158.

40. Chapman, S.C.; Barreto, H.J. Using a chlorophyll meter to estimate specific leaf nitrogen of tropical maize during vegetative growth. *Agron. J.* **1997**, *89*, 557–562. [CrossRef]

41. Sadras, V.O.; Echarte, L.; Andrade, F.H. Profiles of leaf senescence during reproductive growth of sunflower and maize. *Ann Bot.* **2000**, *85*, 187–195. [CrossRef]

42. Peterson, T.A.; Blackmer, T.M.; Francis, D.D.; Schepers, J.S. *G93-1171 Using a Chlorophyll Meter to Improve N Management*; University of Nebraska: Lincoln, NE, USA, 1993.

43. Bremner, J.M.; Mulvaney, C.S. Nitrogen-total. In *Methods of Soil Analysis. Part 2. Chemical and Microbiological Properties*; Page, A.L., Miller, R.H., Keeney, D.R., Eds.; American Society of Agronomy, Soil Science Society of America: Madison, WI, USA, 1982; pp. 595–624.

44. Miller, R.O.; Gavlak, R.; Horneck, D. Soil Nitrate N, NO_3-N Cadmium Reduction S-3.10. In *Soil, Plant, and Water Reference Methods for the Western Region*, 4th ed.; Western Rural Development Center: UT, USA, 2013; p. 39.

45. Miller, R.O.; Gavlak, R.; Horneck, D. Soil Ammonium Nitrogen S-3.50. In *Soil, Plant, and Water Reference Methods for the Western Region*, 4th ed.; Western Rural Development Center: UT, USA, 2013; p. 43.

46. Weiske, A.; Benckiser, G.; Herbert, T.; Ottow, J.C.G. Influence of the nitrification inhibitor 3,4-dimethylpyrazole phosphate (DMPP) in comparison to dicyandiamide (DCD) on nitrous oxide emissions, carbon dioxide fluxes and methane oxidation during 3 years of repeated application in field experiments. *Biol. Fertil. Soils* **2001**, *34*, 109–117.

47. De Antoni Migliorati, M.; Scheer, C.; Grace, P.R.; Rowlings, D.W.; Bell, M.; McGree, J. Influence of different nitrogen rates and DMPP nitrification inhibitor on annual N2O emissions from a subtropical wheat-maize cropping system. *Agric. Ecosyst. Environ.* **2014**, *186*, 33–43. [CrossRef]

48. Yuan, L.; Chen, X.; Jia, J.; Chen, H.; Shi, Y.; Ma, J.; Liang, C.; Liu, Y.; Xie, H.; He, H.; et al. Stover mulching and inhibitor application maintain crop yield and decrease fertilizer N input and losses in no-till cropping systems in Northeast China. *Agric. Ecosyst. Environ.* **2021**, *312*, 107360. [CrossRef]

49. Perez-Castillo, A.G.; Chinchilla-Soto, C.; Elizondo-Salazar, J.A.; Barboza, R.; Dong-Gill, K.I.M.; Muller, C.; Alberto, S.C.; Borzouei, A.; Dawar, K.; Zaman, M. Nitrification inhibitor nitrapyrin does not affect yield-scaled nitrous oxide emissions in a tropical grassland. *Pedosphere* **2021**, *31*, 265–278. [CrossRef]

50. Merino, P.; Menéndez, S.; Pinto, M.; González-Murua, C.; Estavillo, J.M. 3,4-Dimethylpyrazole phosphate reduces nitrous oxide emissions from grassland after slurry application. *Soil Use Manage* **2005**, *21*, 53–57. [CrossRef]

51. Dougherty, W.J.; Collins, D.; Van Zwieten, L.; Rowlings, D.W. Nitrification (DMPP) and urease (NBPI) inhibitors had no effect on pasture yield, nitrous oxide emissions, or nitrate leaching under irrigation in a hot-dry climate. *Soil Res.* **2016**, *54*, 675. [CrossRef]

52. Cardenas, L.M.; Bhogal, A.; Chadwick, D.R.; McGeough, K.; Misselbrook, T.; Rees, R.M.; Thorman, R.E.; Watson, C.J.; Williams, J.R.; Smith, K.A.; et al. Nitrogen use efficiency and nitrous oxide emissions from five UK fertilised grasslands. *Sci. Total Environ.* **2019**, *661*, 696–710. [CrossRef] [PubMed]

53. Nauer, P.A.; Fest, B.J.; Visser, L.; Arndt, S.K. On-farm trial on the effectiveness of the nitrification inhibitor DMPP indicates no benefits under commercial Australian farming practices. *Agric. Ecosyst. Environ.* **2018**, *253*, 82–89. [CrossRef]

54. Asing, J.; Saggar, S.; Singh, J.; Bolan, N.S. Assessment of nitrogen losses from urea and organic manure with and without nitrification inhibitor, dicyandiamide, applied to lettuce under glasshouse conditions. *Aust. J. Soil Res.* **2008**, *46*, 535–541. [CrossRef]

55. Pfab, H.; Palmer, I.; Buegger, F.; Fiedler, S.; Müller, T.; Ruser, R. Influence of a nitrification inhibitor and of placed N-fertilization on N2O fluxes from a vegetable cropped loamy soil. *Agric. Ecosyst. Environ.* **2012**, *150*, 91–101. [CrossRef]

56. Tao, R.; Li, J.; Hu, B.; Shah, J.A.; Chu, G. A 2-year study of the impact of reduced nitrogen application combined with double inhibitors on soil nitrogen transformation and wheat productivity under drip irrigation. *J. Sci. Food Agric.* **2021**, *101*, 1772–1781. [CrossRef]

57. Zheng, J.; Wang, H.; Fan, J.; Zhang, F.; Guo, J.; Liao, Z.; Zhuang, Q. Wheat straw mulching with nitrification inhibitor application improves grain yield and economic benefit while mitigating gaseous emissions from a dryland maize field in northwest China. *Field Crops Res.* **2021**, *265*, 108125. [CrossRef]

58. Borzouei, A.; Mander, U.; Teemusk, A.; Alberto, S.C.; Zaman, M.; Dong-Gill, K.I.M.; Muller, C.; Kelestanie, A.A.; Amin, P.S.; Moghiseh, E.; et al. Effects of the nitrification inhibitor nitrapyrin and tillage practices on yield-scaled nitrous oxide emission from a maize field in Iran. *Pedosphere* **2021**, *31*, 314–322. [CrossRef]

59. Dawar, K.; Sardar, K.; Zaman, M.; Mueller, C.; Alberto, S.C.; Aamir, K.; Borzouei, A.; Perez-Castillo, A.G. Effects of the nitrification inhibitor nitrapyrin and the plant growth regulator gibberellic acid on yield-scale nitrous oxide emission in maize fields under hot climatic conditions. *Pedosphere* **2021**, *31*, 323–331. [CrossRef]

60. Zhang, M.; Fan, C.H.; Li, Q.L.; Li, B.; Zhu, Y.Y.; Xiong, Z.Q. A 2-yr field assessment of the effects of chemical and biological nitrification inhibitors on nitrous oxide emissions and nitrogen use efficiency in an intensively managed vegetable cropping system. *Agric. Ecosyst. Environ.* **2015**, *201*, 43–50. [CrossRef]

61. Abalos, D.; Jeffery, S.; Sanz-Cobena, A.; Guardia, G.; Vallejo, A. Meta-analysis of the effect of urease and nitrification inhibitors on crop productivity and nitrogen use efficiency. *Agric. Ecosyst. Environ.* **2014**, *189*, 136–144. [CrossRef]

62. Zhu, Z.L. Fate and management of fertilizer nitrogen in agro-ecosystems. In *Nitrogen in Soils of China*; Springer: Dordrecht, Germany, 1997; pp. 239–279.

63. Cai, J.; Jiang, D.; Liu, F.; Dai, T.; Cao, W. Effects of split nitrogen fertilization on post-anthesis photoassimilates, nitrogen use efficiency and grain yield in malting barley. *Acta Agric. Scand B Soil Plant Sci.* **2011**, *61*, 410–420. [CrossRef]

64. Rahman, M.A.; Sarker, M.A.Z.; Amin, M.F.; Jahan, A.H.S.; Akhter, M.M. Yield response and nitrogen use efficiency of wheat under different doses and split application of nitrogen fertilizer. *Bangladesh J. Agric. Res.* **2011**, *36*, 231–240. [CrossRef]

65. Chen, Y.; Peng, J.; Wang, J.; Fu, P.; Hou, Y.; Zhang, C.; Fahad, S.; Peng, S.; Cui, K.; Nie, L.; et al. Crop management based on multi-split topdressing enhances grain yield and nitrogen use efficiency in irrigated rice in China. *Field Crops Res.* **2015**, *184*, 50–57. [CrossRef]

66. Zhou, Z.; Plauborg, F.; Liu, F.; Kristensen, K.; Andersen, M.N. Yield and crop growth of table potato affected by different split-N fertigation regimes in sandy soil. *Eur. J. Agron.* **2018**, *92*, 41–50. [CrossRef]

67. Li, H.; Chen, Y.X.; Liang, X.Q.; Lian, Y.F.; Li, W.H. Mineral-nitrogen leaching and ammonia volatilization from a rice–rapeseed system as affected by 3,4-dimethylpyrazole phosphate. *J. Environ. Qual.* **2009**, *38*, 2131–2137. [CrossRef]

68. Coskun, D.; Britto, D.T.; Shi, W.M.; Kronzucker, H.J. Nitrogen transformations in modern agriculture and the role of biological nitrification inhibition. *Nat. Plants* **2017**, *3*, 17074. [CrossRef]

69. Qiao, C.; Liu, L.; Hu, S.; Compton, J.E.; Greaver, T.L.; Li, Q. How inhibiting nitrification affects nitrogen cycle and reduces environmental impacts of anthropogenic nitrogen input. *Glob. Chang Biol.* **2015**, *21*, 1249–1257. [CrossRef]

70. Pereira, J.; Fangueiro, D.; Chadwick, D.R.; Misselbrook, T.H.; Coutinho, J.; Trindade, H. Effect of cattle slurry pre-treatment by separation and addition of nitrification inhibitors on gaseous emissions and N dynamics: A laboratory study. *Chemosphere* **2010**, *79*, 620–627. [CrossRef]

71. Ruser, R.; Schulz, R. The effect of nitrification inhibitors on the nitrous oxide (N2O) release from agricultural soils-a review. *J. Plant. Nutr. Soil Sci.* **2015**, *178*, 171–188. [CrossRef]

72. Singh, A.; Kumar, A.; Jaswal, A.; Singh, M.; Gaikwad, D. Nutrient use efficiency concept and interventions for improving nitrogen use efficiency. *Plant Arch.* **2018**, *18*, 1015–1023.

73. Li, T.; Zhang, X.; Gao, H.; Li, B.; Wang, H.; Yan, Q.; Ollenburger, M.; Zhang, W. Exploring optimal nitrogen management practices within site-specific ecological and socioeconomic conditions. *J. Clean Prod.* **2019**, *241*, 118295. [CrossRef]

74. Janke, C.K.; Moody, P.; Bell, M.J. Three-dimensional dynamics of nitrogen from banded enhanced efficiency fertilizers. *Nutr. Cycl. Agroecosyst.* **2020**, *118*, 227–247. [CrossRef]

75. Souza, E.F.; Soratto, R.P.; Sandaña, P.; Venterea, R.T.; Rosen, C.J. Split application of stabilized ammonium nitrate improved potato yield and nitrogen-use efficiency with reduced application rate in tropical sandy soils. *Field Crop. Res.* **2020**, *254*, 107847. [CrossRef]

76. Nair, D.; Abalos, D.; Philippot, L.; Bru, D.; Mateo-Marín, N.; Petersen, S.O. Soil and temperature effects on nitrification and denitrification modified N2O mitigation by 3, 4-dimethylpyrazole phosphate. *Soil Biol. Biochem.* **2021**, *157*, 108224. [CrossRef]

77. Slangen, J.H.G.; Kerkhoff, P. Nitrification inhibitors in agriculture and horticulture: A literature review. *Fertil. Res.* **1984**, *5*, 1–76. [CrossRef]

78. Raza, S.; Jiang, Y.; Elrys, A.S.; Tao, J.; Liu, Z.; Li, Z.; Chen, Z.; Zhou, J. Dicyandiamide efficacy of inhibiting nitrification and carbon dioxide emission from calcareous soil depends on temperature and moisture contents. *Arch. Agron. Soil Sci.* **2021**, *68*, 1–17. [CrossRef]

79. Pasda, G.; Hähndel, R.; Zerulla, W. Effect of fertilizers with the new nitrification inhibitor DMPP (3, 4-dimethylpyrazole phosphate) on yield and quality of agricultural and horticultural crops. *Biol. Fertil. Soils.* **2001**, *34*, 85–97. [CrossRef]

80. Fisk, L.M.; Maccarone, L.D.; Barton, L.; Murphy, D.V. Nitrapyrin decreased nitrification of nitrogen released from soil organic matter but not amoA gene abundance at high soil temperature. *Soil Biol. Biochem.* **2015**, *88*, 214–223. [CrossRef]

Article

Early and Late Season Nutrient Stress Conditions: Impact on Cotton Productivity and Quality

Solomon Amissah [1], Michael Baidoo [2], Benjamin K. Agyei [1,3], Godfred Ankomah [1], Roger A. Black [4], Calvin D. Perry [5], Stephanie Hollifield [6], Nana Yaw Kusi [7], Glendon H. Harris [1] and Henry Y. Sintim [1,*]

[1] Department of Crop & Soil Sciences, University of Georgia, Tifton, GA 31793, USA
[2] Vision Research Park, United Agronomy, Berthold, ND 58718, USA
[3] Department of Plant, Soil, and Microbial Sciences, Michigan State University, East Lansing, MI 48824, USA
[4] Southeast Georgia Research and Education Center, University of Georgia, Midville, GA 30441, USA
[5] The C. M. Stripling Irrigation Research Park, University of Georgia, Camilla, GA 31730, USA
[6] Southwest District Extension, University of Georgia, Tifton, GA 31794, USA
[7] Research and Extension, South Carolina State University, Orangeburg, SC 29117, USA
*   Correspondence: hsintim@uga.edu

Abstract: Modern cotton (*Gossypium* spp. L) cultivars are efficient in nutrient uptake and utilization, and thus, may potentially tolerate nutrient stress. Early- and late-season nutrient stress (E-stress and L-stress, respectively) effects on cotton productivity and quality were assessed under different production conditions in Camilla and Midville, GA, USA. The E-stress received no nutrient application in the early season, but the full rates were split-applied equally at the initiation of squares and the second week of bloom stages. The L-stress received 30–40% of the full nutrient rates only at the initial stage of planting. The effects of nutrient stress on cotton productivity and fiber quality were not consistent across the different production conditions. Compared to the full nutrient rate, the E-stress did not adversely impact cotton yield, but rather it improved the lint and cottonseed yields under one production condition by 17.5% and 19.3%, respectively. Averaged across all production conditions, the L-stress decreased the lint and cottonseed yields by 34.4% and 36.2%, respectively. The minimal effects of E-stress on cotton suggest nutrient rates at the early season could be reduced and more tailored rates, informed by soil and plant tissue analyses, applied shortly before the reproductive phase.

Keywords: nutrient stress; cotton production; modern cultivars; biomass accumulation; fiber quality

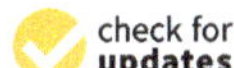

Citation: Amissah, S.; Baidoo, M.; Agyei, B.K.; Ankomah, G.; Black, R.A.; Perry, C.D.; Hollifield, S.; Kusi, N.Y.; Harris, G.H.; Sintim, H.Y. Early and Late Season Nutrient Stress Conditions: Impact on Cotton Productivity and Quality. *Agronomy* **2023**, *13*, 64. https://doi.org/10.3390/agronomy13010064

Academic Editors: Shahram Torabian, Ruijun Qin and Christos Noulas

Received: 28 November 2022
Revised: 20 December 2022
Accepted: 21 December 2022
Published: 25 December 2022

1. Introduction

Cotton (*Gossypium* spp. L) is a valuable industrial crop that contributes substantially to the agricultural economy of many countries. The Food and Agriculture Organization of the United Nations estimated the world's seed cotton (unginned cotton) production in 2020 to be 83.1 million tons, which was valued at $52 billion [1]. The USA ranks third in cotton production in the world, behind China and India. In 2021, about 4.54 million ha of cotton was planted in the USA, and the top three leading producing states were Texas, Georgia, and Arkansas in descending order [2]. Lint and cottonseed are the two valuable industrial products of cotton, with lint being more valuable. In 2021, lint and cottonseed production value in the USA was $7.46 billion and $1.32 billion, respectively. Thus, cotton production management is mainly geared towards enhancing the productivity and quality of lint.

The average cotton lint yield in the USA has increased by 25.6% in the past 30 years, with a yield of 0.73 Mg ha^{-1} in 1991 and 0.92 Mg ha^{-1} in 2021 [2]. The increase in cotton lint yield can largely be attributable to improved agronomic practices and better performance of modern cultivars [3–5]. Rochester and Constable [6] compared cotton cultivars released in 2006 with those released in 1973. The authors observed a 40% increase in lint yield in the 2006 cultivars. In addition, the N, P, and K use efficiencies of the 2006 cultivars increased by 20%, 23%, and 24%, respectively, when compared to those of the 1973 cultivars [6]. In a two-year field study in New Deal, TX, the average lint yields of cotton cultivar PM HS26 (released in 1990), FM 958 (released in 2000), and DP 1646 (released in 2016) were reported to be 1.11 Mg ha^{-1}, 1.26 Mg ha^{-1}, and 1.34 Mg ha^{-1}, respectively [7]. The authors also observed that the 2016 and 2000 cultivars efficiently partitioned and remobilized essential nutrients more than the 1990 cultivar.

The efficient utilization of nutrients by modern cultivars could potentially confer better tolerance to nutrient stress, which would be desirable. There has been instability in the supply and prices of fertilizers over the past couple of years, with most fertilizers exceeding record prices in 2008. Lessons from the 2008 volatility in fertilizer prices suggest that farmers may not be willing to buy the usual tonnage of fertilizers at high price levels [8]. It is therefore important to determine the impact of reduced fertilizer application rates on the productivity and quality of modern cotton varieties. The high fertilizer prices in 2008 may have contributed to the reduced cotton lint yield in succeeding years. The average cotton lint yield in the USA in 2007 was 0.88 Mg ha^{-1}, but it dropped to 0.81 Mg ha^{-1} and 0.78 Mg ha^{-1}, respectively, in 2008 and 2009 [2].

While low fertilizer application could impact cotton productivity, supplying more nutrients than needed could have adverse implications on the environment, such as acidification of soils, eutrophication in aquatic systems, and ozone layer depletion [9,10]. Nutrient uptake and partition studies show that nutrient requirement in cotton is minimal at the vegetative stage and then increases rapidly at the reproductive stage [7,11,12]. Synchronizing nutrient availability with crop demand could potentially increase nutrient use efficiency and reduce nutrient loss through ammonia volatilization, denitrification, runoff, and leaching [13–15]. However, standard nutrient management guidelines suggest the application of all recommended fertilizer rates before or at the initial stages of planting, except for N which is often split-applied. The minimal vegetation cover, coupled with high rainfall and temperature conditions, make fertilizers applied in the early season susceptible to losses.

Determining the response of modern cotton cultivars to no fertilizer application during the early season growth could inform adaptive nutrient management strategies. Residual nutrients in the soil and crop residues could meet the nutritional demand for cotton at the early season growth stage [16–18]. The objective of this study was therefore to assess the impact on the productivity and quality of modern cotton varieties to varying degrees of nutrient stresses under different production conditions.

2. Materials and Methods

2.1. Experimental Site

The research was conducted at the University of Georgia Stripling Irrigation Research Park in Camilla, GA (31°16′45.86″ N, 84°17′29.65″ W) and the Southeast Georgia Research and Education Center in Midville, GA (32°52′54.72″ N, 82°12′54.07″ W). Both sites have a humid subtropical climate, with annual average daily minimum, mean, and maximum air temperatures of 12.8 °C, 19.4 °C, and 26.0 °C, respectively, in Camilla and 11.3 °C, 18.0 °C, and 24.6 °C, respectively, in Midville [19]. The average annual precipitation in Camilla is 1314 mm, with 98 average rainy days, and the average annual precipitation in Midville is 1146 mm, with 102 average rainy days [19].

Air temperature over the two years of the study followed a similar pattern across the two locations, but it was relatively warmer in Camilla compared to Midville, with the average minimum, mean, and maximum air temperatures of 13.7 °C, 19.9 °C, and 26.1 °C, respectively, in Camilla, and 12.4 °C, 18.5 °C, and 24.6 °C, respectively, in Midville (Figure 1). In addition, rainfall received was relatively greater in Camilla, with an annual rainfall of 1378 mm in 2020 and 1384 mm in 2021. Annual rainfall in Midville was 1318 mm in 2020 and 1099 in 2021. Rainfall received between the planting and harvest of cotton was 547 mm and 776 mm in 2020 and 2021, respectively, in Camilla, and 482 mm and 532 mm, respectively, in 2020 and 2021 in Midville.

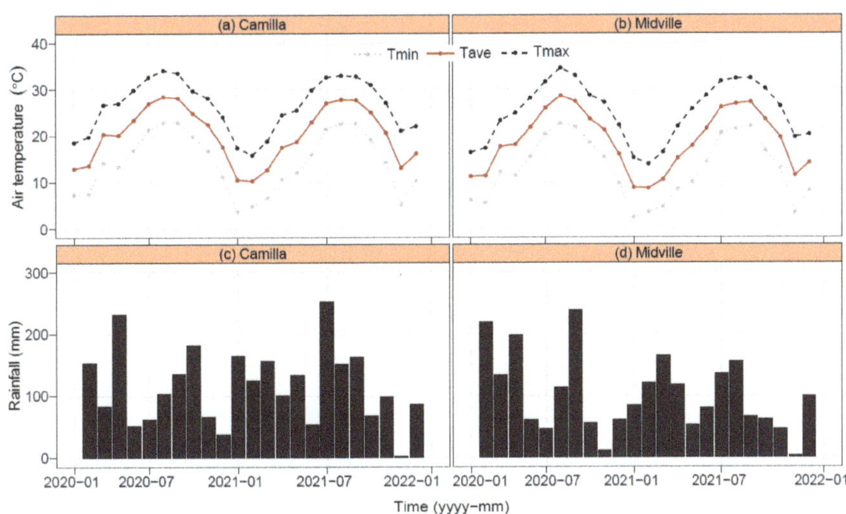

Figure 1. (**a**,**b**) Monthly minimum (Tmin), mean (Tave), and maximum (Tmax) air temperature, as well as (**c**,**d**) monthly total precipitation in Camilla and Midville, GA from 1 January 2020 to 31 December 2021.

The experimental field in Camilla had a Lucy loamy sand, classified as Loamy, kaolinitic, thermic Arenic Kandiudults, whereas the field in Midville had a Dothan loamy sand, classified as fine-loamy, kaolinitic, thermic Plinthic Kandiudults [20]. The average sand, silt, and clay content at the top 15 cm depth of the experimental field soils were 90.7%, 3.2%, and 6.1%, respectively, in Camilla, and 90.9%, 5.1%, and 4.0%, respectively, in Midville. In addition, the average soil pH and organic matter within 0–15 cm depth were 6.51 and 2.90 g kg^{-1} in Camilla, respectively, and 6.40 and 4.67 g kg^{-1} in Midville, respectively.

2.2. Field Experiment

Field experiments were established in 2020 and 2021 to evaluate early- and late-season nutrient stress (E-stress and L-stress) effects on cotton productivity and quality under different production conditions at the two locations. A reduced nutrient stress condition (R-stress) and standard fertility constituted the control treatments. In Camilla, the experiment was established under sub-surface drip irrigation (SSDI) systems in 2020 and 2021, and under an overhead irrigation system in only 2021. In Midville, the experiment was established under rainfed conditions in 2020 and 2021, and under an overhead irrigation system in only 2021, constituting six production conditions across the two locations. The two production conditions at both locations in 2021 were on separate fields (~100 m apart in Camilla and ~300 apart in Midville). The four treatments were assessed under each production condition, except for the Midville 2020 rainfed condition where the standard fertility was not assessed.

The standard fertility treatment referred to nutrient recommendations (Table 1) by the University of Georgia Agricultural and Environmental Services Laboratories to make 1681 kg ha^{-1} lint yield under irrigated conditions and 1121 kg ha^{-1} lint yield under rainfed conditions [21]. The rates were based on the initial soil nutrient status of the experimental fields (Table 2). Nutrients reported as essential for cotton are N, P, K, Ca, Mg, S, Fe, Mn, Zn, B, and Cu [22]. To ensure minimal nutrient stress, the application of some rates of all the essential nutrients was made to the R-stress plots (Table 1). In Camilla, the R-stress plots received 30% of the full nutrient rates at the early stage of planting, another 30% each at square initiation and the second week of bloom (2-WoB) stages, and the remaining 10% at the 6-WoB stage. In Midville, the R-stress plots received 40% of the full nutrient rates at the early stage of planting, and 30% each at square initiation and the 2-WoB stages.

Table 1. Full nutrient application rates (in kg ha^{-1}) for the standard recommendation and reduced nutrient stress (R-stress) treatments imposed in Camilla and Midville, GA, under different production conditions.

Nutrient Elements	SSDI (2020)		Camilla SSDI (2021)		Overhead (2021)		Rainfed (2020)	Midville Rainfed (2021)		Overhead (2021)	
	Standard	R-Stress	Standard	R-Stress	Standard	R-Stress	R-Stress	Standard	R-Stress	Standard	R-Stress
N	84.1	118	106	84.1	118	106	67.3	50.4	78.5	84.1	118
P$_2$O$_5$	0.00	101	44.8	0.00	101	44.8	56.0	33.6	67.3	78.5	101
K$_2$O	112	168	101	112	168	101	112	33.6	84.1	135	140
Mg	0.00	33.6	0.00	0.00	33.6	0.00	28.0	0.00	5.60	0.00	5.60
Ca	0.00	5.60	0.00	0.00	5.60	0.00	5.60	0.00	11.2	0.00	22.4
S	11.2	22.4	11.2	11.2	22.4	11.2	11.2	11.2	11.2	11.2	11.2
B	0.56	2.24	0.56	0.56	2.24	0.56	2.24	0.56	0.56	0.56	1.12
Zn	0.00	2.24	0.00	0.00	2.24	0.00	2.24	0.00	1.12	0.00	2.24
Mn	0.00	1.12	0.00	0.00	1.12	0.00	1.12	0.00	2.24	11.2	5.60
Fe	0.00	1.12	0.00	0.00	1.12	0.00	0.56	0.00	2.24	0.00	2.24
Cu	0.00	0.56	0.00	0.00	0.56	0.00	0.56	0.00	0.56	0.00	0.56

SSDI: Sub-surface drip irrigation; R-Stress: reduced nutrient stress.

Table 2. Initial nutrient status of the experimental field soil in Camilla and Midville under different production conditions.

Soil Depth	N	P	K	Mg	Ca	S	B	Zn	Mn	Fe	Cu
						kg ha^{-1}					
						Camilla					
SSDI (2020)	8.47	82.3	92.0	62.2	955	5.04	0.22	4.6	28.2	36	0.9
SSDI (2021)	2.69	81.6	107	127	907	28	0.45	6.05	13.5	10.1	1.01
Overhead (2021)	0.90	49.3	58.8	95.0	762	28.1	0.45	6.15	15.1	12.3	2.13
						Midville					
Rainfed (2020)	3.49	75.7	81.6	67.7	971	32.4	0.22	5.62	20.2	37.1	0.67
Rainfed (2021)	2.73	68.4	184	166	864	37.9	0.67	7.64	15.6	17.2	0.45
Overhead (2021)	6.97	51.6	71.2	91	716	31.2	0.45	5.9	6.39	19.3	1.01

Soil samples were collected from 0–15 cm depth with a 2.86 cm diameter AMS soil recovery probe (AMS Inc., American Falls, ID, USA). N was measured as Nitrate-N after extraction with 2 M KCl solution, whereas P, K, Ca, Mg, S, Fe, Mn, Zn, B, and Cu were measured after Mehlich I extraction. SSDI: Sub-surface drip irrigation.

The E-stress at both locations was induced by not making any nutrient application until the initiation of squares, after which the full nutrient rates specified for the R-stress were split-applied equally at the initiation of squares and the 2-WoB stages. Thus, the E-stress received the same nutrient application rates as the R-stress. The L-stress in Camilla was induced by applying only 40% of the full nutrient rates specified for the R-stress at the early stage of planting, whereas the L-stress in Midville was induced by applying only 30% of the full nutrient rates specified for the R-stress at the early stage of planting. Table S1 lists the nutrient sources applied. Granular fertilizer sources were used at both locations, except the Camilla SSDI conditions where liquid fertilizer sources were applied via fertigation at the 2-WoB and 6-WoB stages. The treatments were laid out in a randomized complete block design with four replications and plot dimensions (width × length) of 5.5 m × 11.0 m for the Camilla SSDI condition, 7.3 m × 12.2 m for the Camilla overhead irrigation condition, and 7.3 m × 9.1 m for all conditions in Midville.

2.3. Plot Management

Previous cash crops were peanut (*Arachis hypogaea*) and corn (*Zea mays*) for the 2020 and 2021 seasons, respectively, in Camilla and peanut for all seasons in Midville. All sites were under cereal rye (*Secale cereale*) cover crop, and the fields were prepared by strip-tilling to 30.5–45.7 cm depth. Deltapine® cotton variety DP 1646 B2XF (released in 2016) was used in Camilla and Stoneville® cotton variety ST 4550 GLTP (released in 2019) was used in Midville, and they were planted at 107,639 seeds ha^{-1} and 91.4 cm row spacing. The

SSDI system constituted a Netafim Typhoon drip tape (Netafim Irrigation, Inc., Fresno, CA, USA), with 46 cm emitter spacing and 1.5 L h^{-1} discharge rate, installed at the middle of every row (46 cm to the side of plant rows). The drip tapes at every other middle of the plant rows were used for irrigation in this study (one drip tape line serviced two plant rows), as per common grower practice. The overhead irrigation was a lateral irrigation system in Camilla and a center pivot irrigation system in Midville. Irrigation amounts were 200 mm (Camilla 2020 SSDI), 132 mm (Camilla 2021 SSDI), 184 mm (Camilla 2021 overhead), and 95.3 mm (Midville 2021 overhead), which depended on rainfall, location, and irrigation method. Weed and pest control and the use of growth regulators and defoliants followed standard recommendations by the University of Georgia Cooperative Extension [23,24].

2.4. Data Collection

Initial soil nutrient levels within the top 15 cm depth were analyzed following standard protocols by Waters Agricultural Laboratories, Inc. The soil samples were collected with a 2.86 cm diameter AMS soil recovery probe (AMS Inc., American Falls, ID, USA). Nitrate-N was measured, after extraction in a 2 M KCl solution, with the automated flow injection analysis system (FIAlyzer-1000, FIAlab Instruments, Inc., Seattle, WA, USA), and extractable P, K, Ca, Mg, S, Fe, Mn, Zn, B, and Cu were measured, after extraction in Mehlich I solution, with an inductively coupled plasma optical emission spectrophotometer (ICP-OES; iCAP™ 6000 Series, Thermo Fisher Scientific, Cambridge, United Kingdom).

Aboveground plant tissues were sampled at square initiation, 2-WoB, and 7-WoB stages, and shortly before harvest (after defoliation), except for the Midville location where samples were not collected at the 7-WoB stage. At every sampling stage, the aboveground biomass was collected from a 1 m strip of a non-harvest row, ensuring a minimum of 1 m buffer during subsequent sampling. The samples were oven-dried at 78 °C to obtain constant weight, after which the weights were recorded and used to calculate biomass accumulation. Plant height, the number of main stem nodes per plant, the total number of bolls per plant, the number of harvestable bolls per plant, and seed cotton per boll were determined at physiological maturity from five plants selected randomly within non-harvest rows of each plot.

Harvesting was performed mechanically by sampling two entire rows of every plot with a cotton picker, and weights of the seed cotton were measured. Thereafter, the seed cotton samples were ginned at the University of Georgia Micro Gin in Tifton, GA to determine the gin turnout, which was used to calculate the lint and cottonseed yields. Fiber samples were transported to the USDA classing office in Macon, GA to measure fiber quality parameters, including fiber length, fiber strength, uniformity, micronaire, reflectance (RD), and yellowness (+b), following standard protocol [25].

2.5. Statistical Analyses

Separate statistical analyses were performed for each location because of differences in the level of nutrient stress imposed. Plant growth, yield, and fiber quality data, except for the biomass data, were analyzed with the linear mixed model using the "lme4" package in R [26]. The nutrient stress and production conditions were considered fixed effects and block was considered a random effect. The biomass data were analyzed as repeated measure analyses, also using the "lme4" package in R [26]. The sampling time was assigned as a within-plot factor variable, nutrient stress as between plot factor variable, and block as a random term.

Normality of residuals, homoscedasticity of variance, and sphericity assumptions were tested, and appropriate transformations (square root and Box–Cox transformation methods) and corrections (Greenhouse–Geisser and Huynh–Feldt correction methods) were applied as appropriate. Mean separations were performed using the least square means and the adjusted Tukey multiple comparison procedure with the 'emmeans' package in R [27]. The significance level for all analyses was assessed at $p = 0.05$.

3. Results

3.1. Lint Yield, Gin Turnout, and Seed Cotton

The main effects of nutrient stress were significant on cotton lint yield, gin turnout, cottonseed yield, and the seed cotton weight per boll in Camilla, but the effects were only significant on lint yield and cottonseed yield in Midville (Table S2). In addition, the interaction effects of nutrient stress and production conditions were significant on lint yield, cottonseed yield, and seed cotton weight per boll in Camilla, but their effects were not significant on those variables in Midville. Compared to R-stress, the E-stress did not cause a significant reduction in lint yield, but rather, it significantly increased the lint yield by 17.5% under Camilla 2021 SSDI condition (Figure 2a,b). In contrast, the L-stress led to a significant reduction in lint yield in four production conditions when compared to the R-stress. The yield reductions were 41.7%, 33.3%, 69.4%, and 45.4%, respectively, under Camilla 2021 SSDI, Camilla 2021 overhead irrigation, Midville 2020 rainfed, and Midville 2021 rainfed conditions. Compared to R-stress, the standard fertility underperformed but the differences were not significant, except under Camilla 2021 overhead condition (20.6% lower lint yield). Averaged over all production conditions across the two locations, the lint yield was 1.29 Mg ha^{-1}, 0.97 Mg ha^{-1}, 1.30 Mg ha^{-1}, and 1.15 Mg ha^{-1} for the E-stress, L-stress, R-stress, and the standard fertility, respectively.

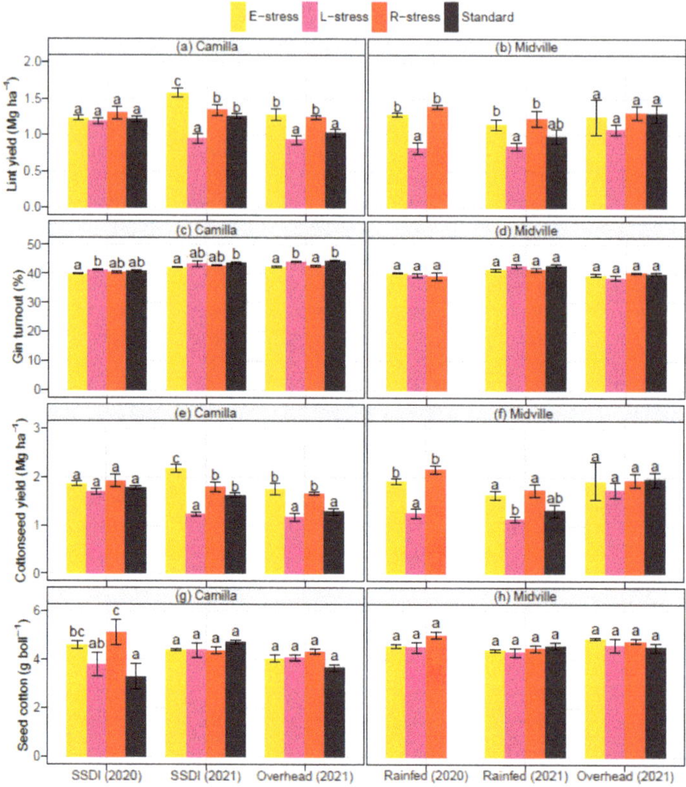

Figure 2. Nutrient stress effects on lint yield (**a,b**), gin turnout (**c,d**), cottonseed yield (**e,f**), and seed cotton yield per boll (**g,h**) in Camilla and Midville, GA under different production conditions. Within location and production conditions, means of nutrient stress treatments not sharing any letter are significantly different using the least squares means and adjusted Tukey multiple comparison procedure ($p < 0.05$). Error bars indicate the standard error of the mean (n = 4). SSDI: Sub-surface drip irrigation.

While the effects of nutrient stress on gin turnout were statistically significant under all production conditions in Camilla, the magnitude of the differences was small, with gin turnout ranging from 39.9% to 41.3% under Camilla 2020 SSDI, 42.2% to 43.7% under Camilla 2021 SSDI, and 42.3% to 44.5% under Camilla 2021 overhead irrigation (Figure 2c,d). Thus, the effects of nutrient stress on cottonseed yield (Figure 2e,f) followed the same trend as the effects on lint yield. Averaged over all production conditions across the two locations, the cottonseed yield was 1.87 Mg ha^{-1}, 1.37 Mg ha^{-1}, 1.86 Mg ha^{-1}, and 1.59 Mg ha^{-1}, respectively, for the E-stress, L-stress, R-stress, and the standard fertility. The seed cotton weight per boll, however, did not follow the same trend as the lint and cottonseed yields, with significant differences between nutrient treatments observed under Camilla 2020 SSDI condition only (Figure 2g,h). The seed cotton weight per boll under Camilla 2020 SSDI condition was least in the standard fertility (3.30 g kg^{-1}) and greatest in R-stress (5.12 g kg^{-1}).

3.2. Plant Height, Nodes, and Boll Development

Compared to R-stress, plant height was significantly reduced by L-stress under Camilla 2021 SSDI (25.0% reduction) and Midville 2021 rainfed (13.6% reduction) conditions (Figure 3a,b and Table S2). The effects of E-stress were minimal on plant height. The E-stress had a similar plant height as the R-stress and standard fertility. In addition, the number of main stem nodes was impacted by L-stress but not by E-stress (Figure 3c,d). A significant reduction in the number of main stem nodes occurred under the overhead irrigation conditions at both locations in 2021. The L-stress reduced the number of main stem nodes by 23.9% and 15.0%, respectively, under the overhead irrigation conditions in Camilla 2021 and Midville 2021. Compared to the E-stress and R-stress, the standard fertility had 21.3% and 17.7% lower number of main stem nodes, respectively, under Camilla 2021 overhead irrigation condition. The total (Figure 3e,f) and harvestable (Figure 3g,h) number of bolls were both significantly reduced by L-stress under Camilla 2021 SSDI conditions, but not under any production conditions in Midville. Compared to the R-stress, the reduction was very severe, 76.8% and 78.8% for the total and harvestable numbers of bolls, respectively. The E-stress, however, did not have a significant impact on the total and harvestable numbers of bolls, and also the R-stress had similar total and harvestable numbers of bolls as the standard fertility.

3.3. Biomass Accumulation

The effects of nutrient stress on biomass accumulation over time are shown in Table S3 and Figure 4. As already mentioned, biomass samples were collected at square initiation, 2-WoB, 7-WoB, and shortly before harvest, except for the Midville location where the samples were not collected at the 7-WoB stage. As expected, biomass accumulation significantly increased over the growth stages under all production conditions at the two locations. However, nutrient stress affected biomass accumulation at only the harvest stage. Significant differences were observed under all conditions in 2021 but not in 2020. Compared to R-stress, the E-stress did not affect biomass accumulation, whereas the L-stress and standard fertility significantly reduced biomass accumulation under four and one conditions, respectively. Averaged over all production conditions across the two locations, the E-stress, L-stress, R-stress, and standard fertility had total aboveground biomass of 11.7 Mg ha^{-1}, 8.82 Mg ha^{-1}, 12.2 Mg ha^{-1}, and 10.4 Mg ha^{-1}, respectively.

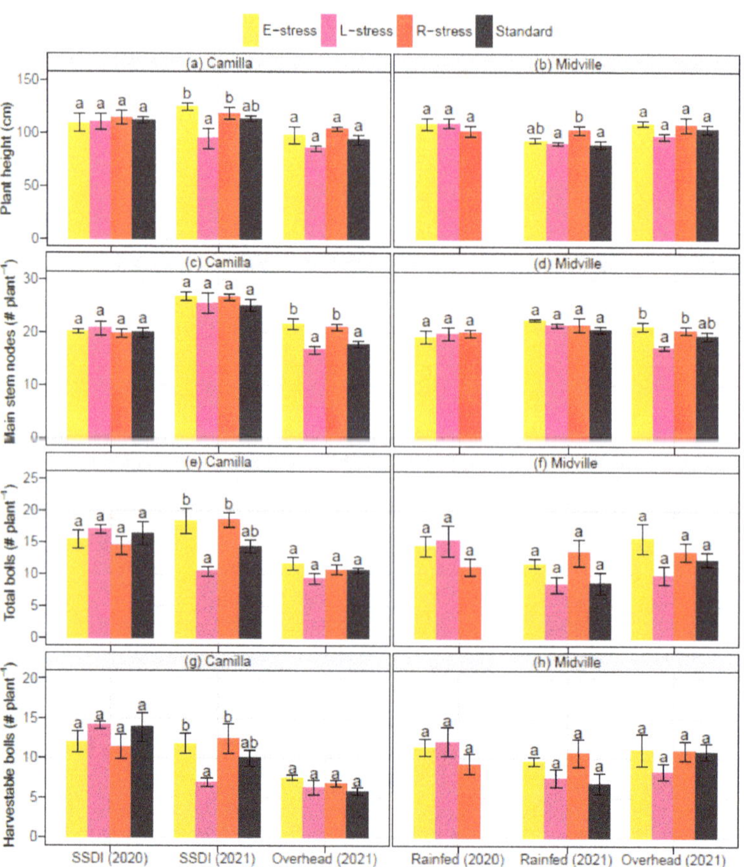

Figure 3. Nutrient stress effects on plant height (**a,b**), main stem nodes (**c,d**), and total (**e,f**) and harvestable (**g,h**) number of bolls in Camilla and Midville, GA under different production conditions. Within location and production conditions, means of nutrient stress treatments not sharing any letter are significantly different using the least squares means and adjusted Tukey multiple comparison procedure ($p < 0.05$). Error bars indicate the standard error of the mean (n = 4). SSDI: Sub-surface drip irrigation.

3.4. Fiber Quality

The effects of nutrient stress on fiber quality indicators across the different production conditions are shown in Table 3 and Table S4. Overall, the magnitude of the differences in fiber quality indicators among the nutrient treatments was minimal even though some of the test statistics were significant. Compared to the R-stress, the E-stress did not affect fiber length and strength, whereas the L-stress significantly reduced the fiber length and strength under Camilla 2021 SSDI condition only. The reduction was 4.23% and 2.02% for the fiber length and strength, respectively. Averaged over all production conditions across the two locations, the E-stress, L-stress, R-stress, and standard fertility had fiber lengths of 3.02 cm, 2.96 cm, 3.01 cm, and 2.97 cm, respectively, and a fiber strength of 31.1 g tex^{-1}, 30.6 g tex^{-1}, 30.9 g tex^{-1}, and 30.7 g tex^{-1}, respectively. The fiber uniformity was significantly increased in E-stress over the R-stress and standard fertility under Camilla 2021 overhead irrigation condition. It was also significantly increased in the E-stress over the R-stress and L-stress under the Midville 2021 rainfed condition. The E-stress, however, tended to decrease the micronaire, with an average reduction of 2.23% when compared to the R-stress. The average micronaire across all production conditions at the two locations was 4.47, 4.57, 4.57, and 4.55 for the E-stress, L-stress, R-stress, and standard fertility, respectively. The RD was not significantly affected by nutrient stress under any production condition. In contrast to the RD, the +b was significantly reduced in the L-stress, and the effect was more obvious under Camilla 2020 SSDI and Midville 2021 rainfed conditions. The average +b under all conditions was 7.60%, 7.39%, 7.55%, and 7.53% for the E-stress, L-stress, R-stress, and standard fertility, respectively.

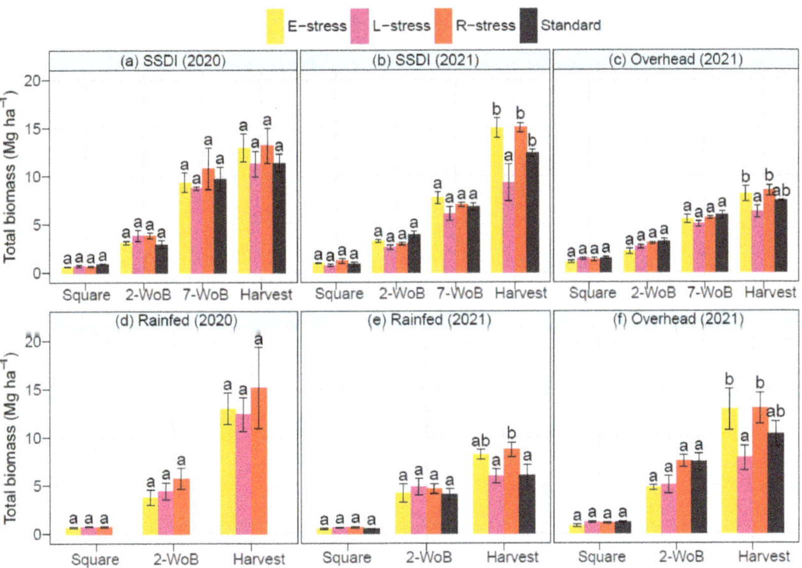

Figure 4. Nutrient stress effects on the total aboveground biomass in Camilla (**a–c**) and Midville (**d–f**), GA under different production conditions. Within production conditions and growth stage, means of nutrient stress treatments not sharing any letter are significantly different using the least squares means and adjusted Tukey multiple comparison procedure ($p < 0.05$). Error bars indicate the standard error of the mean (n = 4). SSDI: Sub-surface drip irrigation; WoB: Week of bloom.

Table 3. Nutrient stress effects on cotton fiber quality in Camilla and Midville under different production conditions.

Nutrient Stress	Fiber Length	Fiber Strength	Uniformity	Micronaire	RD	+b
	cm	g tex^{-1}	%		%	
Camilla (SSDI 2020)						
E-stress	3.20 ± 0.01 [a]	30.5 ± 0.3 [a]	81.9 ± 0.8 [a]	4.08 ± 0.07 [a]	74.9 ± 0.2 [a]	7.80 ± 0.07 [a]
L-stress	3.21 ± 0.01 [a]	30.5 ± 0.3 [a]	81.1 ± 0.1 [a]	4.32 ± 0.07 [b]	75.8 ± 0.2 [a]	7.42 ± 0.15 [b]
R-stress	3.19 ± 0.03 [a]	30.6 ± 0.1 [a]	81.5 ± 0.3 [a]	4.22 ± 0.07 [ab]	73.5 ± 0.4 [a]	7.58 ± 0.09 [b]
Standard	3.17 ± 0.04 [a]	30.1 ± 0.2 [a]	81.6 ± 0.8 [a]	4.20 ± 0.09 [ab]	74.3 ± 0.9 [a]	7.47 ± 0.12 [b]
Camilla (SSDI 2021)						
E-stress	3.10 ± 0.01 [b]	30.5 ± 0.6 [a]	82.3 ± 0.3 [a]	4.55 ± 0.03 [a]	74.0 ± 0.4 [a]	6.92 ± 0.07 [a]
L-stress	2.95 ± 0.02 [a]	29.7 ± 0.3 [ab]	81.6 ± 0.4 [a]	4.67 ± 0.03 [a]	75.9 ± 0.9 [a]	6.83 ± 0.23 [a]
R-stress	3.08 ± 0.02 [b]	30.3 ± 0.1 [a]	81.7 ± 0.2 [a]	4.65 ± 0.03 [a]	74.2 ± 0.3 [a]	6.85 ± 0.06 [a]
Standard	2.96 ± 0.02 [a]	29.3 ± 0.3 [b]	81.6 ± 0.2 [a]	4.60 ± 0.00 [a]	73.6 ± 0.2 [a]	7.03 ± 0.15 [a]

Table 3. *Cont.*

Nutrient Stress	Fiber Length	Fiber Strength	Uniformity	Micronaire	RD	+b
	cm	g tex^{-1}	%		%	
Camilla (Overhead 2021)						
E-stress	3.04 ± 0.01 [c]	27.8 ± 0.2 [a]	82.1 ± 0.2 [a]	4.60 ± 0.04 [ab]	74.7 ± 0.7 [a]	6.92 ± 0.14 [a]
L-stress	2.93 ± 0.02 [ab]	27.0 ± 0.4 [a]	81.7 ± 0.2 [ab]	4.58 ± 0.05 [ab]	74.3 ± 0.6 [a]	6.78 ± 0.05 [a]
R-stress	2.99 ± 0.02 [bc]	27.8 ± 0.3 [a]	81.4 ± 0.2 [b]	4.68 ± 0.03 [b]	73.6 ± 0.9 [a]	6.97 ± 0.12 [a]
Standard	2.90 ± 0.03 [a]	27.3 ± 0.3 [a]	81.3 ± 0.4 [b]	4.47 ± 0.06 [a]	74.7 ± 0.6 [a]	6.95 ± 0.06 [a]
Midville (Rainfed 2020)						
E-stress	2.90 ± 0.03 [a]	32.2 ± 0.4 [a]	83.3 ± 0.3 [a]	4.55 ± 0.12 [a]	77.8 ± 0.2 [a]	7.88 ± 0.22 [a]
L-stress	2.93 ± 0.04 [a]	32.5 ± 0.4 [a]	83.2 ± 0.4 [a]	4.65 ± 0.03 [a]	77.5 ± 0.1 [a]	7.85 ± 0.13 [a]
R-stress	2.98 ± 0.02 [a]	33.0 ± 0.8 [a]	83.5 ± 0.3 [a]	4.47 ± 0.07 [a]	77.7 ± 0.2 [a]	8.18 ± 0.12 [a]
Standard	na	na	na	na	na	na
Midville (Rainfed 2021)						
E-stress	2.94 ± 0.03 [b]	32.6 ± 0.6 [a]	83.3 ± 0.2 [a]	4.62 ± 0.13 [a]	76.7 ± 0.3 [a]	8.10 ± 0.09 [b]
L-stress	2.81 ± 0.04 [a]	30.9 ± 0.5 [a]	82.4 ± 0.2 [b]	4.70 ± 0.00 [a]	76.0 ± 0.2 [a]	7.55 ± 0.06 [a]
R-stress	2.89 ± 0.02 [ab]	31.9 ± 0.2 [a]	82.5 ± 0.3 [b]	4.70 ± 0.07 [a]	75.7 ± 0.1 [a]	7.90 ± 0.08 [ab]
Standard	2.88 ± 0.03 [ab]	31.9 ± 0.5 [a]	83.0 ± 0.4 [ab]	4.72 ± 0.05 [a]	76.2 ± 0.2 [a]	7.75 ± 0.18 [ab]
Midville (Overhead 2021)						
E-stress	2.94 ± 0.03 [a]	32.8 ± 0.2 [a]	83.0 ± 0.4 [a]	4.42 ± 0.09 [a]	77.4 ± 0.2 [a]	7.97 ± 0.23 [a]
L-stress	2.95 ± 0.03 [a]	33.2 ± 0.8 [a]	83.4 ± 0.2 [a]	4.53 ± 0.09 [ab]	77.8 ± 0.0 [a]	7.92 ± 0.25 [a]
R-stress	2.91 ± 0.02 [a]	32.0 ± 0.1 [a]	83.2 ± 0.3 [a]	4.70 ± 0.04 [ab]	78.1 ± 0.2 [a]	7.85 ± 0.09 [a]
Standard	2.92 ± 0.03 [a]	32.3 ± 0.4 [a]	83.5 ± 0.5 [a]	4.80 ± 0.17 [b]	77.6 ± 0.2 [a]	7.82 ± 0.03 [a]

Within location and production conditions, means of nutrient stress treatments not sharing any letter are significantly different using the least squares means and adjusted Tukey multiple comparison procedure ($p < 0.05$). Values represent the mean ± standard error. SSDI: Sub-surface drip irrigation; na: Not available; RD, fiber reflectance; +b, fiber yellowness.

4. Discussion

The residual soil nutrients, which were within the typical range [21], may have met the nutritional needs of the crop by the square stage, as depicted by the lack of significant impact of E-stress on the total aboveground biomass accumulated at the square stage at all production conditions. As an indeterminate crop, cotton can exhibit a high degree of plasticity in growth [11,28,29], which may infer some level of tolerance to partial nutrient stress. Nonetheless, optimum nutrient management is critical for achieving high yield and efficiency in cotton [11]. Nutritional demand for cotton in the early season is reported to be low [7,11,12]. Bassett et al. [12] observed that at the first flower stage, the N, P, K, Ca, and Mg accumulated in the aboveground components of cotton were <15% of the total. In addition, 2–4% of the total seasonal aboveground biomass had accumulated at the square stage [12]. The average aboveground biomass accumulated by the square stage in this study was 8.2% of that accumulated by harvest.

In addition to residual soil nutrients, mineralization of crop residues and organic matter is another good source of nutrients for crops [30–32]. Organic matter at the experimental sites was low to have contributed to any appreciable levels of nutrients (2.90 g kg^{-1} in Camilla and 4.67 g kg^{-1} in Midville). However, residues of the previous crops (corn and peanut) and the use of rye cover crops may have affected the overall nutrient supply. As a biological process, the mineralization of crop residues depends on several abiotic and biotic factors, including temperature, rainfall, soil properties, the chemical composition of the crop residues, and the structure and composition of microbial communities [33–36]. Mineralization of the peanut residues would occur at a greater rate than those of the corn residues or the rye cover crop, as a result of the lower C:N ratio of the peanut residues. Synchronizing fertilizer application and nutrient release from crop residues with plant nutrient demand could enhance crop productivity while reducing over application of mineral fertilizers [33,34].

The E-stress also had no impact on cotton yield, which is consistent with observations made from previous studies that investigated the one-time application of nutrients in cotton [37,38]. In general, the application of N, P, and K at only the first flower stage was reported to maximize nutrient utilization while minimizing the impact on the environment [37,38]. The standard fertility received lower rates of nutrients than the E-stress and R-stress. Compared to the E-stress and R-stress, the standard fertility underperformed, with statistical significance in lint and cottonseed yields observed in Camilla 2021 under both the SSDI and overhead irrigation conditions. While modern cotton cultivars have better nutrient use efficiencies [6,7], the observations of this study indicate they also respond to high nutrient levels. According to Pabuayon et al. [7], genetic improvements to enhance the nutrient efficiency of modern cultivars may have changed their organ nutrient accumulation and requirement rates.

The L-stress received just 30–40% of the nutrient rate of the R-stress, which was applied one time at the early stage of planting. The results showed the L-stress had significantly lower lint and cottonseed yields than the R-stress under four out of the six production conditions tested in this study. Compared to the standard fertility, the L-stress had significantly lower lint and cottonseed yields under just one (Camilla 2021 SSDI condition) out of the five production conditions. As already mentioned, the standard fertility was not tested under the Midville 2020 rainfed condition. Nutrient uptake in cotton peaks from flowering through fruiting, and then slows as the bolls mature [11]. This explains why biomass accumulation was not adversely impacted by L-stress at the square stage, but yield and biomass accumulation were significantly reduced by L-stress at maturity. The 30–40% nutrient rates applied to the L-stress plots may have been depleted by the later growth stages.

Effects of nutrient stress on cotton fiber quality were not consistent across the different production conditions. Where significant, the E-stress tended to increase the fiber length, fiber strength, and uniformity, which was desirable. However, it decreased the micronaire and increased the +b, reflecting poor quality. In contrast, the L-stress tended to decrease the +b. Reported effects of nutrient application on cotton fiber quality are often inconsistent and vary across locations and cultivars [39–41]. Findings from a study, which evaluated seven cotton cultivars under 33 environments in Georgia, showed that production conditions that enhanced yield also led to improved fiber quality [42]. The E-stress and R-stress had the greatest lint yield but had undesirable micronaire and +b properties, which could be due to the high N rates applied. Sui et al. [39] observed a negative correlation between +b and leaf N content. Overall, however, the magnitude of the differences in the fiber quality indicators observed in this study was small and did not affect the grading class.

The global textile market is competitive and fiber quality is critical to ensuring good prices. Moreover, fiber quality affects manufacturing processes and the ultimate use of cotton fiber. Of the fiber quality indicators, color has the highest contribution to the price of cotton [41,43]. Chakraborty et al. [43] reported that color, cleanliness, micronaire, length, and strength contributed 30%, 23%, 22%, 20%, and 5%, respectively, to the price premium paid toward cotton fiber quality. According to Mcveigh [44], a drop from Middling (31) to Strict Low Middling (41) can cause Australian farmers to lose about $760 ha^{-1}. In the USA, the annual cotton price statistics report for the 2021–2022 season by the USDA-AMS showed quotations for color 41, leaf 4, staple 34, micronaire 35–36 and 43–49, strength of 27.0–28.9 g tex^{-1}, and uniformity of 81% to be ~$2.52 kg^{-1} [45]. The quotation for a better cotton fiber quality (color 31, leaf 3, staple 34, micronaire 35–36 and 43–49, strength of 27.0–28.9 g tex^{-1}, and uniformity of 81%) increased by ~2.29 cents kg^{-1} [45].

5. Conclusions

Cotton yield was not adversely impacted by E-stress. However, the L-stress significantly reduced the lint and cottonseed yields under four and one production conditions, when compared to the R-stress and standard fertility, respectively. The E-stress and R-stress had better lint and cottonseed yield than the standard fertility, indicating modern cultivars can respond to high nutrient levels. Significant nutrient stress effects on fiber quality were observed but the magnitude of the differences was small and it did not affect the grading class. The minimal impact of E-stress on cotton yield and quality in this study suggests that the rates of nutrients often applied in the early season can be reduced. More tailored nutrient application rates, based on soil and plant tissue analyses, could then be applied shortly before the reproductive phase of the crop. Such a system will help optimize crop nutrition by synchronizing nutrient availability with crop demand.

Supplementary Materials: The following supporting information can be downloaded at: https://www.mdpi.com/article/10.3390/agronomy13010064/s1, Table S1: Fertilizers applied as the main sources of the different nutrient elements in Camilla and Midville; Table S2: *P*-values of the main effects and interaction effects of nutrient stress and production conditions on the growth and productivity of cotton in Camilla and Midville; Table S3: *P*-values of the main effects and interaction effects of nutrient stress and growth stage on biomass accumulation of cotton in Camilla and Midville under different production conditions; Table S4: *P*-values of the main effects and interaction effects of nutrient stress and production conditions on the fiber quality of cotton in Camilla and Midville.

Author Contributions: Conceptualization: S.A. and H.Y.S.; data curation: S.A., B.K.A., G.A., and H.Y.S.; formal analysis: S.A., M.B., and H.Y.S.; funding acquisition: S.H., G.H.H., and H.Y.S.; investigation: S.A., B.K.A., G.A., R.A.B., C.D.P., and H.Y.S.; methodology: S.A., B.K.A., G.A., and H.Y.S.; project administration: H.Y.S.; Resources: R.A.B., C.D.P., and H.Y.S.; supervision: H.Y.S.; validation: H.Y.S.; visualization: S.A., M.B., and H.Y.S.; writing—original draft preparation: S.A., M.B., N.Y.K., and H.Y.S.; writing—review and editing: S.A., M.B., G.A., R.A.B., C.D.P., S.H., N.Y.K., and H.Y.S. All authors have read and agreed to the published version of the manuscript.

Funding: This research was funded by the Georgia Cotton Commission (Award numbers AWD00012146 and AWD00013224). Further funding was provided by the USDA National Institute for Food and Agriculture through Hatch project 1026085.

Data Availability Statement: The data presented in this study are available on request from the corresponding author. The data are not publicly available due to privacy.

Acknowledgments: We appreciate BASF and Bayer AG for providing cotton seeds. We also thank the staff of Southeast Georgia Research and Education Center and the Stripling Irrigation Research Park of the University of Georgia for assisting with various aspects of the fieldwork.

Conflicts of Interest: The authors declare no conflict of interest.

Abbreviations

E-stress, early-season nutrient stress; L-stress, late-season nutrient stress; R-stress, reduced nutrient stress; SSDI, sub-surface drip irrigation; WoB, week of bloom; RD, fiber reflectance; +b, fiber yellowness

References

1. FAOSTAT. *Food and Agriculture Data*; Statistics Division, Food and Agriculture Organization, United Nations: Rome, Italy, 2022.
2. USDA-NASS. *Quick Stats*; USDA National Agricultural Statistics Service: Washington, DC, USA, 2022.
3. Daystar, J.S.; Barnes, E.; Hake, K.; Kurtz, R. Sustainability Trends and Natural Resource Use in U.S. Cotton Production. *BioResources* **2017**, *12*, 362–392. [CrossRef]
4. Shaheen, M.; Ali, M.Y.; Muhammad, T.; Qayyum, M.A.; Atta, S.; Bashir, S.; Amjad Bashir, M.; Hashim, S.; Hashem, M.; Alamri, S. New Promising High Yielding Cotton Bt-Variety RH-647 Adapted for Specific Agro-Climatic Zone. *Saudi J. Biol. Sci.* **2021**, *28*, 4329–4333. [CrossRef] [PubMed]
5. Constable, G.A.; Bange, M.P. The Yield Potential of Cotton (*Gossypium Hirsutum* L.). F. *Crop. Res.* **2015**, *182*, 98–106. [CrossRef]
6. Rochester, I.J.; Constable, G.A. Improvements in Nutrient Uptake and Nutrient Use-Efficiency in Cotton Cultivars Released between 1973 and 2006. F. *Crop. Res.* **2015**, *173*, 14–21. [CrossRef]
7. Pabuayon, I.L.B.; Lewis, K.L.; Ritchie, G.L. Dry Matter and Nutrient Partitioning Changes for the Past 30 Years of Cotton Production. *Agron. J.* **2020**, *112*, 4373–4385. [CrossRef]
8. Singh, S.; Tan, H.H. *High Natural Gas Prices Could Lead to Spike in Food Costs through Fertilizer Link*; S&P Global Platts: London, UK, 2022.

9. Townsend, A.R.; Howarth, R.W.; Bazzaz, F.A.; Booth, M.S.; Cleveland, C.C.; Collinge, S.K.; Dobson, A.P.; Epstein, P.R.; Holland, E.A.; Keeney, D.R.; et al. Human Health Effects of a Changing Global Nitrogen Cycle. *Front. Ecol. Environ.* **2003**, *1*, 240–246. [CrossRef]
10. Shahzad, K.; Abid, M.; Sintim, H.Y. Wheat Productivity and Economic Implications of Biochar and Inorganic Nitrogen Application. *Agron. J.* **2018**, *110*, 2259–2267. [CrossRef]
11. Rochester, I.J.; Constable, G.A.; Oosterhuis, D.M.; Errington, M. Nutritional Requirements of Cotton during Flowering and Fruiting. In *Flowering and Fruiting in Cotton*; Oosterhuis, D.M., Cothren, J.T., Eds.; The Cotton Foundation: Cordova, TN, USA, 2012; Volume 8, pp. 35–50.
12. Bassett, D.M.; Anderson, W.D.; Werkhoven, C.H.E. Dry Matter Production and Nutrient Uptake in Irrigated Cotton (*Gossypium Hirsutum*). *Agron. J.* **1970**, *62*, 299–303. [CrossRef]
13. Esfandbod, M.; Phillips, I.R.; Miller, B.; Rashti, M.R.; Lan, Z.M.; Srivastava, P.; Singh, B.; Chen, C.R. Aged Acidic Biochar Increases Nitrogen Retention and Decreases Ammonia Volatilization in Alkaline Bauxite Residue Sand. *Ecol. Eng.* **2017**, *98*, 157–165. [CrossRef]
14. Shahzad, K.; Abid, M.; Sintim, H.Y.; Hussain, S.; Nasim, W. Tillage and Biochar Effects on Wheat Productivity under Arid Conditions. *Crop Sci.* **2019**, *59*, 1191–1199. [CrossRef]
15. Mandal, S.; Thangarajan, R.; Bolan, N.S.; Sarkar, B.; Khan, N.; Ok, Y.S.; Naidu, R. Biochar-Induced Concomitant Decrease in Ammonia Volatilization and Increase in Nitrogen Use Efficiency by Wheat. *Chemosphere* **2016**, *142*, 120–127. [CrossRef] [PubMed]
16. Russell, J.S. Evaluation of Residual Nutrient Effects in Soils. *Aust. J. Agric. Res* **1977**, *28*, 461–475. [CrossRef]
17. Torma, S.; Vilček, J.; Lošák, T.; Kužel, S.; Martensson, A. Residual Plant Nutrients in Crop Residues–an Important Resource. *Acta Agric. Scand. Sect. B Soil Plant Sci.* **2018**, *68*, 358–366. [CrossRef]
18. Sintim, H.Y.; Adjesiwor, A.T.; Zheljazkov, V.D.; Islam, M.A.; Obour, A.K. Nitrogen Application in Sainfoin under Rain-Fed Conditions in Wyoming: Productivity and Cost Implications. *Agron. J.* **2016**, *108*, 294–300. [CrossRef]
19. Georgia AEMN. *Georgia Automated Environmental Monitoring Network*; University of Georgia: Griffin, GA, USA, 2022; Available online: www.georgiaweather.net (accessed on 15 January 2022).
20. Soil Survey Staff. *Keys to Soil Taxonomy*, 12th ed.; USDA-Natural Resources Conservation Service: Washington, DC, USA, 2014; ISBN 016085427X.
21. UGA-AESL. *UGFertex: Prescription Lime and Nutrient Guidelines for Agronomic Crops*; UGFertex Version 3.1; University of Georgia Agricultural and Environmental Services Laboratories, University of Georgia: Athens, GA, USA, 2022.
22. Campbell, C.R. *Reference Sufficiency Ranges for Plant Analysis in the Southern Region of the United States*; Southern Cooperative Series Bulletin #394: Raleigh, NC, USA, 2000.
23. Hand, C.; Culpepper, S.; Harris, G.; Kemerait, B.; Liu, Y.; Perry, C.; Porter, W.; Roberts, P.; Smith, A.; Virk, S.; et al. *2021 Georgia Cotton Production Guide*; University of Georgia Extension: Athens, GA, USA, 2021.
24. Whitaker, J.; Culpepper, S.; Freeman, S.; Harris, G.; Kemerait, B.; Perry, C.; Porter, W.; Roberts, P.; Liu, Y.; Smith, A.; et al. *2020 Georgia Cotton Production Guide*; University of Georgia Extension: Athens, GA, USA, 2020.
25. CI. *The Classification of Cotton*; Cotton Incorporated, Agricultural Handbook 566: Cary, NC, USA, 2018.
26. Bates, D.; Maechler, M.; Bolker, B.; Walker, S. Fitting Linear Mixed-Effects Models Using Lme4. *J. Stat. Softw.* **2015**, *67*, 1–48. [CrossRef]
27. Lenth, R.; Love, J.; Herve, M. *Estimated Marginal Means, Aka Least-Squares Means*; Package "Emmeans", Version 1.1.2; The R Foundation: Vienna, Austria, 2018.
28. Atkin, O.K.; Loveys, B.R.; Atkinson, L.J.; Pons, T.L. Phenotypic Plasticity and Growth Temperature: Understanding Interspecific Variability. *J. Exp. Bot.* **2006**, *57*, 267–281. [CrossRef]
29. Li, B.; Tian, Q.; Wang, X.; Han, B.; Liu, L.; Kong, X.; Si, A.; Wang, J.; Lin, Z.; Zhang, X.; et al. Phenotypic Plasticity and Genetic Variation of Cotton Yield and Its Related Traits under Water-Limited Conditions. *Crop J.* **2020**, *8*, 966–976. [CrossRef]
30. De Neve, S. Organic Matter Mineralization as a Source of Nitrogen. In *Advances in Research on Fertilization Management of Vegetable Crops*; Tei, F., Nicola, S., Benincasa, P., Eds.; Springer: Cham, Switzerland, 2017; pp. 65–83.
31. Myers, R.J.K.; Palm, C.A.; Cuevas, E.; Gunatilleke, I.U.N.; Brossard, M. The Synchronisation of Nutrient Mineralisation and Plant Nutrient Demand. In *The Biological Management of Tropical Soil Fertility*; Woomer, P.L., Swift, M.J., Eds.; Wiley: Hoboken, NJ, USA, 1994.
32. Sintim, H.Y.; Zheljazkov, V.D.; Obour, A.K.; Garcia y Garcia, A.; Foulke, T.K. Influence of Nitrogen and Sulfur Application on Camelina Performance under Dryland Conditions. *Ind. Crops Prod.* **2015**, *70*, 253–259. [CrossRef]
33. Whalen, J.K. Managing Soil Biota-Mediated Decomposition and Nutrient Mineralization in Sustainable Agroecosystems. *Adv. Agric.* **2014**, *2014*, 1–13. [CrossRef]
34. Grzyb, A.; Wolna-Maruwka, A.; Niewiadomska, A. Environmental Factors Affecting the Mineralization of Crop Residues. *Agronomy* **2020**, *10*, 1951. [CrossRef]
35. Shahzad, K.; Sintim, H.Y.; Ahmad, F.; Abid, M.; Nasim, W. Importance of Carbon Sequestration in the Context of Climate Change. In *Building Climate Resilience in Agriculture*; Jatoi, W.N., Mubeen, M., Ahmad, A., Cheema, M.A., Lin, Z., Hashmi, M.Z., Eds.; Springer: Cham, Switzerland, 2022; pp. 385–401. ISBN 978-3-030-79407-1.
36. Sintim, H.Y.; Shahzad, K.; Bary, A.I.; Collins, D.P.; Myhre, E.A.; Flury, M. Differential Gas Exchange and Soil Microclimate Dynamics under Biodegradable Plastic, Polyethylene, and Paper Mulches. *Ital. J. Agron.* **2022**, *17*, 1979. [CrossRef]

37. Luo, H.; Wang, Q.; Zhang, J.; Wang, L.; Li, Y.; Yang, G. Minimum Fertilization at the Appearance of the First Flower Benefits Cotton Nutrient Utilization of Nitrogen, Phosphorus and Potassium. *Sci. Rep.* **2020**, *10*, 6815. [CrossRef] [PubMed]
38. Yang, G.; Tang, H.; Tong, J.; Nie, Y.; Zhang, X. Effect of Fertilization Frequency on Cotton Yield and Biomass Accumulation. *F. Crop. Res.* **2012**, *125*, 161–166. [CrossRef]
39. Sui, R.; Byler, R.K.; Delhom, C.D.; Sui, R. Effect of Nitrogen Application Rates on Yield and Quality in Irrigated and Rainfed Cotton. *J. Cotton Sci.* **2017**, *21*, 113–121. [CrossRef]
40. Kusi, N.Y.O.; Lewis, K.L.; Morgan, G.D.; Ritchie, G.L.; Deb, S.K.; Stevens, R.D.; Sintim, H.Y. Cotton Cultivar Response to Potassium Fertilizer Application in Texas' Southern High Plains. *Agron. J.* **2021**, *113*, 5436–5453. [CrossRef]
41. van der Sluijs, M.H.J. Effect of Nitrogen Application Level on Cotton Fibre Quality. *J. Cott. Res.* **2022**, *5*, 9. [CrossRef]
42. Snider, J.L.; Collins, G.D.; Whitaker, J.; Davis, J.W. Quantifying Genotypic and Environmental Contributions to Yield and Fiber Quality in Georgia: Data from Seven Commercial Cultivars and 33 Yield Environments. *J. Cotton Sci.* **2013**, *17*, 285–292.
43. Chakraborty, K.; Ethridge, D.; Misra, S. How Different Quality Attributes Contribute to the Price of Cotton in Texas and Oklahoma? In Proceedings of the Beltwide Cotton Conference, San Antonio, TX, USA, 4–8 January 2000.
44. Mcveigh, M. *The Impact of Colour Discounts to the Australian Cotton Industry*; Project No 1517; Nuffield Australia: North Sydney, Australia, 2017.
45. USDA-AMS. *Cotton Price Statistic 2021–2022*; USDA Agricultural Marketing Service, Cotton and Tobacco Program: Memphis, TN, USA, 2022; Volume 103, p. 13.

Article

Wheat and Faba Bean Intercropping Together with Nitrogen Modulation Is a Good Option for Balancing the Trade-Off Relationship between Grain Yield and Quality in the Southwest of China

Ying-an Zhu [1,†], Jianyang He [2,†], Zhongying Yu [2], Dong Zhou [2], Haiye Li [2], Xinyu Wu [2], Yan Dong [2], Li Tang [2], Yi Zheng [2,3] and Jingxiu Xiao [2,*]

[1] College of Horticulture and Landscape, Yunnan Agricultural University, Kunming 650201, China
[2] College of Resources and Environment, Yunnan Agricultural University, Kunming 650201, China
[3] Department of President Office, Yunnan Open University, Kunming 650599, China
* Correspondence: xiaojingxiuxjx@126.com
† These authors contributed equally to this work.

Abstract: Cereal and legume intercropping could improve cereal yield, but the role of intercropping in grain quality still lacks a full understanding. A two-year bi-factorial trial was conducted to investigate the role of two planting patterns (mono-cropped wheat (MW) and intercropped wheat+faba bean (IW)) and four nitrogen (N) fertilization levels (N0, no N fertilizer applied to both wheat and faba bean; N1, 90 and 45 kg N ha^{-1} applied to wheat and faba bean; N2, 180 and 90 kg N ha^{-1} applied to wheat and faba bean; N3, 270 and 135 kg N ha^{-1} applied to wheat and faba bean), as well as their interaction on the productivity of wheat grain yield (GY) and quality. The results showed that intercropping increased both the yields of wheat grain protein and amino acids (AAs) relative to MW in both years. No difference in Aas content between IW and MW was found but the 9% grain protein content (GPC) of IW was higher than that of MW in 2020. By contrast, wheat gliadin content was increased by 8–14% when wheat was intercropped with faba bean in both years, and some AAs fractions including essential and non-essential AAs were increased under N0 and N1 levels but declined at the N3 level. This means that intercropping increased the grain quality either for protein and AAs content or for fractions. There was no negative relationship between GPC and GY in the present study, and intercropping tended to increase GPC with increasing GY. In conclusion, wheat and faba bean mainly affected GPC and fractions rather than AAs, and intercropping presented a potential to improve both wheat quality and yield concurrently. Modulated N rates benefitted the stimulation of intercropping advantages in terms of grain yield and quality in the southwest of China and similar regions.

Keywords: wheat and faba bean intercropping; grain protein content; protein fractions; profile of amino acids; nitrogen fertilization

Citation: Zhu, Y.-a.; He, J.; Yu, Z.; Zhou, D.; Li, H.; Wu, X.; Dong, Y.; Tang, L.; Zheng, Y.; Xiao, J. Wheat and Faba Bean Intercropping Together with Nitrogen Modulation Is a Good Option for Balancing the Trade-Off Relationship between Grain Yield and Quality in the Southwest of China. *Agronomy* **2022**, *12*, 2984. https://doi.org/10.3390/agronomy12122984

Academic Editors: Christos Noulas, Shahram Torabian and Ruijun Qin

Received: 11 October 2022
Accepted: 22 November 2022
Published: 28 November 2022

Publisher's Note: MDPI stays neutral with regard to jurisdictional claims in published maps and institutional affiliations.

1. Introduction

Traditional planting patterns including intercropping, relay intercropping, and rotation are normally linked with yield increase and sustainability of the agriculture system [1–3]. Legume-based intercropping, a worldwide planting method, always presents increased crop yield and drives higher crude protein yields due to the nitrogen (N) biological fixation of legumes [4–6]. Frequently, improved cereal nutrient was observed because of N and phosphorus (P) transfer from the legume to cereal during their co-growing period in cereal-legume intercropping systems [7], and resulted in better cereal feed/forage quality [8–10]. Thus, the early research argued that the increased protein content of cereals was a result of N fertilization and was linked with legume intercropping [11,12]. Actually, other non-legume-based intercropping was also a benefit for crops yield and quality [13,14].

Protein content and fractions are important for evaluating and determining wheat grain values [15]. Many researchers highlighted the positive effect of cereal and legume intercropping on grain protein content (GPC) [5,16,17], but few studies focused on the effect of intercropping on protein fractions. The content of amino acids (AAs), especially essential amino acids (EAAs), is important to reflect protein quality, but major staple foods including wheat have limited amounts of EAAs for humans [18]. The enhancement of breeding techniques and N topdressing time modulation resulted in improved protein quality [19,20]; however, little attention has focused on the role of planting pattern in grain AAs content and factions.

Wheat and faba bean intercropping, as a typical legume-based intercropping pattern, is widely distributed in many countries either for food or for forage [21]. Tosti and Guiducci observed that wheat temporarily intercropped with faba bean improved both wheat grain yield and protein content [22]. However, De Stefanis et al. found that durum wheat gluten quality, total protein concentration, and monomeric and polymeric protein amounts were significantly increased but wheat grain yield was decreased when durum wheat was temporary intercropped with faba bean [23]. Similarly, wheat temporarily intercropped with clover induced a higher wheat grain protein content but lower grain yield [24]. In fact, a negative relationship or trade-off relationship between the grain yield (GY) and GPC was constantly observed in most cereal grains [25,26], but intercropping was a good strategy to reducing the risk of impairing winter wheat yield and protein content [27].

In the southwest of China, wheat and faba bean had a long co-growing period; thus, the interspecific interaction in this pattern was different from that of wheat temporarily intercropped with faba bean [21]. A previous study illustrated that wheat and faba bean intercropping could increase wheat yield but decrease faba bean yield, and the intercropping yield advantage was decreased with N input [28]. However, there is a lack of comprehensive assessment on the effect of intercropping on grain quality, especially on the content and fractions of wheat protein and AAs, which are tightly related to N input. We hypothesize that wheat and faba bean intercropping could improve wheat grain yield and maintain gain quality simultaneously, and the effect of intercropping on grain quality would vary with N input. Here, we present a two-year field experiment to test the hypothesis: (i) qualifying the effect of intercropping on wheat grain protein and amino acids under different N input conditions, and (ii) identifying the impact of intercropping on the relationship between GY and quality.

2. Material and Methods

2.1. Experimental Site and Growing Conditions

The present study was based on the data collected during 2018/2019 and 2019/2020 cropping seasons in the existing wheat and faba bean intercropping experiment, which was established in 2014. The field experiment was conducted at the Yunnan Agricultural University research station, located in Xundian (23°32′ N, 103°13′ E), Yunnan Province, northwest China. The climate in this region is characterized by a unimodal rainfall pattern with a rainy season from June to September and mean annual rainfall of 1040 mm, and the mean annual air temperature is 14.7 °C. The average monthly temperatures and monthly precipitation amounts during the experiment of 2018/2019 and 2019/2020 are shown in Figure 1. The monoculture corn was planted from May to September for many years before the wheat and faba bean intercropping experiment was established. The soil type in this region is called red soil (Ferralic Cambisol, FAO, 2006) with a bulk density of 1.38 g cm^{-3}, and the content of clay, silt, and sand was 34%, 52%, and 14%, respectively, at a soil depth of 0–30 cm. At the beginning of the multi-year field experiment in 2014, the soil properties were as follows: SOC 12 g kg^{-1}, total N 1.14g kg^{-1}, total P 0.98g kg^{-1}, total K 24.25 g kg^{-1}, available N (NaOH hydrolyzed) 80 mg kg^{-1}, Olsen P 17 mg kg^{-1}, exchangeable K 146 mg kg^{-1}, and pH 7.2 (1:2.5 soil: water). The soil total N and available N contents in each treatment were changed during 2018/2019 and 2019/2020 cropping seasons as compared to the beginning of the field experiment in 2014 due to continuous wheat and faba bean intercropping and different N application rates (data shown in Supplementary Table).

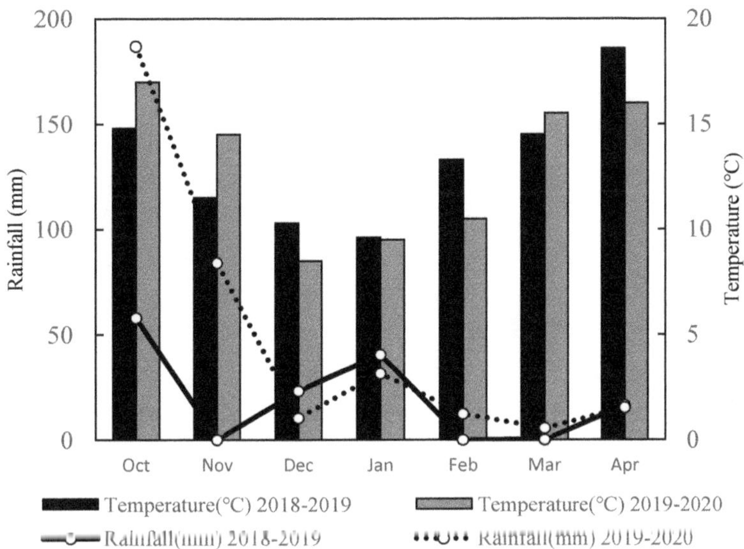

Figure 1. The monthly average temperature and rainfall during experiments of 2018–2019 and 2019–2020.

2.2. Experimental Design

The field experiment was a randomized block design with two factors and three replicates [28]. Factor A was planting patterns (mono-cropped wheat (MW) and intercropped wheat+faba bean (IW)), and factor B was N levels (0 kg N ha^{-1} (N0), 90 kg N ha^{-1}(N1), 180 kg N ha^{-1}(N2), and 270 kg N ha^{-1} (N3) for wheat; 0 kg N ha^{-1} (N0), 45 kg N ha^{-1}(N1), 90 kg N ha^{-1}(N2), and 135 kg N ha^{-1} (N3) for faba bean). In total, the field experiments consisted of 24 plots with eight treatments, and each plot area was 5.4 m × 6.0 m = 32.4 m^2. There were 0.5 m spacings between each plot and 1.0 m spacings between adjacent blocks to avoid water and nutrient interference. The row space of wheat was 0.2 m with a seeding rate at 180 kg ha^{-1}, whereas the faba bean row spacing was 0.3 m and the plant-to-plant spacing was 0.1 m in the present study. The strip intercropping of six rows of wheat intercropped with two rows of faba bean was used in this study based on local farmers practice; thus, there were three strips in each intercropping plot including 18 rows of wheat and six rows of faba bean [28]. The plant density of intercropped wheat and faba bean was identical to that of mono-cropped under the same area, and the row space between wheat and faba bean was 0.25 m in each intercropped plot. Detailed information of a given intercropping plot can be seen in Figure 2.

2.3. Field Experiment Management

The local varieties of Yunmai 52 for wheat (*Triticum aestivum* L.) and Yuxi Dalidou for faba bean (*Vicia faba* L.) were used in the present study since 2014, and the faba bean seed was non-inoculated rhizobium. Wheat and faba bean were sown on the same date normally on 20–30 October with a sowing depth of 10 cm and were harvested in the next year on 10–20 April. After both plants were harvested, all straws were removed from the field and each plot retained fallow from May to September since 2014. The implementation of other crop managements including irrigation and the use of pesticides was according to local farmers' practice.

Urea as N fertilizer was used in the present study. For wheat, one half of the total N application rate for each given treatment was applied as basal fertilizer before sowing by hand, and another half N fertilizer as a topdressing was applied at the wheat elongation stage. For faba bean, all N fertilizers for each treatment were applied as a basal fertilizer before sowing. Amounts of 90 kg P$_2$O$_5$ ha^{-1} (calcium superphosphate) and 90 kg K$_2$O ha^{-1}

(potassium chloride) for each crop were applied as base fertilizers according to local farming practices. In each intercropping plot, topdressing N was only evenly applied to wheat rows by hand.

2.4. Data Collection and Analyses

At maturity, inter- and mono-cropped wheat grains of each whole plot were collected and determined after the grain seeds were fully air-dried during 2018/2019 and 2019/2020 growing seasons, and the experiment of 2019 and 2020 represented two years of experiments, respectively. The wheat grain crude protein; protein fraction contents including albumin, globulin, gliadin, and glutelin; amino acids fraction content were determined in both years.

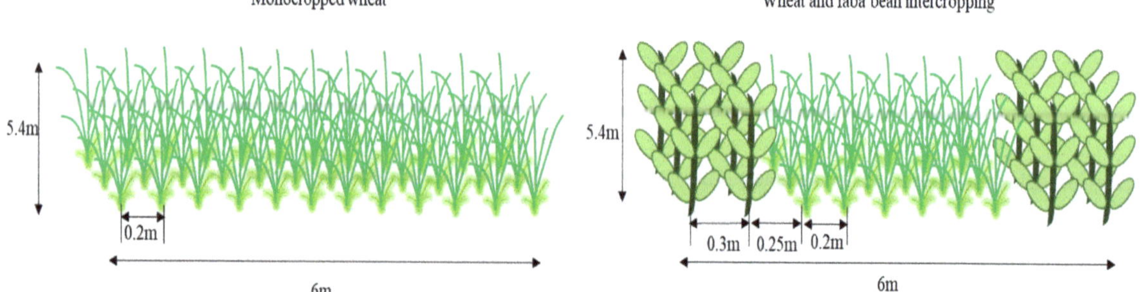

Figure 2. Diagram of the planting pattern for wheat and faba bean intercropping and mono-cropped wheat in the field experiment.

GPC was calculated by multiplying the grain N content with a conversion factor of 5.83 for wheat [29]. Grain N content was analyzed by the Kjeldahl method after the sample digestion with H_2SO_4-H_2O_2. Protein fractions albumin, globulin, gliadin, and glutelin were sequentially extracted from 1 g of wheat grain powder [30,31]. In brief, sequential extraction of albumin and gliadin fractions from the wheat grain sample were carried out by using distilled water and 2% NaCl, followed by extraction with 70% ethanol to obtain the gliadin fraction. The glutelin fraction was extracted from the residue by using 0.05 M NaOH. Protein content was determined using the modified Lowry method of Markwell et al. [32].

Amino acids (AAs) were identified and quantified by a high-performance liquid chromatographer (Agilent 1100) coupled to a DAD detector and a post-column derivatization device. The chromatograph column used was a C18 (250 × 4.6 mm ID) from Thermo Fisher, and the column was operated at a temperature of 40 °C. The chromatograph conditions were set as follows: ultraviolet detector 360 nm; flowrate 1.0 mL min^{-1}; the mobile phase consisted of A = 0.5 M sodium acetate (for HPLC analysis, Sigma Chemical CO., St. Louis, MO, USA) and B = 50% (*v/v*) methanol (Sigma Chemical Co., St. Louis, MO, USA) in water, and the injection volume was 10 μL for all samples. Wheat grain powder was hydrolytic and derivatized before HPLC analysis. Briefly, (1) sample hydrolysis: grain powder was hydrolytic for 24 h at 110 °C in 6M hydrochloric acid; (2) post-column derivatization: the derivatization was performed from a solution containing sodium hydroxide (6 mol L^{-1}), sodium bicarbonate pH 9.0 (0.5 mol L^{-1}), and DNFB. A deviation solution was mixed in a buffer of phosphoric acid pH 7.0 and filtered with a 0.22 μm membrane before HPLC analysis. The identification of amino acids was carried out by comparing retention times of standards and quantification in analytical curves constructed for each amino acid.

The sum content of seventeen AAs was the total AAs (TAAs) content. The seventeen measured AAs were divided into essential amino acids (EAAs) and non-essential amino acid (NEAAs). EAAs are essential for humans and animals but cannot be synthesized in the human body, including Thr, Val, Met, Ile, Leu, Phe, and Lys; NEAAs are non-essential as being synthesized in the human or animal body, including Asp, Ser, Glu, Gly, Ala, Cys, Tyr, His, Arg, and Pro [33].

Protein and TAAs yields represent the yield of protein and/or TAA that can be harvested per unit area of crops [34], which was calculated by protein and AAs content multiplied by each plot grain yield, respectively, in this study.

2.5. Statistic Analysis

A two-way analysis of variance (ANOVA) was performed using the MIXED procedure with SPSS software (IBM SPSS Statistics Version 19.0) to test for significant differences among treatments. Planting patterns and N levels were considered as the fixed factors, and replication was considered the random factor. Significant differences among treatments at each year were investigated using Duncan's multiple range post hoc test when the F-value was significant ($p \leq 0.05$). Linear and quadratic models were used to simulate the relationship among grain yield, grain protein content, and grain AAs content in this study.

3. Results

3.1. Mono- and Inter-Cropped Wheat Grain Protein Content and Yield under Different N Levels

Both the GPC and protein yield were not influenced by the interaction of N levels and planting patterns in the two-year field experiments. Likewise, N levels and planting patterns also had no impact on wheat GPC in the experiment of 2019. However, wheat GPC was increased by 9% when wheat was intercropped with faba bean relative to MW in the experiment of 2020 (Table 1). Similarly, wheat protein yield was increased by 28% and 32% in 2019 and 2020, respectively, when wheat was intercropped with faba bean. In addition, increased protein yield was found with increasing N levels in both years (Table 1).

Table 1. The protein and total amino acids content and yield for inter- and mono-cropped wheat grain under different N levels.

N Levels	Planting Patterns	2019			2020			2019		2020	
(NL)	(PP)	GY	Protein Content	Protein Yield	GY	Protein Content	Protein Yield	TAAs Content	TAAs Yield	TAAs Content	TAAs Yield
		t ha^{-1}	%	g m^{-2}	t ha^{-1}	%	g m^{-2}	mg g^{-1}	g m^{-2}	mg g^{-1}	g m^{-2}
N0		1.69 d	13 a	2.17 d	1.90 c	10 c	1.97 d	92 c	1.55 d	81 d	1.56 d
N1		3.08 c	13 a	4.11 c	3.24 b	10 c	3.41 c	99 b	3.10 c	86 c	2.78 c
N2		4.02 b	14 a	5.45 b	3.72 a	12 b	4.53 b	95 bc	3.81 b	103 b	3.85 b
N3		4.62 a	13 a	6.19 a	3.92 a	14 a	5.37 a	117 a	5.41 a	113 a	4.41 a
	MW	3.08 b	13 a	3.92 b	2.86 b	11 b	3.30 b	100 a	3.20 b	96 a	2.88 b
	IW	3.63 a	14 a	5.04 a	3.53 a	12 a	4.34 a	102 a	3.74 a	95 a	3.42 a
N0	MW	1.41 a	13 a	1.80 a	1.41 a	10 a	1.43 a	93 c	1.32 e	78 e	1.11 e
	IW	1.98 a	13 a	2.54 a	2.39 a	11 a	2.50 a	91 cd	1.79 d	84 d	2.01 d
N1	MW	2.67 a	13 a	3.44 a	2.87 a	9 a	2.70 a	85 d	2.26 c	85 d	2.45 c
	IW	3.49 a	14 a	4.78 a	3.63 a	11 a	4.12 a	113 b	3.95 b	86 d	3.11 b
N2	MW	3.72 a	13 a	4.94 a	3.43 a	11 a	3.90 a	98 bc	3.66 b	100 c	3.42 b
	IW	4.32 a	14 a	5.97 a	4.01 a	13 a	5.15 a	92 cd	3.96 b	107 b	4.29 a
N3	MW	4.50 a	12 a	5.51 a	3.74 a	14 a	5.17 a	124 a	5.56 a	122 a	4.56 a
	IW	4.74 a	14 a	6.87 a	4.09 a	14 a	5.56 a	111 b	5.26 a	105 bc	4.27 a
Sig											
NL		***	ns	***	***	***	***	***	***	***	***
PP		***	ns	***	***	*	***	ns	***	ns	***
NL × PP		ns	ns	ns	ns	ns	ns	***	***	***	***

MW, mono-cropped wheat; IW, inter-cropped wheat; GY, grain yield; TAA, total amino acid. In each column, different letters represent significant differences among treatments at the 0.05 level according to Duncan's multiple range test. * and *** represent significant differences at 0.05 and 0.001 levels, respectively. ns represents no significant difference.

3.2. Mono- and Inter-Cropped Wheat Grain Protein Composition under Different N Levels

Four protein fraction contents including albumin, globulin, gliadin, and glutelin were influenced by N levels, and protein fraction contents were frequently affected by the planting pattern, but they were not influenced by the interaction of N levels and planting patterns (Table 2). In 2019, the increased contents of albumin, gliadin, and glutelin in IW grain were observed as compared to MW, and the increase was 9%, 9%, and 5%, respectively.

In 2020, only the increased content of gliadin in IW grain was observed relative to MW, and the increase was 14%. In addition, all four protein fractions were increased with increasing N levels (Figure 3).

Table 2. Two-way ANOVA analysis of grain protein composition for inter- and mono-cropped wheat under different N levels.

	2019				2020			
	Albumin	Globulin	Gliadin	Glutelin	Albumin	Globulin	Gliadin	Glutelin
N levels (NL)	**	*	***	***	***	***	***	***
Planting patterns(PP)	*	ns	**	*	ns	ns	***	ns
NL×PP	ns	ns	ns	Ns	ns	ns	**	ns

In each column, *, **, and *** represent significant differences at 0.05, 0.01, and 0.001 levels, respectively. ns represents no significant difference.

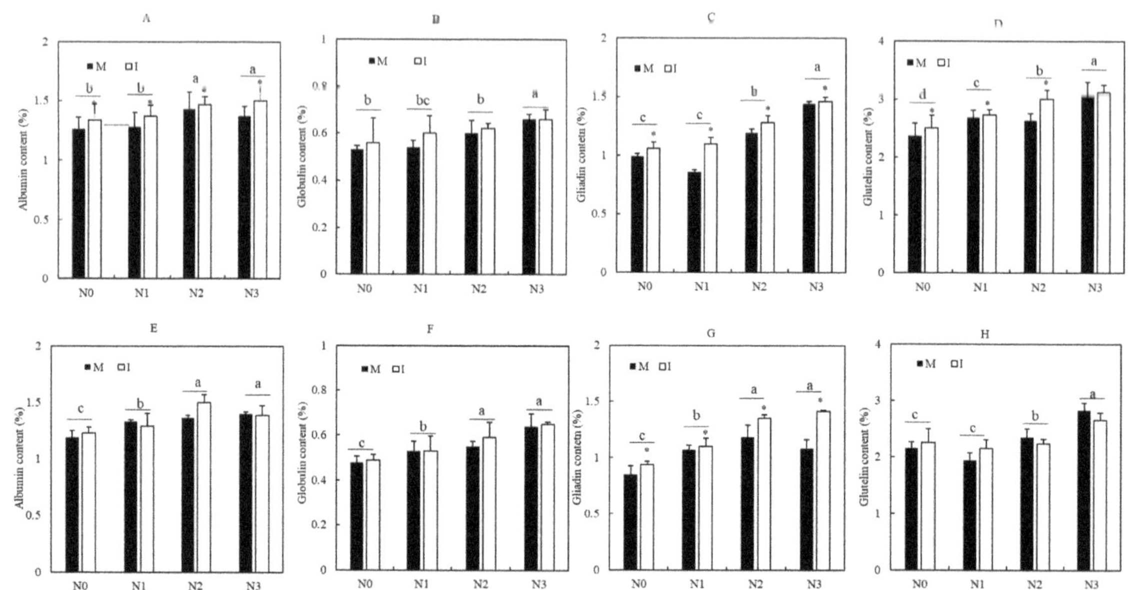

Figure 3. Grain protein fraction content between IW and MW under different N levels. MW, mono-cropped wheat; IW, intercropping wheat. (**A–D**) Albumin, globulin, gliadin, and glutelin content of IW and MW in 2019, respectively; (**E–H**) albumin, globulin, gliadin, and glutelin content of IW and MW in 2020, respectively. Different letters represent significant differences among different N levels ($p < 0.05$); * represents significant differences between IW and MW ($p < 0.05$). Each bar in the figures is the mean value (n = 3), and error bars represent the standard error.

3.3. Mono- and Inter-Cropped Wheat Grain Amino Acids Content and Yield under Different N Levels

Planting patterns had no impact on grain TAAs content in neither year, but TAAs content was influenced by N levels and the interaction of N levels and planting patterns (Table 1). The TAAs content in IW grain was increased by 33% relative to MW at the N1 level in 2019, and wheat grain TAAs content was increased by 7% under N0 and N1 levels when wheat was intercropped with faba bean as compared to MW in 2020. However, wheat grain TAAs content was decreased by 10% and 14% under the N3 level in the experiment of 2019 and 2020, respectively, as compared to MW. Regardless of N levels, the grain TAAs yield was increased by 17–19% when wheat was intercropped with faba bean, whereas no difference in grain TAAs yield between IW and MW was found under the N3 level, due to the interaction between N levels and planting patterns. By contrast, the TAAs yield of IW grain was increased by 35% and 60% under N0 and N1 levels, respectively, in comparison with MW in 2019; the TAAs yield of IW grain was increased by 80%, 27%, and 25% under N0, N1, and N2 levels, respectively, in comparison with MW in 2020 (Table 1).

3.4. Mono- and Inter-Cropped Wheat Grain NEAAs and EAAs Content under Different N Levels

The content of NEAAs and EAAs and the ratio of EAAs and TAAs were not influenced by planting patterns but were affected by N levels and N levels × planting patterns in both years (Table 3). When compared to MW, IW NEAAs content was decreased by 12% and 14% under the N3 level in 2019 and 2020, respectively (Figure 4). By contrast, the NEAAs content of IW was 31% higher than that of MW under the N1 level in 2019; the NEAAs contents of IW were 7% and 5% higher than those of MW under N0 and N2 levels, respectively, in 2020 (Figure 4). Similarly, the IW EAAs content was decreased by 14% at the N3 level in 2020 and was decreased by 9% and 12% at N0 and N2 levels in 2019 as compared to the corresponding MW, respectively. However, grain EAAs was increased by 39% at the N1 level in 2019 and increased by 13% at the N2 level in 2020 when wheat was intercropped with faba bean. As a result, EAAs/TAAs of IW at N0 and N2 levels were decreased by 7% and 6%, respectively, and the EAAs/TAAs of IW at the N3 level was increased by 5% when compared to MW in 2019. In all, we did not find any difference in EAAs/TAAs between MW and IW regardless of N levels (Figure 4).

Table 3. Two-way ANOVA analysis of non-essential amino acids and essential amino acids for inter- and mono-cropped wheat under different N levels.

	2019			2020		
	NEAAs	**EAAs**	**EAAs/TAAs**	**NEAAs**	**EAAs**	**EAAs/TAAs**
N levels (NL)	***	***	**	***	***	***
Planting patterns (PP)	ns	ns	ns	ns	ns	ns
NL × PP	***	***	**	***	***	***

In each column, ** and *** represent significant differences at 0.01 and 0.001 levels, respectively. ns represents no significant difference.

3.5. Mono- and Inter-Cropped Wheat Grain AAs Fraction Content under Different N Levels

The AAs fraction contents including eight EAAs fractions and nine NEAAs fractions were detected in the present study, and they were seldom influenced by planting patterns but were frequently affected by N levels and N levels × planting patterns according to the two-year experiment (Tables 4 and 5). Under N0 and N1 levels, only Met (2019) and Val (2020) contents in IW grain were lower than those in MW; for the other EAAs fractions, wheat and faba bean intercropping either had no impact on EAAs contents or increased EAAs contents. By contrast, under the N3 level, half of the EAAs fraction contents in IW grain were decreased as compared to MW. In the experiment of 2019, Thr, Val, Phe, and Lys contents in the IW grain were decreased by 12%, 40%, 7%, and 9% relative to MW; Val, Met, His, and Lys were decreased by 31%, 28%, 13%, and 26% when compared to MW in the experiment of 2020. On average, the contents of His and Phe in IW grain were higher than those in MW, but the Lys content in IW grain was lower than that in MW in 2019 regardless of N levels. Similarly, no difference in fraction content of EAAs between IW and MW was found except for the His content in IW grain that was higher than that in MW in 2020 (Table 4).

Wheat and faba bean intercropping nearly had no impact on NEAAs fraction contents except for Asp, Pro, Glu, and Tyr in the two-year experiments. Only Asp, Arg, and Cys contents in IW grain at the N0 level and Cys content at the N1 level in 2019 decreased as compared to MW, and the other NEAAs fraction contents in IW grain were either equal to or higher than those in MW. Likewise, only decreased Asp and Ala contents in IW grain in 2019 and decreased Cys in IW grain in 2020 were observed as compared to MW at the

N2 level. By contrast, half of the NEAAs fraction contents in IW grain were decreased in comparison with MW at the N3 level. In all, wheat grain Asp content in 2019 and Gln content in 2020 were decreased when wheat was intercropped with faba bean regardless of N levels, and a similar or higher content for other NEAAs fractions in IW grain was observed relative to MW (Table 5).

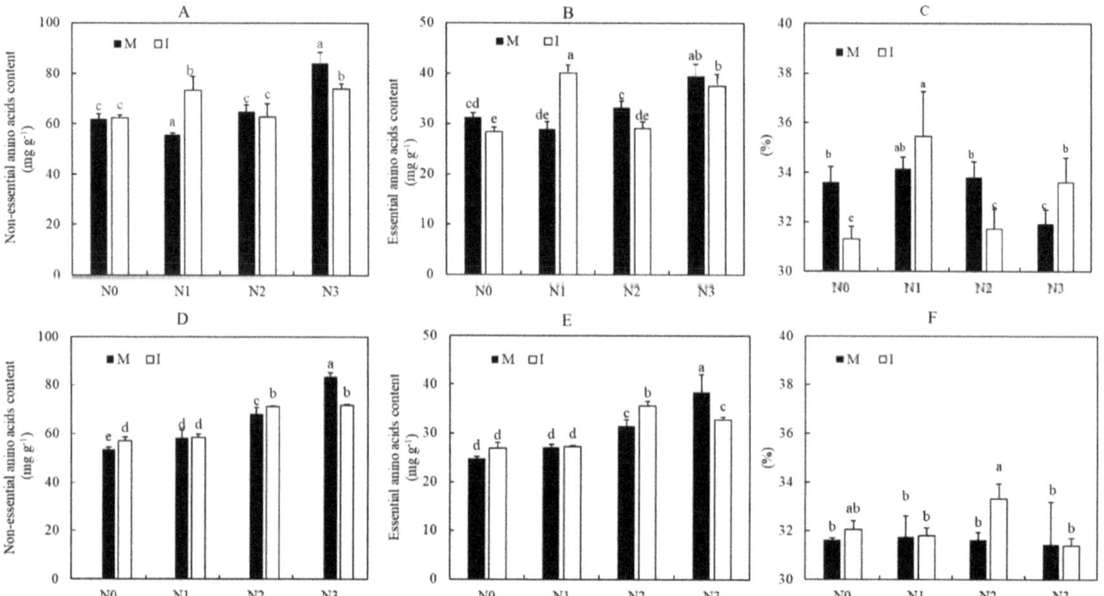

Figure 4. Essential amino acids, non-essential amino acids, and the ratio of essential amino acids to total amino acids between IW and MW under different N levels. (**A,D**) Essential amino acids in 2019 and 2020, respectively; (**B,E**) non-essential amino acids in 2019 and 2020, respectively; (**C,F**) ratio of essential amino acids to total amino acids in 2019 and 2020, respectively. MW, mono-cropped wheat; IW, intercropping wheat; different letters represent significant differences among all treatments. Each bar in the figures is the mean value (n = 3), and error bars represent standard error.

3.6. Co-Relationship of Between Grain Yield, Grain Protein Content, and Amino Acids Content for Mono- and Inter-Cropped Wheat

No relationship between GY and GPC was found for MW, but a quadratic regression was fitted to the relationship between GY and GPC in IW. A positive relationship between GY and AAs content including TAAs, NEAAs, and EAAs was presented, and the AAs content was positively related to GPC (Figure 5).

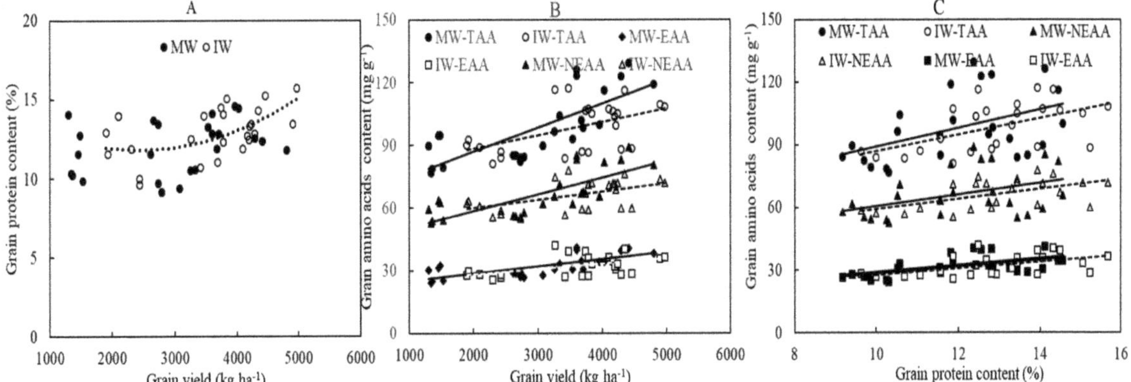

Figure 5. Relationship analysis among grain yield, grain protein content, and grain amino acids fraction. (**A**) Relationship between grain yields and grain protein content. In panel A, grain yield as a function of grain protein content for IW (y = 5 × $10^{-7}x^2$ − 0.0025x + 14.833, R^2 = 0.3228, *p* = 0.017, *n* = 24); (**B**) relationship between grain yields and grain amino acids fraction. In panel B, grain yield as a function of MW-TAA (y = 0.0115x + 63.997, R^2 = 0.5679, *p* = 0.00, *n* = 24), IW−TAA (y = 0.0065x + 75.28, R^2 = 0.2636, *p* = 0.01, *n* = 24), MW-NEAA (y = 0.0081x + 42.344, R^2 = 0.5676, *p* =0.000, *n* = 24), IW-NEAA (y = 0.004x + 51.937, R^2 = 0.2774, *p* = 0.008, *n* = 24), MW-EAA (y = 0.0034x + 21.653, R^2 = 0.5219, *p* = 0.00, *n* = 24), and IW-EAA (y = 0.0025x + 23.343, R^2 = 0.2135, *p* = 0.023, *n* = 24). (**C**) Relationship between grain protein content and grain amino acids fraction. In panel C, grain yield as a function of MW-TAA (y = 4.435x + 44.956, R^2 = 0.2079, *p* = 0.025, *n* = 24), IW-TAA (y = 3.9529x + 47.562, R^2 = 0.2924, *p* = 0.006, *n* = 24), MW-NEAA (y = 2.7833x + 32.919, R^2 = 0.1662, *p* = 0.048, *n* = 24), IW-NEAA (y = 2.4028x + 35.344, R^2 = 0.2968, *p* = 0.006, *n* = 24), MW-EAA (y = 1.6517x + 12.038, R^2 = 0.2976, *p* = 0.006, *n* = 24), and IW-EAA (y = 1.5501x + 12.218, R^2 = 0.2506, *p* = 0.013, *n* = 24). MW, mono-cropped wheat; IW, intercropping wheat. TAA, NEAA, and EAA: total amino acids, non-essential amino acids, and essential amino acids, respectively. The dot-dashed line and solid line represent linear regressions for IW and MW, respectively.

Table 4. The fraction content of each essential amino acids in grain for inter- and mono-cropped wheat grain under different N levels.

N Levels (NL)	Planting Patterns (PP)	2019 Thr	Val	Met	Ile	Leu	Phe	His	Lys	2020 Thr	Val	Met	Ile	Leu	Phe	His	Lys
										%							
N0	MW	3.45 d	0.94 b	7.98 c	3.45 d	6.13 d	4.43 e	2.20 c	2.70 a	2.86 a	0.16 e	5.56 c	3.34 a	5.39 a	4.58 a	1.37 d	1.50 c
	IW	3.51 cd	0.80 bc	5.59 e	3.54 d	6.26 d	4.68 e	2.04 c	1.96 c	3.21 a	0.50 cd	5.72 c	3.36 a	5.97 a	4.29 a	1.92 c	1.94 b
N1	MW	3.44 d	0.65 cd	6.70 d	3.45 d	6.14 d	4.38 e	1.84 d	2.33 b	3.23 a	0.53 cd	6.21 bc	3.26 a	5.91 a	4.46 a	1.53 d	1.96 b
	IW	4.38 a	1.24 a	10.80 a	4.20 abc	8.40 a	6.00 c	2.43 b	2.68 a	3.34 a	0.38 d	5.48 c	3.44 a	6.06 a	4.81 a	1.88 c	1.84 b
N2	MW	3.86 bc	0.65 cd	7.98 c	3.88 bc	7.23 bc	5.39 d	1.73 d	2.46 ab	3.63 a	0.72 b	7.20 b	3.57 a	7.05 a	5.13 a	2.33 b	1.85 b
	IW	3.69 cde	0.49 e	5.70 e	3.70 cd	6.66 cd	5.16 d	1.85 d	1.84 c	3.69 a	0.94 a	9.67 a	3.83 a	7.25 a	5.13 a	2.55 a	2.52 a
N3	MW	4.51 a	0.92 b	9.05 b	4.70 a	8.47 a	7.20 a	2.45 b	2.17 b	4.20 a	0.96 a	9.39 a	4.58 a	7.78 a	6.24 a	2.60 a	2.65 a
	IW	3.98 b	0.55 de	9.21 b	4.33 ab	7.99 ab	6.70 b	2.69 a	1.99 c	3.95 a	0.66 bc	6.74 b	4.14 a	7.26 a	5.88 a	2.25 b	1.94 b
N0		3.48 c	0.87 a	6.79 b	3.49 b	6.20 c	4.55 c	2.12 b	2.33 b	3.04 c	0.33 c	5.64 b	3.35 b	5.68 b	4.43 c	1.65 b	1.72 c
N1		3.91 b	0.95 a	8.75 a	3.82 b	7.27 b	5.19 b	2.14 b	2.50 a	3.28 b	0.45 b	5.84 b	3.35 b	5.99 b	4.63 c	1.70 b	1.90 b
N2		3.78 b	0.57 c	6.84 b	3.79 b	6.95 b	5.28 b	1.79 c	2.15 c	3.66 b	0.83 a	8.44 a	3.70 b	7.15 a	5.13 b	2.44 a	2.19 a
N3		4.25 a	0.73 b	9.13 a	4.51 a	8.23 a	6.95 a	2.57 a	2.08 c	4.07 a	0.81 a	8.06 a	4.36 a	7.52 a	6.06 a	2.43 a	2.29 a
	MW	3.82 a	0.79 a	7.93 a	3.87 a	6.99 a	5.35 b	2.06 b	2.41 a	3.48 a	0.59 a	7.09 a	3.68 a	6.53 a	5.10 a	1.96 b	1.99 a
	IW	3.89 a	0.77 a	7.82 a	3.94 a	7.33 a	5.64 a	2.25 a	2.12 b	3.55 a	0.62 a	6.90 a	3.69 a	6.63 a	5.03 a	2.15 a	2.06 a
Sig																	
NL × PP		***	***	***	*	***	***	***	***	ns	***	***	ns	ns	ns	***	***
NL		***	***	***	***	***	***	***	***	***	***	***	***	***	***	***	***
PP		Ns	ns	ns	Ns	ns	*	***	***	ns	ns	ns	ns	ns	ns	***	ns

MW, mono-cropped wheat; IW, intercropping wheat. Values with different letters are significantly different among the N levels, planting pattern, and interaction of N levels and planting pattern according to Duncan's multiple range test (two-way ANOVA, $p < 0.05$). *, $p < 0.05$; ***, $p < 0.001$; ns, no significance.

Table 5. The fraction content of each non-essential amino acids in grain for inter- and mono-cropped wheat grain under different N levels.

N Levels (NL)	Planting Patterns (PP)	2019 Asp	Glu	Ser	Arg	Gly	Pro	Ala	Cys	Tyr	2020 Asp	Glu	Ser	Arg	Gly	Pro	Ala	Cys	Tyr
N0	MW	5.81 ab	26.30 de	4.47 ef	5.14 bc	3.24 a	9.22 c	5.33 abc	1.07 a	1.22 b	4.03 d	22.64 c	4.08 a	4.27 e	2.80 e	9.81 a	3.72 c	1.32 a	0.87 c
	IW	4.87 cd	26.56 cde	4.50 ef	4.59 ef	3.23 a	10.83 b	5.39 ab	1.09 a	1.17 b	4.67 bc	24.09 cd	4.44 a	4.73 d	3.07 d	9.55 a	4.16 bc	1.03 b	1.30 b
N1	MW	5.16 bcd	21.98 e	4.20 f	4.59 c	3.20 a	9.34 c	5.03 bc	1.13 a	1.10 b	4.54 bc	24.96 c	4.38 a	4.77 cd	3.09 d	10.27 a	4.26 bc	0.77 bc	1.27 b
	IW	6.03 ab	32.97 b	5.87 ab	5.49 ab	3.58 a	11.35 b	6.01 a	0.47 c	1.42 b	4.45 c	24.60 cd	4.34 a	4.92 cd	3.24 cd	10.31 a	3.88 c	1.42 a	1.28 b
N2	MW	6.15 a	25.79 de	5.11 cd	5.13 bc	3.61 a	11.79 b	5.78 a	0.21 d	1.36 b	4.60 bc	31.56 b	4.97 a	5.22 bc	3.39 c	11.43 a	4.63 b	0.98 b	1.30 b
	IW	4.36 d	27.66 cd	4.81 de	5.07 cd	3.45 a	11.16 b	4.42 d	0.70 b	1.14 b	6.27 a	32.50 b	4.92 a	5.09 bcd	3.49 bc	11.72 a	5.60 ab	0.42 d	1.20 b
N3	MW	5.53 abc	38.36 a	6.10 a	5.82 a	4.20 a	16.50 a	5.42 ab	0.50 c	1.76 a	6.42 a	38.62 a	5.59 a	6.19 a	4.19 a	14.40 a	5.82 a	0.76 bc	1.52 a
	IW	5.24 abcd	31.40 bc	5.45 bc	5.63 bc	3.93 a	15.50 a	4.67 cd	0.59 bc	1.43 b	4.97 b	32.23 b	5.36 a	5.47 b	3.72 b	13.32 a	4.64 b	0.57 cd	1.46 a
N0		5.34 a	26.43 b	4.48 c	3.24 b	3.24 c	10.03 c	5.36 a	1.08 a	1.20 b	4.35 b	23.36 c	4.26 c	4.50 d	2.94 d	9.68 c	3.94 b	1.17 a	1.09 c
N1		5.60 a	27.48 b	5.04 b	3.20 b	3.39 bc	10.34 c	5.52 a	0.80 b	1.26 b	4.49 b	24.78 c	4.36 c	4.84 c	3.17 c	10.29 c	4.07 b	0.70 b	1.27 b
N2		5.26 a	26.73 b	4.96 b	3.61 b	3.53 b	11.48 b	5.10 a	0.45 c	1.25 b	5.43 a	32.03 b	4.95 b	5.15 b	3.44 b	11.58 b	5.12 a	0.66 b	1.25 b
N3		5.39 a	34.88 a	5.77 a	4.20 a	4.06 a	16.00 a	5.05 a	0.55 c	1.60 a	5.69 a	35.42 a	5.48 a	5.83 a	3.95 a	13.86 a	5.23 a	0.66 b	1.49 a
	MW	5.66 a	28.11 a	4.97 a	5.17 a	3.56 a	11.71 b	5.39 a	0.73 a	1.36 a	4.90 b	29.45 a	4.76 a	5.11 a	3.37 a	11.48 a	4.61 a	0.96 a	1.24 b
	IW	5.13 b	29.65 a	5.16 a	5.19 a	3.55 a	12.21 a	5.12 a	0.71 a	1.29 a	5.09 a	28.36 b	4.77 a	5.05 a	3.38 a	11.23 a	4.57 a	0.86 a	1.31 a
Sig																			
NL × PP		**	***	***	**	ns	***	***	***	*	***	***	ns	***	***	ns	***	***	***
NL		Ns	***	***	***	***	***	ns	***	**	***	***	***	***	***	***	***	***	***
PP		*	ns	ns	ns	ns	*	ns	Ns	ns	*	*	ns	ns	ns	ns	ns	ns	*

MW, mono-cropped wheat; IW, intercropping wheat. Values with different letters are significantly different among the N levels, planting pattern, and interaction of N levels and planting pattern according to Duncan's multiple range test (two-way ANOVA, $p < 0.05$). *, $p < 0.05$; **, $p < 0.01$; ***, $p < 0.001$; ns, no significance.

4. Discussion

4.1. Effect of Cereal and Legume Intercropping on Grain Protein Content

GPC is an important index to reflect wheat quality; thus, it is of importance to simultaneously achieve high GPC and GY in wheat practice [35]. The present findings are in accordance with a previous study [17] that wheat and faba bean intercropping could simultaneously achieve both high GY and GPC because increased GPC in 2020 and increased protein yield in both years were found, and the intercropping effect was not influenced by N rates (Table 1). Yet, it was noted that GY and N uptake in intercropping depended on the maximum plant height, canopy, radiation use efficiency, interspecies interaction, the period of co-growing season, and so on [36]. Hence, conflict results of the effect of intercropping on GY and GPC were presented in different cereal-legume intercropping systems [22,36]. The wheat N uptake ability from flowering to maturity was one of the main reasons for the high GPC [37] and the N remobilization process was a potential target for improving the quality of wheat grain [20]. Recent studies found that wheat and faba bean intercropping stimulated wheat N uptake during mid- and late- growth stages and induced more N to shift from straws to grain due to intercropping up-regulating the key N assimilation enzyme activity and gene expression during the reproductive growth stages [38,39]. Thus, it could partly explain the reason for intercropping increasing GPC in the present study. Some temporary legume-based intercropping patterns were adopted in many regions due to overcoming some problems including technical and competition in intercropping, and in such conditions, legumes usually improved soil N availability for cereal and finally changed cereal GY and GPC [23,40]. In the present study, we observed that continuous intercropping increased soil N availability especially under low-N-input conditions (Supplementary Table); thus, we could not distinguish the role of the long- and short-term intercropping in improved GPC and GY.

In the present study, increased gliadins in both years and increased glutenins in 2019 were found due to wheat intercropped with faba bean (Figure 3). Gliadins and glutenins content determined the bread-making characteristics of wheat [41], because they play an important role in dough rheology [42]. These results in the present study meant that intercropping could alter the end-use of wheat quality, and more studies are needed to elucidate the mechanism of intercropping modulating protein fractions and their role related to wheat grain quality.

The present finding is partly in accordance with the results of a global meta-analysis that split N had a greater effect on wheat yield and protein content in less fertile soils and at high N rates [43], because GPC was increased by N input in 2020 but was not influenced by N rate in 2019 at the current situation (Table 1). Thus, N management is still a good strategy to improve GPC in the southwest of China. In a previous study, we found that wheat and faba bean had potential to save N input but still maintain wheat grain yield [21]; however, according to the present study, we could not ascertain whether decreased N input in intercropping would affect wheat GPC. This suggests that both GY and GPC should be taken into account when establishing an optimal N rate in the cereal and legume intercropping system.

4.2. Effect of Cereal and Legume Intercropping on Grain Amino Acids Content

In the present study, we found that the effect of intercropping on AAs content including NEAAs and EAAs was dependent on N levels, because some EAAs and NEAAs fractions declined due to intercropping when N was overused (N3 level), but some AAs fractions increased when wheat was intercropped with faba bean at low N levels (N0 and N1 level) (Tables 4 and 5). Taken together, the effect of intercropping on GPC was not affected by N rates in the present study, but it seems that wheat N input should not exceed 180 kg ha^{-1} in intercropping, because intercropping declined wheat AAs content at the N3 level (Table 1). Actually, wheat protein quality is not only dependent on the protein content but also related to the balance of AAs [44]. However, few studies have focused on intercropping on cereal

AAs content. Thus, the findings in the present study suggest that modulating N rates should be imperative to wheat grain quality in the legume-based intercropping system.

High NEAAs, especially high Pro and Glu content, were found in the present study (Table 5), which is in accordance with a previous study [45], whereas NEAAs such as Pro and Glu have a low nutritional value; thus, improvement in EAAs is more important for wheat grain quality. In the present study, it seems that intercropping did not modulate the ratio of EAAs to NEAAs, though there was year's variation (Figure 4), and intercropping had a greater impact on wheat grain protein rather than AAs. These findings should be linked with N remobilization and protein production during grain development. Still, more work on AAs and protein synthesis in intercropping could fully understand the findings.

4.3. Cereal and Legume Intercropping Modulated the Relationship between Grain Yield and Quality

The present study supports a previous study that when agronomic practices were given consideration, there was no trade-off between GY and quality [46], because we found steady GPC (10–15%) with increasing GY for MW, and GPC tended to increase with increasing GY for IW (Figure 5). Actually, wheat GPC content was largely dependent on post-anthesis N uptake [26]. Hence, the shift in the enhanced wheat N from the leaves and the stem to the grain and the stimulated wheat growth rate during the wheat mid-growing season [39,47] should be responsible for the changed correlation between GY and GPC in intercropping. The rainfall and temperature during 2018/2019 and 2019/2020 growing seasons were different (Figure 1), which might induce the effect of intercropping and N levels on GPC, which was different year by year in the present study (Table 1). However, the intercropping yield advantage was stable in the two-year field experiment (Table 1); thus, we thought that grain quality might be more sensitive to temperature and rainfall than grain yield. Hence, no relationship between GY and GPC was found for MW in the present study, but more work should still be conducted in the future to ascertain the correlation between GY and GPC under the current situation.

An early study from Eppendorfer found that correlations between AAs and N content within a variety were similar [48]. However, the correlation between wheat, maize, and soybean GPC and AAs presented a high variation [45,49]. In the present study, linear regression equations were established and significant correlations were found both between AAs and GCP and between AAs and GY for mono- and inter-cropped wheat grain (Figure 5), but we did not analyze the relationship between each AA and GPC and GY. According to our findings, intercropping either increased or decreased some specific AAs content, and intercropping affected the contents of TAAs, EAAs, and NEAAs in wheat grain under different N levels; hence, it could deduce the relationship between GPC and the given AA, which should change due to intercropping.

5. Conclusions

Higher protein yield and AAs yield were obtained when wheat was intercropped with faba bean. Intercropping mainly increased wheat GPC rather than AAs content because intercropping had no impact on AAs content regardless of N levels, but the 9% GPC of IW was higher than that of MW in 2020. Wheat gliadin content was increased on average by 8–14% when intercropped with faba bean. Similarly, some EAAs and NEAAs fraction contents were increased due to intercropping under N0 and N1 levels, but IW presented lower contents of EAAs and NEAAs fractions at the N3 level relative to MW. There was no trade-off relationship between GPC and GY according to regression analysis in the present study, and intercropping was a good option for simultaneously achieving both high GY and GPC. Hence, wheat and faba bean intercropping presented a potential to improve both wheat grain quality and yield, and modulated N rates were important to maximize the intercropping advantage in terms of grain quality. We suggest that the wheat N application rate should not exceed 180 kg ha^{-1} to achieve both intercropping yield and quality advantages in the southwest of China and similar regions.

Supplementary Materials: The following supporting information can be downloaded at: https://www.mdpi.com/article/10.3390/agronomy12122984/s1, Table S1: Soil total nitrogen and available nitrogen contents in the each treatment at soil depths of 0–20cm before the start of the experiment of 2018–2019.

Author Contributions: Conceptualization, J.X. and Y.Z.; methodology, Y.D. and L.T.; formal analysis, Y.-a.Z., J.H. and Z.Y.; investigation, Y.-a.Z., J.H., Z.Y., D.Z., H.L., X.W.; writing—original draft preparation, Y.-a.Z.; writing—review and editing, J.H.; funding acquisition, J.X. All authors have read and agreed to the published version of the manuscript.

Funding: This work was supported by the National Natural Science Foundation of China (32060718 and 31760611) and the Yunnan Agricultural Foundation Joint Project (2018FG001-071).

Data Availability Statement: Not applicable.

Acknowledgments: We thank LetPub (www.letpub.com) for its linguistic assistance during the preparation of this manuscript.

Conflicts of Interest: The authors declare no conflict of interest.

References

1. Beillouin, D.; Ben-Ari, T.; Makowski, D. Evidence map of crop diversification strategies at the global scale. *Environ. Res. Lett.* **2020**, *15*, 19601. [CrossRef]
2. Martin-Guay, M.; Paquette, A.; Dupras, J.; Rivest, D. The new green revolution: Sustainable intensification of agriculture by intercropping. *Sci. Total Environ.* **2018**, *615*, 767–772. [CrossRef] [PubMed]
3. Raseduzzaman, M.; Jensen, E.S. Does intercropping enhances yield stability in arable crop production? A meta-analysis. *Eur. J. Agron.* **2017**, *91*, 25–33. [CrossRef]
4. Aydemir, S.K.; Kızılşimşek, M. Assessing yield and feed quality of intercropped sorghum and soybean in different planting patterns and in different ecologies. *Int. J. Environ. Sci. Technol.* **2019**, *16*, 5141–5146. [CrossRef]
5. Ullah, M.A.; Hussain, N.; Schmeisky, H.; Rasheed, M.; Anwar, M.; Rana, A.S. Foder quality improvement through intercropping and fertilizer applicaiton. *Pak. J. Agri. Sci.* **2018**, *55*, 549–554.
6. Zhang, J.; Yin, B.; Xie, Y.; Li, J.; Yang, Z.; Zhang, G. Legume-cereal intercropping improves forage yield, quality and degradability. *PLoS ONE* **2015**, *10*, e0144813. [CrossRef] [PubMed]
7. Li, C.; Dong, Y.; Li, H.; Shen, J.; Zhang, F. Shift from complementarity to facilitation on P uptake by intercropped wheat neighboring with faba bean when available soil P is depleted. *Sci. Rep.* **2016**, *6*, 18663. [CrossRef] [PubMed]
8. Belel, M.D.; Halim, R.A.; Rafii, M.Y.; Saud, H.M. Intercropping of orn with some selected legumes for improved forage production: A review. *J. Agr. Sci.* **2014**, *6*, 44.
9. Lepse, L.; Sandra, D.; Zeipina, S.; Domínguez-Perles, R.; Eduardo, A.S. Evaluation of vegetable-faba bean (*Vicia faba* L.) intercropping under Latvian agro-ecological conditions. *J. Sci. Food Agr.* **2017**, *97*, 4334–4342. [CrossRef] [PubMed]
10. Salem, A.K.; Fadia, M.; Sultan, F.M.; El-Douby, K.A. Effect of intercropping cowpea (Vigna unguiculata L.) with teosinte (Zea mexicana Schrad) on forage yield productivity and its quality. *Egypt J. Agron.* **2019**, *41*, 183–196. [CrossRef]
11. Putnam, D.H.; Herbert, S.J.; Vargas, A. Intercropped corn-soyabean density studies. II. Yield composition and protein. *Exp. Agric.* **1986**, *22*, 373–381. [CrossRef]
12. Abdel Magid, H.M.; Ghoneim, M.F.; Rabie, R.K.; Sabrah, R.E. Productivity of wheat and alfalfa under intercropping. *Exp. Agr.* **1991**, *27*, 391–395. [CrossRef]
13. Wen, B.; Zhang, X.; Ren, S.; Duan, Y.; Zhang, Y.; Zhu, X.; Wang, Y.; Ma, Y.; Fan, W. Characteristics of soil nutrients, heavy metals and tea quality in different intercropping patterns. *Agroforest Syst.* **2019**, *94*, 1–12. [CrossRef]
14. Qiao, X.; Chen, X.; Lei, J.; Sai, L.; Xue, L. Apricot-based agroforestry system in Southern Xinjiang Province of China: Influence on yield and quality of intercropping wheat. *Agroforest Syst.* **2020**, *94*, 477–485. [CrossRef]
15. Gafurova, D.A.; Tursunkhodzhaev, P.M.; Kasymova, T.D.; Yuldashev, P.K. Fractional and amino-acid composition of wheat grain cultivated in Uzbekistan. *Chem. Nat. Compd.* **2002**, *38*, 462–465. [CrossRef]
16. Stoltz, E.; Nadeau, E. Effects of intercropping on yield, weed incidence, forage quality and soil residual N in organically grown forage maize (*Zea mays* L.) and faba bean (*Vicia faba* L.). *Field Crop Res.* **2014**, *169*, 21–29. [CrossRef]
17. Mthembu, B.E.; Everson, T.M.; Everson, C.S. Intercropping for enhancement and provisioning of ecosystem services in smallholder, rural farming systems in KwaZulu-Natal Province, South Africa: A review. *J. Crop Improv.* **2018**, *33*, 145–176. [CrossRef]
18. Wenefrida, I.; Utomo, H.S.; Blanche, S.B.; Linscombe, S.D. Enhancing essential amino acids and health benefit components in grain crops for improved nutritional values. *Recent Pat. DNA Gene Seq.* **2009**, *3*, 219–225. [CrossRef] [PubMed]
19. Uauy, C.; Brevis, J.C.; Dubcovsky, J. The high protein content gene Gpc-B1 accelerates senescence and has pleiotropic effects on protein content in wheat. *J. Exp. Bot.* **2006**, *57*, 2785–2794. [CrossRef] [PubMed]
20. Zhong, Y.; Xu, D.; Hebelstrup, K.H.; Yang, D.; Cai, J.; Wang, X.; Zhou, Q.; Cao, W.; Dai, T.; Jiang, D. Nitrogen topdressing timing modifies free amino acids profiles and storage protein gene expression in wheat grain. *BMC Plant Biol.* **2018**, *18*, 353. [CrossRef]

21. Xiao, J.; Yin, X.; Ren, J.; Zhang, M.; Tang, L.; Zheng, Y. Complementation drives higher growth rate and yield of wheat and saves nitrogen fertilizer in wheat and faba bean intercropping. *Field Crop Res.* **2018**, *221*, 119–129. [CrossRef]

22. Tosti, G.; Guiducci, M. Durum wheat–faba bean temporary intercropping: Effects on nitrogen supply and wheat quality. *Europ. J. Agron.* **2010**, *33*, 157–165. [CrossRef]

23. De Stefanis, E.; Sgrulletta, D.; Pucciarmati, S.; Ciccoritti, R.; Quarant, F. Influence of durum wheat-faba bean intercrop on specific quality traits of organic durum wheat. *Biol. Agric. Hortic.* **2016**, *33*, 28–29. [CrossRef]

24. Pellegrini, F.; Carlesi, S.; Nardi, G.; Bàrberi, P. Wheat–clover temporary intercropping under Mediterranean conditions affects wheat biomass, plant nitrogen dynamics and grain quality. *Eur. J. Agron.* **2021**, *130*, 126347. [CrossRef]

25. Tari, A. The effects of different deficit irrigation strategies on yield, quality, and water-use efficiencies of wheat under semi-arid conditions. *Agric. Water Manag.* **2016**, *167*, 1–10. [CrossRef]

26. Sieling, K.; Kage, H. Apparent fertilizer N recovery and the relationship between grain yield and grain protein concentration of different winter wheat varieties in a long-term field trial. *Eur. J. Agron.* **2021**, *124*, 126246. [CrossRef]

27. Vrignon-Brenas, S.; Celette, F.; Piquet-Pissaloux, A.; Corre-Hellou, G.; David, C. Intercropping strategies of white clover with organic wheat to improve the trade-off between wheat yield, protein content and the provision of ecological services by white clover. *Field Crop Res.* **2018**, *224*, 160–169. [CrossRef]

28. Xiao, J.; Zhu, Y.; Bai, W.; Liu, Z.; Tang, L.; Zheng, Y. Yield performance and optimal nitrogen and phosphorus application rates in wheat and faba bean intercropping. *J. Integr. Agr.* **2021**, *20*, 3012–3025. [CrossRef]

29. Bereczz, K.; Simon-Sarkadi, L.; Ragasits, I.; Hoffmann, S. Comparison of protein quality and mineral element concentrations in grain of spelt (*Triticum spelta* L.) and common wheat (*Triticum aestivum* L.). *Arch. Agron. Soil Sci.* **2001**, *47*, 389–398. [CrossRef]

30. Triboi, E.; Martre, P.; Triboi-Blondel, A. Environmentally-induced changes in protein composition in developing grains of wheat are related to changes in total protein content. *J. Exp. Bot.* **2003**, *54*, 1731–1742. [CrossRef] [PubMed]

31. Adebiyi, A.P.; Aluko, R.E. Functional properties of protein fractions obtained from commercial yellow field pea (*Pisum sativum* L.) seed protein isolate. *Food Chem.* **2011**, *128*, 902–908. [CrossRef]

32. Markwell, M.A.K.; Haas, S.M.; Bieher, I.I.; Tonbert, N.E. A modification of the Lowry procedure to simplify protein determination in membrane and lipoprotein samples. *Ann. Biochem.* **1978**, *117*, 136. [CrossRef] [PubMed]

33. Wu, G. Amino acids: Metabolism, functions, and nutrition. *Amino Acids* **2009**, *37*, 1–17. [CrossRef]

34. Metho, L.; Taylor, J.; Hammes, P.S.; Randal, P.G. Effects of cultivar and soil fertility on grain protein yield, grain protein content, flour yield and breadmaking quality of wheat. *J. Sci. Food Agric.* **1999**, *79*, 1823–1831. [CrossRef]

35. Ma, J.; Xiao, Y.; Hou, L.; He, Y. Combining protein content and grain yield by genetic dissection in bread wheat under low-input management. *Foods* **2021**, *10*, 1058. [CrossRef] [PubMed]

36. Baumann, D.T.; Bastiaans, L.; Goudriaan, J.; van Laar, H.H.; Kropff, M.J. Analysing crop yield and plant quality in an intercropping system using an eco-physiological model for interplant competition. *Agr. Syst.* **2002**, *73*, 173–203. [CrossRef]

37. Rozbicki, J.; Ceglińska, A.; Gozdowski, D.; Jakubczak, M.; Cacak-Pietrzak, G.; Madry, W.; Golba, J.; Piechocinski, M.; Sobczynski, G.; Studnicki, M.; et al. Influence of the cultivar, environment and management on the grain yield and bread-making quality in winter wheat. *J. Cereal Sci.* **2015**, *61*, 126–132. [CrossRef]

38. Liu, Z.; Wu, X.; Tang, L.; Zheng, Y.; Li, H.; Pan, H.; Zhu, D.; Wang, J.; Huang, S.; Qin, X.; et al. Dynamics of N acquisition and accumulation and its interspecific N competition in a wheat-faba bean intercropping system. *J. Plant Nutr. Fertil.* **2020**, *26*, 1284–1294. (In Chinese)

39. Liu, Z.; Zhu, Y.; Dong, Y.; Tang, L.; Zheng, Y.; Xiao, J. Interspecies interaction for nitrogen use efficiency via up-regulated glutamine and glutamate synthase under wheat-faba bean intercropping. *Field Crop Res.* **2021**, *274*, 108324. [CrossRef]

40. Marcello, G.; Giacomo, T.; Beatrice, F.; Paolo, B. Sustainable management of nitrogen nutrition in winter wheat through temporary intercropping with legumes. *Agron. Sustain. Dev.* **2018**, *38*, 31.

41. Malik, A.H.; Kuktaite, R.; Johansson, E. Combined effect of genetic and environmental factors on the accumulation of proteins in the wheat grain and their relationship to bread-making quality. *J. Cereal Sci.* **2013**, *57*, 170–174. [CrossRef]

42. Barak, S.; Mudgil, D.; Khatkar, B.S. Biochemical and functional properties of wheat gliadins: A review. *Crit. Rev. Food Sci. Nutr.* **2015**, *55*, 357–368. [CrossRef]

43. Hu, C.; Sadras, V.O.; Lu, G.; Zhang, P.; Han, Y.; Liu, L.; Xie, J.; Yang, X.; Zhang, S. A global meta-analysis of split nitrogen application for improved wheat yield and grain protein content. *Soil Till. Res.* **2021**, *213*, 105111. [CrossRef]

44. Jiang, X.; Wu, P.; Tian, J. Genetic analysis of amino acid content in wheat grain. *J. Genet.* **2014**, *93*, 451–458. [CrossRef]

45. Hassan, A.E.; Heneidak, S.; Gowayed, S. Comparative studies of some Triticum species by grain protein and amino acids analyses. *J. Agron.* **2007**, *6*, 286–293.

46. Anderson, W.K.; Shackley, B.J.; Sawkins, D. Grain yield and quality: Does there have to be a trade-off? *Euphytica* **1997**, *100*, 183–188. [CrossRef]

47. Xiao, J.; Dong, Y.; Yin, X.; Ren, J.; Tang, L.; Zheng, Y. Wheat growth is stimulated by interspecific competition after faba bean attains its maximum growth rate. *Crop Sci.* **2018**, *58*, 1–14. [CrossRef]

48. Eppendorfer, W.H. Effects of nitrogen, phosphorus and potassium on amino acid composition and on relationships between nitrogen and amino acids in wheat and oat grain. *J. Sci. Food Agr.* **1978**, *29*, 995–1001. [CrossRef]

49. Sriperm, N.; Pesti, G.M.; Tillman, P.B. The distribution of crude protein and amino acid content in maize grain and soybean meal. *Anim. Feed Sci. Tech.* **2010**, *159*, 131–137. [CrossRef]

agronomy

Article

Determination of Critical Phosphorus Dilution Curve Based on Capsule Dry Matter for Flax in Northwest China

Yaping Xie [1], Yang Li [2], Limin Wang [1], Mir Muhammad Nizamani [3], Zhongcheng Lv [4], Zhao Dang [1], Wenjuan Li [1], Yanni Qi [1], Wei Zhao [1], Jianping Zhang [1,*], Zhengjun Cui [5], Xingrong Wang [1], Yanjun Zhang [1] and Gang Wu [6]

1 Crop Research Institute, Gansu Academy of Agricultural Sciences (GASS), Lanzhou 730070, China
2 College of Agronomy, Shanxi Agricultural University (SXAU), Taiyuan 030006, China
3 School of Life and Pharmaceutical Sciences, Hainan University (HNU), Haikou 570228, China
4 Ordos Institute of Agricultural Sciences (OIAS), Dongsheng 017000, China
5 College of Agronomy, Gansu Agricultural University (GSAU), Lanzhou 730070, China
6 Zhangye Water-Saving Agricultural Experimental Station, Gansu Academy of Agricultural Sciences (GAAS), Zhangye 734000, China
* Correspondence: zhangjpzw3@gsagr.ac.cn; Tel.: +86-0931-761-108-1

Abstract: One of the cores of flax (*Linum usitatissimum* L.) production is to precisely measure the requirement of phosphorus (P) fertilization for optimizing seed yield, grower profits, P-use efficiency, and reducing environmental risk. Therefore, critical P concentration (Pc) was proposed as a suitable analytical tool to assess the flax P nutrition status. Four field experiments, with five P applications (0, 40, 80, 120, 160 kg P_2O_5 ha^{-1}) and four cultivars (Longyaza 1, Longya 14, Lunxuan 2, and Dingya 22) were conducted from the 2017 to 2019 seasons. The capsule Pc dilution curve based on capsule dry matter (CDM) was described by Pc = 2.84 × CDM$^{-0.22}$ (R^2 = 0.87, $p < 0.01$), CDM ranging from 0.60 to 4.17 t ha^{-1}. The P nutrition index (PNI) exhibited a significant positive relationship with P application rate. In addition, the relative seed yield was closely related to PNI. Those results validate that the capsule Pc dilution curve can be an alternative and more rapid tool to diagnose flax P status to support P fertilization precise decisions during the reproductive growth of flax in northwest China.

Keywords: phosphorus dilution curve; flax; phosphorus nutrition index; relative seed yield; phosphorus nutrition status

Citation: Xie, Y.; Li, Y.; Wang, L.; Nizamani, M.M.; Lv, Z.; Dang, Z.; Li, W.; Qi, Y.; Zhao, W.; Zhang, J.; et al. Determination of Critical Phosphorus Dilution Curve Based on Capsule Dry Matter for Flax in Northwest China. *Agronomy* **2022**, *12*, 2819. https://doi.org/10.3390/agronomy12112819

Academic Editors: Christos Noulas, Shahram Torabian and Ruijun Qin

Received: 20 October 2022
Accepted: 7 November 2022
Published: 11 November 2022

Publisher's Note: MDPI stays neutral with regard to jurisdictional claims in published maps and institutional affiliations.

1. Introduction

There has been a growing interest in flax (*Linum usitatissimum* L.), known as oilseed flax or linseed [1], because industries have requested increased quantities of seeds in China. At present, available arable land is declining, so maximizing crop yield per unit area has become a significant aim of agricultural production in China [2,3]. Previous several studies have reported that phosphorus (P) fertilization improved the production of many oil crops, such as flax [4–6], canola [7], safflower and sunflower [8], soybean [9], etc. However, excessive P fertilization in crop production happened occasionally, which resulted in a series of concerns for environmental, ecological, and human health [10–12]. Therefore, the precise management of P fertilization of flax has become a core area of research.

Zamuner et al. [13] indicated that an ideal indicator of a crop P's nutritional status should show P deficiencies and excesses, provide rapid diagnosis, and allow correction during the growing season. In this case, critical P concentration (Pc) was developed, which is defined as the minimum plant P concentration needed to achieve maximum crop biomass, which is a suitable analytical tool to assess the crop P nutrition status [14–16]. Curves of Pc have been generated for many crops, such as potato [13,17], wheat [14,15], timothy [18], mungbean and urdbean [19], rapeseed and maize [15]. Moreover, Pc dilution curves vary among different regions, species, genotypes within species, and practice management [20].

The crop yield and quality are associated with P nutrient status. The Pc dilution curve could be useful as a reference curve to assess flax's nutritional status through the P nutrition index (PNI). Analogously to the N nutrition index, the PNI can be calculated as the ratio between actual plant P concentration and the Pc expected according to that actual crop biomass [14–17].

For diagnosing the P status and estimating the appropriate P fertilizer requirements of flax during the reproductive growth period, it is essential to develop a Pc dilution curve based on the capsule dry matter (CDM) of flax in northwest China. Furthermore, the quantitative assessment of seed yield in response to PNI is highly required to validate the Pc dilution curves as a robust diagnostic tool in flax production. Therefore, the objectives of this study were to establish and validate a Pc dilution curve based on CDM, to assess the relationship between relative seed yield (RY) and PNI in response to flax under different P rates in northwest China for improving P-use efficiency and environmental protection of flax production.

2. Materials and Methods

2.1. Experimental Site

Field experiments 1 and 2 were conducted from 2017 to 2018 at Dingxi Academy of Agricultural Science (34.26° N, 103.52° E, altitude of 2060 m) in Gansu, China, as described in detail in Table 1. Experiments 3 and 4 were carried out from 2018 to 2019 at Yongdeng, Gansu, China (36°02′ N, 103°40′ E, and altitude 2149 m). The soil type is classified as Arenosols [21]. The previous crops for the four experiments were all wheat.

Table 1. Basic information at the Dingxi site in the 2017 and 2018 growing seasons.

Growing Season	Soil Characteristics	Cultivar	P Rate (kg P_2O_5 ha^{-1})	Sampling Data
2017 (Exp 1)	Type: loam	Lunxuan 2	0 (P_0)	86
	Organic matter: 10.2 g kg^{-1}	Dingya 22	40 (P_{40})	92
	Total N: 0.98 g kg^{-1}		80 (P_{80})	98
	Available P: 11.7 mg kg^{-1}		120 (P_{120})	104
	Available K: 122.5 mg kg^{-1}			
	pH: 7.9			110
2018 (Exp 2)	Type: loam	Lunxuan 2	0 (P_0)	86
	Organic matter: 11.0 g kg^{-1}	Dingya 22	40 (P_{40})	92
	Total N: 1.01 g kg^{-1}		80 (P_{80})	98
	Available P: 12.6 mg kg^{-1}		120 (P_{120})	104
	Available K: 135.4 mg kg^{-1}			110
	pH: 8.1			

Exp 1 = experiment 1; Exp 2 = experiment 2; N = nitrogen; P = phosphorus; K = potassium.

2.2. Experimental Design

Data were obtained from four field experiments in which the P rates, flax cultivars, sites, the physicochemical property of pre-planting soil, and varied years were summarized in Tables 1 and 2.

A randomized complete block design with three replicates was used for this study with a plot size of 20 m^2 (4 m × 5 m). Four P application rates (0, 40, 80, and 120 kg P_2O_5 ha^{-1}) were applied to two flax cultivars (Table 1), and five P rates (0, 40, 80, 120, and 160 kg P_2O_5 ha^{-1}) were applied to two other flax cultivars (Table 2). The cultivars were mainly planted in the local agriculture department and farms. Phosphorus fertilizer was applied using calcium superphosphate and broadcast uniformly over the soil surface before seedbed preparation and sowing. Before seedbed preparation and sowing in each site year, 80 kg N ha^{-1} and 120 kg K_2O ha^{-1} were broadcast uniformly over the soil surface using urea and potassium sulfate, respectively. Forty kg N ha^{-1} of urea was top-dressed at the budding stage. To ensure the maximum potential productivity, 40 mm of water was used to irrigate each plot before the flowering of flax. Further crop management procedures followed common agricultural practices to ensure the maximum potential productivity, i.e., no factor other than P was limiting.

Table 2. Basic information at the Yongdeng site in the 2018 and 2019 growing seasons.

Growing Season	Soil Characteristics	Cultivar	P Rate (kg P_2O_5 ha^{-1})	Sampling Data
2018 (Exp 3)	Type: Arenosols	Longyaza 1	0 (P_0)	86
	Organic matter: 9.8 g kg^{-1}	Longya 14	40 (P_{40})	92
	Total N: 1.23 g kg^{-1}		80 (P_{80})	98
	Available P: 10.0 mg kg^{-1}		120 (P_{120})	104
	Available K: 178.3 mg kg^{-1} pH: 7.5		160 (P_{160})	110
2019 (Exp 4)	Type: Arenosols	Longyaza 1	0 (P_0)	86
	Organic matter: 7.6 g kg^{-1}	Longya 14	40 (P_{40})	92
	Total N: 1.05 g kg^{-1}		80 (P_{80})	98
	Available P: 8.7 mg kg^{-1}		120 (P_{120})	104
	Available K: 141.6 mg kg^{-1} pH: 8.2		160 (P_{160})	110

Exp 3 = experiment 3; Exp 4 = experiment 4; N = nitrogen; P = phosphorus; K = potassium.

2.3. Sampling and Measurement

The 30 plants of flax per plot (2 by 1 m) were manually harvested during the reproductive stage (days after sowing 86, 92, 98, 104, and 110 days). Flax capsules (Figure 1) were collected and dried at 75 °C until a constant weight each date. Dried capsules were ground in a sample mill, passed through a 1 mm sieve, and samples were digested using H_2SO_4-H_2O_2, after which the CPC was determined by the Colorimetric Molybdenum-Blue method according to Lithourgidis et al. [22].

Figure 1. The capsules of flax (**A**) at 98 days and capsules of flax (**B**) at 104 days, respectively.

On the harvest date, the crop in each plot was separately harvested using a sickle to determine the seed yield.

2.4. Data Analysis

2.4.1. Construction of the Pc Dilution Curve

The construction of a capsule Pc curve requires identifying critical data points at which the P neither limits growth nor enhances it. The data was collected to determine the Pc dilution curve during the 2018 and 2019 growing seasons at Yongdeng. A P-limiting treatment was defined as a treatment in which additional P led to a significant increase in CDM. A non-P-limiting treatment was defined as one in which P application did not lead to an increase in CDM.

The CDM and CPC values with different P rates were compared using ANOVA (SPSS 19 Software, Inc., Chicago, IL, USA) at a probability level of 5%, and treatment effects were determined using the least significant difference (LSD). A power regression equation was fitted to these theoretical, critical points to determine the equation of the Pc dilution curve. The Pc curve can be established by the following power equation [18]:

$$p_c = \alpha W^{-b} \tag{1}$$

where Pc is the critical P concentration (g kg^{-1}); W is the total dry matter expressed in t ha^{-1}; a and b are positive constants, where a represents the critical plant P concentration in the dry matter (DM) when W = 1 t ha^{-1} and b is a statistical parameter that represents the ratio between the relative decline in plant P concentration and the relative crop growth rate [20].

2.4.2. Phosphorus Nutrition Index

The P nutrition index (PNI) of the capsule in flax, used to characterize crop P status, was calculated according to the following formula [17,20]:

$$\text{PNI} = \frac{p_\alpha}{p_c} \tag{2}$$

where Pa was the actual capsule P concentration and Pc was the expected capsule P concentration. When PNI = 1, P nutrition is considered optimal. When PNI > 1, P nutrition is considered excessive; when PNI < 1, P nutrition is considered insufficient [12].

2.4.3. Relative Yield

The relative seed yield (RY) was obtained by dividing the seed yield at a given P rate by the highest seed yield among all P treatments [13,17]. The RY was calculated as the following equation:

$$\text{RY} = \frac{y_p}{y_h} \tag{3}$$

where Yp is the seed yield of flax in the fertilized plot, and Yh is the seed yield of the highest seed yield treatment (kg ha^{-1}).

The coefficients of determination (R^2) for the relationship between RY and PNI were calculated using SPSS 20.0.

2.4.4. Model Validation

In the current study, the root-mean-squared error (RMSE) and the normalized root-mean-squared error (n-RMSE) in 1:1 plots were used to evaluate the accuracy of the model, which is a common method used to identify the fitness of observed and estimated values [23]. When the n-RMSE $\leq$ 15%, it was looked at as a "good" agreement, 15–30% as a "moderate" agreement, and $\geq$30% as a "poor" agreement [24]. The RMSE and n-RMSE were calculated using Equations (4) and (5):

$$\text{RMSE} = \sqrt{\frac{\sum_{i=1}^{n}(S_i - m_i)^2}{n}} \tag{4}$$

$$\text{n} - \text{RMSE} = \frac{RMSE}{\bar{s}} \tag{5}$$

where n is the number of samples, s_i is an estimated model value; m_i is an observed value, and $\bar{s}$ is the averaged observed value.

3. Results

3.1. Capsule Dry Matter and Capsule P Concentration at Different P Levels

The P application significantly affected the CDM during the reproductive growth period. CDM increased gradually with the increase in P rate, with two years of value CDM averaging from 0.85 to 4.03 t ha^{-1} and from 0.64 to 3.63 t ha^{-1} in 2018 and 2019, respectively. In general, CDM increased significantly from P_0 to P_{80} treatments. Still, there were no significant differences among P_{80}, P_{120}, and P_{160} treatments. The maximum CDM was obtained in the P_{80} treatment for two seasons and two cultivars (Figure 2). During each season, the CDM accorded the following inequality under different P levels:

$$CDM_0 < CDM_{40} < CDM_{80} = CDM_{120} = CDM_{160} \tag{6}$$

where CDM_0, CDM_{40}, CDM_{80}, CDM_{120}, and CDM_{160} present the CDM value of P_0, P_{40}, P_{80}, P_{120} and P_{160}, respectively.

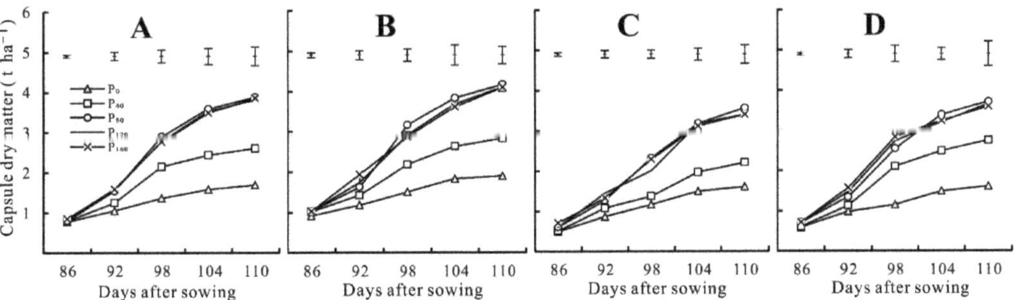

Figure 2. Changes in capsule dry matter for two flax cultivars at five P levels in two growing seasons. (**A**) Lonyaza 1 in 2018; (**B**) Longya 14 in 2018; (**C**) Lonyaza 1 in 2019 and (**D**) Longya 14 in 2019, respectively. The vertical bars indicate the least significant differences (LSDs) with $p \leq 0.05$ among five P levels (n = 3).

The CPC decreased gradually during the reproductive growth period of flax. A higher level of P generally resulted in a higher CPC (Figure 3). The CPC varied between 1.28 g kg^{-1} DM and 3.71 g kg^{-1} DM and from 1.08 g kg^{-1} DM to 3.53 g kg^{-1} DM in 2018 and 2019 across the two cultivars, respectively. The maximum CPC was obtained at the P_{160} treatment for two cultivars and two growing seasons (Figure 3).

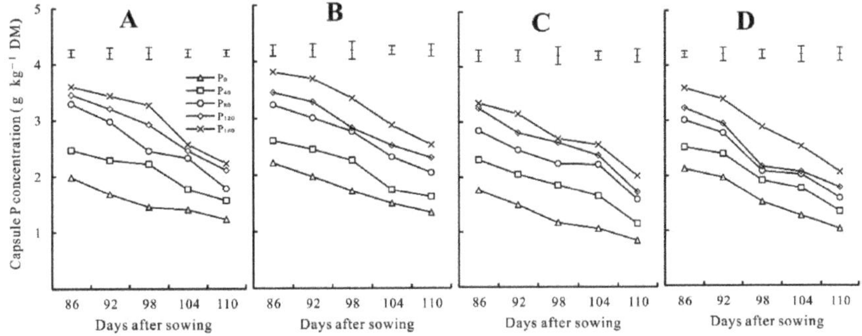

Figure 3. Changes in phosphorus concentration of the capsule for two flax cultivars at five P levels in two growing seasons. (**A**) Lonyaza 1 in 2018; (**B**) Longya 14 in 2018; (**C**) Lonyaza 1 in 2019 and (**D**) Longya 14 in 2019, respectively. The vertical bars indicate the least significant differences (LSDs) with $p \leq 0.05$ among five P levels (n = 3).

3.2. Constructing the Capsule Pc Dilution Curve for Flax

In this study, following the computational procedures of Justes et al. [25], capsule Pc points were established for each sampling date during the reproductive growth period in 2018 and 2019. The theoretical Pc points of CDM were decided for each sampling date, from capsule original to maturity for two cultivars, with 10 data points in the 2018 and 2019 seasons. A reducing trend of Pc points was observed with increasing CDM of flax (Figure 4).

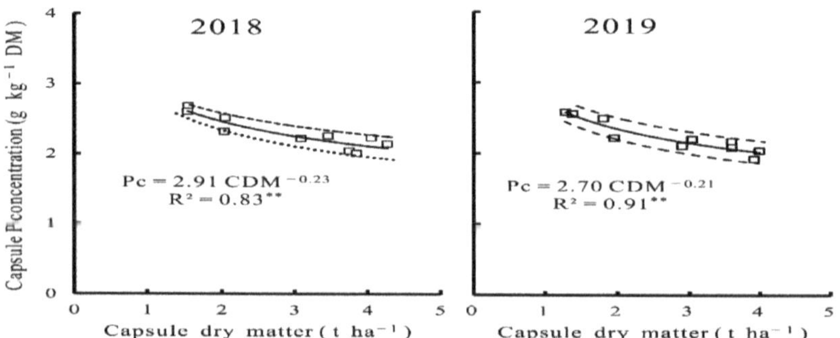

Figure 4. Critical phosphorus (Pc) data points were used to define the critical P dilution curve. The solid line represents the critical P dilution curve, depicting the relationship between the critical P concentration and the capsule dry matter of flax in Gansu, northwest China. The dotted lines represent the confidence intervals ($p = 0.95$). ** Significance at $p < 0.01$ probability level.

The following power equation could match the trend lines:

$$2018: Pc = 2.91\ CDM^{-0.23} \tag{7}$$

$$2019: Pc = 2.70\ CDM^{-0.21} \tag{8}$$

Two years expressed non-significant differences when compared according to calculation procedures recommended by Mead and Curnow [26]. Therefore, data points from two years were pooled to develop the following unified Pc curve (Figure 5):

$$Pc = 2.84\ CDM^{-0.22} \tag{9}$$

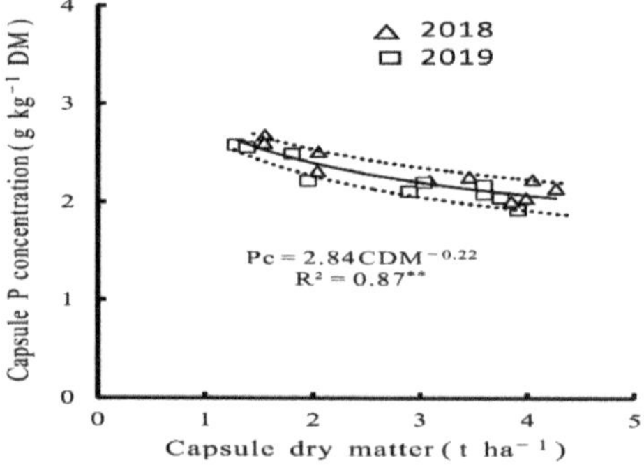

Figure 5. Critical phosphorus (Pc) data points for the capsule Pc curve definition using pooled data from two years. The solid line represents the critical P dilution curve ($Pc = 2.84\ CDM^{-0.22}$, $R^2 = 0.82$ **), which describes the relationship between the critical P concentration and the capsule dry matter of flax in northwest China. The dotted lines represent the confidence intervals ($p = 0.95$). ** Significance at $p < 0.01$ probability level.

3.3. Validation of the Capsule Critical Phosphorus Dilution Curve

The capsule Pc curve was validated by an independent data set from the experiments conducted in 2017–2018 (Experiments 1 and 2, Table 1). The results expressed that the newly established curve can discriminate the P-limiting and non-P-limiting growth conditions of flax during the reproductive growth period. The curve was not affected by cultivar, season, and site. The data points of the P-limiting treatments were below or close to the Pc curve, while the points of the non-P-limiting treatments were near or above this curve (Figure 6). The accuracy of this model was assessed by using the RMSE and n-RMSE, using Equations (4) and (5). Results found that the RMSE of the model was 0.32 g kg^{-1}, and the n-RMSE was 13.86%, indicating "good" agreement between the observed and assessed values (Figure 6). Therefore, in the present study, we constructed that the CDM-Pc model can be used for the diagnosis of plant P nutrition.

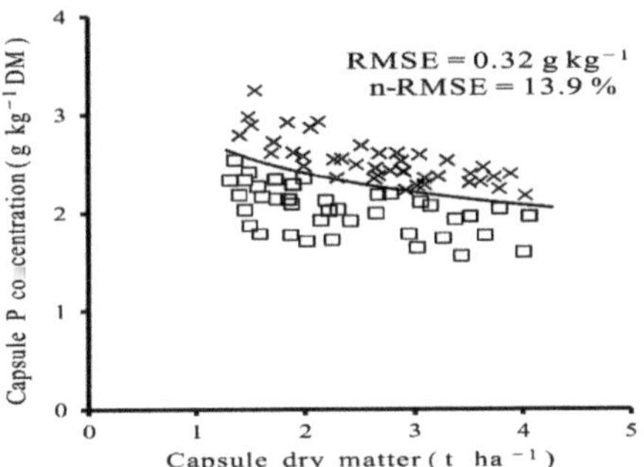

Figure 6. Validation of the critical phosphorus (Pc) curve using the independent data set from Experiments 1 and 2. Data points (□) and (×) represent P-limiting and non-P-limiting treatments, respectively. The solid line depicts the Pc curve based on the capsule dry matter of flax.

3.4. Change of Phosphorus Nutrition Index under Different P Treatments

The P nutrition index (PNI) is useful for diagnosing the crop P nutrition status. The PNI increased with an increasing P rate, ranging from 0.49 to 1.35 (Figure 7). For two years, the PNI values were <1 for the P_0 and P_{40} treatments, indicating the two levels were insufficient for P nutrition. While the values of PNI were >1 for the P_{120} and P_{160} treatments, indicating the presence of a surplus of P uptake in the capsule of flax. The values of PNI were close to 1 for P_{80}, and this indicates that the P rate is optimal for P nutrition for flax growth. Therefore, the optimal P rate was 80 kg P_5O_2 ha^{-1}. These results elucidate that PNI can be used as a robust diagnostic tool for the P status of flax under different P conditions.

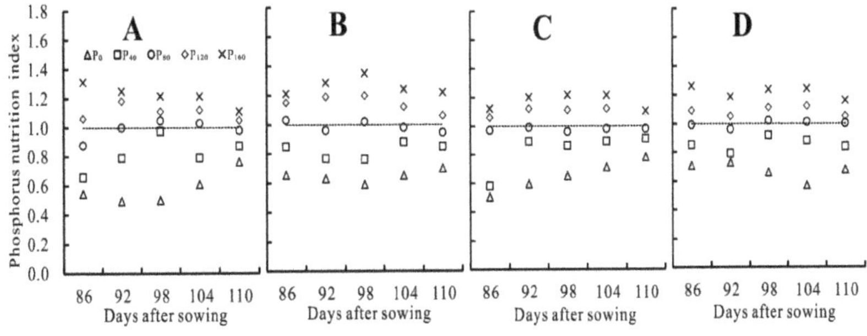

Figure 7. Phosphorus nutrition index (PNI) of two flax cultivars with five P rates in two years. (**A**) Lonyaza 1 in 2018; (**B**) Longya 14 in 2018; (**C**) Lonyaza 1 in 2019 and (**D**) Longya 14 in 2019, respectively. The reference line at PNI = 1 represents optimal P nutrition, while PNI > 1 shows excess P fertilizer application, and PNI < 1 shows P deficiency.

3.5. Relationship between Relative Seen Yield and PNI

The relationship between relative RY and PNI was well illustrated with a second-order polynomial equation (RY = -1.75 PNI^2 + 3.63 PNI-0.88, R^2 = 0.92 **). As seen in Figure 8, for PNI = 1, the relative RY was near 1.0, while for PNI > 1 or PNI < 1, the relative RY decreased. The current study showed that inadequate and excessive P use lowers the relative RY. Only the optimal P application rate results in the maximum relative RY.

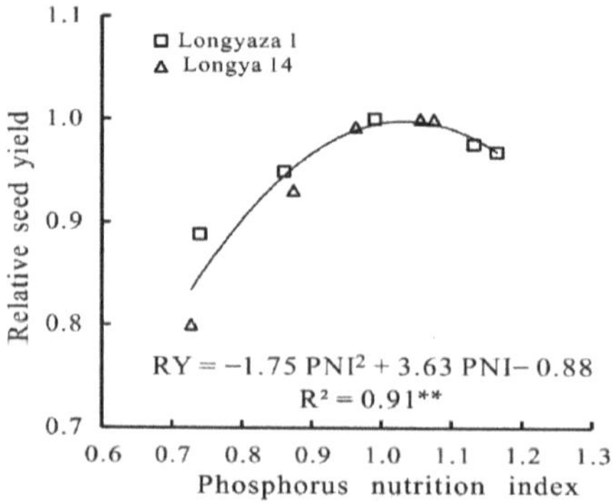

Figure 8. Relationship between relative seed yield (RY) and phosphorus nutrition index (PNI) for two flax cultivars. The PNI data were averaged over two years. ** Significance at $p < 0.01$ probability level.

4. Discussion

This study proposes the idea of the current Pc dilution phenomenon in the capsule of flax firstly. We have constructed a Pc dilution curve based on the CDM and provided a new way of diagnosing and regulating the P status of flax during the reproductive period in northwest China. Phosphate plays a pivotal role in the nexus of photosynthesis [27], energy conservation [28], and carbon metabolism [29] in higher plants and its application to agricultural soils is crucial to achieving optimum crop production [15]. However, excessive P fertilizer cannot ensure a significant increase in crop productivity, yet its abundant application can decrease crop yield and cause environmental damage. In recent years, there has been an increasing demand for simple, accurate, and stable tools for diagnosing the status of crop P, which can provide appropriate on-farm P management. With the advancement of the P dilution principle, the diagnosis model based on this theory has been developed and used to guide agricultural field production.

4.1. Comparison with Other Critical Phosorus Dilution Curves

The Pc dilution curve based on DM (including aboveground DM, vines DM, tubers DM, etc.) has been previously established for different crop species and regions (Table 3). However, this study was the first to analyze the Pc dilution curve in the capsules of flax. Our results show that, as was observed for N [30], there is a dilution of P with increasing CDM. The Pc dilution curve, based on the CDM, was Pc = $2.84 \times$ CDM$^{-0.22}$, where 2.84 represents the Pc when CDM = 1 t, in this study.

Many differences in curve parameters are noticeable between our study and the earlier studies in other crops. In the current study, the value of parameter *a* was lower than the value of parameter a based on plant DM reported in younger timothy and older timothy [18], in potato [13,16,17,20], in wheat [14], and for winter wheat, maize, and rapeseed [15]. The discrepancy was related to genotype, circumstance, and growth stage differences. Firstly, the genotype of crops induces the differences between these curves. Genotype differences in critical P dilution curves have been reported in Switzerland with wheat, maize, and rapeseed [15]. The difference in critical P dilution curves within genotype is probably attributed to the general nature of crop species. Secondly, the growth environment of crops resulted in discrepancies in the Pc dilution curves. This is because the difference from the growth environment may be attributed to differences in pedo-climatic conditions, water supply, soil properties, etc. The growth environment may have affected P uptake from the soil and partitioning by different crop apparatus. However, parameter a in this study was greater than those of parameter a based on tubers DM and vines DM reported in potatoes [17]. Corresponding to the tuber P dilution curve, the main reason may be that the P concentration of a potato tuber was smaller than the P level of a flax capsule, and to the vines' P dilution curve, other factors obviously influence the P dilution curve. Further research is required to be conducted to make a thorough inquiry.

Table 3. Comparison of the parameters of the critical phosphorus concentration (Pc) dilution curve for different regions and measured parameters of timothy, potato, wheat, maize, rapeseed, and flax.

Crop	Region	[e] a	[f] b	Measured parameter	Reference
OT	Canada	3.27	0.20	DM [a]	Bélanger and Ziadi, 2008 [18]
YT		5.23	0.40		
Potato	Argentina	3.92	0.30	DM [b]	Zamuner et al., 2016 [13]
	Colombia	5.23	0.19	DM	Gómez et al., 2016 [16]
	Canada	3.57	0.38	DM	Nyiraneza et al., 2021 [17]
		2.50	0.20	Vines DM	
		2.06	0.14	Tubers DM	
	Brazil	3.91	0.30	DM	Soratto et al., 2020 [20]
Wheat	Canada	4.94	0.49	DM [c]	Bélanger et al., 2015 [14]
	Finland	4.04	0.21	DM	
	Canada	3.62	0.23	DM	
	China	4.40	0.29	DM	
WW	Switzerland	4.44	0.41	DM	Cadot et al., 2018 [15]
Maize		3.49	0.19		
Rapeseed		5.18	0.39		
Flax	China	2.84	0.22	CDM [d]	This article

OT = older timothy; YT= younger timothy; WW = winter wheat; [a] DM, aboveground dry matter; [b] DM, vines + tubers dry matter; [c] DM, shoot dry matter; [d] CDM, capsule dry matter; [e] a, the critical plant P concentration in the capsule dry matter = 1 t ha^{-1}; [f] b, the decline in capsule phosphorus concentration with crop growth.

Parameter b describes the decline in CPC with crop growth and therefore depends on capsule P uptake relative to the CDM increase. The decline in CPC during reproductive growth can be attributed to the decrease in P concentration per unit CDM, similar to the N concentration decline per unit shoot DM in linseed [30]. The decrease in CDC is probably due to (i) a large photosynthetic product of other vegetative parts (stem and leaf) transferred to the seed for yield formation during the reproductive growth period [4,31], which causes the faster rate of CDM accumulation, and (ii) the P nutrition of pericarp and stalk was transferred to the seed for fatty acids and other nutrients formation, which results in a declining P concentration of pericarp and stalk during the reproductive growth period.

Compared with the DM model, the parameter b we established in this study was close to those curves proposed for older timothy in Canada [18], for potatoes in Colombia [16] and in Canada [17], for wheat in Finland [14], and for maize in Switzerland [15]. Whereas the difference existed between the parameter b in the current study and those of the curve developed for younger timothy [18], for potato [13,20] and potato [17], for wheat [14], and for wheat and rapeseed [15]. Those indicated that the rates of plant Pc dilution of the mentioned crops were faster than that of the capsule Pc dilution in flax. These differences were associated with the transfer of large amounts of P to the capsule to satisfy yield and fatty acid formation and a slower CPC decline in the capsule because the capsule was the epicenter of plant growth during the reproductive period.

4.2. Diagnosis of Phosphorus Nutrition Status

The present study validated the Pc curve based on the CDM with independent data. Our results indicated that this curve could accurately distinguish P-limiting and non-P-limiting treatments during the reproductive growth period of flax. For this purpose, the PNI value was calculated based on this Pc curve. This curve has been used to diagnose the P status of timothy [18], potato [13,16,17,20], wheat [14], maize, and rapeseed [15]. In general, those Pc curves were developed during the throughout growth period; only Nyiraneza et al. [17] proposed a tuber Pc dilution curve of potatoes during the reproductive period. Since the reproductive period was another peak of P uptake for flax [32], it is essential to develop a curve for diagnosing the P status of flax after anthesis.

In the study, PNI was the ratio between the measured capsule P concentration of flax and Pc. The PNI can effectively distinguish conditions of P deficiency, optimal levels, or P surplus of flax during the reproductive growth period. Confirmation of P sufficiency during this period of flax can help perform corrective measures, especially under irrigated conditions. Studies showed that foliar P fertilizer applications [33] could be supplied in-season to correct a P deficiency [34], improve photoassimilate production [35], and enhance dry matter distribution [36] and translocation from vegetation organs to seeds [4]; although, only a limited amount of P can be absorbed into the plant through foliage. Therefore, PNI can estimate phosphorus nutrition sufficiency, and this would represent a fast and cost-effective option. This new methodology offers the only reliable way to correct P nutrition during the critical period of flax growth and to ensure maximum yields without having to apply large doses of P at planting. This strategy could help to achieve much more efficient use of P fertilizer and to minimize the cost and environmental threat associated with applying P at high rates.

5. Conclusions

Based on the theory of Pc dilution, this study proposed a Pc dilution curve based on CDM for flax. The curve was described by the following equation, $Pc = 2.84\ CDM^{-0.22}$, for the CDM of flax. The equation was derived from data obtained at five growth stages from two cultivars over five P fertilizer application rates and two growing seasons. The equation was validated using data from two separate cultivars under different sites and the same fertilizer conditions. The PNI values on different sampling dates were generally <1 under P-limiting and >1 under non-P-limiting conditions. There was a significant positive relationship between PNI and the P level across the five development stages, suggesting the Pc dilution curve can be used as a tool for diagnosing the P status of flax. In addition, the relationship between relative RY and PNI was well illustrated with a second-order polynomial equation, in which for PNI = 1, the relative RY was near 1.0, while for PNI > 1 or PNI < 1, the relative RY decreased. Hence, we conclude that the capsule Pc curve can be adopted as a practical diagnostic tool for effective P management during the reproductive growth period of flax in northwest China.

Author Contributions: Conceptualization, J.Z.; investigation, L.W., M.M.N., Z.L., Z.D., W.L., Y.Q. and W.Z.; project administration, Z.C., X.W., Y.Z. and G.W.; and data curation and original draft preparation, Y.X. and Y.L. All authors have read and agreed to the published version of the manuscript.

Funding: This research was funded by the Key Research and Development Projects of Gansu Academy of Agricultural Sciences (2021GAAS20), the National Natural Science Programs of China (31660368; 32060437), the Major Special Projects of Gansu Province (21ZD4NA022), and China Agriculture Research System of MOF and MARA (CARS-17-GW-04).

Institutional Review Board Statement: Not applicable.

Informed Consent Statement: Not applicable.

Data Availability Statement: The datasets used and/or analyzed during the current study are available from the corresponding author upon reasonable request.

Conflicts of Interest: The authors declare no conflict of interest.

References

1. Zuk, M.; Richter, D.; Matuła, J.; Szopa, J. Linseed, the multipurpose plant. *Ind. Crops Prod.* **2015**, *75*, 165–177. [CrossRef]
2. Yao, X.; Zhao, B.; Tian, Y.C.; Liu, X.J.; Ni, J.; Cao, W.X.; Zhu, Y. Using leaf dry matter to quantify the critical nitrogen dilution curve for winter wheat cultivated in eastern China. *Field Crops Res.* **2014**, *159*, 33–42. [CrossRef]
3. Zheng, J.; Fan, J.L.; Zhang, F.C.; Yan, S.C.; Wu, Y.; Lu, J.S.; Guo, J.J.; Cheng, M.H.; Pei, Y.F. Through fall and stem flow heterogeneity under the maize canopy and its effect on soil water distribution at the row scale. *Sci. Total Environ.* **2019**, *660*, 1367–1382. [CrossRef] [PubMed]
4. Xie, Y.P.; Niu, J.Y.; Gan, Y.T.; Gao, Y.H.; Li, A.R. Optimizing phosphorus fertilization promotes dry matter accumulation and P remobilization in oilseed flax. *Crop Sci.* **2014**, *54*, 1729–1736. [CrossRef]
5. Xie, Y.P.; Niu, X.X.; Niu, J.Y. Effect of phosphorus fertilizer on growth, phosphorus uptake, seed yield, yield components and phosphorus use efficiency of oilseed flax. *Agron. J.* **2016**, *108*, 1257–1266. [CrossRef]
6. Xie, Y.P.; Yan, Z.L.; Niu, Z.X.; Coulterd, J.A.; Niu, J.Y.; Zhang, J.P.; Wang, B.; Yan, B.; Zhao, W.; Wang, L.M. Yield, oil content, and fatty acid profile of flax (*Linum usitatissimum* L.) as affected by phosphorus rate and seeding rate. *Ind. Crops Prod.* **2020**, *145*, 112087. [CrossRef]
7. Cheema, M.A.; Malik, M.A.; Hussain, A.; Shah, S.H.; Basra, S.M.A. Effects of time and rate of nitrogen and phosphorus application on the growth and the seed and oil yields of canola (*Brassica napus* L.). *J. Agron. Crop Sci.* **2001**, *186*, 103–110. [CrossRef]
8. Abbadi, J.; Gerendás, J. Effects of phosphorus supply on growth, yield, and yield components of safflower and sunflower. *J. Plant Nutr.* **2011**, *34*, 1769–1787. [CrossRef]
9. Singh, G.; Menon, S. Effect of phosphorus on soybean production—A review. *Plant Cell Biotechnol. Mol. Biol.* **2021**, *22*, 31–34.
10. Stachelek, J.; Ford, C.; Kincaid, D.; King, K.; Miller, H.; Nagelkirk, R. The national eutrophication survey: Lake characteristics and historical nutrient concentrations. *Earth Syst. Sci. Data* **2018**, *10*, 81–86. [CrossRef]
11. Li, H.G.; Huang, G.Q.; Meng, L.; Ma, L.; Yuan, L.; Wang, F.H. Integrated soil and plant phosphorus management for crop and environment in China: A review. *Plant Soil* **2011**, *349*, 157–167. [CrossRef]
12. Li, B.; Boiarkina, I.; Yu, W.; Ming, H.; Munir, T.; Qian, G.; Young, B.R. Phosphorous recovery through struvite crystallization: Challenges for future design. *Sci. Total. Environ.* **2019**, *648*, 1244–1256. [CrossRef] [PubMed]
13. Zamuner, E.C.; Lloveras, J.; Echeverría, H.E. Use of a Critical Phosphorus Dilution Curve to Improve Potato Crop Nutritional Management. *Am. J. Potato Res.* **2016**, *93*, 392–403. [CrossRef]
14. Bélanger, G.; Ziadi, N.; Pageau, D.; Grant, C.; Högnäsbacka, M.; Virkajärvi, P.; Hu, Z.; Lu, J.; Lafond, J.; Nyiraneza, J.A. Model of critical phosphorus concentration in the shoot biomass of wheat. *Agron. J.* **2015**, *107*, 963–970. [CrossRef]
15. Cadot, S.; Bélanger, G.; Ziadi, N.; Morel, C.; Sinaj, S. Critical plant and soil phosphorus for wheat, maize, and rapeseed after 44 years of P fertilization. *Nutr. Cycl. Agroecosyst.* **2018**, *112*, 417–433. [CrossRef]
16. Gómez, M.I.; Magnitskiy, S.; Rodriguez, L.E. Critical Dilution Curves for Nitrogen, Phosphorus, and Potassium in Potato Group Andigenum. *Agron. J.* **2019**, *111*, 419–427. [CrossRef]
17. Nyiraneza, J.; Bélanger, G.; Benjanne, R.; Ziadi, N.; Cambouris, A.; Fuller, K.; Hann, S. Critical phosphorus dilution curve and the phosphorus-nitrogen relationship in potato. *Eur. J. Agron.* **2021**, *123*, 126205. [CrossRef]
18. Bélanger, G.; Ziadi, N. Phosphorus and Nitrogen Relationships during Spring Growth of an Aging Timothy Sward. *Agron. J.* **2008**, *100*, 1757–1762. [CrossRef]
19. Venkatesh, M.S.; Hazra, K.K.; Ghosh, P.K. Determination of Critical Tissue Phosphorus Concentration in Mungbean and Urdbean for Plant Diagnostics. *J. Plant Nutr.* **2014**, *37*, 2017–2025. [CrossRef]
20. Soratto, R.P.; Sandana, P.; Fernandes, A.M.; Martins, J.D.L.; Job, A.L.G. Testing critical phosphorus dilution curves for potato cropped in tropical Oxisols of southeastern Brazil. *Eur. J. Agron.* **2020**, *115*, 126020. [CrossRef]
21. FAO. World Reference Base for Soil Resources 2014. *World Soil Resources Reports No. 106. Rome.* 2015. Available online: http://www.fao.org (accessed on 6 June 2022).
22. Lithourgidis, A.S.; Matsi, T.; Barbayiannis, N.; Dordas, C.A. Effect of liquid cattle manure on corn yield, composition, and soil properties. *Agron. J.* **2007**, *99*, 1041–1047. [CrossRef]
23. Qiang, S.C.; Zhang, F.C.; Miles, D.; Zhang, Y.; Xiang, Y.Z.; Fan, J.L. Determination of critical nitrogen dilution curve based on leaf area index for winter wheat in the Guanzhong Plain, Northwest China. *J. Integr. Agric.* **2019**, *18*, 2369–2380. [CrossRef]
24. Fan, J.L.; Wu, L.F.; Zhang, F.C.; Cai, H.J.; Ma, X.; Bai, H. Evaluation and development of empirical models for estimating daily and monthly mean daily diffuse horizontal solar radiation for different climatic regions of China. *Renew. Sustain. Energy Rev.* **2019**, *105*, 168–186. [CrossRef]
25. Justes, E.; Mary, B.; Machet, J.M. Determination of a critical nitrogen dilution curve for winter wheat crops. *Ann. Bot.* **1994**, *74*, 397–407. [CrossRef]
26. Mead, R.; Curnow, R.N. *Statistical Metahods in Agriculture and Experimental Biology*; Mead, R., Ed.; Champan and Hall: London, UK, 1983; pp. 157–163.
27. Cho, M.H.; Jang, A.; Bhoo, S.H.; Jeon, J.S.; Hahn, T.R. Manipulation of triose phosphate/phosphate translocator and cytosolic fructose-1, 6-bisphosphatase, the key components in photosynthetic sucrose synthesis, enhances the source capacity of transgenic Arabidopsis plants. *Photosy. Res.* **2012**, *111*, 261–268. [CrossRef]
28. Schluepmann, H.; Berke, L. Sanchez-Perez, G.F. Metabolism control over growth: A case for trehalose-6-phosphate in plants. *J. Exp. Bot.* **2012**, *63*, 3379–3390. [CrossRef]

29. Abel, S.; Ticconi, C.A.; Delatorre, C.A. Phosphate sensing in higher plants. *Physiol. Plant* **2002**, *115*, 1–8. [CrossRef]
30. Flénet, F.; Gúerif, M.; Boiffin, J.; Dorvillez, D.; Champolivier, L. The critical N dilution curve for linseed (*Linum usitatissimum* L.) is different from other C3 species. *Eur. J. Agron.* **2006**, *24*, 367–373. [CrossRef]
31. Dordas, C.A. Variation of physiological determinants of yield in linseed in response to nitrogen fertilization. *Ind. Crops Prod.* **2010**, *31*, 455–465. [CrossRef]
32. Xie, Y.P.; Li, A.R.; Yan, Z.L.; Niu, J.Y.; Sun, F.X.; Yan, B.; Zhang, H. Effect of different phosphorus level on phosphorus nutrient uptake, transformation and phosphorus utilization efficiency of oil flax. *Acta Pratacult. Sin.* **2014**, *23*, 158–166. (In Chinese)
33. Hedley, M.J.; McLaughlin, M.J. Reactions of Phosphate Fertilizers and By-Products in Soils. In *Phosphorus: Agriculture and the Environment*; Sharpley, A.N., Ed.; American Society of Agronomy, Crop Science Society of America, and Soil Science Society of America: Madison, WI, USA, 2005; pp. 181–252.
34. Fageria, N.K.; Filho, M.B.; Moreira, A.; Guimaraes, C.M. Foliar fertilization of crop plants. *J. Plant Nutr.* **2009**, *32*, 1044–1064. [CrossRef]
35. ElSayed, A.I.; Weig, A.R.; Sariyeva, G.; Hummel, E.; Yan, S.L.; Bertolini, A.; Komor, E. Assimilate export inhibition in Sugarcane yellow leaf virus-infected sugarcane is not due to less transcripts for sucrose transporters and sucrose-phosphate synthase or to callose deposition in sieve plates. *Physiol. Mol. Plant Pathol.* **2012**, *81*, 64–73. [CrossRef]
36. Fleisher, D.H.; Wang, Q.; Timlin, D.J.; Chun, J.A.; Reddy, V.R. Effects of carbon dioxide and phosphorus supply on potato dry matter allocation and canopy morphology. *J. Plant Nutr.* **2013**, *36*, 566–586. [CrossRef]

Article

Sugarcane Ratoon Yield and Soil Phosphorus Availability in Response to Enhanced Efficiency Phosphate Fertilizer

Clayton Luís Baravelli de Oliveira [1,*], Juliana Bonfim Cassimiro [2], Maikon Vinicius da Silva Lira [2], Ariele da Silva Boni [1], Natália de Lima Donato [1], Roberto dos Anjos Reis, Jr. [3] and Reges Heinrichs [1]

[1] Department of Crop Science, São Paulo State University—Unesp, Rodovia SP 294, km 651, Dracena 17900-000, Brazil
[2] Soil and Fertilizers Lab., University do Oeste Paulista—Unoeste, Rodovia Raposo Tavares, Presidente Prudente 19067-175, Brazil
[3] Wirstchat Polímeros do Brasil, Av. Ayrton Senna da Silva, 300, Torre 01, 8° Andar-Sala 805, Londrina 86050-460, Brazil
* Correspondence: claytonbaravelli@gmail.com

Abstract: The low availability of phosphorus in most Brazilian soils causes a heavy dependence of agricultural production on phosphate fertilizers, which are generally agronomically inefficient in tropical soils. Breeding for increased longevity of sugarcane ratoons is extremely important, but understanding how the efficiency of phosphate fertilization can be improved is equally necessary. The objective of this research was to evaluate the effects of phosphate fertilizers with and without polymer coating on the productivity and nutritional status of sugarcane ratoons and phosphorus availability in the soil. The experiment was carried out on a commercial sugarcane field on a dystrophic Ultisol over two growing seasons in a randomized complete block design with four replications. Two phosphorus sources (monoammonium phosphate (MAP) and MAP + Policote) were tested at four rates (20, 40, 60 and 80 kg P_2O_5 ha^{-1}) in addition to the control (no P fertilization). The Policote-coated phosphate fertilizer induced higher stalk and TRS yields in the first experimental year, while the same effect was not observed in the second year. Nevertheless, with the reapplication of the treatments in the second study year, the mean stalk yield was high in response to the application of 20 kg P_2O_5 ha^{-1} of coated fertilizer and very different from that of the higher rates of the same fertilizer, which yielded 88 Mg ha^{-1}, i.e., 8 Mg ha^{-1} more than the mean of the other rates.

Keywords: *Saccharum* spp.; phosphorus; Policote; polymers

Citation: Oliveira, C.L.B.d.; Cassimiro, J.B.; Lira, M.V.d.S.; Boni, A.d.S.; Donato, N.d.L.; Reis, R.d.A., Jr.; Heinrichs, R. Sugarcane Ratoon Yield and Soil Phosphorus Availability in Response to Enhanced Efficiency Phosphate Fertilizer. *Agronomy* **2022**, *12*, 2817. https://doi.org/10.3390/agronomy12112817

Academic Editors: Christos Noulas, Shahram Torabian and Ruijun Qin

Received: 21 October 2022
Accepted: 9 November 2022
Published: 11 November 2022

Publisher's Note: MDPI stays neutral with regard to jurisdictional claims in published maps and institutional affiliations.

1. Introduction

Sugarcane is an internationally significant crop for the production of renewable energy and is planted on a global acreage of approximately 24.3 million hectares [1]. Brazil is the largest producer, with a cultivated area of around 8 million hectares and an estimated annual output of 521.67 million tons [2]. These data stand for the relevance of sugarcane cultivation in the context of the ongoing expansion of a clean and renewable energy matrix. Under tropical conditions, the yield potential of the crop is enormous. Sugarcane can be grown in approximately 100 countries [3] and due to its versatility of use and high biomass and sucrose production, it has become a focus of global interest [4,5]. All over the world, ways to increase sugarcane yield are being studied, and improving the nutrient supply of the crop may be the answer.

Phosphorus, an essential macronutrient for plants, is often available at insufficient levels, limiting crop yield and productivity [6]. In the case of deficiency of this nutrient, plants cannot complete the production cycle and the structural integrity (nucleic acids, phospholipids) as well as energy production (ATP) for most cellular processes and storage are affected [7]. The importance of P for plants is fundamental and seriously hampered by the reactivity of the nutrient with the soil, making it less available to crops.

Phosphorus in the soil is affected by adsorption and fixation, mainly by binding to Fe and Al oxides, which is intensified in acidic soils, reducing P utilization by plants. In tropical climate regions, soils are deeply weathered and the high complexity of P in relation to the colloidal phase prevents the crop from exploiting more than 15 to 25% of the applied fertilizer P [8,9]. The reason is the high soil P adsorption, depressing the plant available levels, mainly in soils with a predominance of sesquioxides [10].

High phosphorus rates are applied at sugarcane planting, although the residual effect of this initial fertilization is insufficient to meet the crop requirements for subsequent years, causing a decline in ratoon cane yield [11,12]. Phosphate fertilization of ratoon cane is essential to meet the nutritional demand of the crop [13], and more efficient fertilizer sources are being sought, with fixation inhibitors or soil adsorption blockers, as an alternative to increase crop productivity or longevity [14]. Several strategies have been used to increase P fertilization efficiency. Lately, the most frequently used strategy has been the application of enhanced-efficiency fertilizers [9]. These fertilizers contain aggregate technologies that control the nutrient release or stabilize their chemical transformation in the soil, increasing nutrient availability to plants [15].

The need to increase the efficiency of phosphate fertilizers and the lack of information about the issue motivated the hypothesis that the application of polymer-coated fertilizer raises phosphate fertilization efficiency and crop yields. In light of the global importance of sugarcane, the crop requirements during the cycle and low P levels in highly weathered soils, the purpose of this study was to evaluate the effects of phosphate fertilizers with and without polymer coating (fixation inhibitors) on the productivity and nutritional status of sugarcane ratoon and its effects on soil phosphorus availability.

2. Materials and Methods

2.1. Experimental Site and Treatments

The experiment was carried out in Ouro Verde (21°33′15″ S; 51°43′32″ W; 420 m asl) in São Paulo State, Brazil on a commercial sugarcane plantation with variety RB 92579 in the 2018/19 and 2019/2020 growing seasons (Figure 1). The experiment was set up in an area in the third crop cycle (second ratoon) on an Argissolo Vermelho Amarelo soil with sandy texture [16], corresponding to a dystrophic Ultisol [17] and evaluated for two successive growing seasons. The results of soil chemical and particle-size analysis of samples collected after harvesting the first ratoon, from the layers 0.00–0.10 m, 0.10–0.20 m and 0.20–0.40 m are described in Table 1 [18,19].

Table 1. Soil physical and chemical properties at the beginning of the study in 2018.

Layer	pH	OM	P	K	Ca	Mg	Al	H + Al	BS	CEC
m	CaCl$_2$	g kg^{-1}	mg dm^{-3}				mmol$_c$ dm^{-3}			
0–0.10	4.99	10.0	5.19	1.03	10.91	3.82	1	15	15.76	30.76
0.10–0.20	4.86	7.02	7.02	0.36	9.98	3.92	1	15	14.26	29.26
0.20–0.40	5.59	7.23	6.97	0.08	10.65	6.23	0	12	16.96	28.96

	Sand	Silt	Clay	V	m	S	B	Cu	Fe	Mn	Zn
	g kg^{-1}			%				mg dm^{-3}			
0–0.10	822	46	132	51.2	5.97	3.88	0.18	0.77	34.62	8.50	0.58
0.10–0.20	846	32	122	48.7	6.55	5.28	0.17	0.81	43.98	8.91	0.69
0.20–0.40	804	73	123	65.6	0	6.62	0.12	0.81	27.20	5.66	0.53

OM: organic matter; BS: sum of base; CEC: cation exchange capacity; V: base saturation; m: aluminum saturation.

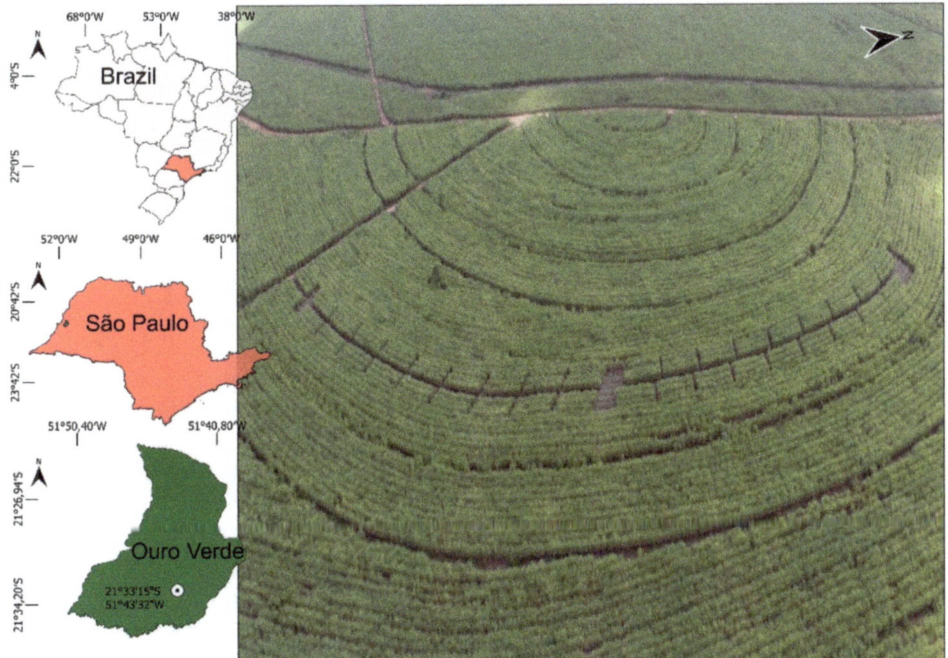

Figure 1. Location of the experimental area under sugarcane, Ouro Verde—SP, Brazil.

The experiment was arranged in a randomized complete block design with four replications, with treatments in a factorial scheme (2 × 4) + 1, represented by two sources, uncoated MAP and Policote-coated MAP (MAP + Policote) at four rates (20, 40, 60 and 80 kg P_2O_5 ha^{-1}) and without P fertilization (control). The N and P_2O_5 concentrations in MAP and coated MAP were 11% and 52%; and 10% and 49%, respectively. To coat the MAP fertilizer, the granules were covered with water-soluble additives based on copolymers with iron and aluminum affinity, called Policote, marketed by Wirstchat Polímeros do Brasil.

The experimental plots consisted of six 20 m long rows with alternating row spacing of 0.90 and 1.5 m on a total area of 144 m^2. Planting of the crop occurred in 2015, with the first cut in 2016 (plant-cane), the second cut in 2017 (first ratoon) and the experiment installed on 16 October 2018 (second ratoon). Phosphate fertilizers were applied on the crop row together with 120 kg N ha^{-1} (34 and 86 kg ha^{-1}, respectively, of ammonium sulfate-N and N urea) and 120 kg K_2O ha^{-1} (potassium chloride), in both growing seasons (2018/20 and 2019/20). After the second cut, 2 Mg ha^{-1} of limestone was applied.

2.2. Weather Conditions

According to the Köppen classification, the climate is Aw, characterized by seasons of a tropical climate with dry winters [20]. Rainfall, temperature and relative humidity data of the experimental period were provided by a meteorological station close to the experimental area and the National Institute of Meteorology (INMET) [21] (Figure 2). The historical average was 1.366 mm, and 1.322 and 1.046 mm in the growing seasons 2018/19 and 2019/20, respectively.

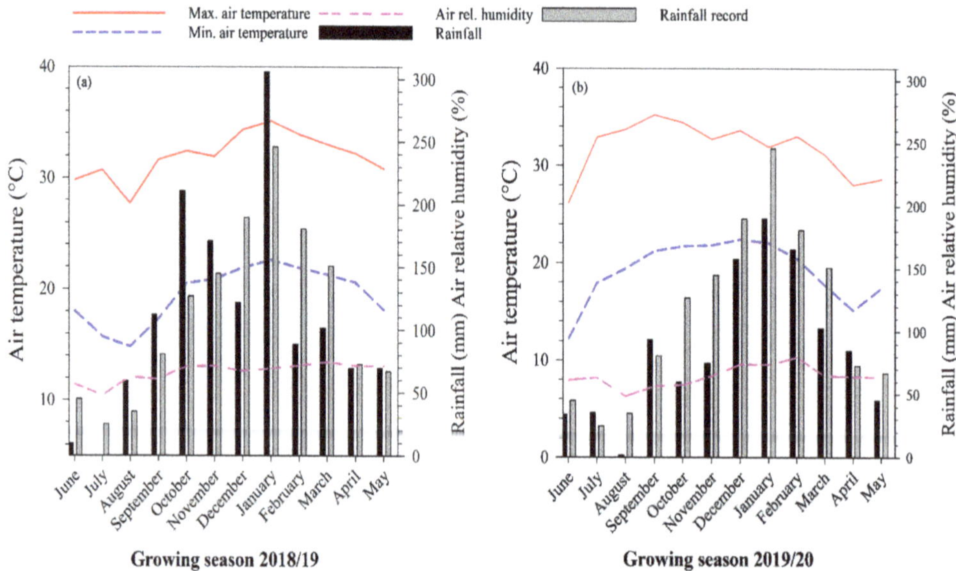

Figure 2. Weather conditions in the growing seasons 2018/19 (**a**) and 2019/20 (**b**).

2.3. Soil Phosphorus

After each harvest, six soil samples per treatment were taken, crumbled, air-dried and sieved (2 mm mesh). Soil P availability was evaluated by the methods of Ion-Exchange Resin (P-Resin) and Mehlich-1 (M1). For P-Resin analysis, the methodology described by Raij et al. [18] was used, in which cationic resin is treated with 1 mol L^{-1} NaHCO$_3$ at pH 8.5. Extractor Mehlich-1 was prepared with a mixture of two dilute acids (0.05 mol L^{-1} HCl and 0.0125 mol L^{-1} H$_2$SO$_4$), as described by Tedesco et al. [22]. The P concentration in the solution of the two extractors was determined by a methodology of recording the phosphomolybdate complex in a UV visible spectrophotometer, with a wavelength reading at 660 nm, proposed by Murphy and Riley [23].

2.4. Plant Analysis

To assess the nutritional status of sugarcane plants, 20 diagnosis leaves (leaf + 1) per plot were randomly collected from the four central rows, leaving a 2 m border. For analysis, the middle third of the leaves were used, excluding the midrib. The N content was determined by the method of sulfuric digestion, titration by micro Kjeldahl and the P, K, Ca, Mg and S levels by nitroperchloric digestion. The P concentration was assessed with a spectrophotometer and the K, Ca, Mg, and S concentrations with an atomic-absorption spectrophotometer [24].

For the technological quality analysis of sugarcane, 12 stalks were randomly sampled at each harvest throughout the experiment. After identification and weighing, the stalks were sent to the laboratory to analyze the following parameters [25]:

Brix (*Bj*): soluble solids content in percent of juice weight, determined by an automatic digital refractometer.

Fiber (*F*): stalk fiber content calculated as: $F = 0.08 * PBU + 0.876$ Where: PBU = Wet bagasse weight.

$$\text{Moisture \% } (U) : \text{ calculated as} : U = \frac{Wwm - Wdm}{Wwm} * 100$$

where: *Wwm* = wet matter weight; *Wdm* = dry matter weight.

Pol in juice (S): apparent sucrose content per juice weight, measured with an automatic digital saccharimeter and calculated as: $S = LPol * (0.26047 - 0.0009882 * Bj)$. Where: $LPol$ = Sucrose reading of clarified juice; and, Bj = Juice Brix.

Pol in cane (PC): calculated as:

$$POL = S * (1 - 0.01 * F) * C$$

where: S = Pol in juice; F = Fiber; C = Coefficient for transformation of pol from juice extracted in press (S) into pol in cane (PC).

Juice purity (Q): apparent purity of cane juice (Q), defined as the ratio of pol to brix expressed as percentage, calculated by:

$$Q = 100 * S / Bj$$

where: S = Pol in juice; Bj = Brix in juice.

Reducing sugars in juice % (RS): percent of reducing sugars (RS) per juice weight was calculated as:

$$RS = 3.641 - 0.0343 * Q$$

where: Q = juice purity.

Total Recoverable Sugar (TRS): computed from pol in cane (PC) and reducing cane sugars (RCS); calculated as:

$$TRS = 9.526 * PC + 9.05 * RS$$

Forage and stalk weight were determined by cutting 15 neighboring plants of the four central rows (excluding plot borders), resulting in a total of 60 plants per plot. After cutting, the plants were weighed immediately on a scale, then husked and shoot tips removed. The material was weighed again and trash weight estimated as the difference between forage and stalk weight. To determine cane yield, the number of stalks within 3 m of the four central rows was counted, discarding 2 m at either end. From these results, the sugarcane yield was calculated.

2.5. Statistical Analysis

Data analysis was performed using Statistical Analysis System software [26]. Residual normality and variance homogeneity were analyzed. To meet the prerequisites, the data were subjected to analysis of variance (ANOVA) at a probability of 5%. In case of significance, the means of the P sources were compared with each other by the F-test and the rates by regression equations [27]. Graphs were plotted using Sigmaplot® version 14.5 (Systat Software, Inc., San Jose, CA, USA, www.sigmaplot.com, accessed on 18 October 2022).

3. Results

3.1. Phosphorus Availability in the Soil

Soil phosphorus levels varied in response to the P sources and rates applied to sugarcane ($p < 0.05$). In the second ratoon crop, extraction by P-Resin detected interaction in both layers (0–0.10 and 0.10–0.20 m) (Figure 3a,c), which became significant in the third ratoon in the lower layer (0.10–0.20 m) (Figure 3g). In the surface layer (0–0.10 m), there was a response to isolated factors, with a linear effect for rates (Figure 3e). Comparing the sources, MAP + Policote made the highest levels of nutrients available (12.72 mg P dm^{-3}) (Table 2). Between the first and second year of evaluation, P-Resin detected a decrease in the mean P concentration of 32% and 41%, respectively, in the 0–0.10 and 0.10–0.20 m layers.

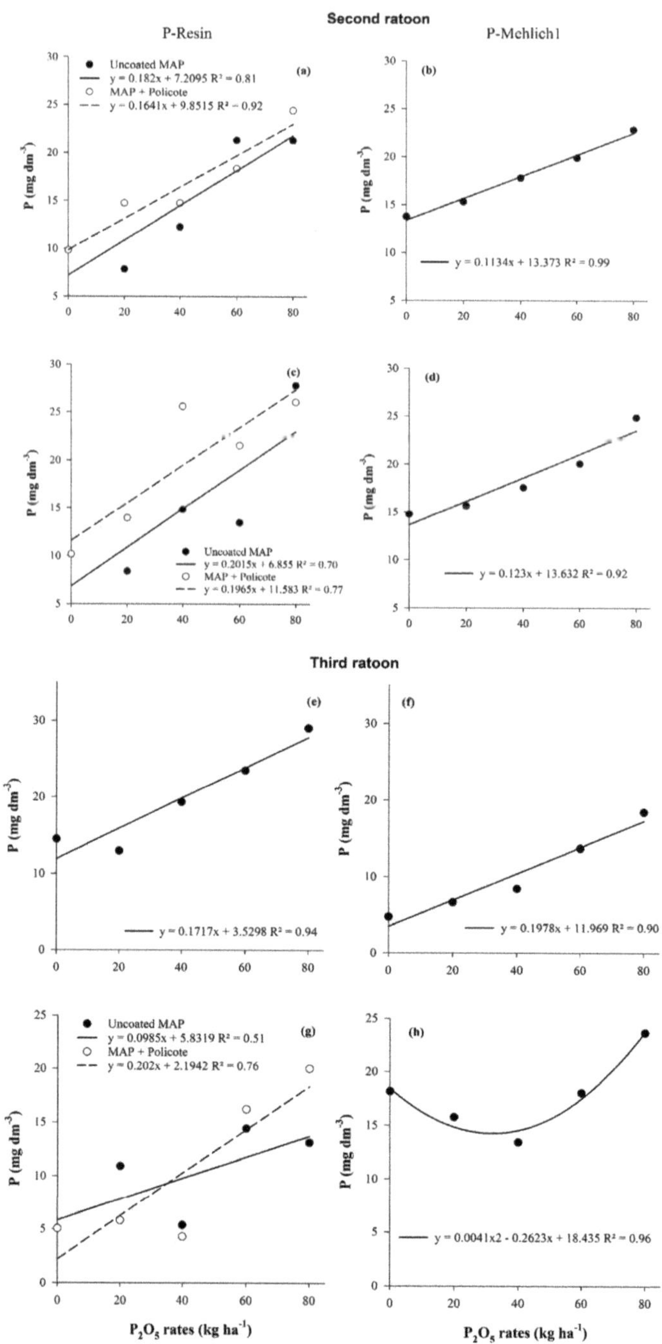

Figure 3. Phosphorus content in soil in response to fertilization with different phosphorus rates with and without Policote coating, extracted by P-Resin from the layer 0–0.10 m (**a**,**e**) and 0.10–0.20 m (**c**,**g**), and by Mehlich-1 from the layer 0–0.10 m (**b**,**f**) and 0.10–0.20 m (**d**,**h**). Growing seasons 2018/2019 and 2019/2020.

Table 2. Mean soil phosphorus contents extracted with P-Resin and Mehlich-1 after sugarcane cultivation in the second and third ratoon crops fertilized with phosphorus sources with and without Policote coating. Growing seasons 2018/2019 and 2019/2020.

Fertilizer Sources	P-Resin	Mehlich-1			
	Third Ratoon	Second Ratoon		Third Ratoon	
	0–0.10 m	0–0.10 m	0.10–0.20 m	0–0.10 m	0.10–0.20 m
			$mg\ dm^{-3}$		
Uncoated MAP	10.89 b	17.24	17.36 b	18.19 b	16.65 b
MAP + Policote	12.72 a	18.61	19.90 a	24.26 a	18.79 a
p-value	0.0149 *	0.332 [ns]	0.0018 **	0.0001 ***	0.0001 ***

Different letters within a column indicate significant differences according to Tukey's test. *, **, *** and ns indicate $p < 0.05$, $p < 0.01$, $p < 0.001$ and [ns]—$p > 0.05$, respectively.

By the Mehlich-1 method, there was no significant interaction between rates and sources ($p > 0.05$). However, there was an isolated effect for the two factors, with a linear response to P rates in the two evaluated years and two layers (Figure 3b,d,f), except in the 0.10–0.20 m layer in the second year when a quadratic response was observed (Figure 3h). Regarding the sources, results in soil P contents were positive in response to Policote-coated fertilizer in the 0–0.10 m layer in the third ratoon and the 0.10–0.20 m layer in both years (Table 2). Between the first and second years of evaluation, Mehlich-1 detected a mean increase of 10.94% in the surface layer and a reduction of 4% in the layer below.

3.2. Plant Nutritional Status

The phosphorus rates and sources had no effect on the nutritional status of sugarcane ($p > 0.05$). However, there was a difference between the years of cultivation (Table 3). Nitrogen and K contents decreased by 29 and 50%, respectively, from the second to the third ratoon crop (Table 3). The levels of P, Ca and Mg soil availability increased from the first to the second year of evaluation, respectively, by 43%, 14% and 39% (Table 3), while S remained constant in the evaluated cycles ($p > 0.05$).

Table 3. Mean levels of macronutrients (leaf + 1) in two sugarcane cycles. Growing seasons 2018/2019 and 2019/2020.

Ratoon	N	P	K	Ca	Mg	S
				$g\ kg^{-1}$		
Second	17.87 a	1.68 b	14.62 a	6.98 b	1.43 b	1.08
Third	12.65 b	2.41 a	7.35 b	7.95 a	2.00 a	1.15
p-value	0.0002 ***	0.003 **	0.0001 ***	0.0132 *	0.005 **	0.5682 [ns]

Different letters within a column indicate significant differences according to Tukey's test. *, **, *** and ns indicate $p < 0.05$, $p < 0.01$, $p < 0.001$ and [ns]—$p > 0.05$, respectively.

Leaf macronutrient contents were within the range considered adequate for the nutritional status of sugarcane [24,28], and in the second and third ratoon crops, S was the only macronutrient below the critical level (1.4 g kg^{-1}), while the levels of the others were within the range considered adequate. In the third ratoon crop, the levels of the macronutrients N, K and S were below the ideal (18, 10 and 1.4 g kg^{-1} respectively), whereas those of the others were adequate.

3.3. Effects on Sugarcane Technological Quality and Yield

The second ratoon stalk yield (Figure 4a) shows that the response to uncoated fertilizer increased linearly up to the rate of 80 kg P_2O_5 ha^{-1}, reaching a production of 105 Mg ha^{-1}. In turn, the response to the Policote-coated phosphate source fitted a quadratic model, with a maximum yield of 106 Mg ha^{-1} at a rate of 58 kg P_2O_5 ha^{-1}. Total reducing sugar (TRS) yield had a similar pattern to that observed for stalk yield (Figure 4c). In the third ratoon, after reapplication of the treatments, no mathematical model could be fitted relating P rates with stalk and TRS yield (Figure 4b,d). However, the rate of 20 kg P_2O_5 ha^{-1} in coated fertilizer stood out among the other treatments, with a yield of 88 Mg ha^{-1} (Figure 4b).

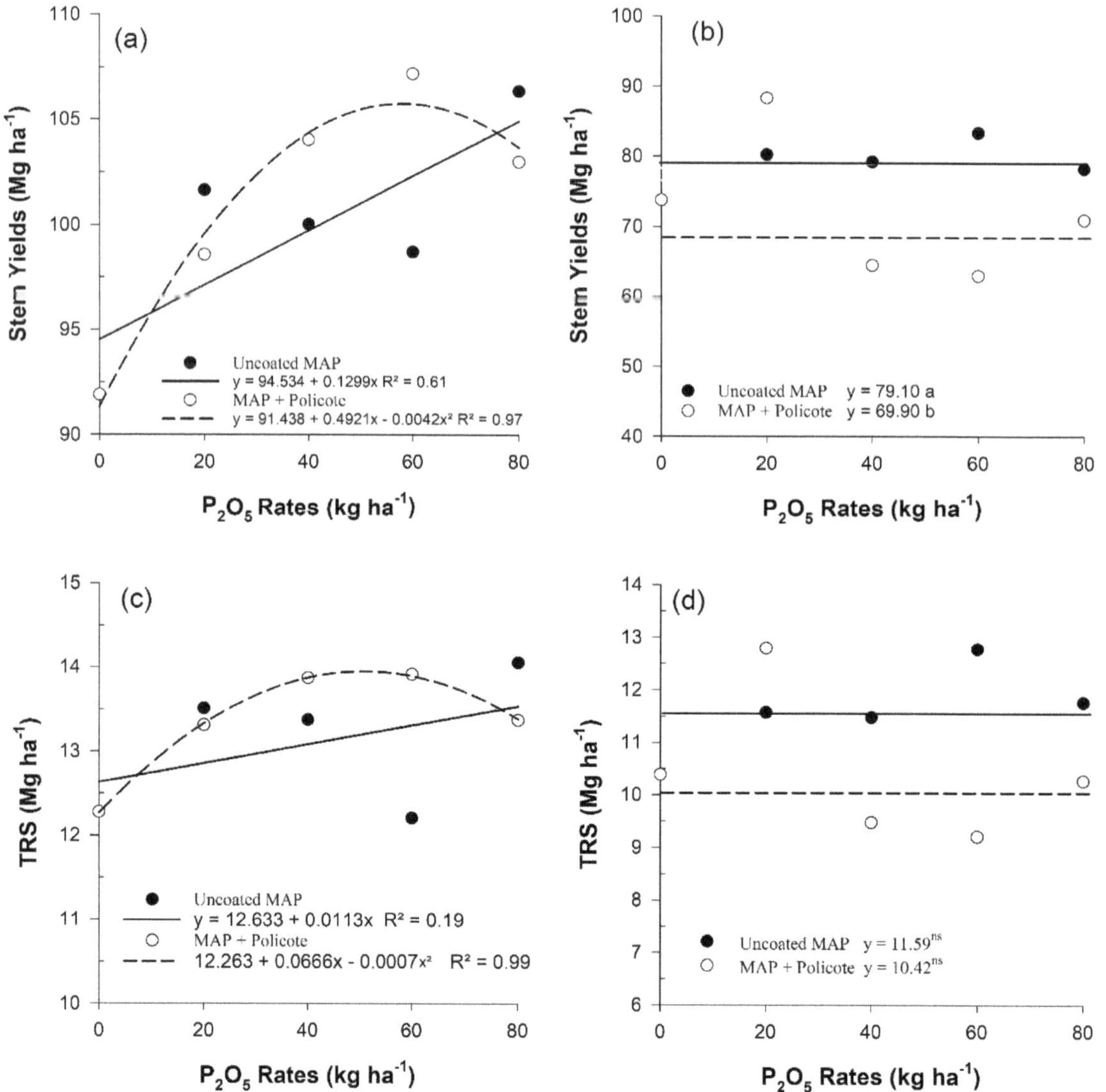

Figure 4. Stalks and TRS yield in the second ratoon (**a,c**) and in the third ratoon crop (**b,d**) of sugarcane fertilized with phosphorus rates and sources without or with Policote coating in two sugarcane production cycles. Growing seasons 2018/2019 and 2019/2020. Different letters indicate significant differences according to Tukey's test ($p < 0.05$), ^{ns}—$p > 0.05$.

Fertilization with P sources and rates had no significant effect on the interaction, nor the separate factors on the technological quality of sugarcane ($p > 0.05$). Comparing the growing seasons, the levels of Brix, Fiber, POL, PC and TRS differed ($p < 0.05$), with an increase of, respectively, 9.5, 9.2, 13.8, 11.7 and 10.8% in the second year of evaluation, while the moisture content was 2.5% higher in the first year (Table 4). For RS and juice purity, no significant differences were observed between the evaluated cultivation cycles.

Table 4. Mean values of stalk technology quality parameters in two sugarcane ratoons. Growing seasons 2018/2019 and 2019/2020.

Ratoon	Brix	Fiber	Moisture	POL	PC	Purity	RS	TRS
				%				kg t^{-1}
Second	15.06 b	11.64 b	73.31 a	15.40 b	13.13 b	87.17 ^{ns}	0.55 ^{ns}	131.47 b
Third	16.49 a	12.71 a	70.80 b	17.53 a	14.66 a	88.91	0.50	145.72 a
p-value	0.003 **	0.001 **	0.0008 ***	0.003 **	0.06 **	0.1262 ^{ns}	0.073 ^{ns}	0.006 **

PC: pol in cane; RS: reducing sugars in juice; TRS: total recoverable sugar. Different letters within a column indicate significant differences according to Tukey's test. **, *** and ns indicate $p < 0.01$, $p < 0.001$ and ^{ns}—$p > 0.05$, respectively.

4. Discussion

In response to soil application of the two phosphorus sources, the methods P-Resin and Mehlich-1 detected increasing concentrations with increasing rates. In the second ratoon crop, the P-Resin extractor better differentiated the fertilizer responses in the 0–0.10 and 0.10–0.20 m layers (Figure 3a,c) based on the principle of ion exchange, affecting the colloidal system. This pattern can be explained by the principle of the polymers used as fertilizer coating, which, when in contact with the soil solution, release charges to saturate positive soil colloid charges. This reduces P adsorption and fixation in the soil colloidal fraction, leaving the nutrient available for plant uptake. In this way, the results can be different depending on the soil characteristic, especially in relation to the colloidal fraction as well as other factors e.g., crop, management system, and climate, among others [29].

In the second year of evaluations (third ratoon), no interaction was observed between rates and sources in phosphorus contents 0–0.10 m by the P-Resin method, showing only a linear effect for rates (Figure 3e). This result can be attributed to factors related to low rainfall in the period (Figure 2b) and intensified by the low water retention capacity due to the sandy texture of the surface layer, which reduces granule solubility and levels out the effect of the two sources on the soil. In the subsurface, an interaction between phosphorus sources and rates was stated, which can be attributed to some residual effect of fertilization applied in the previous year, since rainfall was restricted.

The soil P contents determined by Mehlich-1 were significantly influenced by the rates (Figure 3b,d,f,h), but no difference was identified between the sources in both layers and the two evaluated years, unlike the pattern detected when using P-Resin. These results can be attributed to factors inherent to the Mehlich-1 method, which preferentially extracts P forms bound to Ca, leading to an overestimation of P availability in recently fertilized soils [30]. For the evaluation of fertilizer sources, the results of the extraction methods must be discriminated according to the soil characteristics, especially with regard to the texture class [31].

Several studies correlate the extractors P-resin and Mehlich-1 [32–34]. These authors claimed that the lower the amount of clay, the higher the contents extracted by Mehlich-1. This confirmed the results of this study, which were mostly higher than those obtained by the P-Resin method. However, Mumbach et al. [35] reported contrary results, emphasizing that apart from soil texture, which can be explained by the natural phosphates that are often used for phosphating, the acid extractant predominantly solubilizes Ca-P, resulting in an overestimation of available P.

Table 3 shows the foliar levels of macronutrients, which indicate lower N and K uptake in the second than the first year of evaluation. These results may be related to the lower rainfall in the second year (Figure 2), since N and K movement in the soil is strongly influenced by mass flow, affecting root uptake [36,37], along with the sandy soil with low organic matter content and water retention capacity [38]. Water stress also affects the development of the root system [39,40], reducing the soil volume exploited for nutrient uptake. For both nutrients, the values in the second year were below the critical level considered adequate for the crop [24,28].

The levels of P, Ca and Mg increased in the soil in the second year of evaluation. The difference in P content can be attributed to the increase in the availability of the element in the soil due to phosphate fertilization in the application of treatments in two successive years. While the differences in Ca and Mg observed between the growing seasons must be related to the 2 Mg ha^{-1} limestone applied after the first ratoon harvest, this application may also have affected K uptake due to an imbalance in the K/Ca/Mg ratio [41]. According to the values found, the three nutrients are within the range considered suitable for sugarcane [24,28].

Leaf concentrations of S, although applied in fertilization via ammonium sulfate, were low. This can be attributed to the low amount of soil organic matter, high nutrient mobility in the soil profile, mainly due to the predominant sandy fraction, or to varietal characteristics, as also observed by Calheiros et al. [42] in a study with the same variety.

The concentration of parameters that make up the technological quality of sugarcane stalks was not significant between treatments (Table 4). However, there was a difference between the two years evaluated in some parameters. In the second year, the leaf moisture content (U) was lower than in the first, leading to a concentration effect of brix, PC, Pol and TRS. In turn, the reduction in moisture was a result of the water deficit in the second crop growth cycle (Figure 2). It is worth emphasizing that the plant, even under water stress, continues to synthesize sugar, while photosynthesis is affected if the annual water deficit exceeds 145 mm [43]. According to Araújo et al. [44], the effects of water stress can be beneficial to accumulate TRS, since the increase in TRS is inversely proportional to the moisture decrease up to 51 days of water stress before harvesting the stalks.

Another reason for the positive response in the high levels of technological quality parameters in the second year (Table 4) may be due to the better supply of the system with P since the nutrient influences the apparent sucrose percentage or pol in cane contained in sugarcane juice (PC) and juice purity [45]. Although the difference in purity between the growing seasons was not significant, it increased, confirming the data of Albuquerque et al. (2016) [46] who attributed an increase in Pol to purity to P application.

The cane stalk yield had a quadratic response to Policote-coated fertilizer, while in the absence of the Policote, the response was linear up to the rate of 80 kg P_2O_5 ha^{-1}. The maximum stalk yield (105.85 Mg ha^{-1}) in response to coated fertilizer was reached at a rate of 58.58 kg P_2O_5 ha^{-1}. This yield was similar to that obtained with 80 kg ha^{-1} uncoated P_2O_5, representing a 26% reduction in the applied P rate when using the technology of fertilizer coating (Figure 4). The yield potential is close to that reported by Gava et al. (2011); Abreu et al. (2013) [47,48].

An important parameter in the evaluation of sugarcane fertilization is the production of total reducing sugars, for which the same pattern as for stalk yield was observed, with maximum yield produced at a rate of 47.57 kg P_2O_5 ha^{-1} by applying Policote-coated fertilizer, reaching a TRS yield of 13.85 Mg ha^{-1}. The response to the uncoated monoammonium phosphate source fitted a linear model up to the rate of 80 kg P_2O_5 ha^{-1}, with a TRS yield of 13.34 Mg ha^{-1} (Figure 4). Based on these values, the agronomic efficiency was computed, i.e., 33.21 and 15.87 kg of TRS per kilogram of P_2O_5, respectively, for fertilization with or without Policote coating.

In the evaluations of stalk and TRS yield in the second year (third ratoon), only the difference in the mean of the sources was verified, with higher stalk yield in response to the application of uncoated fertilizer. The difference between fertilizers may be related to the lower biodegradability of the polymer [49,50]. This may explain the lower productivity with the MAP + Policote fertilizer. Clearly, the sparse rainfall directly influenced the dissolution of the coated fertilizer in the soil, hampering the enzymatic action responsible for breaking down the polymers. Along with these factors, the soil SOM levels were low and consequently, the interactions of organisms and enzymes with the fertilizer were reduced.

Although without significant effect, with the reapplication of the treatments in the second study year, the mean stalk yield was high in response to the application of 20 kg P_2O_5 ha^{-1} of coated fertilizer, a very different mean in relation to the higher rates of the same fertilizer. These results suggest the need for further investigation into causes and effects in the application of high rates of fertilizers with technologies for enhanced efficiency, which may allow the use of lower rates due to the use of technology in successive years. Multiple authors describe the beneficial effects of phosphate fertilization with Policote-coated fertilizer and reported no negative effects due to applications of high fertilizer rates [9,31,51–53].

5. Conclusions

Fertilizers with or without Policote coating induced positive responses in soil P, as shown by the extractors P-Resin and Mehlich 1. However, the P-Resin extractor proved to be an adequate detection method of the importance of the polymer in increasing soil phosphorus availability. Leaf contents did not vary in response to phosphate fertilization. The technological quality of cane stalks varied between the studied growing seasons with better results in the second year. The Policote-coated phosphate fertilizer induced higher stalk and TRS yields in the first experimental year, while the same effect was not observed in the second year. Nevertheless, with the reapplication of the treatments in the second study year, the mean stalk yield was high in response to the application of 20 kg P_2O_5 ha^{-1} of coated fertilizer, a very different mean in relation to the higher rates of the same fertilizer.

Further research should be encouraged to understand the dynamics between polymer and the availability of P in soil and the possible effect on the physiology and production of enzymes that may contribute to nutrient use efficiency. These studies will allow the understanding of the physiological phenomena that occur with the highest phosphorus rates in the presence of the polymer.

Author Contributions: Conceptualization, C.L.B.d.O. and R.H.; methodology, C.L.B.d.O., J.B.C., A.d.S.B., N.d.L.D. and R.H.; software, C.L.B.d.O.; validation, C.L.B.d.O. and R.H.; Resource, R.H.; formal analysis, C.L.B.d.O., A.d.S.B. and N.d.L.D.; investigation, C.L.B.d.O., J.B.C. and M.V.d.S.L.; data curation, C.L.B.d.O.; writing—original draft preparation, C.L.B.d.O. and J.B.C. writing—review and editing, C.L.B.d.O., J.B.C., A.d.S.B., R.d.A.R.J. and R.H.; visualization, C.L.B.d.O., R.d.A.R.J. and R.H.; supervision, R.H. All authors have read and agreed to the published version of the manuscript.

Funding: This research received no external funding.

Institutional Review Board Statement: Not applicable.

Informed Consent Statement: Not applicable.

Data Availability Statement: The data presented in this study are available on request from the corresponding author.

Conflicts of Interest: The authors declare no conflict of interest.

References

1. OECD; Food and Agriculture Organization of the United Nations. *OECD-FAO Agricultural Outlook 2022–2031*; OECD: Paris, France, 2022; ISBN 978-92-64-58870-7.
2. de A Conab, C.N. Acompanhamento da safra brasileira cana-de-açúcar. *Acompan. Safra Bras.* **2021**, *8*, 59.
3. Heinrichs, R.; Otto, R.; Magalhães, A.; Meirelles, G. Importance of Sugarcane InBrazilian and World Bioconomy. In *Springer Nature*; S. Dabber: Cham, Switzerland, 2017.
4. Ahorsu, R.; Medina, F.; Constantí, M. Significance and Challenges of Biomass as a Suitable Feedstock for Bioenergy and Biochemical Production: A Review. *Energies* **2018**, *11*, 3366. [CrossRef]
5. de Andrade, A.F.; Flores, R.A.; Casaroli, D.; Bueno, A.M.; Pessoa-de-Souza, M.A.; de Lima, F.S.R.; Marques, E.P. K Dynamics in the Soil–Plant System for Sugarcane Crops: A Current Field Experiment Under Tropical Conditions. *Sugar Tech* **2021**, *23*, 1247–1257. [CrossRef]
6. Gonçalves, V.; Meurer, E.; Tatsch, F.; Carvalho, S.; Neto, O. Biodisponibilidade de Cádmio Em Fertilizantes Fosfatados. *Rev. Bras. Ciência Solo* **2008**, *32*, 2871–2875. [CrossRef]

7. Taiz, L.; Zeiger, E.; Moller, I.M.; Murphy, A. *Fisiologia e Desenvolvimento Vegetal*, 6th ed.; Artmed: Prague, Czech Republic, 2017; ISBN 978-1-60535-255-8.
8. Sanders, J.L.; Murphy, L.S.; Noble, A.; Melgar, R.J.; Perkins, J. Improving Phosphorus Use Efficiency with Polymer Technology. *Procedia Eng.* **2012**, *46*, 178–184. [CrossRef]
9. Zanão, L.A., Jr.; Arf, O.; dos Reis, R.A., Jr.; Pereira, N. Phosphorus Fertilization with Enhanced Efficiency in Soybean and Corn Crops. *Aust. J. Crop Sci.* **2020**, *14*, 78–84. [CrossRef]
10. Fink, J.R.; Inda, A.V.; Bayer, C.; Torrent, J.; Barrón, V. Mineralogy and Phosphorus Adsorption in Soils of South and Central-West Brazil under Conventional and No-Tillage Systems. *Acta Sci. Agron.* **2014**, *36*, 379–387. [CrossRef]
11. Gopalasundaram, P.; Bhaskaran, A.; Rakkiyappan, P. Integrated Nutrient Management in Sugarcane. *Sugar Tech* **2012**, *14*, 3–20. [CrossRef]
12. Costa, D.B.D.; Neto, D.E.S.; Freire, F.J.; De, E.C.A. Adubação fosfatada em cana planta e soca em argissolos do nordeste de diferentes texturas. *Rev. Caatinga* **2014**, *27*, 10.
13. Zambrosi, F.C.B. Phosphorus Fertilizer Reapplication on Sugarcane Ratoon: Opportunities and Challenges for Improvements in Nutrient Efficiency. *Sugar Tech* **2020**, *23*, 704–708. [CrossRef]
14. Guelfi, D.R. Tecnologias e Inovações Para Fertilizantes Fosfatados. *NPTC* **2021**, *10*, 14–33.
15. AAPFCO Association of American Plant Food Control Officials (AAPFCO). Available online: http://www.aapfco.org/publications.html (accessed on 18 February 2021).
16. Santos, E.F. dos Mecanismos de Interação Fósforo-Zinco no Sistema Solo Planta: Disponibilidade No Solo, Avaliações Fisiológicas e Expressão de Transportadores de Fosfato. Ph. D. Thesis, em Biologia na Agricultura e no Ambiente, Universidade de São Paulo, Piracicaba, Brazil, 2018.
17. *Soil Survey Division Keys to Soil Taxonomy*, 12th ed.; United States Department of Agriculture, Natural Resources Conservation Service: Washington, DC, USA, 2014.
18. Raij, B.V.; Andrade, J.C.; Cantarella, H.; Quaggio, J.A. *Análise Química Para Avaliação Da Fertilidade de Solos Tropicais*; Instituto Agronômico de Campinas: Campinas, Brazil, 2001.
19. Teixeira, P.C.; Donagemma, G.K.; Fontana, A.; Teixeira, W.G. *Manual de Métodos de Análise de Solo*; Embrapa Solos: Brasilia, DF, Brazil, 2017; Volume 3, ISBN 978-85-7035-771-7.
20. Alvares, C.A.; Stape, J.L.; Sentelhas, P.C.; de Gonçalves, J.L.M.; Sparovek, G. Köppen's Climate Classification Map for Brazil. *Meteorol. Z.* **2013**, *22*, 711–728. [CrossRef]
21. Instituto Nacional de Meteorologia-INMET. Available online: https://portal.inmet.gov.br/ (accessed on 15 April 2021).
22. Tedesco, M.J.; Gianello, C.; Anghinoni, I.; Bissani, C.A.; Camargo, F.A.O.; Wiethölter, S. *Manual de Adubação e Calagem*.; Comissão de Química e Fertilidade do Solo-RS/SC: Porto Alegre, Brazil, 2004.
23. Murphy, J.; Riley, J.P. A Modified Single Solution Method for the Determination of Phosphate in Natural Waters. *Anal. Chim. Acta* **1962**, *27*, 31–36. [CrossRef]
24. Malavolta, E. Avaliação Do Estado Nutricional Das Plantas: Princípios e Aplicações/Eurípedes Malavolta, Godofredo Cesar Vitti, Sebastião Alberto de Oliveira. Piracicaba Potafos. 1997. Available online: https://books.google.com.sg/books/about/Avalia%C3%A7%C3%A3o_do_estado_nutricional_das_pl.html?hl=pt-BR&id=Lu9EAAAAYAAJ&redir_esc=y (accessed on 18 October 2022).
25. Fernandes, A.C. *Cálculos na Agroindústria da Cana-de-açúcar*; Sociedade dos Técnicos Açucareiros e Alcooleiros do Brasil: Piracicaba, Brazil, 2011.
26. SAS®SAS OnDemand for Academics. Available online: https://welcome.oda.sas.com/home (accessed on 9 February 2022).
27. Pimentel-Gomes, F. *Curso de Estatistica Experimental*, 15th ed.; ESALQ: Piracicaba, Brazil, 2009; ISBN 978-85-7133-055-9.
28. van Raij, B.; Cantarella, H.; Quaggio, J.A.; Furlani, Â.M.C. *Recomendações de Adubação e Calagem para o Estado de São Paulo*; IAC: Campinas, Brazil, 1997; pp. 173–251.
29. de Gazola, R.N.; Buzetti, S.; Dinalli, R.P.; Teixeira Filho, M.C.M.; de Celestrino, T.S. Efeito residual da aplicação de fosfato monoamônio revestido por diferentes polímeros na cultura de milho. *Rev. Ceres* **2013**, *60*, 876–884. [CrossRef]
30. Bortolon, L.; Gianello, C. Simultaneous Multielement Extraction with the Mehlich-1 Solution for Southern Brazilian Soils Determined by ICP-OES and the Effects on the Nutrients Recommendations to Crops. *Rev. Bras. Ciência Solo* **2010**, *34*, 125–132. [CrossRef]
31. Chagas, W.F.T.; Emrich, E.B.; Guelfi, D.R.; Caputo, A.L.C.; Faquin, V. Productive characteristics, nutrition and agronomic efficiency of polymer-coated MAP in lettuce crops. *Rev. Ciência Agronômica* **2015**, *46*, 266–276. [CrossRef]
32. Schlindwein, J.A.; Gianello, C. Nível de Suficiência e Índice de Equivalência Entre o Fósforo Determinado Pelos Métodos da Resina de Troca Iônica e Mehlich-1. *Curr. Agric. Sci. Technol.* **2008**, *14*, 299–306. [CrossRef]
33. Schlindwein, J.A.; Bortolon, L. Soil Phosphorus Available for Crops and Grasses Extracted with Three Soil-Test Methods in Southern Brazilian Soils Amended with Phosphate Rock. *Commun. Soil Sci. Plant Anal.* **2011**, *42*, 283–292. [CrossRef]
34. de Freitas, I.F.; Novais, R.F.; de Villani, E.M.A.; Novais, S.V. Phosphorus Extracted by Ion Exchange Resins and Mehlich-1 from Oxisols (Latosols) Treated with Different Phosphorus Rates and Sources for Varied Soil-Source Contact Periods. *Rev. Bras. Ciênc. Solo* **2013**, *37*, 667–677. [CrossRef]
35. Mumbach, G.L.; de Oliveira, D.A.; Warmling, M.I.; Gatiboni, L.C. Phosphorus Extraction by Mehlich 1, Mehlich 3 and Anion Exchange Resin in Soils with Different Clay Contents. *Rev. Ceres* **2018**, *65*, 546–554. [CrossRef]

36. Crusciol, C.A.C.; Mancuso, M.A.C.; Garcia, R.A.; Castro, G.S.A. Crescimento radicular e aéreo de cultivares de arroz de terras altas em função da calagem. *Bragantia* **2012**, *71*, 256–263. [CrossRef]
37. Lefèvre, I.; Ziebel, J.; Guignard, C.; Hausman, J.-F.; Rosales, R.O.G.; Bonierbale, M.; Hoffmann, L.; Schafleitner, R.; Evers, D. Drought Impacts Mineral Contents in Andean Potato Cultivars. *J. Agron. Crop Sci.* **2012**, *198*, 196–206. [CrossRef]
38. Carneiro, F.M.; Furlani, C.E.A.; Ormond, A.T.S.; Kazama, E.H.; Silva, R.P. da Mechanized Fertilization: Individual Application of Nitrogen, Phosphorus and Potassium in Sugarcane. *Rev. Ciênc. Agron.* **2017**, *48*, 278–287. [CrossRef]
39. de Vasconcelos, A.C.M.; Dionardo-Miranda, L. *Dinâmica do desenvolvimento radicular da cana-de-açúcar e implicações no controle de nematóides*, 2nd ed.; Revista e Ampliada: Campinas, Brazil, 2011; ISBN 978-85-87645-48-7.
40. Simões, W.L.; Calgaro, M.; Guimarães, M.J.M.; de Oliveira, A.R.; Pinheiro, M.P.M.A. Sugarcane crops with controlled water deficit in the submiddle são francisco valley, Brazil. *Rev. Caatinga* **2018**, *31*, 963–971. [CrossRef]
41. Marschner, H. Marschner's Mineral Nutrition of Higher Plants-3rd Edition. Available online: https://www.elsevier.com/books/marschners-mineral-nutrition-of-higher-plants/marschner/978-0-12-384905-2 (accessed on 22 May 2020).
42. Calheiros, A.; Oliveira, M.; Ferreira, V.; Barbosa, G.; Costa, J.; Lima, G.; Aristides, E. Acúmulo de Nutrientes e Produção de Sacarose de Duas Variedades de Cana-de-Açúcar Na Primeira Rebrota, Em Função de Doses de Fósforo. *STAB* **2011**, *29*, 4.
43. Inman-Bamber, N.G. Sugarcane Water Stress Criteria for Irrigation and Drying Off. *Field Crops Res.* **2004**, *89*, 107–122. [CrossRef]
44. Araújo, R.; Alves Junior, J.; Casaroli, D.; Evangelista, A.W.P. Variation in the sugar yield in response to drying-off of sugarcane before harvest and the occurrence of low air temperatures. *Bragantia* **2016**, *75*, 118–127. [CrossRef]
45. Simões Neto, D.E.; de Oliveira, A.C.; Freire, F.J.; dos Freire, M.B.G.S.; do Nascimento, C.W.A.; Rocha, A.T. da Extração de fósforo em solos cultivados com cana-de-açúcar e suas relações com a capacidade tampão. *Rev. Bras. Eng. Agríc. Ambient.* **2009**, *13*, 840–848. [CrossRef]
46. Albuquerque, A.W.; de Sá, L.A.; Rodrigues, W.A.R.; Moura, A.B.; dos Oliveira, M.S. Growth and Yield of Sugarcane as a Function of Phosphorus Doses and Forms of Application. *Rev. Bras. Eng. Agríc. Ambient.* **2016**, *20*, 29–35. [CrossRef]
47. de Gava, G.J.C.; de Silva, M.A.; da Silva, R.C.; Jeronimo, E.M.; Cruz, J.C.S.; Kölln, O.T. Produtividade de três cultivares de cana-de-açúcar sob manejos de sequeiro e irrigado por gotejamento. *Rev. Bras. Eng. Agrícola Ambient.* **2011**, *15*, 250–255. [CrossRef]
48. de Abreu, M.L.; de Silva, M.A.; Teodoro, I.; de Holanda, L.A.; Sampaio Neto, G.D. Crescimento e produtividade de cana-de-açúcar em função da disponibilidade hídrica dos Tabuleiros Costeiros de Alagoas. *Bragantia* **2013**, *72*, 262–270. [CrossRef]
49. Trenkel, M.E. *Slow- and Controlled-Release and Stabilized Fertilizers: An Option for Enhancing Nutrient Use Efficiency in Agriculture*, 2nd ed.; International Fertilizer Industry Association (IFA): Paris, France, 2010; ISBN 978-2-9523139-7-1.
50. Briassoulis, D.; Dejean, C. Critical Review of Norms and Standards for Biodegradable Agricultural Plastics Part I. Biodegradation in Soil. *J. Polym. Environ.* **2010**, *18*, 384–400. [CrossRef]
51. Guelfi, D.R.; Chagas, W.F.T.; Lacerda, J.R.; Chagas, R.M.R.; de Souza, T.L.; Andrade, A.B. Monoammonium Phosphate Coated with Polymers and Magnesium for Coffee Plants. *Ciênc. Agrotec.* **2018**, *42*, 261–270. [CrossRef]
52. Pelá, A.; Ribeiro, M.A.; Bento, R.U.; Cirino, L.H.; Reis Júnior, R.A. Enhanced-Efficiency Phosphorus Fertilizer: Promising Technology for Carrot Crop. *Hortic. Bras.* **2018**, *36*, 492–497. [CrossRef]
53. Pelá, A.; Bento, R.U.; Crispim, L.B.R.; dos Reis, R.A., Jr. Enhanced efficiency of Phosphorus fertilizer in Soybean and Maize. *Aust. J. Crop Sci.* **2019**, *13*, 1638–1642. [CrossRef]

Communication

The Ability of Nitrification Inhibitors to Decrease Denitrification Rates in an Arable Soil

Jie Li [1,*], Wenyu Wang [1,2], Wei Wang [1] and Yaqun Li [1,2]

[1] Institute of Applied Ecology, Chinese Academy of Sciences, Shenyang 110164, China
[2] University of Chinese Academy of Sciences, Beijing 100049, China
* Correspondence: jieli@iae.ac.cn; Tel.: +86-18840608623

Abstract: A nitrification inhibitor is an effective tool that can be used to reduce the loss of nitrogen (N) and improve crop yields. Most studies have focused on the changes in the soil N mineralization process that may influence the dynamics of soil inorganic N and the soil N cycle. However, the effects of the inhibitors on denitrification rates remain largely unclarified. Therefore, in this study, we monitored the dynamics in annual denitrification rates affected by nitrification inhibitors from a maize field for the first time. Treatments included inorganic fertilizer (NPK), cattle manure, a combination of NPK and DMPP (3,4-dimethylpyrazole phosphate), and a combination of manure and DMPP, applied to brown soils in a no-tillage maize field. The findings demonstrated that the denitrification rate and denitrifying enzyme activity (DEA) were highly variable and there were no significant decreases in all treatment groups after the addition of DMPP. Compared to the control soils, the ammonium (NH_4^+-N) concentration was significantly increased, while the nitrate (NO_3^--N) level was significantly decreased in the DMPP-amended soils less than 30 days after treatment application, indicating that nitrification was partially inhibited. The formation of NO_3^--N and the nitrification rates could be markedly reduced by DMPP, while NO_3^--N availability did not affect the denitrification rates. Complete degradation of DMPP was observed in the soil on day 70 after DMPP addition, and its half-life was 10 days. Our study may ultimately help to clarify the characteristics of denitrification rates affected by nitrification inhibitors from different N fertilizer types applied to soils and explore the influencing factors of the dynamics in annual denitrification rates. However, more field studies evaluating the effectiveness of nitrification inhibitors in reducing denitrification under different sites and climate conditions, and the molecular mechanisms driving denitrification rate changes, need to be performed in the future.

Keywords: denitrification process; inhibitors; N transformation; nitrification rates

Citation: Li, J.; Wang, W.; Wang, W.; Li, Y. The Ability of Nitrification Inhibitors to Decrease Denitrification Rates in an Arable Soil. *Agronomy* **2022**, *12*, 2749. https://doi.org/10.3390/agronomy12112749

Academic Editors: Christos Noulas, Shahram Torabian and Ruijun Qin

Received: 18 October 2022
Accepted: 4 November 2022
Published: 5 November 2022

Publisher's Note: MDPI stays neutral with regard to jurisdictional claims in published maps and institutional affiliations.

1. Introduction

In agricultural systems, nitrogen (N) is a critical nutrient that can lead to greater crop yields and higher production of wool, eggs, milk, and animal tissues [1]. The improved N availability has allowed farmers to intensify production and to cultivate low productive soils. However, N may not be fully utilized in soil, as the plant uptake of fertilizer N does not exceed half of the N applied [2]. A significant amount of excess N in the environment originating from fertilizer N is lost from the soil and plant systems through volatilization and denitrification [3]. Thus, considerable attention has been paid to the impact of N on the environment.

As an important anthropogenic greenhouse gas, nitrous oxide (N_2O) has a global warming effect approximately 300 times that of CO_2 [4,5]. N_2O is typically formed in the soils via nitrification–denitrification processes. Nitrification (the oxidation of NH_4^+-N (ammonium) to NO_3^--N (nitrate)) and denitrification (the reduction of NO_3^--N to dinitrogen gas) occur under aerobic and anaerobic conditions, respectively [6]. Denitrification not only leads to the generation of N_2O, but also represents a possible mechanism for the loss of available N in plants [6]. Thus, the development of a novel strategy is needed to protect the environment and guarantee the production of crop products.

Nitrification inhibitors are a potential strategy that can fulfill both environmental and productivity goals [7,8]. By disrupting the roles of *Nitrosomonas bacteria*, nitrification inhibitors can inhibit the conversion of soil NH_4^+-N to NO_3^--N, leading to a decrease in the soil pool of NO_3^- [9]. NH_4^+-N can be more easily absorbed to the soil than NO_3^--N, resulting in the better uptake of NH_4^+-N by plants or immobilization with dissolved organic matter, instead of leaching losses [10]. Most studies have demonstrated that nitrification inhibitors can attenuate nitrate leaching and N_2O lost to the environment, thereby improving N usage and leading to better plant growth [11,12]. By reducing NO_3^--N concentrations, nitrification inhibitors can indirectly regulate other microbial processes (e.g., denitrification rates) [13]. Denitrification can be affected by a variety of factors, such as NO_3^--N concentrations, soil moisture, soil pH, available C, and temperature [14]. However, contradictory results have been obtained regarding the ability of nitrification inhibitors to affect denitrification. It has been reported that nitrification inhibitors can inhibit the oxidation of NH_4^+ to NO_3^-, but do not reduce denitrification rates [15]. On the contrary, nitrification inhibitors have been shown to reduce both nitrification and denitrification in soils after the application of cattle slurry [14]. However, when an inhibitor was added into mineral-fertilized soil, there was no significant decrease in N_2O emissions [16]. It remains unknown whether nitrification inhibitors can decrease denitrification rates in arable soils, where N fertilizer and cattle manure are the major sources of N.

Among the most widely used NIs, 3,4-dimethylpyrazole phosphate (DMPP) is one of the most effective at improving N retention and reducing losses [17]. It has been indicated that DMPP has the advantages of a long-lasting inhibitory effect, low application rate, high persistence, and minor eco-toxicological side effects on plants [18,19]. Previous studies have indicated that the performance of DMPP in reducing the nitrification rate is not constant, as the persistence and effectiveness of DMPP in soil are strongly influenced by the environmental conditions (e.g., soil temperature) [20]. If the widespread use of DMPP is to be encouraged to reduce N losses, it is important to know which other aspects of the N cycle are affected by this compound. Various studies have demonstrated that animal manure applications would input large amounts of metabolizable C, mineral N, and water into soil simultaneously, which may favor both the nitrification and denitrification processes [21]. Previous researchers have mainly focused on the effects of nitrification inhibitors (e.g., DCD, nitripyrin) on changes in the soil N mineralization process that may influence the dynamics of soil inorganic N and the soil N cycle, lacking comparison with denitrification rates from other N fertilizers and nitrification inhibitors (DMPP) [22,23]. Moreover, observations of the denitrification process under various fertilizers with DMPP application during the whole maize growing period are lacking. Hence, a field experiment was performed to determine whether DMPP can affect denitrification rates by limiting NO_3^- availability. Therefore, the specific objectives of this study were to (1) clarify the characteristics of denitrification rates affected by nitrification inhibitors from different N fertilizer types applied to soils; (2) explore the influencing factors of the dynamics in annual denitrification rates (e.g., DMPP concentrations, soil properties). This will provide insightful information for our understanding of the achievement of inhibitors on the mitigation of N losses in arable soil under field conditions.

2. Materials and Methods

2.1. Experimental Fields

The field study was carried out at the Shenyang Experimental Station of the Institute of Applied Ecology, Chinese Academy of Sciences (41°32′ N, 123°23′ E) in Liaoning province. The average annual temperature and precipitation are 7.0–8.0 °C and 700 mm, respectively, with 147–164 frost-free days. According to the soil taxonomy classification, the soil can be classified as Alfisols. The soil properties at 0–20 cm depth are shown in Table 1. Maize (*Zea mays* L.) is continuously planted in early May and harvested in late September.

Table 1. Characteristics of the soils (0–20 cm) and manure used in this study.

Types	pH (H$_2$O)	Total P (g kg^{-1})	Total N (g kg^{-1})	Total K (g kg^{-1})	Available N (mg kg^{-1})	Available P (mg kg^{-1})	Available K (mg kg^{-1})	Organic Carbon (g kg^{-1})
Soil	6.8	0.59	1.49	16.8	91.1	14.7	90.7	12.2
Manure	7.2	6.9	22.35	16.7	-	-	-	2.54

2.2. Experimental Design and Field Management

Five treatments were included and organized in a randomized design with 3 replicates in this study: unfertilized controls (CK); NPK (T2); NPK + DMPP (T3); cattle manure (T4); manure + DMPP (T5). The plots were 20 m^2 (4 m × 5 m) and randomly placed at a distance of 1.5 m apart. In this study, urea was applied at a rate of 200 kg/K/ha; phosphorus (P) and potassium (K) fertilizers, as K$_2$SO$_4$ and KH$_2$PO$_4$, were applied at 58 kg/K/ha and 30 kg/P/ha. The cattle manure was applied at 2 Mg/ha. The manure characteristics are shown in Table 1. The inhibitor DMPP was applied at the rate of 1% of urea N. All fertilizers and inhibitors were added to the topsoil (0–10 cm) before maize sowing. The maize hybrid 'Dandong 1501' was sown on 4 May 2021 at a 25-cm distance between seeds within rows and 60-cm row spacing to reach a target density of 6.67 plants/m^2. All crops were harvested on 28 September. The field management was in accordance with the routine cultivation practices of the local farmers.

2.3. Soil Sampling and Measurement

The soils were sampled on days 0, 5, 10, 15, 20, 30, 45, 70, 100, 130, and 160 after fertilizer application. Four individual samples at each plot were collected using a 3-cm soil auger and then mixed thoroughly to obtain individual bulked plot soil samples. The mineral N concentration (NO$_3$$^-$-N and NH$_4$$^+$-N) was detected after sieving the fresh soils (<2 mm) and extraction with 2 M KCl. A combined electrode pH meter was employed to assess soil pH at a soil: water ratio of 1:2.5. Soil microbial biomass (SMB) was evaluated by substrate-induced respiration. The soil carbon availability was subsequently assessed. Briefly, the soil sample (35 g) was placed in a 1.8-L glass preserving jar, which was then sealed with a septum stopper-fitted lid. After incubation at 25 °C for 7 days, the accumulation of CO$_2$ in the headspace atmosphere of the preserving jar was evaluated as an indicator of soil respiration.

2.4. In Situ Denitrification Rates

The static soil core incubation system was used to measure the in situ denitrification rates according to the acetylene inhibition method [24,25]. Briefly, the intact soil cores consisting of PVC pipes with uniformly distributed holes were isolated and transferred into a glass preserving jar sealed with a septum stopper-fitted lid. Then, 120 mL acetylene (10% v/v of headspace) was placed in the jar, mixed thoroughly, and kept in a temperature-controlled room. Gases (22 mL) were collected from the jars at 30 min and 3, 6, and 24 h following acetylene addition, and then injected into a vacutainer until further use. All specimens were analyzed using a gas chromatography system (Philips, Cambridge, MA, USA) equipped with an electron capture detector at 350 °C. A porous packed column was used to separate the gas samples at 80 °C. The temperature at the injection port was 120 °C. To calculate the total denitrification rate, the solubility of N$_2$O in the soil water was taken into account using the temperature-dependent Bunsen absorption coefficient.

2.5. Denitrifying Enzyme Activity

To measure DEA, 10 g soil samples were weighed into a 100-mL Schott bottle [26,27]. Then, 20 mL nitrate glucose solution (0.1 g KNO$_3$ and 0.2 g glucose in 1 L water) containing 0.125 g chloramphenicol was placed into the Schott bottle. After sealing with rubber septum-fitted lids, the bottles were flushed with N gas for 2 min. Then, 10 mL acetylene was used to prevent the conversion of dinitrogen gas from N$_2$O. All specimens were incubated at

25 °C with shaking. After the removal of 5 mL headspace at 15- and 75-min intervals, the samples were kept in a 3-mL evacuated vacutainer until further use. Finally, all specimens were analyzed using the gas chromatography system (Philips, Cambridge, USA).

2.6. DMPP Extraction and Its Qualification

DMPP extraction and qualification was adapted from previous studies by Benckiser et al. (2013) and Chen et al. (2019) [20,28]. To extract DMPP from the soil sample, 10 g field fresh soil, 10 mL distilled water, and 0.2 mL 1 M K_3PO_4 were mixed together and shaken for 2 h at 30 rpm. Then, 0.2 mL 1 M $CaCl_2$ was added to the soil suspensions and samples were shaken for another 30 min at 30 rmp. Afterwards, 1 mL 1 M NaOH was added and samples were further shaken for 1 h. For transferring DMP into the t-butyl-methyl-ether phase (MTBE), 15 mL of MTBE was added and samples shaken for 1 h. The extract was then centrifuged at 3000 rmp for 5 min. The supernatant was evaporated and filtered through a 0.45-μm Millipore filter. The DMPP concentration was quantified with a Shimadzu HPLC device (Shimadzu, Kyoto, Japan) using a 5 μm, 4.6 × 150 mm Shiseido Spolar C18 column (Shiseido, Tokyo, Japan).

2.7. Statistical Analysis

The differences in soil biochemical parameters were analyzed by two-way ANOVA using SPSS Statistics 16.0 (SPSS Inc., Chicago, IL, USA). The 5% confidence level ($p < 0.05$) was considered statistically different. Pearson correlation analysis was employed to analyze the relationships between soil properties and denitrification rates.

3. Results

3.1. Soil Denitrification Rates

A remarkable seasonal effect was observed for the denitrification rates. The highest denitrification rate was observed in summer, while the lowest denitrification rate was detected in spring and autumn (Figure 1a). The application of urea and manure could lead to an increase in denitrification compared with the control treatment. There was no marked difference between urea- or manure-only soils and DMPP-amended soils, indicating that DMPP may not inhibit denitrification rates (Figure 1a).

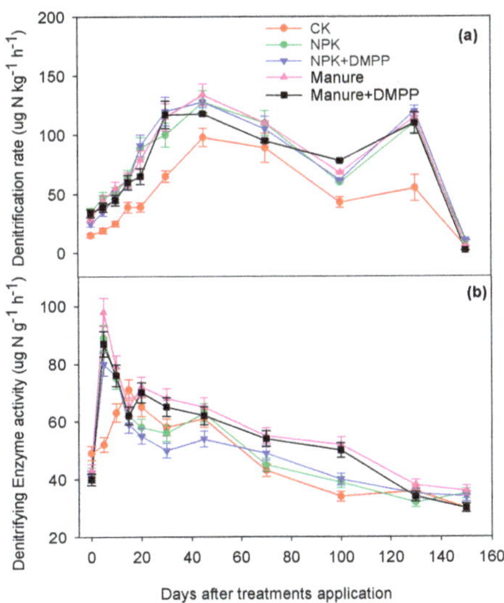

Figure 1. Denitrification rate (**a**) and denitrifying enzyme activity (**b**) at different days following application of different treatments. Data were presented as mean values with standard errors (n = 3).

As with denitrification rates, the highest DEA was found in the spring and summer months, while the lowest DEA was found in the autumn months. The increased DEA in the urea- or manure-amended soils was most obvious on day 5 after the fertilizer application and decreased with time ($p < 0.05$, Figure 1b). DEA was higher in manure-amended soils than other treatments ($p < 0.05$). Application of DMPP to soils could reduce DEA compared to urea- or manure-only treatment, but this difference was not significant (Figure 1b).

3.2. Soil Microbial Mass, C Availability, pH, NH_4^+-N, and NO_3^--N Concentration

The manure-amended soils had an increase in SMB compared with the urea-amended soils on day 5 after manure application ($p < 0.05$). However, the difference was negligible on day 30 (Figure 2a). SMB was not influenced by the application of DMPP in all treatment groups.

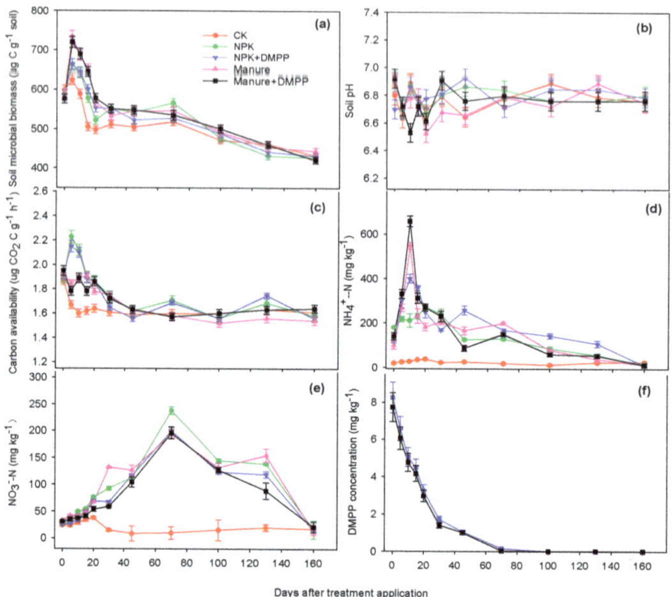

Figure 2. Soil microbial biomass (**a**), soil pH (**b**), carbon availability (**c**), NH_4^+-N concentration (**d**), NO_3^--N concentration (**e**), and DMPP concentration (**f**) at different days following application of different treatments. Data were presented as mean values with standard errors (n = 3).

The manure-amended soils demonstrated an obvious reduction in C availability over the sampling period ($p < 0.05$). However, the C availability was not decreased in both control and urea-amended soils (Figure 2b). No significant differences in carbon availability were observed among all treatment groups, indicating that C availability was not influenced by the addition of nitrification inhibitors in all treatment groups. Likewise, no significant differences in soil pH and C availability were observed between DMPP-amended soils and urea or manure soils (Figure 2c).

A similar variation trend was observed in all the fertilization treatments. After the application of NPK and manure soils, the concentrations of NH_4^+-N and NO_3^--N were first increased and then declined. In the urea and manure treatments, the addition of inhibitors could lead to higher soil NH_4^+-N concentrations compared to the soil without inhibitors (Figure 2d, $p < 0.05$). However, this was only significant before 20 days after inhibitor application, and, by day 30, the difference was negligible (Figure 2d,e). The manure-amended soils with inhibitors also had lower NO_3^--N content compared to those without inhibitors between 45 and 70 days after fertilization ($p < 0.05$).

In this study, Pearson correlation analysis was employed to analyze the relationships between soil properties and denitrification rates. The results revealed that the effect of inhibitors on denitrification rates was not markedly associated with DEA, soil NH_4^+-N, NO_3^--N, soil pH, and carbon availability (Table 2).

Table 2. Pearson correlation analysis of denitrification rates and soil properties.

Variable Factors	NPK		NPK + DMPP		Manure		Manure + DMPP	
	R^2	P	R^2	P	R^2	P	R^2	P
DEA	0.121	0.532	0.247	0.064	0.178	0.074	0.674	0.421
NH_4^+-N	0.378	0.126	0.498	0.079	0.452	0.145	0.546	0.178
NO_3^--N	0.236	0.214	0.216	0.142	0.312	0.126	0.201	0.097
pH	0.145	0.145	0.347	0.078	0.147	0.245	0.394	0.076
Carbon availability	0.189	0.321	0.421	0.231	0.325	0.365	0.414	0.069

3.3. Rate of Inhibitor Loss in the Soils

The degradation rate of DMPP was dramatically increased (Figure 2f). On day 5 after DMPP application, 81% of the applied DMPP was degraded. DMPP was not detectable in the soil on day 70. A half-life of 10 days was detected for DMPP.

4. Discussion

4.1. Impact of DMPP on Soil Denitrification Rates

In all the treatment groups, the denitrification rates increased progressively during the first 40 days after fertilization, with a progressive decrease observed on the following days. Previous research has shown that the denitrification rate is largely dependent on enzyme activity. The reduced denitrification rate over time in each treatment group may be a consequence of the decay of denitrifying microbes—that is, the DEA of soil [29] (Figure 1b). The application of inhibitors did not decrease the denitrification rates compared to soils treated with urea or manure only (Figure 1a). This may be due to the fact that the concentration of NO_3^--N exceeded 5 mg NO_3^-/kg soil, which is regarded as a threshold for denitrification [26]. The high concentration of NO_3^--N in the soil may limit the role of nitrification in supplying nitrate to denitrifiers [30]. Nitrification has become less crucial to ensure the adequate denitrification of NO_3^--N in denitrifying microbes. However, nitrification inhibitors can attenuate the losses of denitrification in soils with low initial NO_3^--N concentrations by affecting the availability of the nitrate pool in denitrifying microorganisms [30]. The limited denitrification by NO_3^- could also be attributed to the rapid degradation of inhibitors in the soil, thus decreasing the effectiveness of the inhibitors, which was supported by the inhibitor degradation data in our study. A remarkable decline (81%) in the concentrations of inhibitors was observed on day 5 after inhibitor application, which was probably due to microbial degradation, leaching, or sorption to soil organic matter (Figure 2f).

Furthermore, our results showed that the application of inhibitors did not affect soil DEA in all treatment groups (Figure 1b). DEA can be used to reflect the population size of denitrifying microorganisms, which is an indicator of the optimum conditions for denitrification [24,26]. The inability of inhibitors to regulate DAE in the soil can explain why the denitrification rates are not influenced by inhibitor application, as the denitrifying microbial communities were not inhibited. Similarly, previous work found that the application of nitrification inhibitors to urea did not inhibit denitrification [26]. Another study demonstrated that the effects of nitrification inhibitors on denitrification rates relied on the levels of available C [31]. The application of manure to the soil slightly increased the quantity of C present in the soil, and therefore the effects of DMPP in manure-amended soils were expected to improve slightly [32]. In our study, the manure-amended soils had higher DEA than the other two treatment groups (Figure 1b). The increased microbial population might be attributed to the manure containing additional sources of nitrate and soluble C for microbial utilization [33].

4.2. Impacts of DMPP on Carbon Availability, Soil Microbial Biomass, and pH

It is crucial to assess whether the application of inhibitors can adversely influence the growth of soil microbial populations, as DMPP can specifically target the nitrifying bacteria [28]. As shown in Figure 2, the application of inhibitors did not affect SMB. These results are in agreement with previous findings reporting that DMPP did not affect the microbial biomass [28]. The non-significant effect of DMPP on SMB may be attributed to the fact that DMPP is bacteriostatic rather than bactericidal in its action [20].

In general, the application of DMPP had no effect on soil pH. A decline in soil pH was noted in all treatment groups between days 5 and 15, which may be related to the nitrification process, as nitrification is a major cause of soil acidification [34]. In this study, multiple stepwise regression analysis revealed that the effect of inhibitors on denitrification rates was not markedly associated with DEA, soil NH_4^+-N, NO_3^--N soil pH, or carbon availability. More studies are needed to clarify the molecular mechanisms underlying the inhibitory effects of nitrification inhibitors on denitrification rates.

5. Summary

The results showed that the denitrification rates and denitrifying enzyme activities were highly variable in different growing periods, but were not affected by the application of inhibitors. Partial inhibition of the nitrification process was observed, as revealed by an increase in the NH_4^+-N concentration and a decrease in the NO_3^--N concentration in the inhibitor treatments compared with the urea- or manure-only treatments. However, the decrease in NO_3^--N was not sufficient to limit NO_3^--N availability to denitrifiers, and, thus, the denitrification rates were found to not decrease. SMB, soil pH, and microbial respiration were not affected by nitrification inhibitors, regardless of whether manure or urea was applied in the soil. Our results concluded that the formation of NO_3^--N and the nitrification rates could be markedly reduced by DMPP, while NO_3^--N availability did not affect the denitrification rates. Furthermore, to confirm the findings of this study, field studies under different sites to explore additional mechanisms driving changes over longer time periods are needed.

Author Contributions: Conceptualization, J.L.; methodology, J.L.; software, J.L.; validation, W.W. (Wenyu Wang), W.W. (Wei Wang), and Y.L.; formal analysis, J.L. and W.W. (Wenyu Wang); investigation, W.W. (Wei Wang) and Y.L.; resources, J.L.; data curation, J.L.; writing—original draft preparation, J.L.; writing—review and editing, J.L.; visualization, J.L.; supervision, J.L., project administration, J.L.; funding acquisition, J.L. All authors have read and agreed to the published version of the manuscript.

Funding: This study was funded by the National Natural Science Foundation of China (42277324); Liaoning Revitalization Talents Program (XLYC2007088); Strategic Priority Research Program of the Chinese Academy of Sciences (XDA28090200); and Guangxi Science and Technology Base and Talent Project (AD 20297090).

References

1. Wang, X.; Wang, G.; Guo, T.; Xing, Y.; Mo, F.; Wang, H.; Fan, J.; Zhang, F. Effects of plastic mulch and nitrogen fertilizer on the soil microbial community, enzymatic activity and yield performance in a dryland maize cropping system. *Eur. J. Soil Sci.* **2021**, *72*, 400–412. [CrossRef]
2. Klimczyk, M.; Siczek, A.; Schimmelpfennig, L. Improving the efficiency of urea-based fertilization leading to reduction in ammonia emission. *Sci. Total Environ.* **2021**, *771*, 13. [CrossRef] [PubMed]
3. Galloway, J.N.; Townsend, A.R.; Erisman, J.W.; Bekunda, M.; Cai, Z.; Freney, J.R.; Martinelli, L.A.; Seitzinger, S.P.; Sutton, M.A. Transformation of the Nitrogen Cycle: Recent Trends, Questions, and Potential Solutions. *Science* **2008**, *320*, 889–892. [CrossRef] [PubMed]

4. Wang, Y.; Yao, Z.; Zhan, Y.; Zheng, X.; Zhou, M.; Yan, G.; Wang, L.; Werner, C.; Butterbach-Bahl, K. Potential benefits of liming to acid soils on climate change mitigation and food security. *Glob. Chang. Biol.* **2021**, *27*, 2807–2821. [CrossRef]
5. Zhou, X.; Wang, S.; Ma, S.T.; Zheng, X.; Wang, Z.; Lu, C. Effects of commonly used nitrification inhibitors-dicyandiamide (DCD), 3,4-dimethylpyrazole phosphate (DMPP), and nitrapyrin on soil nitrogen dynamics and nitrifiers in three typical paddy soils. *Geoderma* **2020**, *380*, 114637. [CrossRef]
6. Dobbie, K.E.; Smith, K.A. Impact of different forms of N fertiliser on N_2O emissions from intensive grassland. *Nutr. Cycl. Agroecosys.* **2003**, *67*, 37–46. [CrossRef]
7. Barton, K.; McLay, C.D.A.; Schipper, L.A.; Smith, C.T. Annual denitrification rates in agricultural and forest soils: A review. *Aust. J. Soil Res.* **1999**, *37*, 1073–1093. [CrossRef]
8. Saud, S.; Wang, D.; Fahad, S. Improved nitrogen use efficiency and greenhouse gas emissions in agricultural soils as producers of biological nitrification inhibitors. *Front. Plant Sci.* **2022**, *13*, 854195. [CrossRef]
9. Ledgard, S.F.; Menneer, J.C. Nitrate leaching in grazing systems and management strategies to reduce losses. *Occas. Rep.* **2005**, *18*, 79–92.
10. Di, H.J.; Cameron, K.C. Reducing environmental impacts of agriculture by using a fine particle suspension nitrification inhibitor to decrease nitrate leaching from grazed pastures. *Agr. Ecosyst. Environ.* **2005**, *109*, 202–212. [CrossRef]
11. Weiske, A.; Benckiser, G.; Ottow, J.C.G. Effect of the new nitrification inhibitor DMPP in comparison to DCD on nitrous oxide (N_2O) emissions and methane (CH4) oxidation during 3 years of repeated applications in field experiments. *Nutr. Cycl. Agroecosys.* **2001**, *60*, 57–64. [CrossRef]
12. Majumdar, D.; Pathak, H.; Kumar, S.; Jain, M.C. Nitrous oxide emission from a sandy loam Inceptisol under irrigated wheat in India as influenced by different nitrification inhibitors. *Agric, Ecosyst, Environ.* **2002**, *91*, 283–293. [CrossRef]
13. Xh, A.; Jme, A.; Ft, B.; Vm, A.; Gm, A.; Fm, A. Dimethylpyrazole-based nitrification inhibitors have a dual role in N_2O emissions mitigation in forage systems under Atlantic climate conditions. *Sci. Total. Environ.* **2022**, *807*, 150670.
14. Hénault, C.; Germon, J.C. NEMIS, a predictive model of denitrification on the field scale. *Eur. J. Soil Sci.* **2000**, *51*, 257–270. [CrossRef]
15. Vallejo, A.L.; Garcia-Torres, J.A.; Diez, A.; Lopez-Fernandez, S. Comparison of N losses (NO_3^-, N_2O, NO) from surface applied, injected or amended (DCD) pig slurry of an irrigated soil in a Mediterranean climate. *Plant Soil* **2005**, *272*, 313–325. [CrossRef]
16. Merino, P.; Menendez, S.; Pinto, M.; Gonzalez-Murua, C.; Estavillo, J.M. 3, 4-Dimethylpyrazole phosphate reduces nitrous oxide emissions from grassland after slurry application. *Soil Use Manag.* **2005**, *21*, 53–57. [CrossRef]
17. Pan, B.; Xia, L.; Lam, S.K.; Wang, E.; Zhang, Y.; Mosier, A.; Chen, D. A global synthesis of soil denitrification: Driving factors and mitigation strategies. *Agr. Ecosyst. Environ.* **2022**, *327*, 107850. [CrossRef]
18. Di, H.J.; Cameron, K.C. Inhibition of ammonium oxidation by a liquid formulation of 3,4-Dimethylpyrazole phosphate (DMPP) compared with a dicyandiamide (DCD) solution in six new Zealand grazed grassland soils. *J. Soils Sediments.* **2011**, *11*, 1032–1039. [CrossRef]
19. Shi, X.Z.; Hu, H.W.; He, J.Z.; Chen, D.L.; Suter, H.C. Effects of 3,4-dimethylpyrazole phosphate (DMPP) on nitrification and the abundance and community composition of soil ammonia oxidizers in three land uses. *Biol. Fertil. Soils.* **2016**, *52*, 927–939. [CrossRef]
20. Chen, H.; Yin, C.; Fan, X.; Ye, M.; Peng, H.; Li, T.; Zhao, Y.; Wakelin, S.A.; Chu, G.; Liang, Y. Reduction of N_2O emission by biochar and/or 3,4-dimethylpyrazole phosphate (DMPP) is closely linked to soil ammonia oxidizing bacteria and nosZI-N_2O reducer populations. *Sci. Total Environ.* **2019**, *694*, 133658. [CrossRef]
21. Luchibia, O.; Suter, H.; Hu, W.; Lam, K.; He, Z. Effects of repeated applications of urea with DMPP on ammonia oxidizers, denitrifiers, and non-targeted microbial communities of an agricultural soil in Queensland, Australia. *Appl. Soil Ecol.* **2020**, *147*, 103392. [CrossRef]
22. Di, H.J.; Cameron, K.C. Sources of nitrous oxide from 15N-labelled animal urine and urea fertilizer with and without a nitri-fication inhibitor, dicyandiamide (DCD). *Aust. J. Soil Res.* **2008**, *46*, 7682.
23. Lui, C.; Mi, X.; Zhang, X.; Fan, Y.; Zhang, W.; Liao, W.; Xie, J.; Gao, Z.; Roelcke, M.; Liu, H. Impacts of slurry application methods and inhibitors on gaseous emissions and N_2O pathways in meadow-cinnamon soil. *J. Environ. Manag.* **2022**, *318*, 115560.
24. Adrian, B.; Mario, C.M.; Luis, M.A.; Pedro, M.A.; Carmen, G.M. Evaluation of a crop rotation with biological inhibition po-tential to avoid N_2O emissions in comparison with synthetic nitrification inhibition. *J. Environ. Sci.* **2023**, *127*, 222–233.
25. Tiedje, J.M.; Simkins, S.; Groffman, P.M. Perspectives on measurement of denitrification in the field including recommended protocols for acetylene based methods. *Plant Soil* **1989**, *115*, 261284. [CrossRef]
26. Luo, J.; White, R.E.; Ball, R.P.; Tillman, R.W. Measuring denitrification activity in soils under pasture: Optimizing conditions for the short-term denitrification enzyme assay and effects of soil storage on denitrification activity. *Soil Biol. Biochem.* **1996**, *28*, 409–417. [CrossRef]
27. Watkins, N.L.; Schipper, L.A.; Sparlinga, G.P.; Thorroldb, B.; Balks, M. Multiple small monthly doses of dicyandiamide (DCD) did not reduce denitrification in Waikato dairy pasture. *N. Z. J. Agri. Res.* **2013**, *56*, 37–48. [CrossRef]
28. Benckiser, G.; Christ, E.; Herbert, T.; Weiske, A.; Blome, J.; Hardt, M. The nitrification inhibitor 3,4-dimethylpyrazole-phosphat (DMPP)—Quantification and effects on soil metabolism. *Plant Soil* **2013**, *371*, 257–266. [CrossRef]
29. Calderon, F.J.; McCarty, G.W.; Reeves, J.B. Nitrapyrin delays denitrification on manured soils. *Soil Sci.* **2005**, *170*, 350–359. [CrossRef]

30. Thompson, R.B. Denitrification in slurry-treated soil: Occurrence at low temperatures, relationship with soil nitrate and reduction by nitrification inhibitors. *Soil Biol. Biochem.* **2005**, *21*, 875882. [CrossRef]
31. Zhang, H.; Hunt, D.E.; Ellert, B.; Maillard, E.; Kleinman, P.J.A.; Spiegal, S.; Angers, D.A.; Bittman, S. Nitrogen dynamics after low-emission applications of dairy slurry or fertilizer on perennial grass: A long term field study employing natural abundance of δ15N. *Plant Soil* **2021**, *465*, 415–430. [CrossRef]
32. David, R.; Wei, S. Nitrapyrin-based nitrification inhibitors shaped the soil microbial community via controls on soil pH and inorganic N composition. *Appl. Soil Ecol.* **2022**, *170*, 104295.
33. Antonio, C.H.; Jesus, G.L.; Antonio, V.; Eulogio, J.B. Effect of urease and nitrification inhibitors on ammonia volatilization and abundance of N-cycling genes in an agricultural soil. *J. Plant Nutr. Soil Sci.* **2020**, *183*, 99–109.
34. Ouyang, Y.; Evans, S.E.; Friesen, M.L.; Tiemann, L.K. Effect of nitrogen fertilization on the abundance of nitrogen cycling genes in agricultural soils: A meta-analysis of field studies. *Soil Biol. Biochem.* **2018**, *127*, 71–78. [CrossRef]

Article

Differences in the Concentration of Micronutrients in Young Shoots of Numerous Cultivars of Wheat, Maize and Oilseed Rape

Jolanta Korzeniowska * and Ewa Stanislawska-Glubiak

Department of Weed Science and Soil Tillage Systems in Wroclaw, Institute of Soil Science and Plant Cultivation-State Research Institute in Pulawy, ul. Orzechowa 61, 50-540 Wroclaw, Poland
* Correspondence: j.korzeniowska@iung.wroclaw.pl

Abstract: Individual species of cultivated plants differ in the content of microelements in the shoots. The aim of our research was to test the hypothesis that the variability of the micronutrient content between cultivars of the same species may be similar or even greater than the differences between species. The research material consisted of shoot samples of 12 wheat, 10 maize and 12 rape varieties collected from production fields in Poland. The smallest number of samples (replicates) within one cultivar was 10. A total of 481 wheat samples, 141 maize samples and 328 rapeseed samples were taken. Wheat samples were taken at the beginning of the stem elongation stage (BBCH 30/31); maize, when the plants reached a height of 25–30 cm (BBCH 14–15); and rape, in the period from the beginning of the main stem elongation stage to the appearance of the first internode (BBCH 30/31). All varieties of the tested crop species were grown in similar soil conditions in terms of pH, texture and TOC content. B, Cu, Fe, Mn and Zn were determined in all plant samples. Wheat showed a significantly lower average concentration of all micronutrients compared to rape and maize (e.g., 10 times less B than rape). On the other hand, among the species tested, rape had the highest concentration of B, Cu and Zn, and maize had the highest concentration of Fe and Mn. In all three tested crops, the differences in the content of B and Zn were greater between species than between cultivars. In the case of Cu, Mn and Fe concentration, the cultivar differences exceeded the species differences. The results suggest that there is no need to take cultivars into account when fertilizing with B and Zn. In contrast, fertilization with Cu, Mn and Fe needs to take into account different requirements of the cultivars for these micronutrients.

Keywords: microelements' diversity; aerial part; crops; species; cultivars

Citation: Korzeniowska, J.; Stanislawska-Glubiak, E. Differences in the Concentration of Micronutrients in Young Shoots of Numerous Cultivars of Wheat, Maize and Oilseed Rape. *Agronomy* **2022**, *12*, 2639. https://doi.org/10.3390/agronomy12112639

Academic Editors: Christos Noulas, Shahram Torabian and Ruijun Qin

Received: 3 October 2022
Accepted: 25 October 2022
Published: 26 October 2022

Publisher's Note: MDPI stays neutral with regard to jurisdictional claims in published maps and institutional affiliations.

1. Introduction

In order to achieve a satisfactory yield, plants should be provided with the proper level of nutrients in the soil. In addition to macronutrients, micronutrients are needed. These elements are needed in small quantities but are absolutely essential for plant life and development. Micronutrients such as boron (B), copper (Cu), iron (Fe), manganese (Mn), molybdenum (Mo) and zinc (Zn) are involved in many metabolic processes in the plant, influencing the optimal use of macronutrients and the overall health and condition of the plant [1].

In addition to being essential for plants, micronutrients are needed by their consumers —humans and animals. Too low levels of these nutrients in food and feed can cause many human and animal diseases. It is estimated that more than two billion people worldwide suffer from a lack of micronutrients, known as "hidden hunger" [2,3]. Iron and zinc deficiencies are the most common [4,5]. These deficiencies occur in people whose daily diet is based on cereal grains, mainly wheat, rice and corn [6]. Such situation occurs in the absence of plant-available forms of micronutrients in the soil, and can be corrected by fertilization.

Different crop species differ in their micronutrient concentration and uptake from the soil. It is generally accepted that plants with higher uptake have higher fertilizer requirements. Many fertilization guides state that the decision to fertilize with micronutrients can be made on the basis of their content in young plants [7–10]. For this purpose, relevant plant parts (usually whole shoots or leaves) at a strictly defined growth stage are taken, micronutrients are determined, and nutrition of the plants is assessed using the respective critical limits/deficiency limits. If the nutritional status of the plants with micronutrients is insufficient, it should be supplemented by foliar fertilization.

Micronutrient deficiency limits in plants are usually set for individual crop species [11–13]. However, within each species there are many cultivars. Breeders, in the search for better yielding, more stress-tolerant or more desirable plants for consumers, create new cultivars every year that meet these expectations. These cultivars may also show significant differences in micronutrient concentration and, thus, require different fertilization.

The literature provides little information on the variation of micronutrient concentrations in staple crop cultivars. Studies performed worldwide have focused mainly on Fe and Zn content in grain. Maganati et al. [14] observed significant differences in the content of Fe and Zn in the grain of 153 rice genotypes. Fe concentrations ranged from 6.9 to 22.3 mg kg^{-1}, while Zn concentrations ranged from 14.5 to 35.3 mg kg^{-1}. Ray et al. [15] showed differences in micronutrient concentration in the grain of many pea, bean, lentil and chickpea cultivars, with significant differences mainly in zinc. Tran et al. [16] showed significant differences in the content of Zn and Fe in the grain of different wheat genotypes. The content of Zn was in the wide range of 86.5–209.0 mg kg^{-1}, and Fe was in the range 51.7–91.8 mg kg^{-1}. Significant differences in B and Cu concentrations in grain and young shoots of several winter wheat cultivars were also observed by Korzeniowska [17] and Korzeniowska and Stanislawska-Glubiak [18]. Genc et al. [19] reported a significant difference in Zn content in young shoots of two winter barley cultivars. According to the literature, the differences in the content of micronutrients in cultivars are genetically and environmentally determined [20]. The main agricultural crops in Poland, along with triticale and rye, are wheat, maize and rape. Wheat is cultivated on 2511 thousand hectares of arable land, maize on 1265 thousand hectares of arable land and rape on 875 thousand hectares of arable land [21]. These three species are the most important representatives of cereal, fodder and oilseed crops in Poland, which together cover more than 40% of the country's sown area.

Based on our previous research and the literature cited, it was hypothesized that the variation in the micronutrient content in plants between cultivars of the same species may be similar or even greater than the differences between species. The aim of our study was to investigate the concentration of microelements in young plants of a dozen varieties of winter wheat (*Triticum aestivum* L.), maize (*Zea mays* L.) and winter rape (*Brassica napus* L.) and to verify our hypothesis on the basis of the data obtained in this way.

2. Materials and Methods

2.1. Sample Collection

In 2016–2018, samples of plants and soil were taken from the fields of winter wheat, winter rape and maize in order to compare the concentrations of micronutrients in species and cultivars. In total, 12 wheat, 10 maize and 12 rape cultivars were included in the study. Samples taken within one cultivar were treated as replications. The number of replications and the characteristics of the cultivars are shown in Tables 1–3. The smallest number of samples (replications) within one cultivar was 10 (Table 2). A total of 481 plant–soil pairs were collected from wheat fields, 141 from maize fields and 328 from rape fields. All the samples were collected by accredited sample takers from 16 Polish provinces, usually one plant–soil pair from each "gmina", the smallest administrative unit. Sampling points were quite evenly distributed throughout Poland (Figure 1).

Table 1. Characteristics of wheat cultivars and the number of samples taken.

No.	Cultivar		Usage [1]	Year [2]	Breeder	No. of Samples
1	Arkadia	Ark	A	2011	DANKO Hodowla Roślin sp. z o.o., Poland	100
2	Bamberka	Bam	A	2009	Hodowla Roślin Strzelce sp. z o.o., Poland	46
3	Bogatka	Bog	B	2004	DANKO Hodowla Roślin sp. z o.o., Poland	14
4	Hondia	Hon	A	2014	DANKO Hodowla Roślin sp. z o.o., Poland	11
5	Julius	Jul	A	-	KWS Lochow GmbH, Germany	88
6	Linus	Lin	A	2011	RAGT 2 n, France	30
7	Muszelka	Mus	B	2008	DANKO Hodowla Roślin sp. z o.o., Poland	11
8	Ostroga	Ost	A	2008	DANKO Hodowla Roślin sp. z o.o., Poland	48
9	Ozon	Ozo	B	2010	KWS Lochow GmbH, Germany	26
10	Sailor	Sai	A	2011	DANKO Hodowla Roślin sp. z o.o., Poland	30
11	Skagen	Ska	A	2009	W. von Borries-Eckendorf GmbH & Co. KG, Germany	44
12	Tonacja	Ton	A	2001	Hodowla Roślin Strzelce sp. z o.o., Poland	33
	Total					481

[1] A: quality, B: bread; [2] year of registration.

Table 2. Characteristics of maize cultivars and the number of samples taken.

No.	Cultivar		FAO	Usage [1]	Year [2]	Breeder	No. of Samples
1	Danubio	Dan	240–250	S	2013	Saatbau Linz eGen, Austria	10
2	Glejt	Gle	230	G	2001	HR Smolice, Poland	16
3	Legion	Leg	260–270	S	2014	HR Smolice, Poland	11
4	Nimba	Nim	260	S	1996	HR Smolice, Poland	10
5	Opoka	Opo	240	S	2006	HR Smolice, Poland	11
6	P8400	P8400	240	G	2013	Pionner, USA	14
7	Reduta	Red	230	G	2000	HR Smolice, Poland	14
8	Rosomak	Ros	250–260	G	2013	HR Smolice, Poland	22
9	Subito	Sub	260	G	2008	HR Smolice, Poland	10
10	Ulan	Ula	270	G	2011	HR Smolice, Poland	23
					Total		141

[1] S: silage, G: grain; [2] year of registration.

Table 3. Characteristics of oilseed rape cultivars and the number of samples taken.

No.	Cultivar		Year [1]	Breeder	No. of Samples
1	Abacus (HY)	Aba	2009	Norddeutsche Pflanzenzucht Hans-Georg Lembke KG, Germany	22
2	Alexander (HY)	But	-	Limagrain Europe, France	21
3	Alvaro (HY)	Alv	2015	KWS Saat SE & Co. KGaA, Germany	20
4	Exquisite (HY)	Exq	2011	Monsanto Technology LLC, USA	14
5	Garou (HY)	Gar	2013	Norddeutsche Pflanzenzucht Hans-Georg Lembke KG, Germany	16
6	Kuga (HY)	Kug	2015	Norddeutsche Pflanzenzucht Hans-Georg Lembke KG, Germany	19
7	Marcopolos (HY)	Mar	2012	KWS Saat SE & Co. KGaA, Germany	23

Table 3. *Cont.*

No.	Cultivar	Year [1]		Breeder	No. of Samples
8	Mercedes (HY)	Mer	2013	Norddeutsche Pflanzenzucht Hans-Georg Lembke KG, Germany	16
9	Monolith (OP)	Mon	2008	Hodowla Roślin Strzelce sp. z o.o. IHAR Group, Poland	71
10	Rohan (HY)	Roh	2008	Norddeutsche Pflanzenzucht Hans-Georg Lembke KG, Germany	30
11	Sherlock (OP)	She	2010	KWS Saat SE & Co. KGaA, Germany	46
12	Visby (HY)	Vis	2008	Norddeutsche Pflanzenzucht Hans-Georg Lembke KG, Germany	30
			Total		328

HY: hybrid cultivar, OP: open-pollinated cultivar,[1] year of registration.

Figure 1. Number of pairs of soil-plant samples taken in Polish provinces: wheat/maize/rape.

Wheat was sampled from an area of 1 m^2, maize was sampled from an area of 8 m^2 and rape was sampled from an area of 4 m^2. Whole shoots of wheat were cut 2 cm above the ground at the beginning of stem elongation stage (BBCH 30/31). The shoots of the other two plants were cut 5 cm above the ground; maize, when the plants reached a height of 25–30 cm (BBCH 14–15); and rape, in the period from the beginning of the main stem elongation stage to the appearance of the first internode (BBCH 30/31) [22]. Each wheat sample consisted of a minimum 80 shoots, and maize and rape of a minimum of 20 shoots. At the same time as the plant samples, corresponding soil samples were taken. Each soil sample was created by mixing five sub-samples taken with a soil sampler to a depth of 20 cm.

2.2. Soil and Climate Characteristic

All plant–soil sample pairs were taken from fields where the pH was in the range 5–7 and the fraction content < 0.02 mm in the range 10–35%. Very acidic and alkaline soils and very light and very heavy soils were not sampled. Extreme conditions were avoided because pH and soil texture have such a strong influence on the uptake of micronutrients by plants that they could distort the picture of their content in the species and cultivars studied [23]. The characteristics of the soil samples taken are shown in Table 4.

Table 4. Characteristic of soil samples.

Soil Feature	Wheat (*n* = 481)			Maize (*n* = 141)			Rape (*n* = 328)		
	Mean	SE	Range	Mean	SE	Range	Mean	SE	Range
pH in KCl	6.1	0.03	5–7	6.0	0.06	5–7	6.1	0.03	5–7
Sand 2.00–0.05 mm, %.	61	0.79	2.5–84.8	64	1.22	13.5–83.6	64	0.73	22.7–83.4
Silt 0.05–0.002 mm, %.	36	0.77	13.8–94.6	33	1.14	14.7–84.0	33	0.68	15.2–72.2
Clay <0.002 mm, %	3	0.05	0.0–6.4	3	0.10	1.3–8.6	3	0.06	1.0–9.8
Fraction <0.02 mm, %	20.5	0.29	10–35	19.5	0.55	10–35	20.0	0.35	10–35
TOC %	1.3	0.03	0.5–9.8	1.2	0.05	0.3–4.1	1.2	0.03	0.3–3.6

SE: standard error.

Poland is located in a temperate transitional climate zone, with an average annual air temperature of 8.7 °C and a total rainfall of 609 mm (1991–2020) [24]. In 2016 and 2017, air temperature and rainfall during the growing season were higher than the climatological normal; in 2018 they were close to the climatological normal.

2.3. Chemical Analysis

B, Cu, Fe, Mn and Zn were determined in all plant and soil samples. Micronutrients in plants were determined by the FAAS method, having first dry ashed the material in a muffle furnace and digested it with 20% nitric acid [25]. The exception was B, which was determined by the ICP-AES technique.

Micronutrients in the soil were determined by Mehlich 3 method [26–28]. During extraction, the ratio of soil to solution was 1: 10, and shaking time on the rotary stirrer was 10 min at 35/40 rpm. The Cu, Fe, Mn and Zn content of the extract was determined using the FAAS technique and the B content using the ICP-AES technique. Moreover, in soil samples the pH was established potentiometrically in 1 mol KCl dm^{-3} [29], total organic carbon (TOC) was determined by Turin method using potassium dichromate [30] and the soil texture was determined by laser diffraction method.

All chemical analyses were performed in state agrochemical laboratories certified by the Polish Centre of Accreditation [31], which ensured high reliability of the analyses.

2.4. Statistical Analysis

Mean micronutrient concentrations for the species tested were calculated from all samples taken, where *n* = 481 for wheat, 141 for maize and 328 for rape. Mean micronutrient concentrations for the cultivars were calculated from replicates within each cultivar.

To test the significance of differences in micronutrient concentrations between species and cultivars, an ANOVA test was performed using Statgraphics v 5.0 software (StatPoint Technologies, Inc., Warrenton, VA, USA). Multiple comparisons among groups were made with Tukey's significant difference test ($p < 0.05$).

3. Results

3.1. Average Soil Micronutrient Content

The soil taken from the fields where wheat, maize and rape were grown differed significantly in the mean concentration of B, Cu and Fe, while there were no differences in the content of Mn and Zn (Figure 2).

The concentration of B in soil from maize fields was more than twice as high as in soil from wheat fields and 50% higher than that from rape fields. Soil Cu levels were the same in wheat and maize fields, while rape fields had about 20% higher soil content of this micronutrient. Similarly to Cu, soil Fe concentration did not differ significantly between wheat and maize fields, and soil from rape fields was about 20% richer in Fe.

Despite differences in micronutrient content in the fields where the studied plants grew, no deficiency was found anywhere. The concentration of B, Cu, Fe, Mn and Zn in the soil was sufficient for all three species according to Polish standards [32].

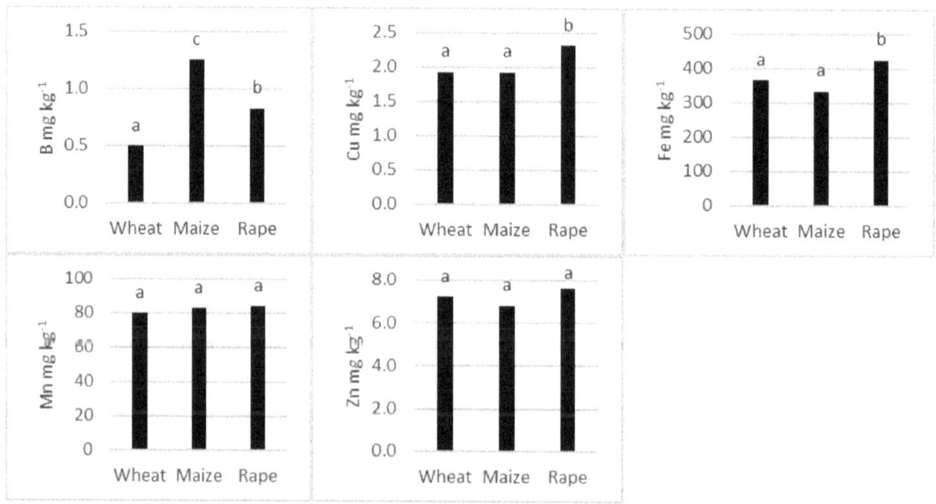

Figure 2. Concentration of micronutrients in the soil determined with the Mehlich 3 extractant—average of all collected samples. Bars marked with the same letters indicate no significant difference according to Tukey's test ($p < 0.05$).

3.2. Average Concentration of Micronutrients in the Shoots of the Studied Species

The average concentration of micronutrients in wheat, maize and rape shoots differed significantly, with wheat containing the least of each element tested (Figure 3). The greatest interspecies differences occurred in B and Zn contents. B concentration was ten times higher in oilseed rape and two times higher in maize than in wheat. In addition, Zn was twice as high in oilseed rape and 1.5 times higher in maize than in wheat.

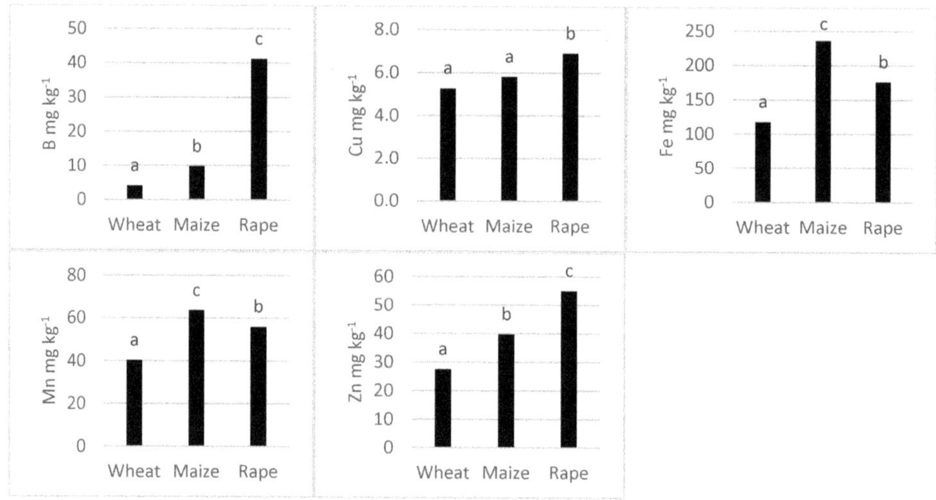

Figure 3. Concentration of micronutrients in shoots of the tested species—average of all collected samples. Bars marked with the same letters indicate no significant difference according to Tukey's test ($p < 0.05$).

The smallest differences between the plants were in the Cu content. There was no statistically significant difference between wheat and maize, and rape contained only 1/3 more of this element than wheat.

The concentration of Fe and Mn was the highest in maize, which contained twice as much Fe and 60% more Mn than wheat. On the other hand, rape contained 50% more Fe and 40% more Mn than wheat.

3.3. Average Concentration of Micronutrients in the Different Cultivars of the Species Tested

3.3.1. Wheat

The wheat cultivars tested were sufficiently supplied with all micronutrients (Figure 4). No lower concentrations were found in the shoots than the deficiency limits set by Korzeniowska et al. (2020). However, the individual cultivars differed in their micronutrient levels. For each micronutrient, it was possible to distinguish groups of cultivars with similar contents of this element, i.e., cultivars between which there were no significant differences. In contrast, significant differences were observed among groups. In some cases, one cultivar was classified in two or even three groups (Figure 4).

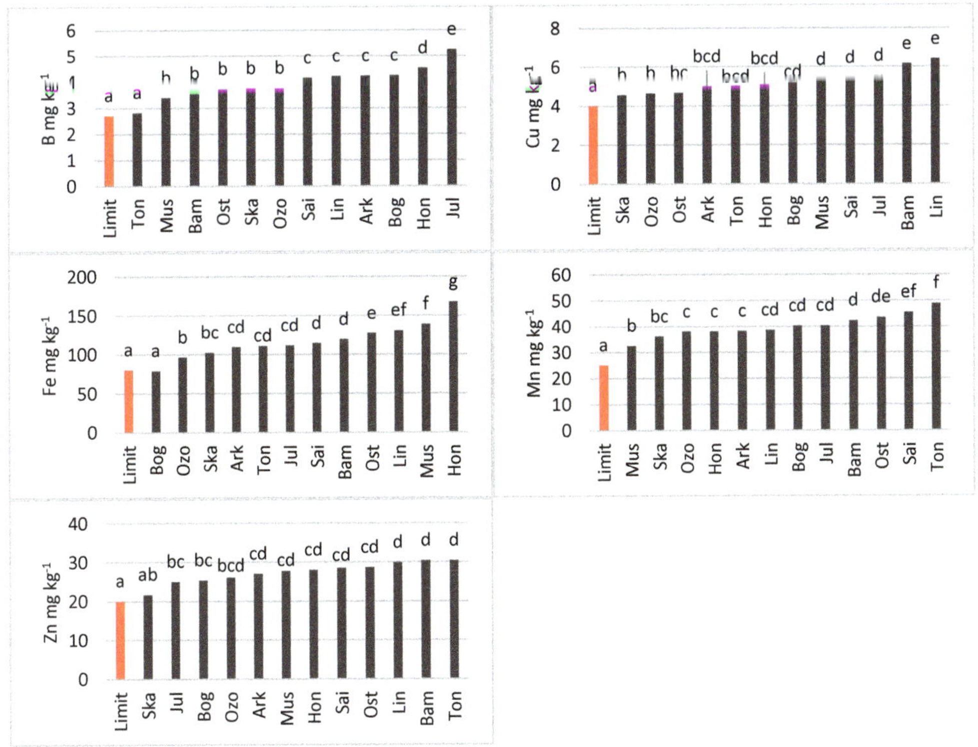

Figure 4. Concentration of microelements in shoots of wheat cultivars. Bars marked with the same letters indicate no significant difference according to Tukey's test ($p < 0.05$).

The lowest concentration of B was recorded in the shoots of the cultivar Tonacja (2.8 mg kg^{-1}), which alone formed the first group (a). The average concentration of B in the second group of cultivars (b), which included Muszelka, Bamberka, Ostka, Skagen and Ozon, was 3.6 mg kg^{-1}. Sailor, Linus, Akcadia and Bogatka formed the third cultivar group (c) with an average B concentration in shoots of 4.2 mg kg^{-1}. Hondia as an independent group (d) contained 4.5 mg kg^{-1}, and Julius, with the highest B concentration (5.2 mg kg^{-1}), belonged to the last group (e).

The group with the lowest Cu concentration included the cultivars Skagen and Ozon (average concentration of 4.6 mg kg^{-1}), while Linus and Bamberka contained the highest amount of this element (6.2 mg kg^{-1}).

The greatest significant difference in Fe concentration occurred between the first group (a), whose sole representative was Bogatka (78 mg kg^{-1}) and the last group (g), represented by Hondia (167 mg kg^{-1}). The remaining cultivars formed seven groups (b, bc, cd, d, e, ef and f), with Fe concentrations ranging from 96 to 139 mg kg^{-1}.

In the case of Mn, the Muszelka cultivar alone formed the group with the lowest concentration of this nutrient (33 mg kg^{-1}), while Tonacja formed the group with the highest concentration (49 mg kg^{-1}). The other cultivars formed six groups in which Mn concentration ranged from 36 to 45 mg kg^{-1}.

The lowest concentration of Zn in the shoots was presented by the cultivar Skagen (22 mg kg^{-1}), while the group which included the cultivars Linus, Bamberka and Tonacja contained the highest amount of this nutrient (average 30 mg kg^{-1}). The Zn concentration in the other cultivars ranged from 25–29 mg kg^{-1}.

3.3.2. Maize

Evaluation of the supply of micronutrients to maize using the limits developed by Korzeniowska et al. (2020) showed that it was sufficient, except for one case concerning Cu in the cultivar Nimba (Figure 5).

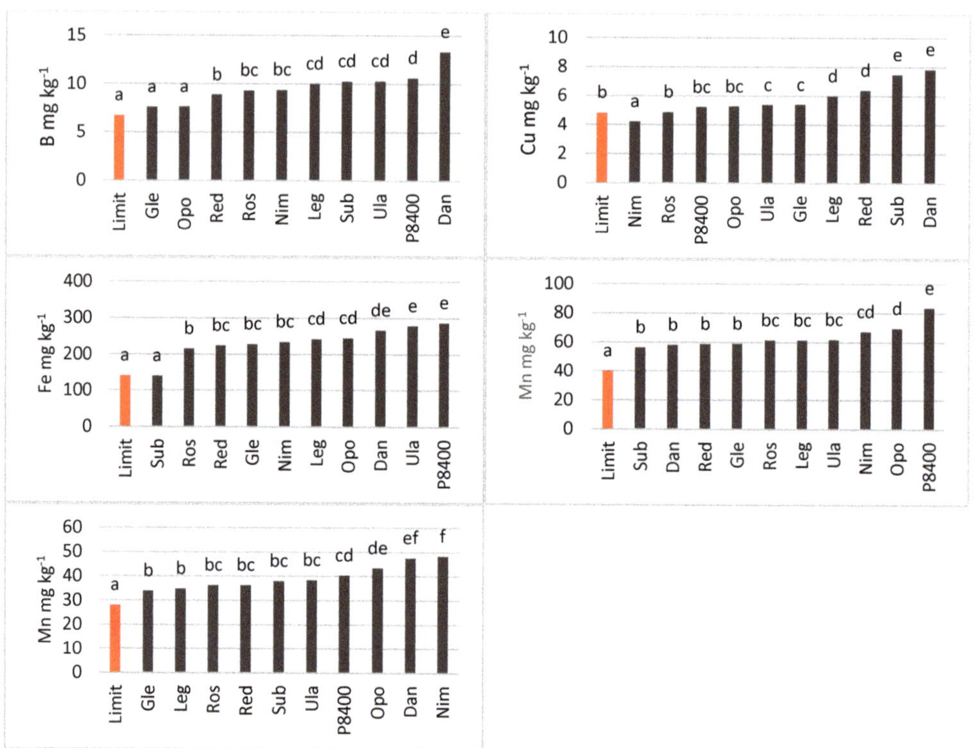

Figure 5. Concentration of microelements in shoots of maize cultivars. Bars marked with the same letters indicate no significant difference according to Tukey's test ($p < 0.05$).

The concentration of B in maize shoots ranged from 7.5 mg kg^{-1} for the first group of cultivars (a), represented by Glejt and Opoka, to 13.3 mg kg^{-1} in the cultivar Danubio, which alone formed the last group (e).

The Cu concentration in maize shoots was lowest in the aforementioned Nimba cultivar at 4.2 mg kg^{-1}. Nimba's Cu supply was too low, as the limit is 4.8 mg kg^{-1}. The highest Cu concentration (average 7.6 mg kg^{-1}) was in the group formed by Subito and Danubio (e).

The Fe concentration in plants of the first group (a), which was formed by one cultivar, Subito, was 139 mg kg^{-1} and was much lower compared to the other cultivar groups. The cultivars Ulan and P8400 belonged to the group with the highest concentration of this element, averaging 282 mg kg^{-1} (e).

The variability of Mn in the shoots of the maize cultivars tested was relatively low. As many as seven cultivars (Subito, Danubio, Reduta, Glejt, Rosomak, Legion and Ulan) did not differ significantly (groups b and bc). In contrast, the cultivar P8400 stood out as having the highest concentration of Mn (83 mg kg^{-1}).

The concentration of Zn in maize shoots ranged on average from 34.5 mg kg^{-1} for Glejt and Legion (group b) to 47.5 mg kg^{-1} for Danubio and Nimba (group f).

3.3.3. Oilseed Rape

All rape cultivars were characterized by shoot micronutrient contents above the critical limits provided in Korzeniowska et al. (2020) (Figure 6).

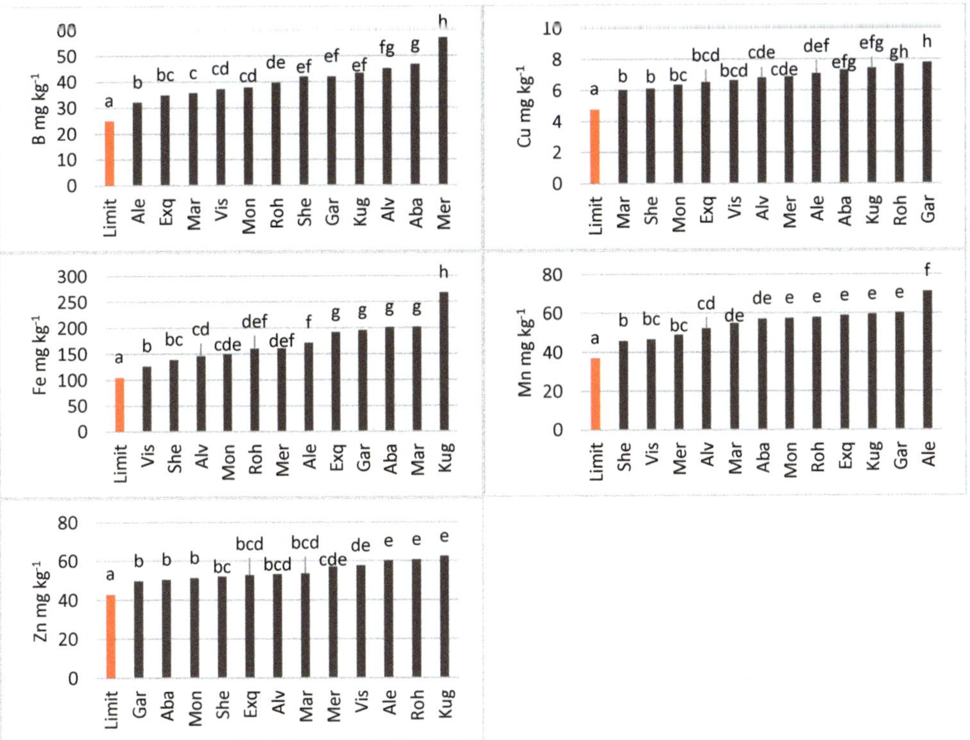

Figure 6. Concentration of microelements in shoots of oilseed rape cultivars. Bars marked with the same letters indicate no significant difference according to Tukey's test ($p < 0.05$).

The B concentration ranged from 32 mg kg^{-1} for the cultivar Alexander (b) to 57 mg kg^{-1} for the cultivar Mercedes, which formed the last group (h) on its own. The other cultivars belonged to as many as seven groups, indicating a wide variation of B in rape shoots.

Cu content varied less than B content. In the group of cultivars that included Marcopolos and Sherlock (b), an average Cu concentration was 6.1 mg kg^{-1}. The highest concentration of this element (7.8 mg kg^{-1}) was presented by Garou (h).

The Fe concentration ranged from 126 mg kg^{-1} for Visby (b) to 267 mg kg^{-1} for Kuga (h). The Fe concentration in the Kuga cultivar was as much as 35 % higher than in the cultivars of the previous group (g), which included Exquisite, Garou, Abakus and Marcopolos (average 197 mg kg^{-1}).

The Sherlock cultivar (b) had the lowest concentration of Mn (46 mg kg^{-1}), while the Alexander cultivar, which on its own formed a group that differed significantly from the other groups, contained the most of this micronutrient (71 mg kg^{-1}).

Zn concentration ranged from 51 mg kg^{-1} for the Garou, Abacus and Monolith (b) cultivar group to 61 mg kg^{-1} for the Alexander, Rohan and Kuga (e) cultivars.

3.4. Comparison of Cultivar and Species Diversity

The ranges of microelements in the shoots of the studied species presented in Figure 7 were determined on the basis of the micronutrient concentrations in cultivars. It was observed that for some micronutrients, the ranges for wheat, maize and rape partially overlapped, and for other micronutrients were completely divergent. The greatest difference between species was found for B. The range of B in rape (32–57 mg kg^{-1}) differed significantly from that found in wheat and maize. At the same time, the concentration of B in all maize cultivars (7.5–13.3 mg kg^{-1}) was higher than in wheat cultivars (2.8–5.2 mg kg^{-1}).

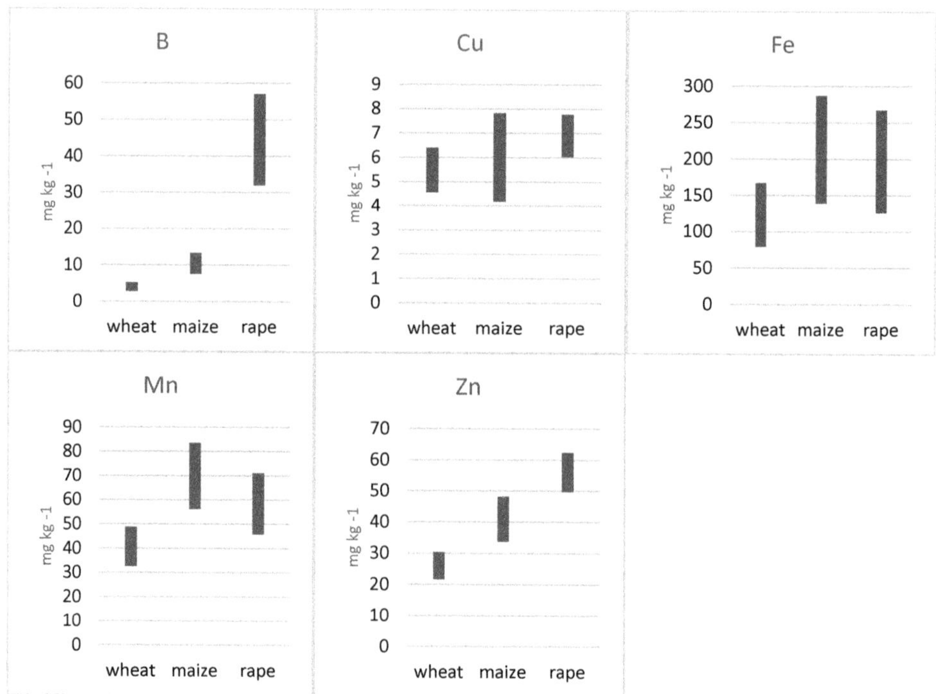

Figure 7. The range of micronutrient concentration in shoots of the tested species.

The species tested were much less variable in Cu concentration than B. The range of Cu in maize (4.2–7.8 mg kg^{-1}) was broad enough to include the ranges found in wheat (4.6–6.4 mg kg^{-1}) and rape (6.0–7.8 mg kg^{-1}).

The ranges of Fe in maize (139–287 mg kg^{-1}) and rape (126–267 mg kg^{-1}) largely overlapped with each other, but only slightly with wheat (79–167 mg kg^{-1}). The concentration of Mn found in wheat cultivars (33–49 mg kg^{-1}) differed from the concentrations observed in maize (56–83 mg kg^{-1}) and rape (46–81 mg kg^{-1}), which were higher than in wheat and largely overlapped with each other.

The ranges of Zn concentration were separate for each species and did not overlap. The lowest range of Zn was found in wheat shoots (22–30 mg kg^{-1}) and the highest range was found in canola shoots (50–62 mg kg^{-1}).

Table 5 shows the differences between the cultivars with the lowest and the highest average concentrations of the five micronutrients tested. These differences depended on the plant species. Undoubtedly, the smallest differences in micronutrient concentration among cultivars were observed for wheat.

Table 5. Differences among cultivars in the concentration of micronutrients in shoots—based on cultivars with the lowest and highest mean concentration.

Element	Wheat				Maize				Oilseed Rape			
	Cultivar	*n*	Mean mg kg^{-1}	Difference mg kg^{-1}	Cultivar	*n*	Mean mg kg^{-1}	Difference mg kg^{-1}	Cultivar	*n*	Mean mg kg^{-1}	Difference mg kg^{-1}
B	Ton	33	2.8		Gle	16	7.5		Alex	21	32	
	Jul	88	5.2	2.4	Dan	10	13.3	5.8	Mer	16	57	25
Cu	Sko	44	4.6		Nim	10	4.2		Mar	22	6	
	Lin	30	6.4	1.8	Dan	10	7.8	3.6	Gar	16	7.8	1.7
Fe	Bog	14	79		Sub	10	139		Vis	30	126	
	Hon	11	167	88	P8400	14	287	148	Kug	19	267	141
Mn	Mus	11	33		Sub	10	56		She	46	46	
	Ton	33	49	16	P8400	14	83	27	Alex	21	71	25
Zn	Ska	44	22		Gle	16	34		Gor	16	50	
	Ton	33	30	9	Nim	10	48	14	Kug	19	62	12

n number of observations.

For rape and maize, the differences among cultivars were clearly greater than for wheat. For both of these species, the differences among cultivars for Fe, Mn and Zn were similar. Furthermore, it was observed that rape showed greater variation in B concentration and smaller Cu concentration compared to maize.

Table 6 shows the differences between the average micronutrient concentrations in the plants for each pair of species tested. The greatest differences in B, Cu and Zn concentrations were observed for the wheat–maize pair, while the greatest variation in Fe and Mn concentrations occurred between wheat and maize.

Table 6. Differences among species in the concentration of micronutrients in shoots—based on the mean of all cultivars *.

Element	Species	Mean mg kg^{-1}	Difference	Species	Mean mg kg^{-1}	Difference	Species	Mean mg kg^{-1}	Difference
B	Wheat	4.0	5.7	Wheat	4.0	37.1	Maize	9.7	31.4
	Maize	9.7		Rape	41.1		Rape	41.1	
Cu	Wheat	5.2	0.6	Wheat	5.2	1.7	Maize	5.8	1.1
	Maize	5.8		Rape	6.9		Rape	6.9	
Fe	Wheat	117	119	Wheat	117	58	Maize	236	61
	Maize	236		Rape	175		Rape	175	
Mn	Wheat	40	24	Wheat	40	16	Maize	64	8
	Maize	64		Rape	56		Rape	56	
Zn	Wheat	27	13	Wheat	27	28	Maize	40	15
	Maize	40		Rape	55		Rape	55	

* number of observations: wheat, 481; maize, 141; rape, 328.

Using the data from Tables 5 and 6, it can be confirmed whether greater differences in micronutrient concentrations occurred between species, or between cultivars within a species. For this purpose, the values of these differences for a given pair of species were compared with the values for the cultivars of each species in the pair. For example, the difference in average Cu concentration between species for the wheat–maize pair is 0.6 mg kg^{-1} (Table 6), while the difference among wheat cultivars is 1.8 mg kg^{-1}, and among maize cultivars it is 3.6 mg kg^{-1} (Table 5). From this comparison, we conclude that there are greater differences among cultivars than between species of the wheat–maize pair.

4. Discussion

4.1. Soil and Weather Conditions

All cultivars of the three plant species tested grew under similar conditions of pH, texture and TOC content (Table 4), those soil features that have a strong influence on the bioavailability of micronutrients to plants [33,34]. The same was also true of the concentration of bioavailable forms of Mn and Zn in the soil from the fields of all three species. However, for B, Cu and Fe there were some differences between the growth sites of each species, with the greatest variation in B in soil (Figure 2). Wheat had the least available B forms in soil, followed by rape, and maize had the most. The concentrations of Cu and Fe in the soil were the same in the fields of wheat and maize, and significantly higher in fields of rape. It should be noted, however, that no soil micronutrient deficit was shown for any of the species tested. The concentrations of bioavailable B, Cu, Fe, Mn and Zn were sufficient for all three species according to Polish current standards for assessing soils in micronutrients [32].

The lack of micronutrient deficiency in the soil was also confirmed by their concentration in plant shoots. The concentration of micronutrients in plants, compared with the respective deficiency limits, showed a sufficient supply of these nutrients for wheat, maize and rape cultivars. (Figures 4–6).

Of the three species studied, rape is the most sensitive to B deficiency and has the highest demand for this nutrient [35]. Although the soil B concentration in rape fields was significantly lower than in maize fields, no B deficiency was found in the shoots of any rape cultivar (Figure 5).

Similar pH, texture and TOC in the soils sampled and the absence of micronutrient deficits in both the soils and the shoots of the cultivars leads us to believe that soil conditions were not a factor that significantly influenced the differences between species and cultivars in the concentration of micronutrients in the shoots. This is confirmed by the correlation between the soil features and the concentration of microelements in shoots, which was insignificant or low ($r \leq 0.19$) (Table 7). The exception was the content of Mn in wheat and maize shoots, which was dependent on the soil pH at the level of $r = -0.30$.

In addition to soil properties, precipitation and temperature during the growing season have an impact on the uptake and concentration of microelements in plants. Abundant rainfall and optimal temperature favor the production of large biomass, which may be associated with a reduction in the content of microelements due to the so-called dilution effect, especially in the case of a deficiency of micronutrients in the soil. The local weather conditions at the sampling sites were certainly a factor that also influenced the variability of the micronutrient concentration in plants. Nevertheless, it was not possible to eliminate this factor from the research. It was assumed that, despite some variation in weather conditions, the average micronutrient concentrations calculated from several hundred samples reliably reflect the real differences in micronutrient concentrations between species and cultivars.

Table 7. Pearson correlation coefficient (r) between micronutrient concentration in shoots and soil features.

Crop	Micronutrient	pH	Fraction <0.02 mm	Corg
Wheat $n = 481$	B	ns	−0.16 ***	−0.11 *
	Cu	ns	ns	ns
	Fe	ns	ns	ns
	Mn	−0.30 ***	ns	ns
	Zn	ns	ns	ns
Maize $n = 141$	B	ns	ns	ns
	Cu	0.19 **	0.18 *	ns
	Fe	ns	ns	−0.18 *
	Mn	−0.31 **	ns	ns
	Zn	−0.18 *	0.17 *	ns
Rape $n = 328$	B	ns	0.19 ***	ns
	Cu	ns	ns	ns
	Fe	ns	ns	ns
	Mn	ns	−0.16 **	−0.14 *
	Zn	ns	ns	ns

*, **, *** significant level $p < 0.05$; 0.01; 0.001, respectively; ns: nonsignificant; n: number of samples; Corg: organic carbon.

4.2. Concentration of Micronutrients in the Plant Species Studied

In the present study, very extensive research material was used, which influenced the high reliability of the results. The average concentration of micronutrients in the shoots of the plant species studied was calculated on the basis of many samples taken for a dozen cultivars: for wheat 12 cultivars where used (481 samples); for maize 10 culivars (141 samples); and for rape 12 culitvars (328 samples) (Tables 1–3).

In general, rape and maize showed significantly higher concentrations of micronutrients in the shoots than wheat (Figure 2). The high concentrations of B and Zn in rape and Fe in maize are particularly noteworthy. This corresponds to some extent to the nutritional requirements of these species. The known high sensitivity of rape to B deficiency and the fairly high sensitivity of maize to Fe deficiency [36] translates into a frequent need to fertilize rape with boron and maize with iron. However, the high Zn concentration in rape is not related to its high sensitivity to deficiency of this micronutrient. Rape, unlike maize, is not considered a crop with high sensitivity to Zn deficiency [27].

There are not many opportunities to compare our results with studies by other authors because there are no publications that compare micronutrients in shoots at the same growth stages in the species we studied. Only Korzeniowska et al. [37] report the average concentration of micronutrients in winter wheat shoots calculated on the basis of 357 samples taken in 2010–2011 from fields located in Poland: B was 3.9, Cu was 5.3, Fe was 171, Mn was 45, Zn was 37 mg kg^{-1}. The values of B, Cu and Mn reported by these authors are very similar to ours, while Fe and Zn are higher by 45 and 37%, respectively. In addition, Bergmann [11] gives optimum ranges for B, Cu, Mn and Zn concentrations in wheat and maize shoots and rape leaves taken at the same growth stages as ours. In general, all mean concentrations of microelements calculated from the optimal ranges of Bergmann were higher than the concentrations observed in our study. The greatest differences, up to twofold, were found for B and Cu in wheat shoots. It can be assumed that the differences between our results and Bergmann's are due to the different cultivars used now and in the 1990s. This suggests that the B and Cu ranges provided by Bergman for wheat have become obsolete and should not be used to assess plant nutritional status.

4.3. Differences in Micronutrient Content among Cultivars and Species

Our research hypothesis was that differences in plant micronutrient content may be greater among cultivars within a species than between species. Previous findings indicated large differences in the content of microelements in cultivars and their different response to micronutrients fertilization. Korzeniowska [38] separated three distinct groups of wheat cultivars among the 10 studied, which showed high, medium and low demand for Cu fertilization. These groups differed significantly in both response to fertilization and Cu concentration in shoots. Stanislawska-Glubiak and Sienkiewicz [39] studied micronutrient concentrations in seven spring barley cultivars. These authors showed that the maximum difference in concentration among cultivars was 22% in Cu, 40% in Mn and 49% in Fe. In addition, Wrobel and Korzeniowska [40] observed significant differences in the concentration of B in the cob leaf in the seven maize cultivars studied.

Despite previous results, the present extensive research has shown that our thesis of greater differences in micronutrient content among cultivars than species is only true for Cu, Fe and Mn, and does not apply to B and Zn.

In the case of Cu, this difference for the wheat–maize pair was 0.6 mg kg^{-1}, for the wheat–rape pair it was 1.7 mg kg^{-1} and for the maize–rape pair was 1.1 mg kg^{-1} (Table 6). At the same time, the difference in Cu concentration among cultivars was clearly greater than between species and was 1.8 mg kg^{-1} for wheat, 3.6 mg kg^{-1} for maize and 1.7 mg kg^{-1} for rape (Table 5).

Differences in Fe and Mn content were also often greater among cultivars within a species than between species. The difference in Fe and Mn concentration between maize cultivars was 148 and 27 mg kg^{-1}, respectively, and between rape cultivars was 141 and 25 mg kg^{-1} (Table 5). At the same time, for the maize–rape pair the difference was 61 mg kg^{-1} Fe and 8 mg kg^{-1} Mn. Among wheat cultivars, the difference in Fe and Mn content was 88 and 16 mg kg^{-1}, respectively, and for the wheat–rape pair it was 58 and 16 mg kg^{-1} (Table 6).

In contrast, differences in plant B and Zn content were greater between species than among cultivars within a single species (Tables 5 and 6). The difference in B concentration between species was as high as 37.1 mg kg^{-1} for the wheat–rape pair, while it was only 2.4 mg kg^{-1} among wheat cultivars and 25 mg kg^{-1} between rape cultivars. Larger differences between species compared to cultivars were also found in Zn content, although not as large as for B. The largest difference was found for the wheat–rape pair (28 mg kg^{-1}), while for the wheat cultivars the difference was only 9 mg kg^{-1} and for the rape cultivars it was 12 mg kg^{-1}.

The results suggest that cultivar should be taken into account when assessing the need to fertilize wheat, maize and rape with Cu, Fe and Mn, while the assessment of the need for fertilization of these species with B and Zn can be carried out independently of the cultivar used.

When fertilizing certain crops with micronutrients, it would be advisable to take into account not only the nutritional needs of the individual species, but also to adapt micronutrient doses to the requirements of the cultivars within the species. Such a measure could contribute to a more efficient use of fertilizers, in line with sustainable agriculture.

5. Conclusions

The highest average concentrations of B, Cu and Zn were observed in rape shoots and the highest average concentrations of Fe and Mn were observed in maize shoots. Wheat showed significantly lower concentrations of all micronutrients than rape and maize.

All the wheat, rape and maize cultivars tested had sufficient average micronutrient concentrations in the shoots, equal to or above the deficiency limit. The exception was one maize cultivar (Nimba), in which a concentration below the limit was observed.

For B and Zn concentrations, greater differences were found between species than cultivars for all three plants tested. On the contrary, for Cu concentration, varietal differences always exceeded species differences. In contrast, for Mn and Fe, varietal differences exceeded species differences for wheat–maize and maize–maize pairs, excluding the wheat–maize pair.

The results suggest that the fertilization of wheat, maize and rape with Cu, as well as Mn and Fe, needs to take into account different requirements of the cultivars for these micronutrients. In contrast, there is no need to take cultivars into account when fertilizing with B and Zn. Nevertheless, further research should confirm to what extent the concentration of micronutrients in the early stage of growth affects the size of the final crop yield.

Author Contributions: Conceptualization, J.K. and E.S.-G.; investigation, J.K. and E.S.-G.; methodology, J.K. and E.S.-G.; writing—original draft, J.K. and E.S.-G.; writing—review & editing, J.K. All authors have read and agreed to the published version of the manuscript.

Funding: We acknowledge funding from the Polish Ministry of Agriculture and Rural Development under 2.33 Scientific Research Program of the Institute of Soil Science and Plant Cultivation in Pulawy.

Informed Consent Statement: Not applicable.

Data Availability Statement: Presented data in this study are available upon request from the corresponding author.

Acknowledgments: The authors would like to thank National Agrochemical Station in Warsaw for organizing field sampling and chemical analysis.

Conflicts of Interest: The authors declare no conflict of interest.

References

1. Bell, R.W.; Dell, B. *Micronutrients for Sustainable Food, Feed, Fibre and Bioenergy Production*; International Fertilizer Industry Association: Paris, France, 2008; pp. 1–175.
2. Lowe, N.M. The global challenge of hidden hunger: Perspectives from the field. *Proc. Nutr. Soc.* **2021**, *80*, 283–289. [CrossRef] [PubMed]
3. Morgounov, A.; Gomez-Becerra, H.F.; Abugalieva, A.; Dzhunusova, M.; Yessimbekova, M.; Muminjanov, H.; Zelenskiy, Y.; Ozturk, L.; Cakmak, I. Iron and zinc grain density in common wheat grown in Central Asia. *Euphytica* **2007**, *155*, 193–203. [CrossRef]
4. Lockyer, S.; White, A.; Buttriss, J.L. Biofortified crops for tackling micronutrient deficiencies–what impact are these having in developing countries and could they be of relevance within Europe? *Nutr. Bull.* **2018**, *43*, 319–357. [CrossRef]
5. Ramzan, Y.; Hafeez, M.B.; Khan, S.; Nadeem, M.; Batool, S.; Ahmad, J. Biofortification with zinc and iron improves the grain quality and yield of wheat crop. *Int. J. Plant Prod.* **2020**, *14*, 501–510. [CrossRef]
6. Kenzhebayeva, S.; Abekova, A.; Atabayeva, S.; Yernazarova, G.; Omirbekova, N.; Zhang, G.; Turasheva, S.; Asrandina, S.; Sarsu, F.; Wang, Y. Mutant lines of spring wheat with increased iron, zinc, and micronutrients in grains and enhanced bioavailability for human health. *Biomed. Res. Int.* **2019**, *2019*, 1–10. [CrossRef]
7. Hochmuth, G.; Maynard, D.; Vavrina, C.; Hanlon, E.; Simonne, E. Plant tissue analysis and interpretation for vegetable crops in Florida. In *Nutrient Management of Vegetable and Row Crops Handbook*; University of Florida press: Gainesville, FL, USA, 2012; pp. 45–92.
8. Meyer, R.D.; Marcum, D.B.; Orloff, S.B. Understanding micronutrient fertilization in alfalfa. In Proceedings of the Western Alfalfa and Forage Symposium, Sparks, NV, USA, 11–13 December 2002; Available online: https://alfalfa.ucdavis.edu/+symposium/proceedings/2002/02-087.pdf (accessed on 5 October 2022).
9. Sarangthem, I.L.; Sharma, D.; Oinam, N.; Punilkumar, L. Evaluation Of Critical Limit Of Zinc In Soil And Plant. *Int. J. Curr. Res. Life Sci.* **2018**, *7*, 2584–2586.
10. Schulte, E.E.; Kelling, K.A. *Plant Analysis: A diagnostic Tool*; Cooperative Extension Service; Purdue University: West Lafayette, IN, USA, 1991; Available online: https://www.extension.purdue.edu/extmedia/nch/nch-46.html (accessed on 5 October 2022).
11. Bergmann, W. *Nutritional Disorders of Plants-Development, Visual and Analytical Diagnosis*; Gustav Fischer Verlag: Jena, Germany; Stuttgart, Germany; New York, NY, USA, 1992; pp. 343–361.
12. Jones, J.B.; Wolf, B.; Mills, H.A. *Plant Analysis Handbook*; Micro-Macro Publishing Inc.: Athens, GA, USA, 1991; pp. 1–213.
13. Korzeniowska, J.; Stanislawska-Glubiak, E.; Lipinski, W. New limit values of micronutrient deficiency in soil determined using 1 M HCl extractant for wheat and rapeseed. *Soil Sci. Annu.* **2020**, *71*, 205–214. [CrossRef]
14. Maganti, S.; Swaminathan, R.; Parida, A. Variation in iron and zinc content in traditional rice genotypes. *Agr. Res.* **2020**, *9*, 316–328. [CrossRef]
15. Ray, H.; Bett, K.; Tar'an, B.; Vandenberg, A.; Thavarajah, D.; Warkentin, T. Mineral micronutrient content of cultivars of field pea, chickpea, common bean, and lentil grown in Saskatchewan, Canada. *Crop. Sci.* **2014**, *54*, 1698–1708. [CrossRef]
16. Tran, B.T.T.; Cavagnaro, T.R.; Able, J.A.; Watts-Williams, S.J. Bioavailability of zinc and iron in durum wheat: A trade-Off between grain weight and nutrition? *Plants People Planet* **2021**, *3*, 627–639. [CrossRef]
17. Korzeniowska, J. Response of ten winter wheat cultivars to boron foliar application in a temperate climate (South-West Poland). *Agron. Res.* **2008**, *6*, 471–476.

18. Korzeniowska, J.; Stanislawska-Glubiak, E. The effect of foliar application of copper on content of this element in winter wheat grain. *Pol. J. Agron.* **2011**, *4*, 3–6.

19. Genc, Y.; McDonald, G.K.; Graham, R.D. Critical deficiency concentration of zinc in barley genotypes differing in zinc efficiency and its relation to growth responses. *J. Plant Nutr.* **2002**, *25*, 545–560. [CrossRef]

20. Becerra, H.F.G.; Yazici, A.; Ozturk, L.; Budak, H.; Peleg, Z.; Morgounov, A.; Fahima, T.; Saranga, Y.; Cakmak, I. Genetic variation and environmental stability of grain mineral nutrient concentrations in Triticum dicoccoides under five environments. *Euphytica* **2010**, *171*, 39–52. [CrossRef]

21. Statistics Poland (GUS). Land Use and Sown Area in 2019. 2020. Available online: https://stat.gov.pl/obszary-tematyczne/rolnictwo-lesnictwo/rolnictwo/uzytkowanie-gruntow-i-powierzchnia-zasiewow-w-2019-roku,8,15.html (accessed on 5 October 2020).

22. Meier, U. *Growth Stages of Mono- and Dicotyledonous Plants, BBCH Monograph*, 2nd ed.; Blackwell Science: Berlin, Germany, 2001; pp. 1–158.

23. Kabata-Pendias, A.; Pendias, H. *Trace Elements in Soils and Plants*; CRC Press: Boca Raton, FL, USA, 2001; pp. 1–413.

24. *Climate of Poland 2020*; Institute of Meteorology and Water Management-National Research Institute: Warsaw, Poland, 2021; pp. 1–43. Available online: https://www.imgw.pl/sites/default/files/2021-04/imgw-pib-klimat-polski-2020-opracowanie-final-rozkladowki-min.pdf (accessed on 5 October 2020).

25. *PN-R 04014:1991*; Agrochemical Plant Analyse. Methods of Mineralization of Plant Material for Determination Macro- and Microelements. Polish Committee for Standardization: Warsaw, Poland, 1991. (In Polish)

26. Korzeniowska, J.; Stanislawska-Glubiak, E. Comparison of 1 M HCl and Mehlich 3 for assessment of the micronutrient status of polish soils in the context of winter wheat nutritional demands. *Commun. Soil Sci. Plan.* **2015**, *46*, 1263–1277. [CrossRef]

27. Korzeniowska, J.; Stanislawska-Glubiak, E.; Lipinski, W. Development of the limit values of micronutrient deficiency in soil determined using Mehlich 3 extractant for Polish soil conditions. Part, I. Wheat. *Soil Sci. Annu.* **2019**, *70*, 314–323. (In Polish) [CrossRef]

28. Mehlich, A. Mehlich 3 soil test extractant: A modification of Mehlich 2 extractant. *Commun. Soil Sci. Plan.* **1984**, *15*, 1409–1416. [CrossRef]

29. *ISO 10390:2005*; Soil quality: Determination of pH. International Standardization Organization: Geneva, Switzerland, 2005.

30. *PN-ISO 14235:2003*; Soil quality: Determination of Organic Carbon in Soil by Sulfochromic Oxidation. Polish Committee for Standardization: Warsaw, Poland, 2003. (In Polish)

31. *PN-EN ISO/IEC 17025:2018-02*; General Requirements for the Competence of Testing and Calibration Laboratories. Polish Committee for Standardization: Warsaw, Poland, 2018. (In Polish)

32. Korzeniowska, J.; Stanislawska-Glubiak, E.; Jadczyszyn, J.; Lipinski, W. *Fertilization of Agricultural Crops with Microelements. New Limit Values for the Assessment of Micronutrients in Soil*; IUNG-PIB: Pulawy, Poland, 2021; pp. 1–26, (In Polish). Available online: https://schr.gov.pl/download/NAWOZENIE-MIKROELEMENTY.pdf (accessed on 5 October 2022).

33. Kabata-Pendias, A.; Mukherjee, A.B. *Trace Elements from Soil to Human*; Springer: Berlin, Heidelberg, 2007; pp. 1–550.

34. Wall, D.P.; O'Sullivan, L.; Creamer, R.; McLaughlin, M.J. Soil Fertility and Nutrient Cycling. In *The Soils of Ireland*; Creamer, R., O'Sullivan, L., Eds.; World Soils Book Series; Springer: Cham, Switzerland, 2018; pp. 223–234.

35. Alexander, A.; Gondolf, N.; Orlovius, K.; Paeffgen, S.; Trott, H.; Wissemeier, A.H. *Mikronährstoffe in der Landwirtschaft Und Im Gartenbau Bedeutung-Mangelsymptome-Düngung*; Bundesarbeitskreis Düngung (BAD): Frankfurt am Main, Germany, 2013; pp. 1–64.

36. Katyal, J.C.; Randhawa, N.S. Micronutrient. *FAO Fertil. Plant Nutr. Bull* **1983**, *7*, 1–82.

37. Korzeniowska, J.; Stanislawska-Glubiak, E.; Kantek, K.; Lipinski, W.; Gaj, R. Micronutrient status of winter wheat in Poland. *J. Cent. Eur. Agric.* **2015**, *16*, 54–64. [CrossRef]

38. Korzeniowska, J. Comparison of different winter wheat cultivars in respect to their copper fertilization requirements. *Zesz. Probl. Postępów Nauk. Rol.* **2009**, *541*, 255–263. (In Polish). Available online: https://www.researchgate.net/profile/Jolanta-Korzeniowska-2/publication/259100834_Comparison_of_different_winter_wheat_cultivars_with_respect_to_their_copper_fertilization_demand/links/0c960529f737f2537c000000/Comparison-of-different-winter-wheat-cultivars-with-respect-to-their-copper-fertilization-demand.pdf (accessed on 5 October 2022).

39. Stanislawska-Glubiak, E.; Sienkiewicz, U. Response of spring barley cultivars to soil acidity, liming and molybdenum fertilization. *Zesz. Probl. Postępów Nauk. Rol.* **2004**, *502*, 349–356. (In Polish). Available online: https://agro.icm.edu.pl/agro/element/bwmeta1.element.agro-article-a9b5dc8b-975f-430c-a2b2-bb9a7c5ce240/c/Zeszyt_Probl_Poste_Nau_Rol_r.2004_t.502_s.349-356.PDF (accessed on 5 October 2022).

40. Wróbel, S.; Korzeniowska, J. Assessment of the need to fertilize maize with boron. *Stud. I. Rap. IUNG-PIB* **2007**, *8*, 127–142. (In Polish). Available online: http://iung.pulawy.pl/sir/zeszyt08_11.pdf (accessed on 5 October 2022).

 *agronomy*

Article

Improving Nutrients Uptake and Productivity of Stressed Olive Trees with Mono-Ammonium Phosphate and Urea Phosphate Application

Hamdy A. Z. Hussein [1], Ahmed A. M. Awad [2],* and Hamada R. Beheiry [1]

[1] Horticulture Department, Faculty of Agriculture, Fayoum University, Fayoum 63514, Egypt
[2] Soils and Natural Resources, Faculty of Agriculture and Natural Resources, Aswan University, Aswan 81528, Egypt
* Correspondence: ahmed.abdelaziz@agr.aswu.edu.eg; Tel.: +20-1000421124

Abstract: Nutritional status improvement is a surrogate approach to overcoming undesirable soil conditions. This study was performed in sandy clay loam soil that was characterized by certain undesirable parameters (ECe = 6.4 vs. 7.2 dS m^{-1}, CaCO$_3$ = 8.8 vs. 9.2%, and pH = 7.78 vs. 7.89) on olive (*Olea europaea*, Arbequina cv.) in the 2020 and 2021 seasons to investigate the influence of two highly soluble phosphorus fertilizers, mono-ammonium phosphate (MAP) and urea phosphate (UP). The treatments included 0.336, 0.445, and 0.555 kg tree^{-1} for MAP$_1$, MAP$_2$, and MAP$_3$ and 0.465, 0.616, and 0.770 kg tree^{-1} for UP$_1$, UP$_2$, and UP$_3$, respectively, in comparison to granular calcium super-phosphate (GCSP) at the recommended rate (0.272 kg P$_2$O$_5$ equal 1.75 kg tree^{-1}). This experiment was established according to a randomized complete block design. Generally, our results indicated that both MAP and UP applications surpassed GCSP for all studied parameters except leaf copper uptake in the 2021 season. Moreover, among the HSPFs applied, it was found that applying the maximum levels gave the best results. However, MAP$_3$ gave the maximum values for shoot length, SPAD reading, and dry fruit matter. Moreover, UP$_3$ produced the best results for the leaf area, olive tree yield, total olive yield, total fresh weight, flesh weight (FlW), fruit length (FrL), and leaf Fe content in both seasons.

Keywords: *Olea europaea* trees; nutrients uptake; phosphorus fertilizers; growth and physiological parameters; yield and fruit quality

Citation: Hussein, H.A.Z.; Awad, A.A.M.; Beheiry, H.R. Improving Nutrients Uptake and Productivity of Stressed Olive Trees with Mono-Ammonium Phosphate and Urea Phosphate Application. *Agronomy* **2022**, *12*, 2390. https://doi.org/10.3390/agronomy12102390

Academic Editors: Christos Noulas, Shahram Torabian and Ruijun Qin

Received: 29 August 2022
Accepted: 29 September 2022
Published: 2 October 2022

Publisher's Note: MDPI stays neutral with regard to jurisdictional claims in published maps and institutional affiliations.

1. Introduction

Abiotic stresses (ABSs), including salinity, calcification, and high soil pH, are major constraints affecting the agricultural sector in many parts of the world. However, calcareous soils are characterized by high calcium carbonate (CaCO$_3$) content, which, in turn, affects soil properties—for example, causing a low cation exchange capacity (CEC), high pH, and decreased availability of most essential nutrients, in addition to low content of soil organic matter (SOM) and loss of nutrients through deep percolation, causing a nutritional imbalance among different nutrients [1,2]. According to [3], most calcareous soils exist in arid and semi-arid regions and cover more than 30% of the Earth's surface. Thus, soil salinity is no less important than calcareous soil; however, approximately 4 × 10^4 ha becomes unsuitable for cultivation every year owing to salinization [4]. Based on reports published by specialized agencies of the United Nations, it was revealed that approximately half of the irrigated area is either salinized or has the possibility of developing salinity in the future. Soil salinity occurs owing to soluble salt accumulation in the root zone, resulting in abnormal plant growth and development, which, in turn, affects productivity. Generally, saline soil is identified by the electrical conductivity (ECe) of the saturated soil paste in the root zone exceeding 4 dsm^{-1} at 25 °C and an exchangeable sodium percentage (ESP) $\geq$ 15% [5]. The total cultivated area in arid and semi-arid regions is estimated at around 831 million ha across the world, and it is expected that more than 50% of arable land will be saline by 2050 [6,7]. Given the aforementioned information, this issue requires more attention and further efforts among researchers to overcome these undesirable characteristics that hinder nutrient uptake, causing the abnormal growth and development of plants, which, in turn, influences crop productivity.

Balanced fertilization is the best agronomic practice for soil management in plants under stressed conditions. Among the essential macronutrients, phosphorus (P) is one of the most important, along with nitrogen (N) and potassium (K), as it is considered the most influential for root development and thus increases the plant's ability to absorb water and nutrients from the soil. Moreover, it has a crucial role in several metabolic processes, including protein synthesis, cell division and elongation, respiration, the consumption of energy-rich compounds (adenosine tri-, di-, and monophosphate, ATP, ADP, and AMP), the photosynthesis process and nutrients' movement within plants [8–10]. Furthermore, P is an essential integrated element of nucleic acids and phospholipids and plays a central role in sugar assimilation [11,12]. Besides these vital roles, P plays a fundamental role in phosphoprotein and fat metabolism, sulfur metabolism, biological oxidation, and several other metabolisms dependent on the application of P [13]; however, both saline and calcareous soils suffer from the unavailability of P and other micronutrients as a result of high pH, in addition to chemical reactions that affect these nutrients, whether by loss or fixation, due to the reaction of P anions with calcium (Ca) and magnesium (Mg) to form insoluble phosphate complex compounds with limited solubility, besides decreasing the organic matter below the critical level [14–16]. Generally speaking, P is absorbed in the form of $H_2PO_4^-$ and HPO_4^{--} through root hairs and root tips; although the total amount of P may be high, the majority is often restricted [17], in addition to the loss of P from the soil due to its negative charge. However, more than 80% of added P converts into an unavailable form due to its fixation and adsorption processes [18,19]. As is well known, either a deficiency or excess of P in the soil can negatively affect plant performance. P causes stunted plants and root diameter decrease [20,21], as well as disturbances to chlorophyll pigment production and the accumulation of anthocyanins, resulting in purple discoloration [22–24]. On the other hand, the overapplication of P at levels that exceed crop demands could increase P losses to the subsurface and groundwater [9] and decrease the absorption of zinc (Zn), manganese (Mn), copper (Cu), and iron (Fe); consequently, the symptoms of their deficiency appear on the crop, which, in turn, affects the productivity [25].

Recently, attention has turned towards applying highly soluble phosphorus fertilizers (HSPF) including mono-ammonium phosphate (MAP), urea phosphate (UP), and mono-potassium phosphate (MKP) as an alternative surrogate to overcome the fixing and retaining of phosphate ions. Both MAP and UP are acidic phosphorus fertilizers that markedly enhance phosphorus use efficiency (PUE) by lowering soil pH in saline and calcareous soils with high pH values. However, a decreasing pH enhances micronutrients' availability, thus improving the solubility of calcium and preventing its association with P [26,27]. UP is an amino-structured complex and a highly acidic fertilizer produced by the reaction of phosphoric acid (H_3PO_4) with urea $CO(NH_2)_2$, and its chemical structure is $H_3PO_4.CO(NH_2)_2$ [28,29]. Moreover, MAP is an acidic fertilizer, but is manufactured via the reaction of H_3PO_4 with ammonia (NH_3), and its chemical structure is $NH_4H_2PO_4$. Despite the little information available about HSPF, its positive influences were an important factor for the generalization of its application instead of traditional fertilizers such as calcium super-phosphate. The authors of [30] reported that spraying P in different forms, such as MAP, UP, and MKP, increased nitrogen and potassium accumulation. Similarly, the results of [31] indicated that applying MAP, UP, and MKP as foliar treatment improved the flowering, fruit set, yield, and oil content of picual and kalamata cultivars. These results were confirmed by [32,33], which stated that the increases obtained with N and P application could be due to increases in hermaphrodite flowers, thus improving the flowering set, fruiting, fruit quality, and yield. Some studies [34,35] stated that MAP application was the best treatment to improve P availability compared with traditional P fertilizers such as calcium super-phosphate. These results have been confirmed by [36,37]; however, they indicated that the application of MAP improved the chemical constituents and productivity of potatoes.

By 2018, olive (*Olea europaea* L.) cultivation had reached approximately 11 million ha throughout the world, with more than 90% concentrated in Mediterranean countries [38]. Olive trees are cultivated to produce oil and table olives. In Egypt, olive cultivation is considered among the most important commercial cultivation practices, and it ranks fourth in Africa after citrus, mango, and table grapes [39]. Egypt is responsible for more than 13% of the world's production; however, the total cultivation area reached 101,326 ha, with total production reaching 874,748 tons, in 2017, according to the Ministry of Agriculture. However, the majority is cultivated in newly reclaimed lands; most of these lands are sandy soils that suffer from some negative characteristics. P fertilization is one of the most important factors in its annual growth cycle; however, P is essential to enhance flower formation, cell division and elongation, the development of new growth tissues, and the photosynthesis process and root growth, which in turn increase the productivity [40,41]. However, P is the most important basic nutrient determining the oil yield and its components; moreover, the quality parameters of oil can be altered due to the influence of P on phospholipid formation. Despite all the positive effects of P fertilizer, some previous studies indicated that P did not cause any increases in the yield or its attributes [42,43]. Under these ABSs, some types of phosphorus fertilizers, such as mono-ammonium phosphate (MAP) and urea phosphate (UP), are applied instead of calcium super-phosphate (CSP), whereas both MAP and UP may be more effective and easier to apply via fertigation and foliar spray.

The main objective of this research was to evaluate the potential performance of two types of highly soluble phosphorus fertilizers, namely MAP and UP, with low pH (<7.0), due to the nature of Egyptian soils with high soil pH. To do so, three levels of P_2O_5—0.205, 0.272, and 0.339 tree^{-1}—were applied with both fertilizers in a comparative study with one level (0.272 P_2O_5 tree^{-1}) of granular calcium super-phosphate (GCSP), with a high pH (>7.0), in an attempt to improve the nutrient uptake of olive trees (Arbequina cv.) grown under multi-stress conditions, which in turn affects the growth and productivity characteristics.

2. Materials and Methods

2.1. Study Location, Weather Conditions, and Plant Materials

This study was accomplished through the Egyptian–Spanish Project in Kawm Ushim district (29°55′ N; 30°88′ E), located on Cairo–Fayoum Desert Road, Egypt, during the seasons of 2020 and 2021. It was performed on olive (*Olea europaea* L. Arbequina cv.) trees grown on sandy clay loam soil to investigate the influence of two types of highly soluble phosphorus fertilizers, mono-ammonium phosphate (MAP) and urea phosphate (UP), which were applied five times, in comparison with granular calcium superphosphate (GCSP) with chemical structure $Ca(H_2PO_4)_2$ as a control treatment.

The trees were around 15 years old, propagated by leaf cutting, and planted at a distance of 5×8 m^2 from one another under a drip irrigation system, and the selected trees were visually free from diseases. The arbequina olive cultivar was chosen for its characteristics of self-pollination, an abundant yield, and a strong ability to resist drought and high temperatures. Accordingly, it is considered the most suitable for the Mediterranean countries; its olives are distributed as food products or used to produce oils rich in antioxidants. All horticultural practices, including irrigation and weed, pest, and disease control, were applied according to the recommendations of the Egyptian Ministry of Agriculture and Soil Reclamation. The selected trees were as uniform in shape and size as possible, and similar in vigor and growth. The weather data of the study region are presented in Table 1.

Table 1. Average climate data for Kawm Ushim region (29°55′ N; 30°88′ E), Fayoum, Egypt in 2020 and 2021 growing seasons.

Month	AD	AN	ARH	AWS	AM-PEC-A	AP
	(°C)		(%)	(ms^{-1})	(mmd^{-1})	$(mm\ d^{-1})$
January	25.04	2.94	61.81	2.44	3.43	0.08
February	26.74	3.87	60.63	2.35	4.32	0.96
March	32.58	5.00	55.56	2.81	5.04	0.46
April	37.45	7.48	45.13	3.26	5.58	0.04
May	43.86	13.89	35.22	3.54	6.87	0.00
June	41.92	16.84	35.60	3.78	7.56	0.00
July	42.15	19.68	37.03	3.42	6.88	0.00
August	41.32	20.83	38.84	3.30	6.78	0.00
September	42.32	18.84	45.35	3.64	8.64	0.00
October	37.30	15.52	50.85	3.25	6.61	0.02
November	30.47	10.22	58.60	2.36	4.63	0.28
December	25.22	5.75	61.72	2.30	3.49	0.15

AD °C = Average day temperature, AN °C = Average night temperature, ARH = average relative humidity, AWS = average wind speed, AM-PEC-A = average measured pan evaporation class A and AP = average precipitation. Source: https://power.larc.nasa.gov/index.php, accessed on 22 August 2022.

2.2. Treatment and Experimental Design

According to technical bulletin No. 2 of 2016, issued by the General Administration of Agriculture, the recommended fertilization program for olive trees aged over 6 years is 394, 500, 810, and 400 g of N, P_2O_5, K_2O, and MgO, respectively. Both experiments included three levels of P, namely 0.205, 0.272, and 0.339, which were calculated as $P_2O_5\%$ from two highly soluble phosphorus fertilizers (HSPFs) (MAP at total $MAP_1 = 0.336$, $MAP_2 = 0.445$, and $MAP_3 = 0.555$ kg $tree^{-1}$ in five equal doses at rate 67.2, 89.0, and 111.0 g $tree^{-1}$) and (UP at rate $UP_1 = 0.465$, $UP_2 = 0.616$, and $UP_3 = 0.770$ kg $tree^{-1}$ in five equal doses at rate 93.0, 123.2, and 154.0 g $tree^{-1}$) in comparison with the recommended level of P_2O_5 (0.272) at GCSP, 1.75 kg $tree^{-1}$.

The experimental plots were colonized and identified by the three levels of MAP and three levels of UP in addition to one level of GCSP, which were allocated in 7 treatments, and each treatment was repeated five times in the middle of March, April, May, June, and July in both growing seasons as a soil application in four plots, as described in Table 2. Each treatment consisted of three trees.

Both fertilizers applied, MAP and UP, were purchased from the ICL and SQM companies via their distributors in Egypt. Meanwhile, GCSP was produced by the Suez company that produces fertilizers in Egypt. The field experiment was established according to a randomized complete block design (RCBD). The chemical analysis of the applied PFs in this study is shown in Table 3.

Table 2. Details of the treatments applied in this study: phosphorus fertilizers applied, composition of treatments, replications, and application times on olive trees (*Olea europaea* L. arbequina cv.) in 2020 and 2021.

Symbol	Phosphorus Fertilizer Applied	Composition Treatment (kg tree^{-1})	Replication	Applying Time
GCSP	Granular calcium super-phosphate	5.0 kg of AS + 1.75 kg GCSP + 1.5 kg K_2SO_4 + 0.4kg $MgSO_4.7H_2O$		
MAP_1		4.81 kg of AS + 0.336 kg MAP + 1.5 kg K_2SO_4 + 0.4kg $MgSO_4.7H_2O$		
MAP_2	Mono-ammonium phosphate	4.74 kg of AS + 0.445 kg MAP + 1.5 kg K_2SO_4 + 0.4kg $MgSO_4.7H_2O$	These quantities are equally added five times in four plots	All treatments were performed five times in the middle of March, April, May, June, and July
MAP_3		4.67 kg of AS + 0.555 kg MAP + 1.5 kg K_2SO_4 + 0.4kg $MgSO_4.7H_2O$		
UP_1		4.60 kg of AS + 0.465 kg UP + 1.5 kg K_2SO_4 + 0.4kg $MgSO_4.7H_2O$		
UP_2	Urea phosphate	4.47 kg of AS + 0.616 UP + 1.5 kg K_2SO_4 + 0.4kg $MgSO_4.7H_2O$		
UP_3		4.34 kg of AS + 0.770 UP + 1.5 kg K_2SO_4 + 0.4kg $MgSO_4.7H_2O$		

GSCP = granular calcium super-phosphate, $Ca(H_2PO_4)_2 \approx 15.5\%P_2O_5$, MAP = mono-ammonium phosphate $NH_4H_2PO_4 \approx 61\%P_2O_5$, UP = urea-phosphate $H_2N-C = NH_2.H_2PO_4$ $44\%P_2O_5$, AS = ammonium sulfate $(NH_4)_2SO_4 \approx 20.6\%N$.

Table 3. Chemical analysis of phosphorus fertilizers applied in this study.

Properties	GCSP	MAP	UP
Chemical formula	$Ca(H_2PO_4)_2$	$NH_4H_2PO_4$	$CO(NH_2)_2 \cdot H_3PO_4$
pH (1% solution)	7.5	4.5	1.8
N (%)	0.0	12.00	17.72
P_2O_5 (%)	15.5	61.00	44.00

2.3. Soil sampling and Determination

Soil samples were randomly taken from the surface layer at a depth of 0–25 cm, before the application of treatments, and transferred to the Soil, Water, and Plant Analysis Laboratory (SWPAL) at the Faculty of Agriculture and Natural Resources, Aswan University, to determine some soil chemical and physical properties (Table 4) Particle size distribution was evaluated using the hydrometer method [44], soil pH was measured in soil paste using a pH meter [45], electrical conductivity (EC) was measured in soil paste extract using an EC meter, and calcium carbonate content ($CaCO_3$%) was determined using a calcimeter, as described by [46].

Table 4. Some soil chemical and physical properties.

Soil Property	2020	2021
Particle size distribution (%)		
Sand	47.32	48.49
Silt	19.56	20.20
Clay	33.12	31.31
Soil texture	Sandy clay loam	Sandy clay loam
pH (in soil paste)	7.78	7.89
ECe (dS m^{-1})	6.4	7.2
Organic matter (%)	0.63	0.52
$CaCO_3$ (%)	8.8	9.2
Soluble ions (mmol L^{-1})		
CO_3^{--}	-	-
HCO_3^{-}	2.8	3.7
Cl^{-}	53.4	55.3
SO_4^{--}	19.3	21.1
Ca^{++}	39.6	41.2
Mg^{++}	7.8	8.4
Na^{+}	22.4	24.3
K^{+}	5.7	6.2
Macronutrients (mg kg^{-1})		
Total N	414	640
Extractible P NaHCO$_3$ pH = 8.5	4520	4830
Extractible K NH$_4$OAC pH = 7.0	1337	1415
DTPA Extractible micronutrients (mg kg^{-1})		
Fe	10.7	11.2
Mn	4.5	6.3
Zn	0.15	0.14
Cu	0.48	0.38

In addition, soil organic matter (SOM) was determined according to the Walkley–Black method [47]. Regarding the determination of soluble ions, the soluble cations, sodium (Na^{+}), potassium (K^{+}), calcium (Ca^{++}), and magnesium (Mg^{++}) were extracted with 1N NH_4AC; however, Na^{+} and K^{+} were determined with a flame photometer [48], whereas Ca^{++} and Mg^{++} were measured with the EDTA titration method. Soluble anions, carbonate (CO_3^{--}), bicarbonate (HCO_3^{-}), chloride (Cl^{-}), and sulfate (SO_4^{--}) were determined with the titration method [45]. Nitrogen (N), phosphorus (P), and potassium (K) extracted were determined by the modified micro Kjeldahl method, as in [49–51], respectively.

Some available micronutrients, including iron (Fe), manganese (Mn), zinc (Zn), and copper (Cu), were extracted with DTPA [52] and determined using inductively coupled plasma–optical emission spectrometry (ICP-EOS, PerkinElmer OPTIMA 2001 DV, Norwalk, CT, USA), as described in [53].

2.4. Physiological and Growth Parameters

Twenty shoots at one year old were randomly selected on each side of the ten olive orchard trees in mid-September (after growth cycle) and spotted for every replicate to measure some attributes, including shoot length (ShL), which was measured in cm; number of leaves per shoot (NLSh); average number of leaves per meter and leaf area (LA, cm^2) of the third and fourth leaves from the top of new spring shoots, which were estimated using a digital planimeter device (Planx 7 Tamaya). Relative chlorophyll content (SPAD) was determined using a SPAD-502 m device (Minolta, Osaka, Japan).

2.5. Leaf Nutrient Measurements

Leaf samples were collected from the twenty selected shoots from ten trees, washed with distilled water, oven-dried at 70 °C for 72 h, and crushed to determine N, P, K, Ca, Mg, and Na according to the method described in [50]. Micronutrients (Fe, Mn, Zn, and Cu) were determined using inductively coupled plasma–optical emission spectrometry (ICP-EOS, PerkinElmer OPTIMA 2001 DV, Norwalk, CT, USA) as described in [53].

2.6. Total Olive Yield (kg tree^{-1})

In mid-October (harvesting time) in 2020 and 2021, the average yield was recorded (in kg tree^{-1}) for each tree under each treatment, and the total olive yield (TOY) per hectare was calculated based on the number of trees in a hectare.

2.7. Fruits' Physical and Chemical Characteristics

Samples of 100 fruits from each treated tree were randomly picked in both seasons, and we examined shoots from each replicate to study their physical and chemical characteristics, namely fruit length (FrL, cm), fruit diameter (FrD, cm), fruit shape index (LD), flesh weight (FlW), fruit weight (TFrW, g), and flesh/fruit ratio, according to [54]. Fruit oil percentage as a dry weight was determined according to [55] by extracting the oil from the dried flesh samples using a Soxhlet fat extraction apparatus and petroleum ether of (60–80 °C) boiling point as a solvent, and the percentage of oil was determined on a dry weight basis. Regarding the determination of dry weight and moisture content (%), a sample of 50 fruits was dried at 70 °C in an electric oven until a constant weight was reached. The average dry weight was determined and the percentage of moisture per fruit was calculated.

2.8. Statistical Analysis

Analysis of variance (ANOVA) and Duncan's test were performed on three replicates for nutrient determinations and five replicates for physiological and growth parameters and yield and its attributes using the InfoStat statistical package, version 2011 (InfoStat Microsoft) [56]; here, the replicate was considered the random variable, whereas the treatment was the fixed variable. The standard of error (±SE) was calculated for each treatment. A stepwise regression test was performed to identify the extent of the relationships between the olive tree yield (OTY, kg) and olive oil content (OOC, %) with the nutrients, growth, physiological parameters, and yield attributes under multi-abiotic stresses.

3. Results

3.1. Leaf Nutrient Contents

As presented in Table 5, we found that the application of 0.770 kg tree^{-1} of urea phosphate (UP$_3$) was the superior treatment; it recorded the highest values (0.23 and 1.67%)

for phosphorus (LPU) and calcium uptake (LCaU), respectively, in the 2020 season, and (0.72%) for leaf potassium uptake (LKU) in the 2021 season. Moreover, the trees fertilized by UP with 0.616 (UP_2) and 0.465 kg tree^{-1} (UP_1) displayed the maximum leaf magnesium uptake (LMgU) in the first season and leaf sodium uptake (LNaU) in the second season, respectively. On the other hand, the influence of the applied MAP was no less important than that of UP, whereas the trees treated with 0.336 kg tree^{-1} of MAP (MAP_1) produced the greatest values (2.92 and 0.72%) for LNU and LKU in the first growing season, as well as 0.26 for LPU and 1.19% for LMgU in the second season, whereas applying UP at 0.465 kg (UP_1) and MAP at 0.445 kg tree^{-1} (MAP_2) gave the best values (2.27 and 1.48%) for LNU and LCaU, respectively, in the second season. It can be seen in Table 5 that the percentage increases of the greatest and lowest values were 66.91 vs. 149.58 for N, 76.92 vs. 85.71 for P, 18.03 vs. 22.03 for K, 96.47 vs. 74.00 for Ca, 72.50 vs. 41.67 for Mg, and 85.37 vs. 103.03 for Na in the two growing seasons, respectively.

The obtained data listed in Table 5 showed that the application of two different HSPFs, MAP and UP, irrespective of their applied levels, appreciably outperformed the traditional phosphorus fertilizer, GCSP, in improving the olive leaf nutrient content.

The results of the ANOVA indicated that all treatments had significant effects (at $p \leq 0.01$) on the LNU, LPU, LMgU, and LNaU in both seasons; in addition, LKU in the first season and LCaU in the second season experienced significant effects, whereas there was a significant impact (at $p \leq 0.05$) on LKU in 2020 and no significant influence on LCaU in 2021.

The influence of MAP and UP application on leaf micronutrient content in the 2020 and 2021 seasons are illustrated in Table 6. However, the highest values (234.42 vs. 239.00) for leaf iron uptake (LFeU) in both seasons and (22.00 mgkg^{-1}) for leaf manganese uptake (LMnU) in the first season were recorded with the application of UP_3, whereas the highest values (28.42 vs. 4.02 mgkg^{-1}) of LMnU and leaf copper uptake (LCuU) were achieved via UP_1.

Concerning MAP impacts, our results showed that MAP_1 gave the maximum values (49.86 vs. 49.36 mgkg^{-1}) for leaf zinc uptake (LZnU) in the 2020 and 2021 seasons, respectively. Moreover, LCuU recorded the greatest values (3.50 mgkg^{-1}) in trees treated with MAP_3. In contrast, dissimilar data were obtained regarding the lowest values. The UP_2 treatment was the least effective, as it recorded the lowest values (169.51 mgkg^{-1}) for LFeU in the first season and (24.49 vs. 28.99 mgkg^{-1}) for LCuC in the two growing seasons, respectively. Meanwhile, the lowest LFeU was obtained with UP_1. Similar data were observed for LMnU and LCuU, however, with the lowest values (21.09 vs. 2.50 mgkg^{-1}) in fertilized trees with GCSP in the second season, whereas MAP_3 was the least effective on LMnU in the first season, which reached 15.92 mgkg^{-1}. Based on the comparison between the highest and lowest values, the percentages of increase reached 44.48 vs. 41.24% for Fe, 38.19 vs. 34.76 for Mn, 103.59 vs. 70.27% for Zn, and 109.58 vs. 60.80% for Cu in the 2020 and 2021 seasons, respectively.

The general trend of the data presented in Table 6 indicated that UP application was slightly more beneficial than MAP. Analysis of variance showed that the treatments had a significant influence (at $p \leq 0.01$) on all studied micronutrient uptake in both seasons.

3.2. Physiological and Growth Attributes

The results pertaining to the influence of both phosphorus fertilizers applied, namely MAP and UP, in comparison with GCSP as a soil application on some physiological and growth parameters of olive trees grown under multi-abiotic stresses in the 2020 and 2021 seasons are graphically illustrated in Figures 1–4. The obtained results indicated marked improvements for all studied physiological and growth parameters in both seasons; however, the highest values for shoot length (ShL) and leaf area (LA) were obtained via the applying of UP_3 treatment in the two growing seasons.

Table 5. Influence of different levels of MAP and UP in a comparison with the recommended GCSP level on leaf macronutrient uptakes of olive (*Olea europaea* L. arbequina cv.) trees grown in sandy loam clay soil under multi-abiotic stresses ($CaCO_3$ = 8.8 vs. 9.2%, ECe = 6.4 vs. 7.2 dS m^{-1}, and pH = 7.78 vs. 7.89) during 2020 and 2021 seasons.

Treatment	LNU	LPU	LKU	LCaU	LMgU	LNaU
			(%, in DM of Leaves)			
			2020 Season			
GCSP	1.63d ± 0.03	0.13e ± 0.03	0.66bc ± 0.01	0.85b ± 0.20	0.47cd ± 0.01	0.41f ± 0.01
MAP$_1$	1.36d ± 0.05	0.14f ± 0.06	0.72a ± 0.03	1.41ab ± 0.01	0.63b ± 0.01	0.45e ± 0.01
MAP$_2$	2.04ab ± 0.03	0.17d ± 0.16	0.71ab ± 0.04	1.39ab ± 0.01	0.41e ± 0.01	0.66c ± 0.01
MAP$_3$	1.53e ± 0.03	0.19c ± 0.12	0.70ab ± 0.02	1.18ab ± 0.01	0.52c ± 0.03	0.76a ± 0.01
UP$_1$	2.27a ± 0.03	0.17d ± 0.03	0.61c ± 0.01	1.42ab ± 0.24	0.43de ± 0.03	0.73b ± 0.01
UP$_2$	1.55e ± 0.03	0.21b ± 0.03	0.71ab ± 0.01	1.12ab ± 0.01	0.69a ± 0.01	0.56d ± 0.01
UP$_3$	1.76c ± 0.04	0.23a ± 0.06	0.63c ± 0.01	1.67a ± 0.17	0.40e ± 0.01	0.64c ± 0.01
			2021 season			
GCSP	2.04d ± 0.09	0.14f ± 0.05	0.69b ± 0.04	0.74f ± 0.01	0.84f ± 0.02	0.39e ± 0.01
MAP$_1$	2.66b ± 0.05	0.26b ± 0.06	0.64c ± 0.09	0.95e ± 0.01	1.19a ± 0.01	0.37f ± 0.01
MAP$_2$	1.19e ± 0.06	0.21de ± 0.06	0.61c ± 0.09	1.48a ± 0.03	0.91e ± 0.01	0.47d ± 0.01
MAP$_3$	2.97a ± 0.4	0.20e ± 0.09	0.65bc ± 0.07	0.96de ± 0.05	0.80g ± 0.01	0.62b ± 0.01
UP$_1$	1.98b ± 0.07	0.22cd ± 0.12	0.71a ± 0.09	1.38b ± 0.01	1.06c ± 0.01	0.67a ± 0.01
UP$_2$	2.23c ± 0.05	0.31a ± 0.08	0.71a ± 0.07	1.06c ± 0.01	1.13b ± 0.01	0.33g ± 0.01
UP$_3$	2.55b ± 0.06	0.24bc ± 0.06	0.72a ± 0.06	1.03cd ± 0.02	0.94d ± 0.02	0.51c ± 0.02

Mean values (±SE) with different letters in each column are significant (at $p \leq 0.05$). GCSP represents granular calcium super-phosphate applied at 1.75 kg tree^{-1}, MAP = mono-ammonium phosphate, MAP$_1$, MAP$_2$, and MAP$_3$ represent MAP applied at 0.336, 0.445, and 0.555 kg tree^{-1}, UP = urea phosphate, UP$_1$, UP$_2$, and UP$_3$ represent UP applied at 0.465, 0.616, and 0.770 kg tree^{-1}, all treatments were applied as a soil application. According to Duncan's multiple ange test, Means sharing the same letter in each column are not significantly different.

Table 6. Influence of different levels of MAP and UP in comparison with the recommended GCSP level on leaf micronutrient uptakes of olive (*Olea europaea* L. arbequina cv.) trees grown in sandy loam clay soil under multi-abiotic stresses ($CaCO_3$ = 8.8 vs. 9.2%, ECe = 6.4 vs. 7.2 dS m^{-1}, and pH = 7.78 vs. 7.89) during 2020 and 2021 seasons.

	2020 Season			
Treatment	LFeU	LMnU	LZnU	LCuU
	Leaves (mgkg^{-1})			
GCSP	195.00c ± 2.89	19.59cd ± 0.63	37.10b ± 0.62	6.33a ± 0.03
MAP$_1$	168.84e ± 1.59	18.34d ± 0.72	49.86a ± 0.16	1.67d ± 0.07
MAP$_2$	208.83b ± 2.17	20.42b–d ± 0.38	31.48c ± 0.25	2.34c ± 0.10
MAP$_3$	177.42d ± 2.55	15.92e ± 0.58	31.78c ± 0.13	3.50b ± 0.02
UP$_1$	162.25e ± 1.06	23.58a ± 0.77	30.81cd ± 0.43	3.42b ± 0.04
UP$_2$	165.75e ± 2.02	21.17bc ± 0.82	24.49e ± 0.29	2.34c ± 0.02
UP$_3$	234.42a ± 4.28	22.00ab ± 0.05	29.90d ± 0.68	3.34b ± 0.14
	2021 season			
GCSP	196.00c ± 1.44	21.09d ± 0.63	38.10b ± 0.53	2.50d ± 0.29
MAP$_1$	184.34d ± 2.50	22.34d ± 0.77	49.36a ± 0.15	3.09b–d ± 0.24
MAP$_2$	218.77b ± 2.08	25.77bc ± 0.82	34.48c ± 0.58	2.84cd ± 0.10
MAP$_3$	188.92d ± 8.90	21.28d ± 0.34	34.78c ± 1.83	4.00a ± 0.29
UP$_1$	176.50e ± 2.89	28.42a ± 0.72	32.81c ± 1.00	4.02a ± 0.24
UP$_2$	169.509a ± 0.72	24.57c ± 0.67	28.99d ± 1.12	3.34a–c ± 0.67
UP$_3$	239.42a ± 3.99	27.50ab ± 0.87	33.06c ± 0.42	3.67ab ± 0.10

Mean values (±SE) with different letters in each column are significant (at $p \leq 0.05$). GCSP = granular calcium super phosphate, MAP = mono-ammonium phosphate, UP = urea phosphate, MAP$_1$, MAP$_2$, and MAP$_3$ represent MAP applied as a soil application at 0.336, 0.445, and 0.555 kg tree^{-1}, UP$_1$, UP$_2$, and UP$_3$ represent UP applied as a soil application at 0.465, 0.616, and 0.770 kg tree^{-1}, control represent GCSP applied as a soil application at 1.75 kg tree^{-1}. According to Duncan's multiple range test, Means sharing the same letter in each column are not significantly different.

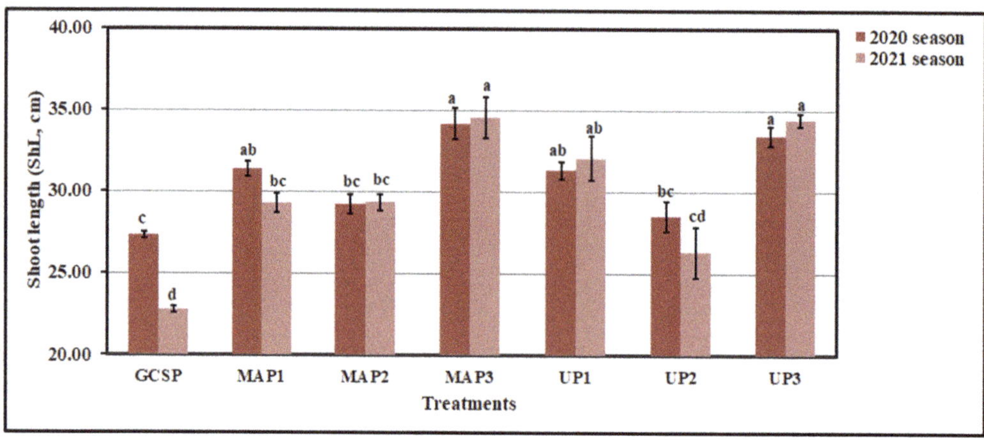

Figure 1. Influence of two phosphorus fertilizers; mono-ammonium phosphate (MAP) and urea-phosphate (UP) in comparison with granular calcium phosphate (GCSP) applied to shoot length (ShL, cm) of olive (arbequina cv.) trees grown in sandy loam clay soil under multi-abiotic stresses ($CaCO_3$ = 8.8 vs. 9.2%, ECe = 6.4 vs. 7.2 dS m^{-1}, and pH = 7.78 vs. 7.89) during 2020 and 2021 seasons. GSCP applied represents 1.75 kg tree^{-1}, MAP$_1$, MAP$_2$, and MAP$_3$ represent MAP applied at 0.336, 0.445, and 0.555 kg tree^{-1}, and UP$_1$, UP$_2$, and UP$_3$ represent UP applied at 0.465, 0.616, and 0.770 kg tree^{-1}. Bars in the same years with a different letter indicate significant differences between treatments at $p \leq 0.01$. According to Duncan's multiple range test, bars sharing the same letter in each column are not significantly different.

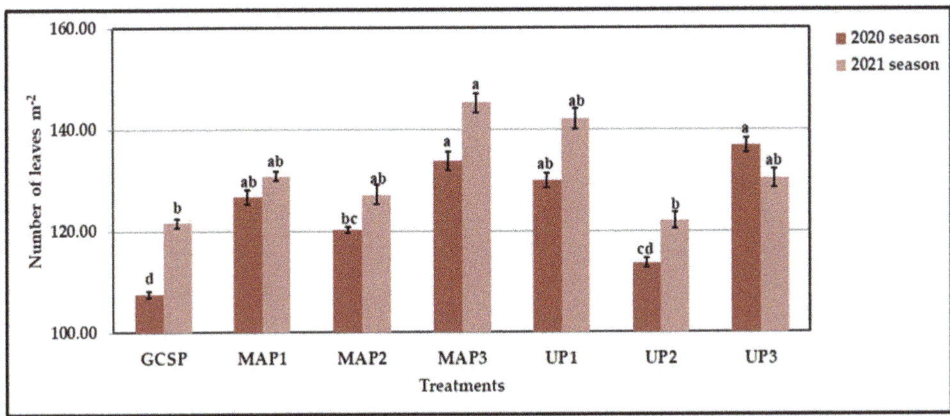

Figure 2. Influence of phosphorus fertilizers; mono-ammonium phosphate (MAP) and urea phosphate (UP) applied to number of leaves m^{-2} of olive (arbequina cv.) trees grown in sandy loam clay soil under multi-abiotic stresses (CaCO$_3$ = 8.8 vs. 9.2%, ECe = 6.4 vs. 7.2 dS m^{-1}, and pH = 7.78 vs. 7.89) during 2020 and 2021 seasons. GSCP applied represents 1.75 kg tree^{-1}, MAP$_1$, MAP$_2$, and MAP$_3$ represent MAP applied at 0.336, 0.445, and 0.555 kg tree^{-1}, and UP$_1$, UP$_2$, and UP$_3$ represent UP applied at 0.465, 0.616, and 0.770 kg tree^{-1}. Bars in the same years with a different letter indicate significant differences between treatments at $p \leq 0.01$. According to Duncan's multiple range test, bars sharing the same letter in each column are not significantly different.

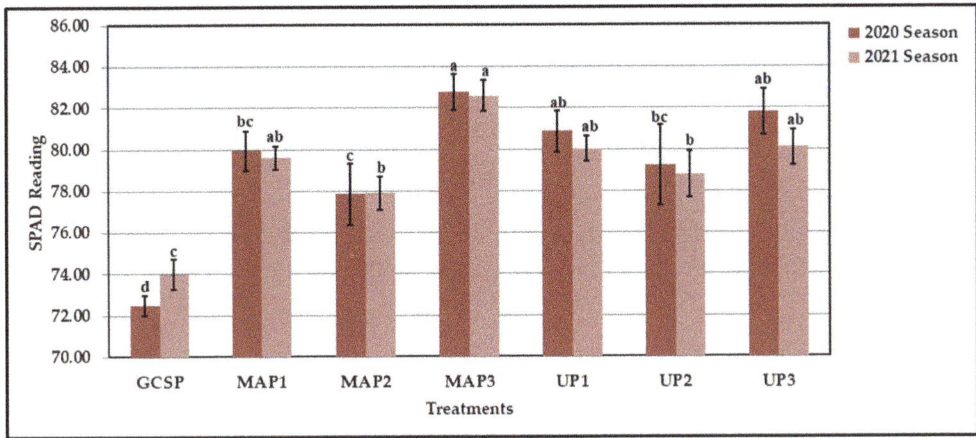

Figure 3. Influence of phosphorus fertilizers; mono-ammonium phosphate (MAP and urea phosphate (UP) applied to SPAD reading of olive (arbequina cv.) trees grown in sandy loam clay soil under multi-abiotic stresses (CaCO$_3$ = 8.8 vs. 9.2%, ECe = 6.4 vs. 7.2 dS m^{-1}, and pH = 7.78 vs. 7.89) during 2020 and 2021 seasons. GSCP applied represents 1.75 kg tree^{-1}, MAP$_1$, MAP$_2$, and MAP$_3$ represent MAP applied at 0.336, 0.445, and 0.555 kg tree^{-1}, and UP$_1$, UP$_2$, and UP$_3$ represent UP applied at 0.465, 0.616, and 0.770 kg tree^{-1}. Bars in the same years with a different letter indicate significant differences between treatments at $p \leq 0.01$. According to Duncan's multiple range test, bars sharing the same letter in each column are not significantly different.

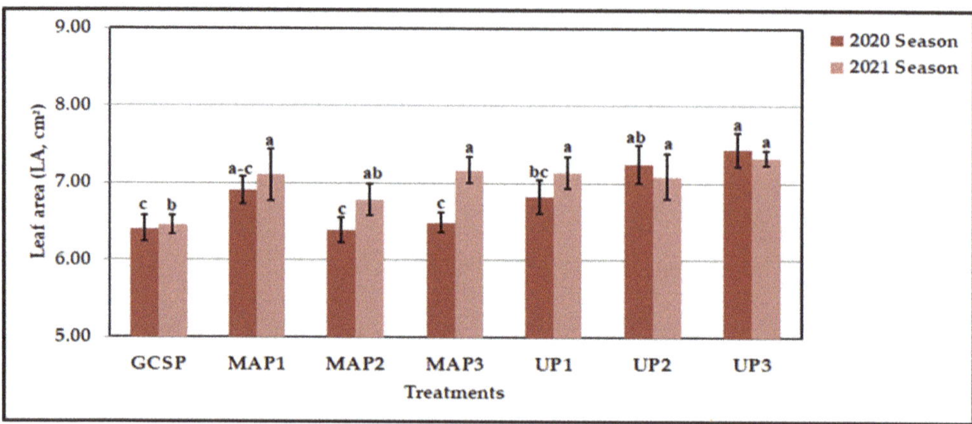

Figure 4. Influence of phosphorus fertilizers; mono-ammonium phosphate (MAP and urea phosphate (UP) applied to leaf area (LA, cm^2) of olive (arbequina cv.) trees grown in sandy loam clay soil under multi-abiotic stresses (CaCO$_3$ = 8.8 vs. 9.2%, ECe = 6.4 vs. 7.2 dS m^{-1}, and pH = 7.78 vs. 7.89) during 2020 and 2021 seasons. GSCP applied represents 1.75 kg tree^{-1}, MAP$_1$, MAP$_2$, and MAP$_3$ represent MAP applied at 0.336, 0.445, and 0.555 kg tree^{-1}, and UP$_1$, UP$_2$, and UP$_3$ represent UP applied at 0.465, respectively. 0.616 and 0.770 kg tree^{-1}. Bars in the same years with a different letter indicate significant differences between treatments at $p \leq 0.01$. According to Duncan's multiple range test, bars sharing the same letter in each column are not significantly different.

Meanwhile, olive trees fertilized with 0.555 kg tree^{-1} of MAP (MAP$_3$) showed the highest number of leaves (in area unit m^2) and SPAD reading in both seasons, whereas the values reached 136.77 vs. 145.31 and 82.76 vs. 82.58 for both aforementioned attributes in the two growing seasons. On the other hand, we noted that the minimum values for all aforementioned parameters, with the exception of LA in the second season, were recorded in trees fertilized with the recommended level of GCSP (1.75 kg tree^{-1}) in both seasons, whereas the lowest values of LA in the first season were recorded with MAP$_2$ treatment. According to the comparison between the maximum and minimum values, the percentages of increase reached 22.37 vs. 51.19 for ShL, 27.17 vs. 19.54 for NLf, 14.18 vs. 11.64 for SPAD reading, and 16.43 vs. 13.47 for LA in the first and second seasons, respectively. The results of the ANOVA indicated that all treatments had significant effects (at $p \leq 0.01$) for all aforementioned parameters except ShL in both seasons. However, there were significant (at $p \leq 0.05$) and non-significant impacts for ShL in the 2020 and 2021 seasons, respectively.

3.3. Olive Fruit Quality

The results presented in Table 7 indicated that the olive trees fertilized with UP$_3$ gave the maximum values (1.63 vs. 1.65 g) for total fruit weight (TFrW) and (1.32 vs. 1.31 g) for flesh weight (FlW) in the 2020 and 2021 seasons, respectively. Dissimilar results were obtained for seed weight (SeW), where the trees fertilized with GCSP gave the best values (0.29 g) in the first season, and trees treated with MAP$_1$ and UP$_1$ in the second season, since both of them gave the same value (0.30 g). As shown in Table 7, based on the obtained values for TFrW, SeW, and FlW, we found that the maximum values (81.26 vs. 4.35 and 81.32 vs. 4.36) for both fruit flesh weight (FrFlW%) and flesh/pit ratio (FPR), respectively, were achieved by applying UP$_3$ in the growing season of 2020 and MAP$_1$ in the growing season of 2021.

Table 7. Influence of different levels of MAP and UP in comparison with the recommended GCSP level on some fruit quality of olive (*Olea europaea* L. arbequina cv.) trees grown in sandy loam clay soil under multi-abiotic stresses ($CaCO_3$ = 8.8 vs. 9.2%, ECe = 6.4 vs. 7.2 dS m^{-1}, and pH = 7.78 vs. 7.89) during 2020 and 2021 seasons.

Treatment	TFrW	SeW	FlW	FrFlW	FPR	FrL	FrD	LD
	(g)			(%)		(mm)		
				2020 Season				
GCSP	1.47b ± 0.01	0.29c ± 0.01	1.18b ± 0.01	80.39ab ± 0.70	4.11ab ± 0.18	13.33cd ± 0.07	11.55cd ± 0.19	1.16cd ± 0.02
MAP$_1$	1.52b ± 0.03	0.32bc ± 0.01	1.20b ± 0.03	78.99ab ± 0.75	3.77ab ± 0.17	14.24b ± 0.16	11.62b–d ± 0.19	1.23ab ± 0.01
MAP$_2$	1.36c ± 0.01	0.31bc ± 0.02	1.05c ± 0.02	77.14b ± 0.93	3.40b ± 0.23	13.17d ± 0.06	11.28d ± 0.12	1.19a–d ± 0.01
MAP$_3$	1.55b ± 0.06	0.35b ± 0.01	1.20b ± 0.06	77.47b ± 0.89	3.45b ± 0.17	15.02a ± 0.24	12.14a ± 0.18	1.21a–c ± 0.03
UP$_1$	1.52b ± 0.02	0.32bc ± 0.04	1.20b ± 0.04	78.85ab ± 0.71	3.84ab ± 0.49	14.16b ± 0.08	11.91a–c ± 0.14	1.17b–d ± 0.02
UP$_2$	1.52b ± 0.03	0.44a ± 0.02	1.07c ± 0.02	70.75c ± 0.85	2.43c ± 0.10	13.68c ± 0.18	12.00a–c ± 0.12	1.14d ± 0.01
UP$_3$	1.63a ± 0.06	0.30bc ± 0.01	1.32a ± 0.01	81.26a ± 0.65	4.35a ± 0.19	15.11a ± 0.10	12.13ab ± 0.37	1.24a ± 0.03
				2021 season				
GCSP	1.50c ± 0.03	0.40a ± 0.01	1.10b ± 0.04	73.35d ± 0.82	2.77c ± 0.17	14.09d ± 0.06	11.90b–d ± 0.25	1.19bc ± 0.03
MAP$_1$	1.61a ± 0.04	0.30d ± 0.01	1.31a ± 0.03	81.32a ± 0.29	4.36a ± 0.08	15.05bc ± 0.10	12.35a–c ± 0.18	1.22a–c ± 0.02
MAP$_2$	1.49c ± 0.02	0.35bc ± 0.01	1.14b ± 0.06	76.40c ± 0.33	3.24bc ± 0.06	13.79d ± 0.06	11.47d ± 0.21	1.21a–c ± 0.03
MAP$_3$	1.61a ± 0.03	0.36b ± 0.02	1.25a ± 0.03	77.73bc ± 0.56	3.50b ± 0.12	15.43b ± 0.15	12.20a–d ± 0.43	1.29a ± 0.05
UP$_1$	1.53bc ± 0.01	0.30d ± 0.02	1.24a ± 0.03	80.63ab ± 0.87	4.24a ± 0.42	14.63c ± 0.16	11.62cd ± 0.27	1.26ab ± 0.02
UP$_2$	1.58ab ± 0.01	0.33b–d ± 0.01	1.25a ± 0.02	79.01a–c ± 0.94	3.78a ± 0.22	14.76c ± 0.31	12.58ab ± 0.17	1.17c ± 0.01
UP$_3$	1.65a ± 0.01	0.32cd ± 0.01	1.31a ± 0.01	80.65ab ± 0.54	4.17a ± 0.01	16.08a ± 0.15	12.86a ± 0.33	1.25a–c ± 0.03

Mean values (±SE) with different letters in each column are significant (at $p \leq 0.05$). GCSP represents granular calcium super-phosphate applied at 1.75 kg tree^{-1}, MAP = mono-ammonium phosphate, MAP$_1$, MAP$_2$, and MAP$_3$ represent MAP applied at 0.336, 0.445, and 0.555 kg tree^{-1}, UP = urea phosphate, UP$_1$, UP$_2$, and UP$_3$ represent UP applied at 0.465, 0.616, and 0.770 kg tree^{-1}. TFrW = total fruit weight, SeW = seed weight, FlW = flesh weight, FrFlW = TFrW/FlW, FPR = flesh/pit ratio, FrL = fruit length, FrD = fruit diameter, and LD = fruit shape index. All treatments were applied as soil applications. According to Duncan's multiple range test. Means sharing the same letter in each column are not significantly different.

Despite the improvements achieved with MAP and UP, MAP_2 treatment was the least influential on the TFrW in both seasons, with values of 1.63 vs. 1.65 g, and on FlW, recording 1.05 in the first season. Meanwhile, UP_2 treatment had the weakest influence on SeW, FrFlW%, and FPR, recording values of 0.44, 70.75, and 2.43 in the first season. Accordingly, the lowest values (1.10, 73.35, and 2.77) for FlW, FrFlW, and FPR, respectively, were obtained in trees fertilized with GCSP in the second season. The obtained data indicated that the increment rates were 19.85 vs. 10.74 for TFrW, 25.71 vs. 19.09 for FlW, 14.86 vs. 10.87% for FrFlW, and 79.01 vs. 57.40% for FPR. Meanwhile, the rate of decline reached 34.09 vs. 25% for SeW in the two growing seasons, respectively. Analysis of variance indicated that the treatments had a significant impact (at $P \leq 0.01$) on all studied attributes.

It is clear from Table 7 that UP_3 and MAP_3 led to appreciable improvements in fruit length (FrL) and fruit diameter (FrD), which in turn impacted the fruit shape index (LD). Our obtained results showed that the trees fertilized with UP_3 achieved the highest values (15.11 vs. 16.08 mm) for FrL in both seasons, with 12.86 mm for FrD in the second season. Meanwhile, the greatest value (12.14 mm) for FrD was produced in trees treated with MAP_3 in the first season. Based on the obtained values for FrL and FrD, the highest values (1.24 vs. 1.29) for LD were determined as a result of applying UP_3 and MAP_3 in the two growing seasons, respectively. On the contrary, trees fertilized with the MAP_2 treatment yielded the minimum values of 13.17 vs. 13.79 mm for FrL and 11.28 vs. 11.47 mm for FrD in the 2020 and 2021 seasons, respectively. Meanwhile, the lowest values of 1.14 vs. 1.17 for LD were produced via UP_2 treatment in both seasons, respectively. The obtained results in Table 7 show that the percentages of increase were 14.73 vs. 16.61, 7.62 vs. 12.12, and 8.77 vs. 10.26 for FrL, FrD, and LD in the growing seasons of 2020 and 2021, respectively. As displayed in Table 7, some parameters related to the flesh and seeds of olive fruits were significantly improved due to the application of the phosphorus fertilizers (MAP and UP) in comparison with GCSP. In our investigation, UP_3 was the superior treatment for these fruit quality parameters. Although the improvements in the studied attributes were slight, the statistical analysis indicated that all treatments had significant effects (at $p \leq 0.01$) on all studied parameters in the 2020 and 2021 seasons, respectively. Meanwhile, LD had a significant influence (at $p \leq 0.05$) in the growing season of 2020.

3.4. Table and Oil Olive Yield

The impacts of different levels of MAP and UP in comparison with GCSP on fruit dry matter (FrDrM%), total olive yield (TOY, $tree^{-1}$, and ha^{-1}), and olive oil content (OOC, %) in the 2020 and 2021 seasons are presented in Table 8. The UP application was more effective compared with MAP treatment. The trees fertilized with UP_3 produced the maximum total yield values, (42.67 vs. 42.83 kg $tree^{-1}$), and (10.75 vs. 10.79 ton ha^{-1}), in the 2020 and 2021 seasons. Moreover, it was the best treatment for FrDrM% and OOC% in the second season, which reached 31.39 and 42.71%, respectively. Meanwhile, the trees treated with MAP_3 recorded the highest values (32.99%) for FrDr% and (41.18%) for OOC in the first season, respectively.

Regarding the lowest values, the general trends indicated that the olive trees fertilized with GCSP recorded the minimum values (38.67 vs. 37.67 kg) for OTY and (35.92 vs. 35.45%) for OOC% in both seasons, respectively, as well as (29.22%) for FrDrM in the second season. In addition, the minimum values (9.66 vs. 9.41 ton ha^{-1}) for TOY in the 2020 and 2021 seasons, respectively, and (29.22%) for FrDrM% in the second season were produced using 0.445 kg $tree^{-1}$ (MAP_2). The overall trends of our study showed that the trees fertilized with either MAP or UP outperformed their counterparts fertilized with GCSP.

As presented in Table 8, the percentage increases amounted to 12.90 vs. 7.43% for FrDrM%, 10.34 vs. 13.70% for OTY, 11.28 vs. 14.67% for TOY, and 14.64 vs. 20.48% for OOC in the growing seasons of 2020 and 2021, respectively. The results obtained from the statistical analysis revealed significant differences (at $p \leq 0.01$) between treatments for all studied parameters in the first and second seasons, respectively.

Table 8. Influence of different levels of MAP and UP in comparison with the recommended GCSP level on fresh matter %, total olive yield (both tree and ha), and olive oil content of olive (*Olea europaea* L. arbequina cv.) trees grown in sandy loam clay soil under multi-abiotic stresses ($CaCO_3$ = 8.8 vs. 9.2%, ECe = 6.4 vs. 7.2 dS m^{-1}, and pH = 7.78 vs. 7.89) during 2020 and 2021 seasons.

Treatment	FrDrM	OTY	TOY	OOC
	(%)	(kg tree^{-1})	(ton ha^{-1})	(%, DM)
2020 Season				
GCSP	29.84e ± 0.45	38.67b ± 0.67	9.74b ± 0.13	35.92d ± 0.24
MAP$_1$	30.88d ± 0.85	39.33b ± 0.67	9.91b ± 0.14	37.43cd ± 0.36
MAP$_2$	29.22e ± 0.38	38.33d ± 0.88	9.66b ± 0.14	36.15cd ± 0.31
MAP$_3$	32.99a ± 0.26	42.33a ± 0.58	10.67a ± 0.12	41.18a ± 0.47
UP$_1$	32.49ab ± 0.46	41.67a ± 0.67	10.50a ± 0.12	38.99b ± 0.50
UP$_2$	31.98bc ± 0.57	39.00b ± 0.58	9.83b ± 0.13	37.46c ± 0.42
UP$_3$	31.39cd ± 0.51	42.67a ± 0.58	10.75a ± 0.13	40.72a ± 0.28
2021 season				
GCSP	29.22e ± 0.30	37.67d ± 0.67	9.49d ± 0.14	35.45d ± 0.62
MAP$_1$	30.88de ± 0.11	38.33cd ± 0.58	9.66cd ± 0.11	36.01cd ± 0.22
MAP$_2$	29.84de ± 0.27	37.33d ± 0.67	9.41d ± 0.11	37.96bc ± 0.44
MAP$_3$	32.99cd ± 0.39	41.67a ± 0.33	10.50a ± 0.13	41.42a ± 0.26
UP$_1$	32.49bc ± 0.70	39.33bc ± 0.67	9.91bc ± 0.14	37.10b–d ± 0.89
UP$_2$	31.98ab ± 0.46	39.67b ± 0.58	10.00b ± 0.14	38.87b ± 0.26
UP$_3$	31.39a ± 0.26	42.83a ± 0.33	10.79a ± 0.14	42.71a ± 0.23

Mean values (±SE) with different letters in each column are significant (at $p \leq 0.05$). GCSP represent granular calcium super-phosphate applied at 1.75 kg tree^{-1}, MAP = mono-ammonium phosphate, MAP$_1$, MAP$_2$, and MAP$_3$ represent MAP applied at 0.336, 0.445, and 0.555 kg tree^{-1}, UP = urea phosphate, UP$_1$, UP$_2$, and UP$_3$ represent UP applied at 0.465, 0.616, and 0.770 kg tree^{-1}, all treatments were applied as a soil application. According to Duncan's multiple range test, Means sharing the same letter in each column are not significantly different.

3.5. Regression and Stepwise Analysis

The results obtained from the stepwise regression, shown in Table 9, indicate the relationship of the olive tree yield (OTY, kg) and olive oil content (OOC, %) with the leaf nutrient content, growth parameters, and yield attributes under multi-abiotic stresses in the 2020 and 2021 growing seasons. In both seasons, these factors made highly significant contributions to the OTY and OOC. In our results, the adjusted R^2 values were 0.637 and 0.840 (r = 0.821 and 0.934) for OTY and 0.909 and 0.388 (r = 0.960 and 0.647) for OOC in the two seasons, respectively. The fitted equation then obtained demonstrated that the variation in OTY was explained by the variation in attributes such as FrL and LNC in 2020 and FrL, FrDrM%, FrFlW, and LD in 2021. Meanwhile, FrL, FrDrM%, and LMgC in 2020 and FrL in 2021 contributed to the OOC variation.

Table 9. Proportional contribution in predicting olive tree yield (TOY, kg) and olive oil content (OOC, %) using stepwise multiple linear regression for multi-stressed olive trees fertilized by mono-ammonium phosphate (MAP) and urea phosphate (UP) in three levels in comparison with the recommended granular calcium super phosphate (GCSP) level in 2020 and 2021 seasons.

r	R^2	Adjusted R^2	SEE	Significance	Fitted Equation
				2020 season	
0.821	0.673	0.637	1.192	***	OTY = 8.008+2.172FrL + 16.209LNC
0.960	0.922	0.909	0.626	***	OOC = −8.245 + 2.237FrL + 0.443FrDrM% + 2.906LMgC
				2021 season	
0.934	0.872	0.840	0.810	***	OTY = 0.298 + 1.838FrL + 0.642FrDrM% − 0.203FrFlW + 8.541LD
0.647	0.419	0.388	2.197	***	OOC = 3.758 + 2.343FrL

4. Discussion

This manuscript describes work that was carried out under multi-abiotic stresses through the application of two highly soluble phosphorus fertilizers (HSPFs) differing in their content of nitrogen (N%) and phosphorus (P_2O_5%), namely, MAP and UP, compared with GCSP as one of the most widely used phosphate fertilizers in Egypt, in an attempt to overcome the problem of P fixation and the unavailability of micronutrients under some abiotic stresses in olive trees (*Oleaeuropaea* L. arbequina cv.). As shown in Table 4, the tested soil suffered from more than one undesirable property, such as $CaCO_3$ = 8.8 vs. 9.2%, ECe = 6.4 vs. 7.2 dS m^{-1}, and pH = 7.78 vs. 7.89, in the two growing seasons of 2020 and 2021, respectively, which hindered the optimal growth of the olive trees. All these undesirable characteristics combined to negatively affect the absorption of nutrients and thus lead to different physiological and growth attributes, which in turn affect the table and olive oil yield and its components. Generally speaking, the obtained results revealed that the applied HSPFs, either MAP or UP, irrespective of their applied levels, significantly affected all studied nutrients. Our obtained data indicated that LPU, LFeU, and LMnU in both seasons; LNU, LCaU, and LMgU in the first season; and LKU in the second season were significantly increased with the UP application, irrespective of the use level. Additionally, LKU and LNaU in the 2020 season, in addition to LNU, LCaU, and LMgU in the 2021 season and LZnU in both growing seasons, were obtained in plants fertilized with MAP, regardless of the applied levels. In this context, the influences of MAP and UP were somewhat similar in terms of the availability of nutrients compared with GCSP. Furthermore, the remarkable superiority of the application of UP over MAP was demonstrated. These results may be attributed to the improved effects of MAP and UP in reducing soil pH; however, the mean pH values of MAP and UP were 4.5 and 1.8, in comparison with GCSP, whose pH was 7.5, as presented in Table 3. This pH value can improve the availability of nutrients and make them more soluble for uptake by olive tree roots. Very recently, some results were reported by [57]; they mentioned the positive impact of phosphoric acid (H_3PO4) in reducing soil pH. The obtained results are in accordance with the results of [58–60]. In this regard, similar results reported that the simulative influence of MAP and UP may be due to their vital role in reducing soil pH, which in turn markedly influences nutrient availability and plays a fundamental role in fixing atmospheric nitrogen, which is beneficial to enhancing LNU [61,62]. The notable declines in LNU, as shown in the MAP_1, MAP_3, and UP_1 treatments, could be due to the translocation of N from leaves to fruit during the pollination stage. As shown in Table 3, irrespective of the applied level, applying ammonium sulfate with GCSP under a high soil pH encouraged the occurrence of mineralization in both 2020 and 2021. Then, the ammonium (NH_4^+) ions were converted into nitrate (NO_3^-) ions, which were lost by leaching; this could be due to the negative charge and increased water requirements, regardless of the nature of the dry climate. Although these results are not in agreement with the findings of [63,64], in which decreases in LNU were proposed to be due to the translocation of N to form young shoots, these results were in accordance with those obtained by [63,65]. They were not in line with [64], especially regarding LNU, wherein the recorded lower values may be due to the high $CaCO_3$ content in the tested soil, in addition to the prevalent climatic conditions related to the ARH and AP, as presented in Table 1. In other words, both MAP and UP enhanced the root hair system, thus increasing the absorption efficiency of roots in the growing olive trees. These results were further explained based on the soil's chemical and physical properties; HPO_4^{--} and $H_2PO_4^-$ ions from both MAP and UP were absorbed quickly by root trees compared to GCSP. These ions were fixed in soil due to the high pH of dicalcium phosphate ($CaHPO_4$) and tricalcium phosphate [$Ca_3(PO_4)_2$], and their solubility was limited according to the following equation: $Ca(H_2PO_4) + 2Ca^{+2} \rightarrow Ca_3(PO_4)_2 + 4H^+$. The precipitation of HPO_4^- on the surface of $CaCO_3$ can be expressed by the following equation: $Ca(H_2PO_4) + 2CaCO_3$ $Ca_3(PO_4)_2 + 2CO_2 + 2H_2O$. Moreover, HPO_4^{--} ions are fixed by an absorption reaction with Fe, Zn, and Mn. The only exception was that the highest LCuU was produced in plants treated with GCSP in the 2020 season. Similar results were reported by [66], who observed that using GCSP as a foliar application on eggplants in a high dose (2%) enhanced plant growth, which in turn affected nutrient uptake; alternatively, the result may have been due to an antagonistic effect between Cu and Fe, Mn, and Zn. In other words, the results could indicate that applying MAP or UP is better than applying GCSP, due to the fact that the presence of N and P in one chemical structure is better for the absorption of both nutrients compared to adding them individually with GCSP treatment. In summary, nutritional status is the basis upon which to evaluate physiological and growth parameters. It could be noticed that the maximum values were produced when applying the maximum level of the applied HSPF, irrespective of its type. However, the ShL values and SPAD readings were obtained with the MAP_3 treatment in both seasons, in addition to the number of leaves per m^2 ($NLfm^2$) in the second season. Meanwhile, the maximum values of LA were produced in trees fertilized with UP_3 in both seasons. These results also explain that the P reaction products differ from each other in their solubility. This confirms that the different sources of phosphate fertilizers are not equally effective, due to the presence of NH_4^+ ions in MAP and their conversion into NO_3^- ions. Similarly, the presence of amide groups ($-NH_2$) in UP and their conversion into NH_4^+ ions and then into NO_3^- ions lead to a lowered soil pH in the rhizosphere zone [67]; in addition, the absorption of NO_3^- enhanced the dissolution of precipitated Ca-P compounds and P availability [68,69]. These results could be attributed to the vital role of MAP and UP in reducing soil pH and increasing the levels of available P, which, in turn, markedly affect several metabolic and physiological processes, such as protein synthesis [9] and phosphorprotein, fat, and sulfur metabolism [13]. In addition, it is an essential element in energy-rich compounds such as ATP, ADP, and AMP and in the photosynthesis process [8]; in turn, it significantly influences cell division and elongation. To confirm the role of soil pH in nutrient availability, some studies have been reported [70,71] regarding the influence of organic manure on reducing soil pH, which in turn positively impacted nutrient availability in Jerusalem artichoke plants. On the other hand, applying GCSP yielded the lowest values for all of the studied attributes. This is clear evidence of the difference in the solubility of $H_2PO_4^-$ and HPO_4^{--} ions in the three studied phosphorus fertilizers, and thus their different behaviors in the soil. These results indicated that the P utilization from MAP and UP was higher than P in GCSP in the vegetative growth stage [72–74]. However, under high $CaCO_3$ content conditions, the P in GCSP fertilizer converts from available to unavailable forms such as Ca_2-P, Ca_8-P, and Ca_{10}-P.

The beneficial effects of P in MAP and UP, which, in turn, were reflected in the total olive yield and its attributes, are presented in Table 8; however, they could be a result of nutritional status improvement. These results are in agreement with the results of [58,61], who reported that absorbed N, P, and K act as cofactors to increase the total carbohydrates and their assimilation, which causes an increase in the assimilation products, which is consequently reflected in the studied yield attributes, such as TFrW, FlW, FrL, FrD, and FrDrM. In other words, these enhancements may be due to the improved impact of MAP and UP on the leaf K and Zn content [75–78]. Regarding the maximum OTY and TOY, it could be observed from our results that the maximum values of OTY and TOY were recorded for trees fertilized with the UP_3 treatment, followed by the MAP_3 treatment, in both seasons. It is evident that the TOY depends on the high level of the applied highly soluble phosphorus fertilizer. These results could be due to P and its synergistic effects on the translocation of different nutrients' availability. Additionally, the increases in P application might cause improvements in the root system [79], consequently enabling plants to absorb more water and nutrients from the depths of the soil. Furthermore, the N present within the chemical structure of both MAP and UP played a cooperative role with P in enhancing the plant growth and the ability to increase flowering due to their direct influence on growth and on the promotion the chlorophyll formation [10]. These obtained results are in line with the previous results of [80,81].

5. Conclusions

Under saline calcareous alkaline soils, phosphorus and other micronutrients are fixed in unavailable forms. This work was conducted on olive (*Olea europaea*, Arbequina cv.) trees grown in sandy clay loam soil characterized by multiple undesirable properties ($CaCO_3$ = 8.8 vs. 9.2%, ECe = 6.4 vs. 7.2 dS m^{-1}, and pH = 7.78 vs. 7.89) in the 2020 and 2021 seasons, respectively, under a drip irrigation system. Generally speaking, from our results, three main points could be concluded: (1) The application of highly soluble phosphorus fertilizers (HSPFs), mono-ammonium phosphate (MAP), and urea phosphate (UP), irrespective of the use level, was the most influential compared with granular calcium super-phosphate (GCSP) for all studied characteristics except leaf copper uptake. (2) Regardless of the applied level, plants subjected to the application of UP yielded superior results to their counterparts fertilized with MAP. (3) The application of the maximum level of either MAP (0.555 kg tree^{-1}) or UP (0.770 kg tree^{-1}) gave the best results for most of the studied traits. However, the trees fertilized with MAP$_3$ gave the maximum values for shoot length, SPAD reading, and dry fruit matter. Meanwhile, the plants fertilized with UP$_3$ produced the best results for the leaf area, olive tree yield, total olive yield, total fresh weight, flesh weight (FlW), fruit length (FrL), and leaf Fe content in both seasons. In short, the application of HSPFs under these conditions might be an alternative surrogate to improve nutrient efficiency and thus improve productivity.

Author Contributions: Conceptualization, H.A.Z.H. and H.R.B.; data curation, A.A.M.A., H.A.Z.H. and H.R.B.; formal analysis, A.A.M.A. and H.A.Z.H.; investigation, H.A.Z.H., A.A.M.A. and H.R.B.; methodology, A.A.M.A. and H.A.Z.H.; resources, H.A.Z.H. and H.R.B.; software A.A.M.A. and H.R.B.; writing—original draft, A.A.M.A.; writing—review and editing, A.A.M.A. and H.A.Z.H. All authors have read and agreed to the published version of the manuscript.

Funding: This research received no external funding.

Institutional Review Board Statement: Not appreciable.

Informed Consent Statement: Not appreciable.

Data Availability Statement: The data presented in this study are available upon request from the corresponding author.

Conflicts of Interest: The authors declare no conflict of interest.

References

1. Aboukila, E.F.; Nassar, I.N.; Rashad, M.; Hafez, M.; Norton, J.B. Reclamation of calcareous soil and improvement of squash growth using brewers' spent grain and compost. *J. Saudi Soc. Agric. Sci.* **2018**, *17*, 390–397. [CrossRef]
2. Wahba, M.M.; Labib, F.; Zaghloul, A. Management of calcareous soils in arid region. *Int. J. Environ. Pollut. Environ. Model.* **2019**, *2*, 248–258.
3. FAO. *Soils Portal: Management of Calcareous Soils*; Food and Agriculture Organization of United Nations: Rome, Italy, 2016. Available online: http://www.fao.org/soils-portal/soilmanagement/managementof-someproblem-soils/calcareous-soils/ar/ (accessed on 1 April 2016).
4. Saad El-Dein, A.A.; Galal, M.E. Prediction of Reclamation Processes in Some Saline Soils of Egypt. *Egypt. J. Soil Sci.* **2017**, *57*, 293–301.
5. Agbna, G.H.M.; Ali, A.B.; Bashir, A.K.; Eltoum, F.; Hassan, M.M. Influence of biochar amendment on soil water characteristics and crop growth enhancement under salinity stress. *Int. J. Eng. Works* **2017**, *4*, 49–53.
6. Sun, J.; Yang, R.; Li, W.; Pan, Y.; Zheng, M.; Zhang, Z. Effect of biochar amendment on water infiltration in a coastal saline soil. *J. Soil Sediment* **2018**, *18*, 3271–3279. [CrossRef]
7. Jamil, A.; Riaz, S.; Ashraf, M.; Foolad, M.R. Gene expression profiling of plants under salt stress. *Crit. Rev. Plant Sci.* **2011**, *30*, 435–458. [CrossRef]
8. Vance, C.P.; Uhde-Stone, C.; Allan, D.L. Phosphorus acquisition and use: Critical adaptations by plants for securing a nonrenewable resource. *New Phytol.* **2003**, *157*, 423–447. [CrossRef] [PubMed]
9. Von Wandruszka, R. Phosphorus retention in calcareous soils and the effect of organic matter on its mobility. *Geochem. Trans.* **2006**, *7*, 6–14. [CrossRef]
10. Taiz, L.; Zeiger, E.; Møller, I.M.; Murphy, A. *Plant Physiology and Development*, 6th ed.; Sinauer Associates: Sunderland, MA, USA, 2015.
11. Barłóg, P.; Grzebisz, W.; Feć, M.; Łukowiak, R.; Szczepaniak, W. Row method of sugar beet (*Beta vulgaris* L.) fertilization with multicomponent fertilizer based on urea-ammonium nitrate solution as a way to increase nitrogen efficiency. *J. Cent. Eur. Agric.* **2010**, *11*, 225–234.

12. Salimpour, S.; Khavazi, K.; Nadian, H.; Besharati, H.; Miransari, M. Enhancing phosphorous availability to canola (*Brassica napus* L.) using P solubilizing and sulfur oxidizing bacteria. *Aust. J. Crop Sci.* **2010**, *4*, 330–334.
13. Uygar, V.; Şen, M. The effect of phosphorus application on nutrient uptake and translocation in wheat cultivars. *Int. J. Agric. For. Life Sci.* **2018**, *2*, 171–179.
14. Faye, I.; Diouf, O.; Guisse, A.; Sene, M.; Diallo, N. Characterizing root responses to low phosphorus in pearl millet [*Pennisetum glaucum* (L.) R. Br.]. *Agron. J.* **2006**, *98*, 1187–1194. [CrossRef]
15. Awad, A.A.M.; Sweed, A.A.A. Influence of organic manures on soil characteristics and yield of Jerusalem artichoke. *Commun. Soil Sci. Plant Anal.* **2020**, *51*, 1101–1113. [CrossRef]
16. Farrag, H.M.; Bakr, A.A.A. Biological reclamation of a calcareous sandy soil with improving wheat growth using farmyard manure, acid producing bacteria and yeast. *Int. J. Agric. Sci.* **2021**, *3*, 53–71. [CrossRef]
17. Ringeval, B.; Augusto, L.; Monod, H.; van Apeldoorn, D.; Bouwman, L.; Yang, X.; Achat, D.L.; Chini, L.P.; Van Oost, K.; Guenet, B.; et al. Phosphorus in agricultural soils: Drivers of its distribution at the global scale. *Glob. Chang. Biol.* **2017**, *23*, 3418–3432. [CrossRef] [PubMed]
18. Lopez-Bucio, J.; Vega, O.M.; Guevara-García, A.; Herrera-Estrella, L. Enhanced phosphorus uptake in transgenic tobacco plants that overproduce citrate. *Nat. Biotechnol.* **2000**, *18*, 450–453. [CrossRef]
19. Adesemoye, A.; Kloepper, J. Plant–microbes interactions in enhanced fertilizer use efficiency. *Appl. Microbiol. Biotechnol.* **2009**, *85*, 1–12. [CrossRef]
20. Fageria, N.K.; Moreira, A.; and Castro, C. Response of soybean to phosphorus fertilization in Brazilian Oxisol. *Commun. Soil Sci. Plant Anal.* **2011**, *42*, 2716–2723. [CrossRef]
21. Bargaz, A.; Noyce, G.L.; Fulthorpe, R.; Carlsson, G.; Furze, J.R.; Jensen, E.S.; Dhiba, D.; Isaac, M.E. Species interactions enhance root allocation, microbial diversity and P acquisition in intercropped wheat and soybean under P deficiency. *Appl. Soil Ecol.* **2017**, *120*, 179–188. [CrossRef]
22. Osborne, S.L.; Schepers, J.S.; Francis, D.D.; Schlemmer, M.R. Detection of phosphorus and nitrogen deficiencies in corn using spectral radiance measurements. *Agron. J.* **2002**, *94*, 1215. [CrossRef]
23. Choi, J.M.; Lee, C.W. Influence of elevated phosphorus levels in nutrient solution on micronutrient uptake and deficiency symptom development in strawberry cultured with fertigation system. *J. Plant Nutr.* **2012**, *35*, 1349–1358. [CrossRef]
24. Viégas, I.J.M.; Cordeiro, R.A.M.; Almeida, G.M.; Silva, D.A.S.; De Silva, B.C.; Okumura, R.S.; Junior, M.L.S.; De Silva, S.P.; De Freitas, J.M.N. Growth and visual symptoms of nutrients deficiency in Mangosteens (*Garcinia mangostana* L.). *Am. J. Plant Sci.* **2018**, *9*, 1014–1028. [CrossRef]
25. Bagayoko, M.; George, E.; Romheld, V.; Buerkert, A.B. Effects of mycorrhizae and phosphorus on growth and nutrient uptake of millet, cowpea and sorghum on a west African soil. *J. Agric. Sci.* **2000**, *135*, 399407. [CrossRef]
26. Tang, J.W.; Mu, R.Z.; Zhang, B.L.; Fan, X.S. Solubility of urea phosphate in water + phosphoric acid from (277.00 to 354.50) K. *J. Chem. Eng. Data* **2007**, *52*, 1179–1181. [CrossRef]
27. Navizaga, C.; Boecker, J.; Sviklas, A.M.; Galeckiene, J.; Baltrusaitis, J. Adjustable N:P$_2$O$_5$ ratio urea phosphate fertilizers for sustainable phosphorus and nitrogen use: Liquid phase equilibria via solubility measurements and raman Spectroscopy. *ACS Sustain. Chem. Eng.* **2017**, *5*, 1747–1754. [CrossRef]
28. Kucharski, J.; Wyszkowska, J.; Borowik, A. Biological effects produced by urea phosphate in soil. *J. Res. Appl. Agric. Eng.* **2013**, *58*, 25–28.
29. Zhang, H.Y.; Huang, Z.H.; Wang, J.; Zhang, J.; Meng, C.R.; Wei, C.Z. Effects of different acidifiers on pH and phosphorus availability in calcareous soil. *Soils Fert. Sci. China* **2019**, *1*, 145–150.
30. Diego, B.; Ercan, H.; Munoz, C.; Arquero, O. Factors influencing the efficiency of mono-potassium phosphate in the olive. *Intl. J. Plant Prod.* **2010**, *4*, 235–240.
31. Shaheen, A.A. Effect of Using Some Sources of Phosphorus on Flowering, Fruiting and Productivity of Olive Trees. *World J. Agric. Sci.* **2019**, *15*, 103–113.
32. Barone, E.; Mantia, M.; La-Marchese, A.; Marra, F.P. Improvement in yield and fruit size and quality of table olive cvs. *Sci. Agric.* **2014**, *71*, 52–57. [CrossRef]
33. Ran, E.U.; Yasuor, H.C.; Schwartz, D.A.; Gall, A.B.; Dag, A. Phosphorous nutritional level, carbohydrate reserves and flower quality in olives. *J. Plant Sci. Genet. Isr.* **2016**, *16*, 1–19.
34. Hopkins, B.G. Russet Burbank potato phosphorus fertilization with dicarboxylic acid copolymer additive (AVAIL®). *J. Plant Nutr.* **2011**, *33*, 1422–1434. [CrossRef]
35. Rosen, C.J.; Kelling, A.K.; Stark, J.C.; Gregory, A.P. Optimizing Phosphorus Fertilizer Management in Potato Production. *Am. J. Potato Res.* **2014**, *91*, 145–160. [CrossRef]
36. Carl, J.R.; Mcnearney, M.; Peter, B. Effect of liquid fertilizer sources and avail on potato yield and quality. In Proceedings of the Idaho Potato Conference, Poccatelo, ID, USA, 19–20 January 2011.
37. Abdel-Razzak, H.S.; Moussa, A.G.; Abd El-Fattah, M.A.; El-Morabet, G.A. Response of sweet potato to integrated effect of chemical and natural phosphorus fertilizer and their levels in combination with mycorrhizal Inoculation. *J. Biol. Sci.* **2013**, *13*, 112–122. [CrossRef]
38. FAO. *FAOSTAT*; Food and Agriculture Organization of the United Nations: Rome, Italy, 2017. Available online: http://www.fao.org/faostat/en/ (accessed on 10 December 2019).

39. Mahmoud, T.M.; Emam, S.; Mohamed, A.; El-Sharony, T.F. Influence of foliar application with potassium and magnesium on growth, yield and oil quality of "Koroneiki" olive trees. *Am. J. Food Technol* **2017**, *12*, 209–220. [CrossRef]
40. Stan, K.; David, H. *Producing Table Olives*; Land links Press: Collingwood, Australia, 2007; p. 3066.
41. Abo Arab, D.E.A.A. Effects of Bio and Mineral Phosphorus Fertilization on the Quality and Yield of Melon (Cantaloupe). M.Sc. Thesis, High Institute of Public Health, Alexandia University, Alexandria, Egypt, 2014.
42. Gregoriou, C.; El-Kholy, M. Fertilization. In *Olive GAP Manual: Good Agricultural Practices for the Near East & North Africa Countries*; FAO: Rome, Italy, 2010.
43. Fernandez-Escobar, R. Fertilization. In *El Cultivo del Olive*, 7th ed.; Barranco, D., Fernandez-Escobar, R., Rallo, I., Eds.; Mundi-Prensa: Madrid, Spain, 2017; pp. 419–460.
44. Bouyoucos, C.J. Hydrometer method improved for making particle size analysis of soil. *Soil Sci. Soc. Proc.* **1981**, *26*, 446–465.
45. McLean, E.O. Soil pH and lime requirement. In *Methods of Soil Analysis. Part 2. Chemical and Microbiological Properties*; Page, A.L., Ed.; American Society of Agronomy: Madison, WI, USA, 1982; pp. 199–224.
46. Page, A.L.; Miller, R.H.; Keeney, D.R. *Method of Soil Analysis. Part 2. Chemical and Microbiological Methods*; American Society of Agronomy: Madison, WI, USA, 1982; pp. 225–246.
47. Allison, L.E. Organic carbon. In *Methods of Soil Analysis. Part II*; Dans, C., Black, A., Eds.; American Society of Agronomy: Madison, WI, USA, 1965; Chapter 90; pp. 1372–1376.
48. Jackson, M.L. *Soil Chemical Analysis*; Constable and Co. Ltd.: London, UK, 1962.
49. Pregl, F. *Quantitative Organic Micro-Analysis*, 4th ed.; Churchill, A., Ed.; P. Blakiston's Son & Company: London, UK, 1945; pp. 126–129.
50. Olsen, S.R.; Cole, C.V.; Watanabe, F.S.; Dean, L.A. *Estimation of Available Phosphorus in Soils by Extraction with Bicarbonate*; Department of Agriculture: Washington, DC, USA, 1954.
51. Chapman, H.D. Cation-exchange capacity. In *Methods of Soil Analysis. Part II*; Dans, C., Black, A., Eds.; American Society of Agronomy Inc.: Madison, WI, USA, 1965; Chapter 57–58; pp. 891–903.
52. Lindsay, W.L.; Norvell, W.A. Development of ADTPA-Soil test for zinc, iron, manganese and cooper. *Soil Sci. Soc. Am. J.* **1978**, *42*, 421–428. [CrossRef]
53. Baird, R.B.; Eaton, E.D.; Rice, E.W. *Standard Methods for the Examination of Water and Waste Water*, 23rd ed.; American Public Health Association: Washington, DC, USA, 2017.
54. Fouad, M.M.; Omima, A.K.; El-Said, M. Comparative studies on flowering fruit set and yield of some olive cultivars under Giza conditions. *Egypt J. Appl. Sci.* **1992**, *7*, 630–644.
55. A.O.A.C. Association of Agriculture Chemicals. *Official Methods of Analysis of Association of Official Analytical Chemicals*, 17th ed.; Association of Agriculture Chemicals: Washington, DC, USA, 2000.
56. Di Rienzo, J.A.; Casanoves, F.; Balzarini, M.G.; Gonzalez, L.; Tablada, M.; Robledo, C.W. InfoStat versión. Group InfoStat, FCA, Universidad Nacional de Córdoba, Argentina. 2011. Available online: http://www.infostat.com.ar (accessed on 29 September 2020).
57. Zaki, M.E.; Mohamed, M.H.M.; Abd El-Wanis, M.M.; Glala, A.A.A.; Hamoda, A.H.M.; Shams, A.S. Implications of applied P-Sources with calcium super phosphate, phosphoric acid and rock phosphate, and phosphate dissolving bacteria on snap bean grown under greenhouses conditions. In Proceedings of the 5th International Conference on Biotechnology Applications in Agriculture (ICBAA), Benha University, Benha, Egypt, 8 April 2021.
58. El-zeiny, O.A.H. Effect of bio-fertilizers and root exudates of two weeds as a source of natural growth regulators on growth and productivity of bean plants (*phaseolus vulgaris* L.). *J. Agric. Biol. Sci.* **2007**, *3*, 440–446.
59. Gharib, A.A.; Shahen, M.M.; Ragab, A.A. Influence of rhizobium inoculation combined with *Azotobacter chrococcum* and *Bacillus megaterium* var phosphaticum on growth, nodulation, yield and quality of two snap bean (*Phasealus vulgaris*, L.) Cultivers. In Proceedings of the 4th Conference on Recent Technologies in Agriculture, Giza, Egypt, 3–5 November 2009; pp. 650–660.
60. Massoud, O.N.; Morsy, E.M.; Nadia, H.E. Field response of snap bean (*Phaseolus vulgaris* L.) to N_2-fixers bacillus circulans and arbuscular mycorrhizal fungi inoculation through accelerating rock phosphate and feldspar weathering. *Aust. J. Basic Appl. Sci.* **2009**, *3*, 844–852.
61. Ahmed, A.M.; Gheeth, R.H.M.; Galal, R.M. Influence of organic manures and rock phosphate combined with feldspar on growth, yield and yield components of bean (*Phaseolus vulgaris* L.). *Assiut J. Agric. Sci.* **2013**, *44*, 71–89.
62. Sabry, M.Y.; Gamal, S.R.; Salama, A.A. Effect of phosphorus sources and arbuscular mycorrhizal inoculation on growth and productivity of snap bean (*Phaseolus vulgaris* L.). *Gesunde Pflanzen.* **2017**, *69*, 139–148.
63. Bouhafa, K.; Moughli, L.; Bouabid, R.; Douaik, A.; Taarabt, Y. Dynamics of macronutrients in olive leaves. *J. Plant Nutr.* **2018**, *41*, 956–968. [CrossRef]
64. Gimenez, M.; Nieves, M.; Gimeno, H.; Martinez, J.; Martinez-Nicola, J.J. Nutritional diagnosis norms for three olive tree cultivars in superhigh density orchards. *Int. J. Agric. Nat. Resour.* **2021**, *48*, 34–44. [CrossRef]
65. Zipori, I.; Yermiyahu, U.; Ben-Gal, A.; Dag, A. Sustainable Management of Olive Orchard Nutrition: A Review. *Agriculture* **2020**, *10*, 11. [CrossRef]
66. El-Tohamy, W.A.; Ghoname, A.A.; Abou-Hussein, S.D. Improvement of pepper growth and productivity in sandy soil by different fertilization treatments under protected cultivation. *J. Appl. Sci. Res.* **2006**, *2*, 8–12.
67. Nardi, S.; Tosoni, M.; Pizzeghello, D.; Provenzano, M.R.; Cilenti, A.; Sturaro, R.; Vianello, A. A comparison between humic substances extracted by root exudates and alkaline solution. *Soil Sci. Soc. Am. J.* **2005**, *69*, 2012–2019. [CrossRef]

68. Chien, S.H.; Prochnow, L.I.; Tu, S.; Snyder, C.S. Agronomic and environmental aspects of phosphate fertilizers varying in source and solubility: An update review. *Nutr. Cycl. Agroecosyst.* **2011**, *89*, 229–255. [CrossRef]
69. Leytem, A.B.; Dungan, R.S.; Moore, A. Nutrient availability to corn from dairy manures and fertilizer in a calcareous soil. *Soil Sci.* **2011**, *176*, 426–434. [CrossRef]
70. Awad, A.A.M.; Ahmed, H.M.H. Impact of organic manure combinations on performance and rot infection of stressed-Jerusalem artichoke plants. *Egypt J. Soil Sci.* **2018**, *58*, 417–433. [CrossRef]
71. Awad, A.A.M.; Ahmed, H.M.H. Response of Jerusalem artichoke plants grown under saline calcareous soil to application of different combined organic manures. *Egypt J. Soil Sci.* **2019**, *59*, 117–130. [CrossRef]
72. Pramanik, K.; Bera, A.K. Effect of seedling age and nitrogen fertilizer on growth, chlorophyll content, yield and economics of hybrid rice. *Intl. J. Agron. Plant Prod.* **2013**, *4*, 34893499.
73. Tari, D.B.; Daneshian, J.; Amiri, E.; Rad, A.H.S.; Moumeni, A. Investigation chlorophyll condition at different nitrogen fertilization methods in rice by applied mathematics relations. *J. Sci. Res.* **2013**, *14*, 1056–1058.
74. Pradhan, P.P.; Dash, A.K.; Panda, N.; Samant, P.K.; Mishra, A.P. Effect of graded doses of neem coated urea on productivity of low land rice. *ORYZA Int. J. Rice* **2019**, *56*, 68–74. [CrossRef]
75. Awad, A.A.M.; Ahmed, A.I.; Abd Elazem, A.H.; Sweed, A.A.A. Mitigation of $CaCO_3$ influence on *Ipomoea batatas* plants using *Bacillus megaterium* DSM 2894. *Agronomy* **2022**, *12*, 1571. [CrossRef]
76. Sarrwy, S.M.A.; Enas, A.M.; Hassan, H.S.A. Effect of Foliar Sprays with Potassium Nitrate and Mono-potassium Phosphate on leaf mineral contents, fruit set, yield and fruit quality of picual olive trees grown under sandy soil conditions. *Am. Eurasian J. Agric. Environ. Sci.* **2010**, *8*, 420–430.
77. Ben Mimoun, M.; Loumi, O.; Ghrab, M.; Latiri, K.; Hellali, R. Foliar potassium application on olive tree. In Proceedings of the Regional Workshop on Potassium and Fertigation Development in West Asia and North Africa, Rabat, Morocco, 24–28 November 2004.
78. Ramezani, S.; Shekafandeh, A. Role of gibberellic acid and zincsulphate in increasing size and weight of olive fruit. *Afr. J. Biotechnol.* **2009**, *8*, 6791–6794.
79. Gobarah, M.E.; Mohamed, M.H.; Tawfik, M.M. Effect of phosphorus fertilizer and foliar spraying with zinc on growth, yield and quality of groundnut under reclaimed sandy soils. *J. Appl. Sci. Res.* **2006**, *2*, 491496.
80. El-Habbasha, S.F.; Kandil, A.A.; Abu-Hagaza, N.S.; Abdel-Haleem, A.K.; Khalafallah, M.A.; Behiary, T.G. Effect of phosphorus levels and some biofertilizers on dry matter, yield and yield attributes of groundnut. *Bull. Fac. Agric. Cairo Univ.* **2005**, *56*, 237–252.
81. Kumar, S.; Dhar, S.; Kumar, A.; Kumar, D. Yield and nutrient uptake of maize (*Zea mays*)–wheat (*Triticum aestivum*) cropping system as influenced by integrated potassium management. *Indian J. Agron.* **2015**, *60*, 511–515.

Article

Enhancement of Rose Scented Geranium Plant Growth, Secondary Metabolites, and Essential Oil Components through Foliar Applications of Iron (Nano, Sulfur and Chelate) in Alkaline Soils

Amany E. El-Sonbaty [1], Saad Farouk [2,*], Hatim M. Al-Yasi [3], Esmat F. Ali [3,*], Atef A. S. Abdel-Kader [4] and Seham M. A. El-Gamal [5]

1   Soil, Water and Environment Research Institute, Agriculture Research Centre, El-Gama St., Giza 3725004, Egypt
2   Agricultural Botany Department, Faculty of Agriculture, Mansoura University, Mansoura 35516, Egypt
3   Department of Biology, College of Science, Taif University, Taif 21944, Saudi Arabia
4   Department of Medicinal and Aromatic Plants, Horticulture Research Institute, Agricultural Research Center, Giza 12619, Egypt
5   Medicinal and Aromatic Plants Research Department, Horticulture Research Institute, Agricultural Research Center, Giza 12619, Egypt
*   Correspondence: gadalla@mans.edu.eg (S.F.); a.esmat@tu.edu.sa (E.F.A.); Tel.: +20-102-172-1645 (S.F.)

Citation: El-Sonbaty, A.E.; Farouk, S.; Al-Yasi, H.M.; Ali, E.F.; Abdel-Kader, A.A.S.; El-Gamal, S.M.A. Enhancement of Rose Scented Geranium Plant Growth, Secondary Metabolites, and Essential Oil Components through Foliar Applications of Iron (Nano, Sulfur and Chelate) in Alkaline Soils. *Agronomy* **2022**, *12*, 2164. https://doi.org/10.3390/agronomy12092164

Academic Editors: Christos Noulas, Shahram Torabian and Ruijun Qin

Received: 21 August 2022
Accepted: 7 September 2022
Published: 11 September 2022

Publisher's Note: MDPI stays neutral with regard to jurisdictional claims in published maps and institutional affiliations.

Abstract: Iron (Fe) deficiency exists as a widespread nutritional disorder in alkaline and calcareous soils; therefore, Fe-enriching strategies may be used to overcome this issue. Field experiments were conducted with a randomized complete design with three replicates for evaluating the effectiveness of iron oxide nanoparticles (Fe-NPs) against traditional Fe compounds (sulfate or chelate), which have various shortcomings on Rose-scented geranium (RSG) herb in terms of plant growth, phytopharmaceuticales, essential oil (EO), and its constituents. Supplementation of Fe-sources considerably improved RSG plant growth and EO yield in the 1st and 2nd cut throughout the two seasons over non-treated control plants. A total of 11 compounds of RSG-EO were identified; the main constituents were citronellol, geraniol, and eugenol. The results indicate that EO composition was significantly affected by Fe-sources. Amendments of Fe-sources considerably augmented photosynthetic pigments, total carbohydrates, nitrogen, phosphorous, potassium, iron, manganese, zinc, phenols, flavonoids, and anthocyanin. Commonly, Fe-NPs with humic acid (Fe-NPs-HA) supplementation was superior to that of traditional sources. The highest values were recorded with spraying Fe-NPs-HA at 10 mg L^{-1} followed by 5 mg L^{-1}, meanwhile, the lowest values were recorded in untreated control plants. Current findings support the effectiveness of nanoparticle treatment over Fe-sources for improving growth and yield while also being environmentally preferred in alkaline soil. These modifications possibly will be applicable to EO quality and its utilization in definite food and in medical applications.

Keywords: chlorophyll; essential oil; nano-iron; phytopharmaceutical; rose-scented geranium; yield

1. Introduction

Rose-scented geranium (RSG, *Pelargonium graveolens* L. Her. ex Ait. 'Synonym *Prasophyllum roseum* Willd.'; Geraniaceae) is a highly valued perennial aromatic shrub worldwide [1]. The chief RSG production takes place in China and the Middle East, i.e., Egypt [2]. Its EO is extensively used in the perfumery, cosmetic, and aromatherapy industries [1,3,4]. Additionally, they are becoming increasingly popular for several human disorders, i.e., relieving dysentery, cancer, sterility, urinary stones, and liver complications [5,6]. The main constituents of RSG-EO are citronellol (19.28–40.23%), geraniol (6.45–18.40%), linalool (3.96–12.90%), iso-menthone (5.20–7.20%), citronellyl formate (1.92–7.55%), Guaia-6,9-diene

(0.15–4.40%), and bits of more than 100 constituents [1,4]. Accordingly, EO composition is strongly affected by environmental factors and micronutrients including iron [7,8].

Iron (Fe) represents the 4th supreme plentiful element in the earth's crust, which participates in several species' physio-biochemical pathways [9,10]. It is a co-factor for approximately 140 enzymes elaborated in photosynthesis, gas exchange, nitrogen fixation, and nucleic acid assimilation [11,12]. It is also involved in chlorophyll biosynthesis, chloroplast development, and electron transport systems [13,14]. Iron deficiency (FDS) is a widespread threat affecting 30–50% of cultivated alkaline soils in dry regions, i.e., Egyptian soil [15,16]. Considering the soil–plant–animal–human food chain FDS not only affects plant growth and development but can also accelerate anemia in animals and humans [1]. Therefore, usage of the proper amount and forms of Fe is a prerequisite to extra studies, to lessen FDS, and to increase nutrient-use efficiency. Presently, several products were applied to overcome FDS. The EU Directive No. 2003/2003 [17,18] comprises chelates i.e., ethylene diaminetetraacetic acid (Fe-EDTA) and ethylenediamine-*N*, *N'*-biso-hydroxyphenyl acetic acid (Fe-o,o-EDDHA) complexes; and inorganic salts as a promising method for improving Fe uptake and lessens Fe-chlorosis. The effectiveness of inorganic and chelated Fe fertilizers in mitigating FDS is exceedingly variable depending on their solubility, constancy, infiltration capacity via leaf cuticle and translocation into the plant tissues [19,20]. The usage of Fe chelates does not represent a viable approach for agronomists to avoid Fe chlorosis as a result of the excessive cost and ecological hazards [21]. Furthermore, most of these chelates are recalcitrant products in soils and waters, and there has been developing anxiety recently about the ecological threat of their amendment to soils [22].

Recently, there has been a thrust to develop innovative nanoparticle (NP) fertilizer formulations including iron nano-oxide (Fe-NPs), for reducing the quantity of conventional fertilizers owing to (1) their unique physical and chemical attributes (small size, huge surface area, pureness, and steadiness), and (2) the interface amongst nanoparticles and biomolecules possibly will provoke metabolic pathways in treated plants [8,18,23]. The stimulating impact of Fe-NPs on the growth and economic yield of different herbs has been reported previously [8,23,24]. In this regard, El-Khateeb et al. [8] on sweet marjoram found that Fe-NPs foliar spraying augmented plant growth, chlorophyll concentration, carbohydrates, EO %, and yield as well as their constituents. Nejad et al. [25] found that Fe-spraying increased the photosynthetic pigments, phenols, and EO % of RSG plants. Gutierrez-Ruelas et al. [18] recorded that Fe-NPs spraying increased green bean plant biomass, total chlorophyll, and Fe content as well as nitrate reductase activity.

However, it is unclear how Fe-NP supplementation affects RSG plant development and some biochemical characteristics when used in place of conventional Fe-sources. As a result, the main goal of the current study is to determine the effects of Fe-sources (chelate, sulfur, and nano) on the growth of RSG, EO content, and their constituents, as well as their phytochemicals. We hypothesized that various Fe sources have varying effects on plant growth, EO yield, and composition as well as phytopharmaceuticals production. As a novel Fe source, Fe-NPs were also very successful in eliciting the accumulation of phytopharmaceuticals, as well as boosting EO yield and plant antioxidant activity.

2. Materials and Methods

2.1. Assimilation of Fe-NPs

The synthesis of magnetic iron oxide nanoparticles (Fe-NPs) was created with an eco-friendly adapted scheme [26]. The co-precipitation method synthesized the Fe-NPs in situ, which is a classical method for Fe_3O_4 generation. Concisely, 6.1 g of ferric chloride was dissolved in 100 mL of distilled water, subsequently, the addition of an aliquot of concentrated HCl to evade $Fe(OH)_3$ precipitation, afterward 4.2 g of $FeSO_4 \cdot 7H_2O$ were dissolved in a mix, and heated to 90 °C, then 10 mL of NH_4OH (25%) was poured quickly, and pH of the solution was sustained at 10. The mixture was stirred at 90 °C for 30 min and then cooled to lab temperature. The black substance was collected via centrifugation at $600 \times g$, and then washed with ethanol and distilled water.

2.2. Characterization of Fe-NPs

The dimension and shape of Fe-NPs were detected by transmission electron microscopy (TEM, JEOL Ltd. Tokyo, Japan). The TEM samples were prepared via dropping solution on a carbon-coated copper grid and then exposed to the infra-light for 30 min (Okenshoji Co., Ltd., Tokyo, Japan microgrid B). The micrograph was examined by JEOL-JEM 6510 at 70 kV in the RCMB, Mansoura University, Egypt.

2.3. Experimental Location, Climate Data, and Soil Properties

Two field experiments were done at a private farm in El-Serw City (31°14′19.21″ N, 31°39′13.64″ E; 16 m ASL), Damietta, Egypt, during the 2018 and 2019 seasons for assessing the response of RSG plant growth, yield, and EO content to foliar applications of Fe-sources. Physical-chemical examination of the soil surface (0–60 cm) was employed before transplanting [27]. The soil texture was clay, and its properties were recorded in Table 1. Diurnal experimental site ecological information involved temperature, solar radiation, relative humidity, and wind speed of the 1st and 2nd seasons as presented in Supplementary Materials Table S1.

Table 1. Physical and chemical analyses of the experimental soil in two seasons.

Soil Properties		Values	
		1st Season	**2nd Season**
Particle size distribution (%)	Sand (%)	21.00	21.19
	Silt (%)	35.92	34.82
	Clay (%)	43.08	44.08
Some physical and chemical trials	Electrical conductivity (dSm^{-1})	4.070	4.060
	pH (soil paste)	7.630	7.700
	Calcium carbonate (%)	3.730	3.810
	Nitrogen (mg kg^{-1} soil)	20.32	21.03
	Phosphorus (mg kg^{-1} soil)	16.72	17.63
Cations (meq 100 g^{-1} soil)	Calcium	2.000	4.000
	Magnesium	11.33	12.12
	Sodium	2.740	2.720
	potassium	2.060	2.090
Anions (meq 100 g^{-1} soil)	Carbonate	0.000	0.000
	Bicarbonate	0.370	0.360
	Chloride	4.690	4.650
	sulfate	5.630	5.690

2.4. Experimental Layout

The experimental soil was mechanically plowed twice prior to transplantation until the soil surface was steady and established in the plots. Uniform seedlings of 25–30 cm length (from the Dept. of Medicinal and Aromatic Plants, Ministry of Agric., Egypt) were individually transplanted on 1st March, during the 2018 and 2019 seasons, in 3 × 3.5 m plots, rows with 60 cm apart and 60 cm amongst the seedlings. In both seasons, the plants were received the recommended doses of mineral fertilizers (ammonium sulfate '20.5%', calcium superphosphate '15.5%', and potassium sulfate '52%' at 200, 100, and 55 kg/fed. '4200 m^{2}', correspondingly) before planting and once first cut in both seasons. Entirely agricultural practices of plants were carried out following the endorsements of the Ministry of Agriculture, Egypt. The design of the experiment was completely randomized that contained 11 treatments at three replicates, and they are displayed in Table 2. The preliminary study provided the basis for choosing this concentration. The Fe-forms were sprayed directly on the plants four times at 45 and 60 days (for the 1st cut), and 135 and 150 days (for the 2nd cut) from transplanting (15 days prior to flowering and at the start of the flowering stage in both cuts).

Table 2. The experimental treatments and their identifications.

No.	Treatments	Abbreviation
1	Control (Spraying with tap water)	T1
2	Spraying with 5 mg L^{-1} Fe-NPs	T2
3	Spraying with 10 mg L^{-1} Fe-NPs	T3
4	Spraying with 5 mg L^{-1} Fe-NPs with humic (Fe-NPs-HA)	T4
5	Spraying with 10 mg L^{-1} Fe-NPs with humic (Fe-NPs-HA)	T5
6	Spraying with 100 mg L^{-1} ferric sulfate	T6
7	Spraying with 200 mg L^{-1} ferric sulfate	T7
8	Spraying with 100 mg L^{-1} EDDHA	T8
9	Spraying with 200 mg L^{-1} EDDHA	T9
10	Spraying with 100 mg L^{-1} EDTA	T10
11	Spraying with 200 mg L^{-1} EDTA	T11

2.5. Measurements and Data Collection

Plants were harvested (cuts) 10 cm above the soil two times at full bloom (after 90 and 180 days from transplanting) in each season for determining growth characteristics (plant height 'cm', branches number/plant, shoot fresh and dry weights 'g/plant') and EO (%, yield/plant, yield/fed.), meanwhile both cuts in the second season was used for determining photosynthetic pigments, ions, phytopharmaceuticals, and EO composition.

2.6. Determination of Essential Oil

Using a modified Clevenger apparatus for three hours, the EO was hydro-distilled from the air-dried plants that had been in the shade for 48 h [28]. After distillation, the EO was dried by a glass separator, filtered two times, kept in the fridge at 4 °C, and preserved in dark closed bottles for preventing light and oxygen exposure. EO % = (EO volume/shoot fresh weight) × 100. The EO yield (mL/plant) was calculated following the current equation; EO yield = shoot fresh weight (g) × EO%.

The EO components were recognized, with a Varian Chrompack CP-3800 gas chromatograph (Varian Company, California, USA) with a mass detector (4000 GC-MS/MS). Helium served as the gas carrier at a flow rate of 2 mL min^{-1} with a linear velocity of 32 cm s^{-1}. The flame ionization detector temperature was 265 °C and the injector temperature was 250 °C. Detection of the constituents was dependent on a judgment of their mass spectra with those of a computer library or with realistic composites and validation of compound individualities was also gained via Retention index (RI) assessed regarding a homologous series of C5–C24 (n-alkanes) as designated by Adams [29].

2.7. Ion Concentration

Nitrogen (N), and phosphorus (P) were extracted and estimated [30] from the plant dry shoot. Roughly 0.2 g shoot dry mass was cautiously moved to a digestion flask with 5 mL of concentrated H_2SO_4, at 100 °C for 2 h; then, the combinations were cool for 15 min in lab temperature. An aliquot of H_2SO_4/$HClO_3$ mix was poured dropwise. Total N was assessed with the micro-Kjeldahl scheme. The outline of Cooper [31] was followed for the assessment of P alongside the phosphate standard curve. In the meantime, the potassium (K), Fe, manganese (Mn), and zinc (Zn) were extracted by acid digestion (70% nitric acid and 35% hydrochloric acid) in a Milestone MLA 1200 Mega microwave digestion device, then estimated using iCAPTM 7000 Plus Series ICP-OES (Thermo Scientific[TM], Boston, MA, USA, Boston) following Bettinelli et al. [32] protocol.

2.8. Photosynthetic Pigment

Chlorophylls and carotenoids were assessed by Lichtenthaler and Wellburn [33] procedure. Generally, 0.2 g FW from the 4th upper leaves was extracted overnight in pre-cooled methanol (96%) accompanied by 0.05% sodium bicarbonate. The optical density (OD) was read at 470, 653, and 666 nm spectrophotometrically (T60 UV–Visible spectrophotometer, Leicestershire, UK). Pheophytin (Pheo) and Chlorophyllide (Chlide) were assessed in the 4th upper leaves according to Radojevic and Bashkin [34] and Harpaz-Saad et al. [35], respectively. On the other hand, the protocol described by Sarropoulou et al. [36] was applied for the estimation of protoporphyrin (Proto), Mg-protoporphyrin (Mg-Proto), and protochlorophyllide (Pchlide).

2.9. Total Carbohydrates

The colorimetric technique designated by Zhang et al. [37] was used to estimate total carbohydrate concentrations in plant shoots using 3, 5-dinitrosalicylic acid (DNS), after extraction with hot ethanol (80%). An aliquot of shoot extract (3 mL) was mixed with 3 mL DNS reagent in a test tube, then heated in a boiling water bath for 5 min. Consequently, 40% Rochelle salt solution (1 mL) was quickly added, to the mix, and placed in a water bath at lab temperature for about 25 min., subsequently; the OD at 510 nm is recorded with a spectrophotometer (T60 UV–Visible spectrophotometer, Leicestershire, UK).

2.10. Total Phenolic Compounds, Total Flavonoids, and Total Anthocyanin

The Folin–Ciacolteu procedure was utilized spectrophotometrically (T60 UV–Visible spectrophotometer, Leicestershire, UK) to estimate the total phenolic concentration [38]. Concisely, the ethanolic plant extract was added to the Folin–Ciocalteu reagent and sodium carbonate solution (20%), homogenized, and incubated in the dark for 30 min. The OD was then measured at 650 nm. A calibration curve for gallic acid was used to estimate their concentration (mg gallic g^{-1} DW).

The technique established by Meda et al. [38] was employed to assess the total concentration of flavonoids (mg quercetin g^{-1} DW) using the aluminum chloride colorimetric scheme. An aliquot of ethanolic extract, 0.1 mL of aluminum chloride, 0.1 mL of sodium acetate, and 2.8 mL of distilled water was combined and stirred. The mixture's OD was deliberate spectrophotometrically (T60 UV–Visible spectrophotometer, Leicestershire, UK) at a wavelength of 415 nm.

Total anthocyanin concentration was determined according to the method of Abdel-Aal and Hucl [39], in which the OD of each pre-chilled acidified methanolic extract was assessed spectrophotometrically (T60 UV–Visible spectrophotometer, UK) at 530 nm. The concentration (mg 100 g^{-1} FW) was expressed as cyaniding-3-glucoside using a molar extinction coefficient of 27.900.

2.11. Statistical Analysis

The similarity of variables error variance was performed earlier in the analysis of variance (ANOVA). The outputs demonstrated that all data satisfied the uniformity to accomplish further ANOVA checks. The data acquired were exposed to one way-ANOVA at a 95% confidence level by CoHort Software, 2008 statistical package (CoHort software, 2006; Raleigh, NC, USA). The mean values of treatments were compared via Tukey's HSD-MRT test at $p \leq 0.05$. Values attended by diverse letters were significantly different at $p \leq 0.05$. The data presented are mean values $\pm$ standard error (SE). The levels of significance were denoted by * $p < 0.05$, ** at $p < 0.01$, *** $p < 0.001$ and NS, no significant.

3. Results

3.1. Magnetite Nanoparticles Characterization

By using TEM imaging, the physicochemical properties of Fe-NPs were considered (Figure 1). The images of synthesized magnetite nanoparticles with an average particle size of 9–14 nm and a large number of diffraction rings characteristic of crystalline spherical Fe-NPs. The nanoparticles used in this study have a mean diameter of 12.6 nm, suggesting that the particles can cross bio-membranes.

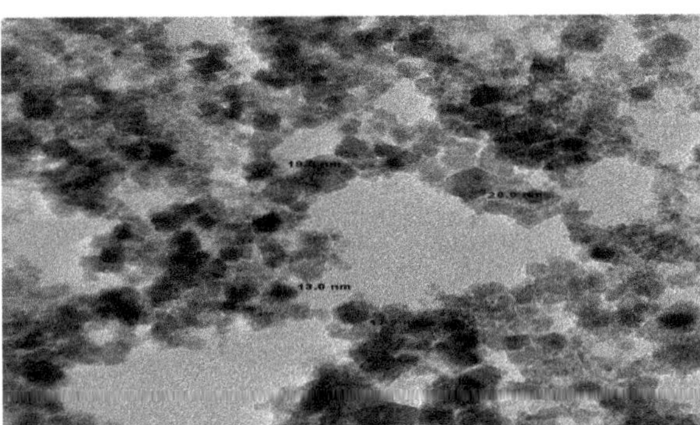

Figure 1. TEM imaging of the prepared magnetite nanoparticles revealed spherical shape of particles, with an average size of 9–14 nm.

3.2. Morphological Characterization

Data in Table 3 shows that application of Fe-sources significantly increased the growth parameters in both the 1st and 2nd cut in both seasons over control plants. The highest morphological values were significantly associated with RSG treated with Fe-NPs HA at 10 mg L^{-1}, followed by 5 mg L^{-1} Fe-NPs HA, correspondingly, and mostly, they produced equivalent effects in both seasons. Meanwhile, the lowest values were usually detected in a non-treated plant, with statistical significance.

Table 3. Effect of iron (nano, sulfate, and chelated) foliar spray on some vegetative growth parameters of Sweet Scented geranium during the 2018 and 2019 experimental seasons. Means of three replicates are presented with ± SE.

| | First Season | | | | | | | |
| | Cut 1 | | | | Cut 2 | | | |
Treatments	Plant Height (cm)	Branches No/Plant	Shoot Fresh Weight (g)	Shoot Dry Weight (g)	Plant Height (cm)	Branches No/Plant	Shoot Fresh Weight (g)	Shoot Dry Weight (g)
T1	34.6 ± 0.88 [h]	13.0 ± 0.57 [h]	617.6 ± 5.48 [k]	108.5 ± 0.94 [k]	43.3 ± 0.88 [h]	18.0 ± 0.57 [g]	838.3 ± 6.64 [k]	164.0 ± 1.36 [k]
T2	66.3 ± 0.88 [c]	30.6 ± 0.88 [c]	1179 ± 8.50 [d]	238.5 ± 1.72 [d]	75.6 ± 1.20 [c]	43.0 ± 1.15 [c]	1852 ± 7.53 [d]	438.61.78 [d]
T3	71.3 ± 0.88 [b]	33.3 ± 0.88 [bc]	1261 ± 4.61 [c]	258.1 ± 0.94 [c]	81.0 ± 1.15 [b]	46.0 ± 0.57 [bc]	2024 ± 7.83 [c]	483.6 ± 1.87 [c]
T4	75.0 ± 0.57 [ab]	35.0 ± 0.57 [ab]	1433 ± 8.14 [b]	303.6 ± 1.72 [b]	84.3 ± 1.20 [ab]	48.0 ± 1.15 [ab]	2262 ± 8.14 [b]	545.5 ± 1.96 [b]
T5	78.3 ± 0.88 [a]	38.0 ± 0.57 [a]	1560 ± 6.35 [a]	341.1 ± 1.39 [a]	89.0 ± 1.15 [a]	51.3 ± 0.88 [a]	2469 ± 6.08 [a]	616.0 ± 1.51 [a]
T6	42.6 ± 0.88 [fg]	17.0 ± 0.57 [fg]	719.3 ± 7.51 [i]	125.2 ± 1.30 [i]	49.6 ± 0.88 [fg]	20.6 ± 0.88 [fg]	1020 ± 7.83 [i]	217.5 ± 1.67 [i]
T7	39.6 ± 0.88 [g]	15.0 ± 0.57 [gh]	665.0 ± 6.08 [j]	116.5 ± 0.68 [j]	45.6 ± 0.88 [gh]	19.0 ± 0.57 [fg]	965.0 ± 4.72 [j]	198.5 ± 1.00 [j]
T8	46.3 ± 0.88 [f]	20.0 ± 0.57 [ef]	783.6 ± 6.93 [h]	137.6 ± 1.21 [gh]	53.6 ± 0.88 [ef]	23.0 ± 0.57 [f]	1123 ± 6.11 [h]	243.2 ± 1.32 [h]
T9	51.3 ± 0.88 [e]	23.0 ± 0.57 [de]	834.3 ± 6.11 [g]	148.3 ± 1.08 [g]	58.3 ± 0.88 [e]	28.6 ± 0.88 [e]	1332 ± 11.4 [g]	291.2 ± 2.49 [g]
T10	57.6 ± 0.88 [d]	24.0 ± 0.57 [d]	928.3 ± 5.78 [f]	174.1 ± 1.08 [f]	65.60.88 [d]	32.3 ± 0.88 [de]	1483 ± 7.93 [f]	336.2 ± 1.79 [f]
T11	61.3 ± 0.88 [d]	26.0 ± 0.57 [d]	990.3 ± 4.33 [e]	187.2 ± 0.81 [e]	72.0 ± 1.15 [c]	36.0 ± 0.57 [d]	1597 ± 7.05 [e]	365.8 ± 1.61 [e]
ANOVA *p*	***	***	***	***	***	***	***	***

Table 3. *Cont.*

	Second season							
	Cut 1				Cut2			
Treatments	Plant Height (cm)	Branches No/Plant	Shoot fresh weight (g)	Shoot dry weight (g)	Plant Height (cm)	Branches No/Plant	Shoot fresh weight (g)	Shoot dry weight (g)
T1	36.6 ± 0.88 [h]	15.0 ± 0.57 [f]	648.6 ± 5.23 [k]	114.0 ± 0.90 [k]	44.6 ± 0.88 [g]	19.0 ± 0.57 [f]	851.3 ± 4.94 [k]	166.8 ± 1.00 [k]
T2	67.6 ± 0.88 [c]	32.6 ± 0.88 [b]	1192 ± 4.33 [d]	241.4 ± 0.87 [d]	77.0 ± 0.57 [c]	44.6 ± 0.88 [b]	1886 ± 6.93 [d]	447.3 ± 1.64 [d]
T3	72.3 ± 0.88 [b]	34.3 ± 0.88 [ab]	1283 ± 3.60 [c]	263.0 ± 0.73 [c]	82.3 ± 0.88 [b]	47.6 ± 0.88 [ab]	2052 ± 6.38 [c]	491.0 ± 1.52 [c]
T4	75.6 ± 0.88 [b]	37.0 ± 1.15 [ab]	1457 ± 3.84 [b]	309.1 ± 0.812 [b]	86.6 ± 0.88 [ab]	50.0 ± 1.15 [a]	2282 ± 6.08 [b]	551.3 ± 1.46 [b]
T5	81.3 ± 0.88 [a]	39.0 ± 1.15 [a]	1583 ± 4.05 [a]	346.6 ± 0.88 [a]	91.0 ± 1.15 [a]	51.6 ± 1.20 [a]	2489 ± 6.42 [a]	621.7 ± 1.60 [a]
T6	44.6 ± 0.88 [fg]	18.6 ± 1.20 [ef]	734.6 ± 4.91 [e]	128.0 ± 0.85 [i]	52.6 ± 0.88 [f]	23.0 ± 0.57 [ef]	1055 ± 6.80 [i]	225.2 ± 1.45 [i]
T7	41.6 ± 0.88 [g]	16.0 ± 0.57 [ef]	688.0 ± 3.78 [j]	120.3 ± 0.66 [j]	47.6 ± 0.88 [g]	20.0 ± 0.57 [f]	890.3 ± 4.97 [j]	186.2 ± 1.06 [j]
T8	47.3 ± 0.88 [f]	20.3 ± 1.20 [de]	810.0 ± 4.35 [h]	142.5 ± 0.76 [h]	55.6 ± 0.88 [f]	25.3 ± 0.88 [e]	1154 ± 4.35 [h]	250.1 ± 0.94 [h]
T9	52.3 ± 0.88 [e]	24.0 ± 1.15 [cd]	849.3 ± 6.11 [g]	151.3 ± 1.09 [g]	61.3 ± 0.88 [e]	31.0 ± 1.15 [d]	1377 ± 7.00 [g]	301.4 ± 1.53 [g]
T10	60.3 ± 0.88 [d]	25.3 ± 0.88 [c]	952.0 ± 5.50 [f]	178.8 ± 1.03 [f]	67.6 ± 0.88 [d]	34.3 ± 0.88 [cd]	1509 ± 5.50 [f]	342.4 ± 1.24 [f]
T11	63.3 ± 0.88 [cd]	27.6 ± 0.88 [c]	1019 ± 4.91 [e]	193.0 ± 0.93 [e]	72.6 ± 0.88 [c]	37.6 ± 0.88 [c]	1624 ± 6.35 [e]	372.3 ± 1.45 [e]
ANOVA *p*	***	***	***	***	***	***	***	***

Levels of significance are represented by *** $p < 0.001$. For each parameter in the year, different letters within the column show significant differences between the treatments and control according to Tukey's HSD test at $p < 0.05$. T1, T2, T3, T4, T5, T6, T7, T8, T9, T10, and T11 are control, 5 mg L^{-1} Fe-NPs, 10 mg L^{-1} Fe-NPs, 5 mg L^{-1} Fe-NPs HA, 10 mg L^{-1} Fe-NPs HA, 100 mg L^{-1} $FeSO_4$, 200 mg L^{-1} $FeSO_4$, 100 mg L^{-1} EDDHA, 200 mg L^{-1} EDDHA, 100 mg L^{-1} EDTA, and 200 mg L^{-1} EDTA, respectively.

3.3. Essential Oil Yield

Data presented in Figure 2A–F indicate that Fe-sources supplementation significantly raised EO %, accompanied by increasing EO yield per plant and per fed. in both cuts relative to control plants (water spraying plants). The highest EO %, EO yield per plant, and EO yield per fed. were recorded by spraying Fe-NPs-HA at 10 mg L^{-1} followed by 5 mg L^{-1}, meanwhile, the lowest values were recorded in control plants. In this regard, EO% of the 1st cut ranged from 0.132 to 0.293% based on air-dry weight, meanwhile it was 0.101 to 0.192% in the 2nd cut in the first season. On the other hand, it was from 0.137 to 0.295% in the 1st cut and from 0.103 to 0.209% in the second cut, respectively, in the second season.

Regarding EO yield per plant and per fed., the results showed that Fe-sources spraying had a significant impact on EO yield at both harvests in the first and second seasons. In most cases, the yield was slightly higher in the 1st cut than in the 2nd cut in both seasons. In the first season, the EO yield per plant and fed. in the first cut was 0.819–4.577 mL/plant and 13.374–74.737 L/fed. meanwhile the 2nd cut recorded 0.849–4.740 mL/plant and 13.869–77.396 L/fed. respectively (Figure 2). Additionally, in the second season, the EO yield/plant recorded 0.888–4.676 mL/plant in the first cut and 0.876–5.201 mL/plant in the second cut. Meanwhile, the EO yield/fed. was 14.510–76.354 and 14.313–84.924 in the 1st and 2nd cut, respectively. The highest EO yield per plant and per fed. In the 1st and 2nd cut throughout both seasons was obtained in plants sprayed with 10 mg L^{-1} Fe-NPs-Ha and the lowest values were detected in untreated plants.

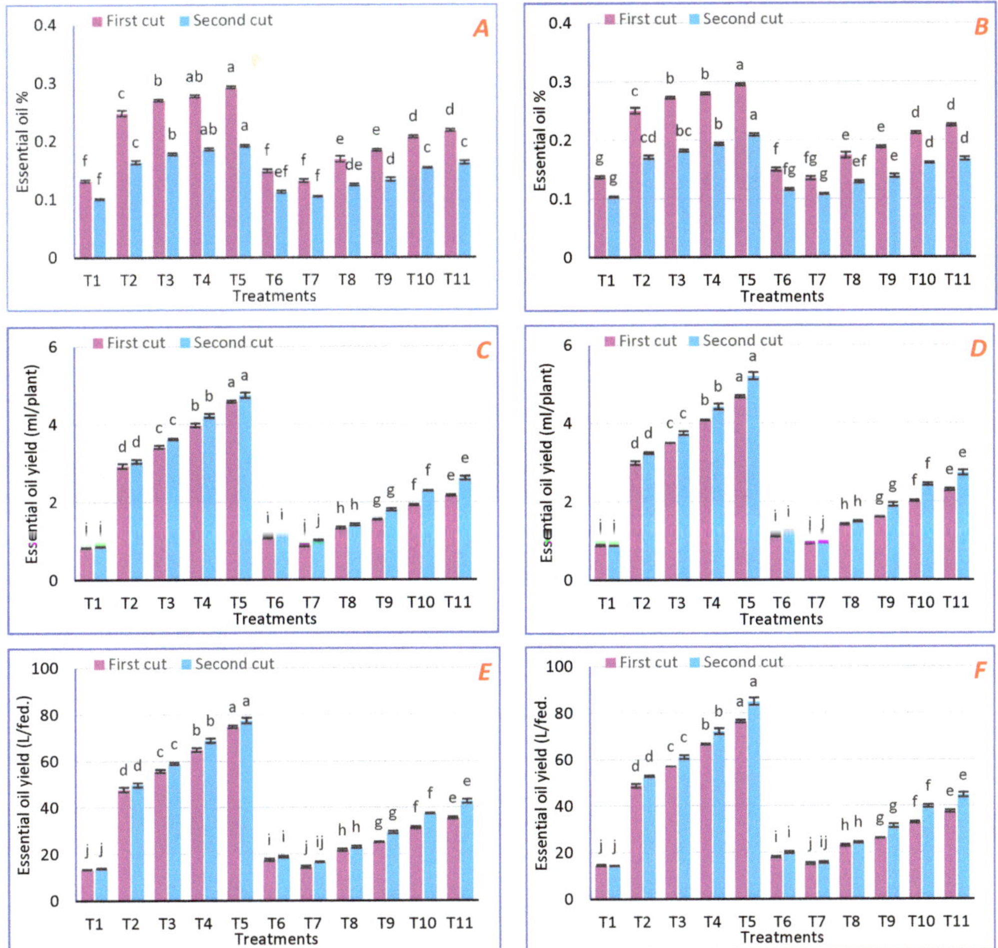

Figure 2. Effect of iron (nano, sulfate, and chelated) foliar spray on EO oil yield of Sweet Scented geranium during experimental seasons. (**A**) EO % in two cuts of the 1st season, (**B**) EO % in two cuts of the 2nd season, (**C**) EO yield (mL/plant) in two cuts of the 1st season, (**D**) EO yield (mL/plant) in two cuts of the 2nd season, (**E**) EO yield (L/fed.) in two cuts of the 1st season, (**F**) EO yield (L/fed.) in two cuts of the 2nd season. Means of three replicates are presented with ± SE. For each parameter in the year, different letters within the column show significant differences between the treatments and control according to Tukey's HSD test at $p < 0.05$. T1, T2, T3, T4, T5, T6, T7, T8, T9, T10, and T11 are control, 5 mg L^{-1} Fe-NPs, 10 mg L^{-1} Fe-NPs, 5 mg L^{-1} Fe-NPs HA, 10 mg L^{-1} Fe-NPs HA, 100 mg L^{-1} FeSO$_4$, 200 mg L^{-1} FeSO$_4$, 100 mg L^{-1} EDDHA, 200 mg L^{-1} EDDHA, 100 mg L^{-1} EDTA, and 200 mg L^{-1} EDTA, respectively.

3.4. Chemical Composition of Essential Oils

Rose-scented geranium EO was slightly light green with a 0.889 g/mL density. The data belonging to qualitative and quantitative constituents of EO, collected from the 1st and 2nd cuts during the 2019 season of RSG herbs subjected to Fe-sources foliar application were identified (Tables 4 and 5). In total, 11 constituents were detected in EO accounting for 86.04% and 91.55% of the total EO in the 1st and 2nd cut respectively. A comparison of the entire set of EO analytical data showed significant variations in the EO's qualitative and quantitative composition as a result of the use of Fe-sources.

Table 4. Effect of iron (nano, sulfate, and chelated) foliar spray on essential oil active constituent's retention time (RT) and percentage (area %) of Sweet Scented geranium in the first cut during the 2019 experimental season. Means of three replicates are presented with ± SE.

Treatments	α–Pinene		Myrcene		Isomenthone		Linalool		Citronelyl Formate		Geranyl Formate		Citronelol		Geraniol		Geranyl Butrate		Eugenol		β-Caryophyllene		Unknown Constituents	C/G Ratio
	RT	Area%	RT	Area%	RT	Area%	RT	Area%	RT	Area%	RT	Area%	RT	Area%	RT	Area%	RT	Area%	RT	Area%	RT	Area%	Area%	
T1	2.10	0.36	3.81	1.46	5.12	4.29	5.34	4.21	6.06	7.92	6.83	8.00	7.48	21.43	8.22	21.02	8.50	1.66	10.50	13.23	11.43	2.46	13.96	0.933
T2	2.03	0.35	3.72	0.82	5.01	3.99	5.25	5.98	5.93	5.91	6.70	5.33	7.33	22.42	8.07	19.09	8.35	9.23	10.31	8.26	10.95	4.20	14.42	0.817
T3	2.18	0.54	3.89	0.73	5.20	5.99	5.41	5.21	6.12	8.75	6.87	7.78	7.50	25.29	8.23	23.79	8.50	1.75	10.48	10.51	11.09	1.38	8.28	1.005
T4	1.80	0.53	3.45	1.04	4.71	6.60	4.95	8.01	5.62	9.11	6.37	7.04	6.98	25.58	7.71	23.05	7.98	1.15	9.90	6.21	10.33	1.49	10.19	1.107
T5	2.77	1.37	4.20	1.28	5.36	4.00	5.57	5.09	6.29	5.59	7.07	7.41	7.71	24.93	8.47	27.58	8.98	1.70	10.70	10.01	11.30	2.25	3.48	0.814
T6	2.24	0.87	3.99	0.36	5.33	4.72	5.55	6.38	6.25	5.58	7.03	5.48	7.63	18.29	8.42	30.21	8.67	2.57	10.69	8.61	11.32	2.75	14.18	0.621
T7	2.24	0.59	4.19	0.54	5.32	4.20	5.54	6.01	6.26	6.15	7.04	7.55	7.64	19.15	8.41	26.56	8.68	3.07	10.70	11.56	11.33	5.01	9.51	0.678
T8	2.13	0.13	3.53	0.60	4.62	5.50	4.88	5.41	5.58	8.72	6.34	7.54	7.00	22.53	7.75	26.37	7.99	1.53	9.91	8.67	10.52	2.58	9.92	0.882
T9	1.73	0.50	3.34	0.45	4.64	4.94	4.90	5.94	5.60	7.87	6.36	5.88	7.02	20.54	7.80	29.54	8.03	1.64	9.93	9.79	10.50	1.99	10.92	0.766
T10	2.06	0.30	3.83	0.12	5.16	3.36	5.41	6.03	6.15	7.09	6.93	9.14	7.59	23.89	8.37	27.73	8.87	2.35	10.61	10.82	11.55	1.90	7.27	0.787
T11	2.05	0.72	3.49	0.19	5.12	4.90	5.36	6.55	6.06	5.43	6.84	6.17	7.46	21.71	8.25	36.38	8.74	0.43	10.47	6.47	11.08	1.43	9.12	0.624

T1, T2, T3, T4, T5, T6, T7, T8, T9, T10, and T11 are control, 5 mg L^{-1} Fe-NPs, 10 mg L^{-1} Fe-NPs, 5 mg L^{-1} Fe-NPs HA, 10 mg L^{-1} Fe-NPs HA, 100 mg L^{-1} FeSO$_4$, 200 mg L^{-1} FeSO$_4$, 100 mg L^{-1} EDDHA, 200 mg L^{-1} EDDHA, 100 mg L^{-1} EDTA, and 200 mg L^{-1} EDTA, respectively.

Table 5. Effect of iron (nano, sulfate, and chelated) foliar spray on essential oil active constituent's retention time (RT) and percentage (area %) of Sweet Scented geranium in the second cut during the 2019 experimental season. Means of three replicates are presented with ± SE.

Treatments	α-Pinene		Myrcene		Isomenthone		Linalool		Citronelyl Formate		Geranyl Formate		Citronelol		Geraniol		Geranyl Butrate		Eugenol		β-Caryophyllene		Unknown Constituents	C/G Ratio
	RT	Area %	RT	Area %	RT	Area %	RT	Area %	RT	Area %	RT	Area %	RT	Area %	RT	Area %	RT	Area %	RT	Area %	RT	Area %	Area %	
T1	2.06	0.62	3.80	1.09	5.19	6.20	5.40	5.11	6.15	9.35	6.87	4.34	7.63	35.55	8.30	17.28	8.79	0.85	10.52	8.36	11.08	2.85	8.45	2.029
T2	2.11	0.48	3.88	0.77	5.22	6.66	5.44	6.31	6.16	8.75	6.91	6.27	7.55	26.07	8.28	24.41	8.54	1.83	10.50	5.25	11.11	0.76	12.44	1.071
T3	2.05	0.38	3.75	0.35	4.93	4.33	5.19	12.78	5.86	6.99	6.63	2.76	7.27	28.93	8.02	22.63	8.51	2.02	10.22	5.20	10.86	1.17	12.46	1.307
T4	2.03	0.34	4.12	0.73	5.52	5.36	5.72	5.53	6.46	8.56	7.23	5.63	7.92	34.46	8.63	17.55	8.92	2.81	10.95	9.63	11.60	3.27	6.13	2.809
T5	2.27	1.31	4.07	1.22	5.48	5.59	5.67	3.47	6.44	8.53	7.20	3.19	7.93	32.19	8.62	14.06	8.91	2.15	10.97	10.24	11.61	2.92	15.13	2.098
T6	2.26	0.42	3.99	1.03	5.39	4.20	5.56	2.69	6.35	7.31	7.07	3.09	7.86	32.94	8.51	12.50	8.81	3.61	10.85	13.33	11.43	2.46	16.42	2.096
T7	2.35	0.32	4.08	0.95	5.42	4.53	5.58	3.33	6.37	8.51	7.04	3.55	7.81	36.27	8.47	14.96	8.75	2.56	10.72	11.41	11.28	3.30	10.31	2.125
T8	2.15	0.30	3.87	0.79	5.22	4.98	5.39	2.48	6.14	7.63	6.85	3.22	7.61	34.50	8.23	11.28	8.54	2.95	10.53	13.23	11.44	4.23	14.41	2.478
T9	2.16	0.68	3.88	1.31	5.21	5.71	5.40	3.09	6.12	7.69	6.86	3.18	7.53	31.44	8.21	15.28	8.51	2.13	10.50	12.21	11.09	4.48	12.8	1.896
T10	2.04	1.07	3.47	1.41	4.62	6.98	4.86	2.76	5.55	8.04	6.29	4.91	6.96	36.97	7.63	11.89	7.92	1.88	9.85	7.78	10.80	2.49	13.82	2.409
T11	2.17	0.52	3.81	0.43	5.10	4.51	5.30	3.87	6.02	7.69	6.73	4.50	7.42	26.53	8.14	22.78	8.39	2.66	10.31	10.64	11.17	2.56	13.31	1.142

T1, T2, T3, T4, T5, T6, T7, T8, T9, T10, and T11 are control, 5 mg L^{-1} Fe-NPs, 10 mg L^{-1} Fe-NPs, 5 mg L^{-1} Fe-NPs HA, 10 mg L^{-1} Fe-NPs HA, 100 mg L^{-1} FeSO$_4$, 200 mg L^{-1} FeSO$_4$, 100 mg L^{-1} EDDHA, 200 mg L^{-1} EDDHA, 100 mg L^{-1} EDTA, and 200 mg L^{-1} EDTA, respectively.

Citronellol and geraniol were the main ingredients of RSG-EO with treatments in the 1st cut, accounting for 18.29–25.58% and 19.09–36.88% of the total. There were also moderate amounts of eugenol (6.21–13.23%), geranyl formate (5.33–9.14%), citronelyl formate (5.43–9.11%), linalool (4.21–8.01%), and isomenthone (3.36–6.60%), as well as very variable amounts of α-pinene (0.13–1.37%), myrcene (0.12–1.46%), geranyl butyrate (0.43–9.23%), and β-caryophyllene (1.38–5.01%). According to Table 4's findings, 5 mg L^{-1} Fe-NPs-HA was used to produce the maximum levels of citronellol, citronely formate, linalool, and isomenthone. Meanwhile, the application of 10 mg L^{-1} Fe-NPs-HA, 100 mg L^{-1} EDTA, 200 mg L^{-1} EDTA, and 200 mg L^{-1} $FeSO_4$ correspondingly resulted in the greater amount of α–pinene, geranyl formate, geraniol, and β–caryophyllene.

Citronellol (26.07–36.97%) and geraniol (11.28–24.41%) made up the majority of RSG-EO in the second cut with all treatments (Table 5). There were also moderate amounts of eugenol (5.20–13.33%), geranyl formate (2.76–6.27%), citronelyl formate (6.99–9.35%), linalool (2.48–12.78%), isomenthone (4.20–6.98%), and very variable amounts of α-pinene (0.30–1.31%), myrcene (0.35–1.41%), geranyl butyrate (0.85–3.61%), and β-caryophyllene (0.76–4.48%). The results in Table 4 demonstrate that 5 mg L^{-1} Fe-NPs were necessary to produce the greatest amount of geranyl formate and geraniol. Meanwhile, myrcene, isomenthone, citronellol (100 mg L^{-1} EDTA), geranyl butyrate, eugenol (100 mg L^{-1} $FeSO_4$), and β-caryophyllene (200 mg L^{-1} EDDHA) are present in larger concentrations.

Citronellol (C), geraniol (G), and their esters are the quality features in RSG-EO. Different C/G ratio was established in RSG herbs at the 1st and 2nd cut (Tables 4 and 5). In the 1st cut, the C/G ratio (from 0.621 to 1.107), additionally, the maximum C/G ratio (1.107) was recorded in T4 after that T3 (1.005) as compared with T1 (0.933). Similarly, in the 2nd cut, the C/G ratio varied from 1.071 to 2.809, with the maximum C/G ratio documented in T4 (2.809) followed by T8 (2.478) relative to T1 (2.029).

3.5. Measurement of Chlorophyll and Its Assimilation and Chlorophyll Precursor

Foliar spraying of Fe-forms significantly improved total chlorophyll and carotenoid concentrations in RSG leaves above the control plants. It is observed from the data also that Fe-NPs in special with humic acid were most effective than other Fe-forms. The greatest chlorophyll and carotenoid concentrations were obtained after 10 mg L^{-1} Fe-NPs-HA spraying, which increased by 136 and 70% in the first cut and by 118 and 98% in the second cut respectively, over control plants (Table 6).

Table 6 shows that application of Fe-sources especially 10 mg L^{-1} Fe-NPs-HA significantly increased Pheo, Achl a, Chl a/Chlide, and Chl b/Chlide comparative to non-treated herbs. Additionally, Table 6 designates that porphyrin intermediate assimilation (Mg-proto, proto, and Pchlide) was considerably decreased by Fe-sources supplementation.

3.6. Measurement of Ion Levels

Data existing in Table 7 display that Fe sources supplementation significantly amplified the level of ions in plant shoots in both cuts over untreated control plants. Additionally, the data also indicate that the usage of nano-forms of iron was superior to traditional sources in increasing the ion level on plant shoots. The greatest values of nitrogen (3.33 and 3.97%), phosphorous (0.222 and 0.222%), potassium (2.07 and 2.45%), iron (443 and 534 mg g^{-1}), manganese (89.7 and 43 mg g^{-1}), and zinc (87 and 89.7 mg g^{-1}) in the first cut and second cut, respectively, were recorded when plant sprayed twice with 10 mg L^{-1} Fe-NPs-HA relative to other treatments or control plants.

Table 6. Effect of iron (nano, sulfate, and chelated) foliar spray on chlorophyll of Sweet Scented Geranium in the first and second cuts during the 2019 experimental season. Means of three replicates are presented with $\pm$ SE.

Treatments	Total Chlorophyll (mg g^{-1} FW)	Total Carotenoids (mg g^{-1} FW)	Chl A (mg g^{-1} FW)	Pheo A (mg g^{-1} FW)	Chl a/Child	Chl b/Child	Mg Proto (µg g^{-1} FW)	Proto (µg g^{-1} FW)	Pchilde (mg g^{-1} FW)
				First Cut					
T1	0.862 ± 0.031 [g]	0.190 ± 0.006 [f]	0.530 ± 0.079 [e]	0.317 ± 0.014 [b]	0.940 ± 0.023 [f]	0.895 ± 0.036 [c]	0.274 ± 0.015 [a]	0.461 ± 0.004 [a]	0.977 ± 0.005 [a]
T2	1.679 ± 0.038 [cd]	0.313 ± 0.010 [a–c]	1.148 ± 0.027 [ab]	0.476 ± 0.034 [a]	1.926 ± 0.062 [c]	1.677 ± 0.066 [b]	0.209 ± 0.001 [bc]	0.293 ± 0.001 [f]	0.542 ± 0.033 [e]
T3	1.810 ± 0.052 [bc]	0.317 ± 0.008 [ab]	1.204 ± 0.022 [a]	0.481 ± 0.004 [a]	2.281 ± 0.020 [b]	2.226 ± 0.092 [a]	0.188 ± 0.002 [cd]	0.265 ± 0.001 [g]	0.538 ± 0.009 [e]
T4	1.955 ± 0.055 [ab]	0.319 ± 0.008 [ab]	1.217 ± 0.075 [a]	0.508 ± 0.021 [a]	2.428 ± 0.039 [ab]	2.328 ± 0.127 [a]	0.175 ± 0.002 [cd]	0.244 ± 0.001 [h]	0.530 ± 0.008 [e]
T5	2.038 ± 0.008 [a]	0.323 ± 0.001 [a]	1.248 ± 0.091 [a]	0.538 ± 0.016 [a]	2.577 ± 0.014 [a]	2.355 ± 0.068 [a]	0.160 ± 0.001 [e]	0.226 ± 0.001 [i]	0.503 ± 0.018 [e]
T6	1.116 ± 0.008 [f]	0.237 ± 0.031 [d–f]	0.830 ± 0.016 [cd]	0.436 ± 0.035 [ab]	0.987 ± 0.026 [ef]	0.923 ± 0.041 [c]	0.223 ± 0.002 [b]	0.314 ± 0.001 [c]	0.785 ± 0.027 [bc]
T7	0.957 ± 0.005 [g]	0.214 ± 0.005 [ef]	0.767 ± 0.010 [de]	0.429 ± 0.035 [ab]	0.975 ± 0.035 [f]	0.878 ± 0.082 [c]	0.253 ± 0.002 [a]	0.354 ± 0.001 [b]	0.815 ± 0.035 [b]
T8	1.256 ± 0.025 [f]	0.255 ± 0.011 [c–e]	0.892 ± 0.053 [b–d]	0.435 ± 0.017 [ab]	1.157 ± 0.037 [e]	1.028 ± 0.045 [c]	0.217 ± 0.001 [b]	0.303 ± 0.001 [de]	0.686 ± 0.028 [cd]
T9	1.261 ± 0.015 [f]	0.263 ± 0.004 [be]	0.936 ± 0.010 [b–d]	0.453 ± 0.032 [a]	1.698 ± 0.037 [d]	1.544 ± 0.046 [b]	0.216 ± 0.002 [b]	0.305 ± 0.001 [d]	0.580 ± 0.036 [de]
T10	1.448 ± 0.003 [e]	0.288 ± 0.005 [a–d]	1.017 ± 0.055 [a–d]	0.476 ± 0.018 [a]	1.725 ± 0.028 [d]	1.552 ± 0.051 [b]	0.213 ± 0.002 [bc]	0.299 ± 0.001 [d–f]	0.562 ± 0.011 [e]
T11	1.562 ± 0.007 [de]	0.298 ± 0.004 [a–c]	1.086 ± 0.032 [a–d]	0.469 ± 0.030 [a]	1.773 ± 0.024 [cd]	1.618 ± 0.054 [b]	0.211 ± 0.002 [bc]	0.296 ± 0.001 [ef]	0.556 ± 0.007 [e]
ANOVA *p*	***	***	***	***	***	***	***	***	***
				Second Cut					
T1	0.883 ± 0.011 [h]	0.144 ± 0.002 [g]	0.430 ± 0.000 [f]	0.333 ± 0.011 [f]	0.706 ± 0.019 [g]	0.645 ± 0.026 [h]	0.298 ± 0.002 [a]	0.415 ± 0.001 [a]	0.913 ± 0.019 [a]
T2	1.650 ± 0.008 [c]	0.263 ± 0.004 [ab]	1.054 ± 0.053 [ab]	0.561 ± 0.026 [bc]	2.213 ± 0.065 [c]	2.032 ± 0.085 [cd]	0.198 ± 0.002 [d]	0.278 ± 0.001 [e]	0.567 ± 0.018 [c]
T3	1.813 ± 0.030 [b]	0.274 ± 0.004 [ab]	1.148 ± 0.043 [a]	0.573 ± 0.025 [ab]	2.404 ± 0.057 [bc]	2.237 ± 0.106 [bc]	0.193 ± 0.001 [d]	0.269 ± 0.001 [f]	0.562 ± 0.011 [c]
T4	1.802 ± 0.010 [b]	0.283 ± 0.001 [a]	1.167 ± 0.022 [a]	0.676 ± 0.021 [a]	2.526 ± 0.002 [b]	2.400 ± 0.054 [b]	0.149 ± 0.001 [e]	0.209 ± 0.001 [g]	0.426 ± 0.017 [d]
T5	1.933 ± 0.010 [a]	0.286 ± 0.001 [a]	1.210 ± 0.068 [a]	0.676 ± 0.018 [a]	3.671 ± 0.069 [a]	3.399 ± 0.126 [a]	0.100 ± 0.001 [f]	0.140 ± 0.001 [h]	0.424 ± 0.004 [d]
T6	1.132 ± 0.024 [g]	0.179 ± 0.005 [f]	0.674 ± 0.032 [e]	0.444 ± 0.033 [e]	1.087 ± 0.038 [ef]	1.012 ± 0.052 [fg]	0.248 ± 0.003 [b]	0.349 ± 0.001 [b]	0.749 ± 0.009 [b]
T7	1.096 ± 0.013 [g]	0.163 ± 0.012 [fg]	0.667 ± 0.016 [e]	0.437 ± 0.018 [ef]	0.917 ± 0.022 [fg]	0.845 ± 0.037 [gh]	0.250 ± 0.002 [b]	0.348 ± 0.001 [b]	0.887 ± 0.004 [a]
T8	1.272 ± 0.012 [f]	0.193 ± 0.001 [ef]	0.767 ± 0.010 [de]	0.455 ± 0.009 [de]	1.142 ± 0.034 [ef]	1.032 ± 0.037 [fg]	0.248 ± 0.002 [b]	0.346 ± 0.001 [b]	0.624 ± 0.015 [c]
T9	1.398 ± 0.040 [e]	0.210 ± 0.008 [de]	0.861 ± 0.043 [cd]	0.458 ± 0.024 [c–e]	1.302 ± 0.021 [e]	1.214 ± 0.039 [ef]	0.243 ± 0.001 [b]	0.338 ± 0.001 [c]	0.606 ± 0.005 [c]

Table 6. *Cont.*

Treatments	Total Chlorophyll (mg g^{-1} FW)	Total Carotenoids (mg g^{-1} FW)	Chl A (mg g^{-1} FW)	Pheo A (mg g^{-1} FW)	Chl a/Child	Chl b/Child	Mg Proto (µg g^{-1} FW)	Proto (µg g^{-1} FW)	Pchilde (mg g^{-1} FW)
				Second Cut					
T10	1.495 ± 0.023 ^{de}	0.230 ± 0.004 ^{cd}	0.955 ± 0.010 ^{bc}	0.556 ± 0.018 $^{b-d}$	1.560 ± 0.046 ^{d}	1.399 ± 0.060 ^{e}	0.219 ± 0.001 ^{c}	0.309 ± 0.001 ^{d}	0.575 ± 0.005 ^{c}
T11	1.539 ± 0.006 ^{d}	0.249 ± 0.007 ^{bc}	1.086 ± 0.032 ^{ab}	0.569 ± 0.002 ^{b}	2.186 ± 0.067 ^{c}	1.852 ± 0.071 ^{d}	0.202 ± 0.002 ^{d}	0.280 ± 0.001 ^{e}	0.573 ± 0.014 ^{c}
ANOVA *p*	***	***	***	***	***	***	***	***	***

Levels of significance are represented by *** $p < 0.001$. For each parameter in the year, different letters within the column show significant differences between the treatments and control according to Tukey's HSD test at $p < 0.05$. T1, T2, T3, T4, T5, T6, T7, T8, T9, T10, and T11 are control, 5 mg L^{-1} Fe-NPs, 10 mg L^{-1} Fe-NPs, 5 mg L^{-1} Fe-NPs HA, 10 mg L^{-1} Fe-NPs HA, 100 mg L^{-1} FeSO$_4$, 200 mg L^{-1} FeSO$_4$, 100 mg L^{-1} EDDHA, 200 mg L^{-1} EDDHA, 100 mg L^{-1} EDTA, and 200 mg L^{-1} EDTA, respectively.

Table 7. Effect of iron (nano, sulfate, and chelated) foliar spray on nutrients content of Sweet Scented Geranium in the first and second cut during the 2019 experimental season. Means of three replicates are presented with ± SE.

Treatments	Cut 1						Cut2					
	N%	P%	K%	Fe (mg L^{-1})	Mn (mg L^{-1})	Zn (mg L^{-1})	N%	P%	K%	Fe (mg L^{-1})	Mn (mg L^{-1})	Zn (mg L^{-1})
T1	1.98 ± 0.026 ^{e}	0.150 ± 0.001 ^{g}	1.13 ± 0.014 ^{g}	144 ± 0.352 ^{k}	21.7 ± 0.161 ^{k}	21.0 ± 0.282 ^{k}	2.35 ± 0.014 ^{f}	0.150 ± 0.001 ^{g}	1.59 ± 0.011 ^{e}	152 ± 1.00 ^{k}	17.5 ± 0.178 ^{i}	35.6 ± 0.294 ^{i}
T2	2.83 ± 0.017 ^{b}	0.179 ± 0.001 ^{c}	1.65 ± 0.017 ^{c}	275 ± 0.889 ^{d}	43.0 ± 0.280 ^{d}	41.7 ± 0.115 ^{d}	3.37 ± 0.011 ^{b}	0.179 ± 0.001 ^{c}	2.28 ± 0.015 ^{b}	451 ± 0.542 ^{d}	25.4 ± 0.121 ^{d}	52.3 ± 0.161 ^{d}
T3	2.87 ± 0.017 ^{b}	0.180 ± 0.001 ^{c}	1.72 ± 0.014 ^{b}	400 ± 1.39 ^{c}	48.5 ± 0.060 ^{c}	47.1 ± 0.282 ^{c}	3.42 ± 0.014 ^{b}	0.180 ± 0.001 ^{c}	2.43 ± 0.011 ^{a}	458 ± 0.069 ^{c}	28.6 ± 0.103 ^{c}	58.9 ± 0.219 ^{c}
T4	2.89 ± 0.026 ^{b}	0.205 ± 0.001 ^{b}	2.03 ± 0.012 ^{a}	411 ± 0.987 ^{b}	77.5 ± 0.092 ^{b}	75.3 ± 0.057 ^{b}	3.43 ± 0.014 ^{b}	0.205 ± 0.001 ^{b}	2.44 ± 0.020 ^{a}	524 ± 0.744 ^{b}	36.7 ± 0.127 ^{b}	75.9 ± 0.173 ^{b}
T5	3.33 ± 0.014 ^{a}	0.222 ± 0.001 ^{a}	2.07 ± 0.014 ^{a}	443 ± 1.40 ^{a}	89.7 ± 0.083 ^{a}	87.0 ± 0.271 ^{a}	3.97 ± 0.017 ^{a}	0.222 ± 0.001 ^{a}	2.45 ± 0.014 ^{a}	534 ± 0.600 ^{a}	43.0 ± 0.196 ^{a}	89.7 ± 0.132 ^{a}
T6	2.18 ± 0.020 ^{d}	0.168 ± 0.001 ^{e}	1.22 ± 0.014 ^{f}	184 ± 0.606 ^{i}	26.0 ± 0.190 ^{i}	25.3 ± 0.127 ^{i}	2.60 ± 0.011 ^{de}	0.168 ± 0.001 ^{e}	2.04 ± 0.014 ^{c}	176 ± 0.519 ^{i}	19.4 ± 0.063 ^{h}	39.6 ± 0.132 ^{h}
T7	2.14 ± 0.017 ^{d}	0.161 ± 0.001 ^{f}	1.19 ± 0.011 ^{fg}	157 ± 0.467 ^{j}	23.2 ± 0.176 ^{j}	22.5 ± 0.161 ^{j}	2.54 ± 0.020 ^{e}	0.161 ± 0.001 ^{f}	1.78 ± 0.018 ^{d}	161 ± 0.404 ^{j}	17.8 ± 0.109 ^{i}	36.6 ± 0.225 ^{i}
T8	2.20 ± 0.023 ^{d}	0.172 ± 0.001 ^{de}	1.23 ± 0.011 ^{f}	191 ± 0.623 ^{h}	28.2 ± 0.242 ^{h}	27.4 ± 0.167 ^{h}	2.62 ± 0.014 ^{de}	0.172 ± 0.001 ^{de}	2.06 ± 0.020 ^{c}	198 ± 0.877 ^{h}	20.3 ± 0.161 ^{g}	41.7 ± 0.305 ^{g}
T9	2.23 ± 0.017 ^{d}	0.173 ± 0.001 ^{de}	1.24 ± 0.011 ^{f}	222 ± 0.207 ^{g}	30.6 ± 0.383 ^{g}	29.7 ± 0.063 ^{g}	2.64 ± 0.020 ^{d}	0.173 ± 0.001 ^{de}	2.08 ± 0.068 ^{c}	229 ± 0.831 ^{g}	21.9 ± 0.167 ^{f}	45.2 ± 0.254 ^{f}
T10	2.43 ± 0.014 ^{c}	0.177 ± 0.001 ^{cd}	1.35 ± 0.014 ^{e}	253 ± 0.900 ^{f}	33.5 ± 0.228 ^{f}	32.5 ± 0.242 ^{f}	2.90 ± 0.017 ^{c}	0.177 ± 0.001 ^{cd}	2.11 ± 0.014 ^{c}	388 ± 0.906 ^{f}	24.5 ± 0.225 ^{e}	50.6 ± 0.155 ^{e}
T11	2.49 ± 0.023 ^{c}	0.178 ± 0.001 ^{c}	1.49 ± 0.012 ^{d}	268 ± 1.03 ^{e}	40.5 ± 0.167 ^{e}	39.3 ± 0.150 ^{e}	2.97 ± 0.020 ^{c}	0.178 ± 0.001 ^{c}	2.26 ± 0.014 ^{b}	444 ± 0.456 ^{e}	24.7 ± 0.103 ^{de}	50.8 ± 0.069 ^{e}
ANOVA *p*	***	***	***	***	***	***	***	***	***	***	***	***

Levels of significance are represented by *** $p < 0.001$. For each parameter in the year, different letters within the column show significant differences between the treatments and control according to Tukey's HSD test at $p < 0.05$. T1, T2, T3, T4, T5, T6, T7, T8, T9, T10, and T11 are control, 5 mg L^{-} Fe-NPs, 10 mg L^{-1} Fe-NPs, 5 mg L^{-1} Fe-NPs HA, 10 mg L^{-1} Fe-NPs HA, 100 mg L^{-1} FeSO$_4$, 200 mg L^{-1} FeSO$_4$, 100 mg L^{-1} EDDHA, 200 mg L^{-1} EDDHA, 100 mg L^{-1} EDTA, and 200 mg L^{-1} EDTA, respectively.

3.7. Total Carbohydrate

Data in Table 8 displayed that, in general, the spraying of Fe sources increased significantly total carbohydrate concentration in the plant shoot over untreated control plants. The highest carbohydrate concentration was documented under the treatment of foliar application with 10 mg L^{-1} Fe-NPs-HA as compared with other treatments or untreated control plants.

Table 8. Effect of iron (nano, sulfate, and chelated) foliar spray on carbohydrates and phytopharmaceuticals of Rose Scented Geranium in the first and second cut during the second season. Means of three replicates are presented with $\pm$ SE.

Treatments	Carbohydrates (mg g^{-1} FW)		Phenol (mg gallic acid g^{-1} DW)		Flavonoids (mg quercetine g^{-1} DW)		Anthocyanin (mg 100 g^{-1} FW)	
	Cut 1	Cut 2	Cut 1	Cut 2	Cut 1	Cut 2	Cut 1	Cut 2
T1	3.041 ± 0.439 [b]	3.295 ± 0.124 [e]	8.084 ± 0.157 [g]	10.91 ± 0.199 [d]	0.989 ± 0.007 [d]	0.998 ± 0.008 [g]	2.156 ± 0.028 [c]	2.167 ± 0.012 [f]
T2	5.091 ± 0.143 [a]	5.143 ± 0.081 [a–c]	12.99 ± 0.199 [a–c]	13.82 ± 0.124 [ab]	2.690 ± 0.067 [a]	2.719 ± 0.054 [b–d]	3.815 ± 0.047 [ab]	3.959 ± 0.009 [b–d]
T3	5.424 ± 0.097 [a]	5.532 ± 0.016 [ab]	13.73 ± 0.264 [ab]	13.98 ± 0.356 [ab]	2.690 ± 0.044 [a]	2.787 ± 0.040 [bc]	4.114 ± 0.053 [a]	4.339 ± 0.049 [a–c]
T4	5.557 ± 0.025 [a]	5.604 ± 0.047 [ab]	13.98 ± 0.242 [a]	14.43 ± 0.227 [a]	2.736 ± 0.033 [a]	2.851 ± 0.041 [ab]	4.146 ± 0.024 [a]	4.828 ± 0.115 [ab]
T5	5.965 ± 0.416 [a]	5.971 ± 0.020 [a]	14.27 ± 0.264 [a]	14.83 ± 0.530 [a]	2.762 ± 0.047 [a]	3.007 ± 0.012 [a]	4.238 ± 0.224 [a]	5.183 ± 0.023 [a]
T6	4.369 ± 0.136 [ab]	4.104 ± 0.261 [de]	10.95 ± 0.318 [ef]	12.61 ± 0.264 [bc]	1.658 ± 0.073 [b]	2.478 ± 0.022 [e]	3.318 ± 0.113 [b]	2.954 ± 0.026 [ef]
T7	4.315 ± 0.063 [ab]	3.978 ± 0.060 [de]	10.37 ± 0.446 [i]	11.89 ± 0.448 [cd]	1.425 ± 0.042 [c]	1.429 ± 0.040 [i]	3.250 ± 0.018 [b]	2.786 ± 0.032 [ei]
T8	4.529 ± 0.079 [ab]	4.184 ± 0.052 [c–e]	11.31 ± 0.246 [d–f]	12.70 ± 0.338 [bc]	1.840 ± 0.038 [b]	2.559 ± 0.023 [de]	3.361 ± 0.292 [b]	3.249 ± 0.079 [de]
T9	4.645 ± 0.929 [ab]	4.441 ± 0.351 [cd]	11.60 ± 0.369 [d–f]	13.33 ± 0.136 [a–c]	2.550 ± 0.007 [a]	2.584 ± 0.011 [de]	3.557 ± 0.248 [ab]	3.475 ± 0.263 [c–e]
T10	4.749 ± 0.073 [ab]	4.737 ± 0.356 [b–d]	11.96 ± 0.102 [c–e]	13.66 ± 0.408 [ab]	2.593 ± 0.025 [a]	2.669 ± 0.042 [cd]	3.674 ± 0.008 [ab]	3.462 ± 0.044 [c–e]
T11	4.883 ± 0.033 [a]	4.785 ± 0.323 [b–d]	12.59 ± 0.213 [b–d]	13.69 ± 0.220 [ab]	2.609 ± 0.015 [a]	2.703 ± 0.038 [b–d]	3.704 ± 0.063 [ab]	3.655 ± 0.539 [c–e]
ANOVA p	***	***	***	***	***	***	***	

Levels of significance are represented by *** $p < 0.001$. For each parameter in the year, different letters within the column show significant differences between the treatments and control according to Tukey's HSD test at $p < 0.05$. T1, T2, T3, T4, T5, T6, T7, T8, T9, T10, and T11 are control, 5 mg L^{-1} Fe-NPs, 10 mg L^{-1} Fe-NPs, 5 mg L^{-1} Fe-NPs HA, 10 mg L^{-1} Fe-NPs HA, 100 mg L^{-1} FeSO$_4$, 200 mg L^{-1} FeSO$_4$, 100 mg L^{-1} EDDHA, 200 mg L^{-1} EDDHA, 100 mg L^{-1} EDTA, and 200 mg L^{-1} EDTA, respectively.

3.8. Phytopharmaceuticals

As shown in Table 8, the spraying of Fe-forms significantly increased the leaf phytopharmaceutical concentrations (phenol, flavonoid, and anthocyanin) in relation to non-treated plants. The supreme of phenols (14.27 and 14.83 mg gallic acid g^{-1} DW), flavonoids (2.762, and 3.007 mg quercetin g^{-1} DW), and anthocyanin (4.238, 5.183 mg 100 g^{-1} FW) concentrations in both cuts were recorded in the plant shoot treated with 10 mg L^{-1} Fe-NPs-HA. On the other hand, the lower levels of phytopharmaceuticals were recorded in untreated control plants in either the 1st and 2nd cuts.

4. Discussion

Around the world, iron deficiency (FED) is a significant issue that may have the desired effect on plant productivity in alkaline and calcareous soil. As a result, FED may be overcome via Fe-enriching methods, which involved conventional (sulphate or chelated) and nano-compounds supplementation. According to the results of the present investigation, foliar application of Fe-sources modifies the composition of EO, phytopharmaceuticals, and RSG-plant growth. It was also noted that the use of nano-sources specifically designed for humic acid Fe-NPs-HA offered the highest values of all examined attributes and enhanced the composition and quality of EO. Kah et al. [40] conveyed that nanofertilizers application had up to 30% more effective than traditional products. The peculiar characteristics of nano-particles, i.e., their large surface area, quick mass allocation, small size, high purity, and stability may be the cause of this observation [41]. In addition to accelerating enzymatic activities, nanoparticles also have the ability to reduce the accumulation of reactive oxygen species and oxidative damage that is improved plant development [23]. Moreover, it is attributed to their functions in modifying gene expression linked to several plant metabolic pathways [42].

Compared to untreated control plants, plant growth was dramatically boosted by the application of Fe-sources. These results were supported by previous investigations [8,18,24]. In this regard, the performance, root growth, and leaf count of sweet basil were all enhanced by the application of Fe_3O_4-NPs (1, 2, and 3 mg L^{-1}) concentration [43]. Additionally, ryegrass and pumpkin showed improved root elongation with Fe supplementation [44]. Similar findings indicating the improved influence of Fe_3O_4 NPs on a shoot and root elongation were gathered by Zahra et al. [45]. The improvement of photosynthetic processes and nucleic acid assimilation, which is reflected in an increase in photoassimilates needed for cell division and enlargement and improved plant development, may be the cause of Fe-sources' beneficial effects on plant growth [8,18,46].

It has been demonstrated that the use of Fe-sources significantly increased RSG-EO yield. Additionally, in both cuts during the first and second seasons, Nano-Fe in particular with HA (10 mg L^{-1}) was the most successful treatment (Figure 2). Previous studies have also observed an increase in EO caused by the use of Fe-sources [7,47]. On sweet marjoram, El-Khateeb et al. [8] discovered that applying Fe-NPs boosted EO% and EO production. According to Nejad et al. [25], applying Fe-sources significantly raised EO% when compared to untreated RSG plants. The generation of carbohydrates and the buildup of plant EO were positively correlated [48]. According to the results of the current study, foliar application of Fe-sources led to a greater accumulation of total carbohydrates in the herb than the control (Table 8). As a result, $FeSO_4$ application enhanced the content of total carbohydrates in coriander plants, according to Abou-Sreea et al. [7]. Fe-NPs foliar treatments considerably boosted the photosynthetic rate and chemical contents (carbohydrate, flavonoids, crude protein, total fatty acids, IAA), as well as oil yield, according to Abdel Wahab and Taha [49]. Additionally, El-Khateeb et al. [8] demonstrate that the total carbohydrates concentration in plant shoots of sweet marjoram treated with Fe-NPs was markedly elevated.

Essential oils, as a secondary metabolite, are highly complex mixtures of volatile compounds. Fe-sources applications affected not only EO yield but also EO constituents. In the present study, 11 constituents were identified in RSG-EO and the main components were citronellol, geraniol, and eugenol (Tables 4 and 5). A widespread study was done on the constituents of RSG-EO, which distinguished considerable variations in their constituents worldwide. In this regard, Sharopov et al. [50] in Tajikistan identified 95.1% of RSG-EO constituents, including 79 components including citronellol (37.5%), geraniol (6.0%), caryophyllene oxide (3.7%), menthone (3.1%), linalool (3.0%), β-bourbonene (2.7%), isomenthone (2.1%), and geranylformate (2.0%).

Citronellol (C), geraniol (G), and their esters are the prime components in RSG-EO as per the prerequisites of perfumery productions [1]. The C/G is the main aspect that regulates the standard of RSG-EO for fragrance manufacturing [51]. Commonly, C/G proportion of 1:1–3:1 is satisfactory; nonetheless, the best ratio is 1:1 [1,52]. Oil of C/G ratio of over 3:1 is deliberated to be of deprived quality for fragrance manufacturing nonetheless still, it can be used for the manufacture of creams, toiletries, and fragrance-based objects at a lesser price [53,54]. The variance in the C/G ratio is probably associated with environmental factors at the harvesting, which eventually influences the assimilation of citronellol and geraniol. It is described that citronellol concentrations were greater in the warm season relative to the winter season [55].

The data herein revealed that the application of Fe sources significantly raised chlorophyll above untreated plants. In line with the current results, several researchers recognized that the application of Fe-NPs [24]; Fe-sulphate [18], EDDHA [18], and EDTA [56] increased leaves chlorophyll concentration over untreated plants. The encouragement roles of Fe on chlorophyll accumulation resulted from regulating Fe, Mg, and N uptake and increase Fe availability (Table 7), as well as regulate Chl assimilation gene expression [57], stimulation chlorophyll assimilation pathways [58] and encouraging the transformation of Mg-Proto to Pchlide and consequently Chl a and b. Moreover, Fe-sources application interferes Chl degradation as indicated in the present study (Table 6), by Pheo production and avoids the change of Mg-prototp Pchlide [59], as well as hastein ALA assimilation [60] due to declining Mg-proto and proto accumulation. As indicated previously, Fe-NPs were superior to other Fe sources in increasing chlorophyll concentration due to: (1) accelerating a dramatic upregulation of photosystem marker genes [61,62] formation of a complex with phytoferritin (leaves iron-binding protein), leading to greater involvement in chlorophyll assimilation [63]; (2) Improving thylakoid and chloroplast metabolic pathways that sequentially rise photosynthetic activities and lessening of chloroplast ROS [58,64].

The most recent results showed that spraying with Fe sources significantly increases the levels of N, P, K, Fe, Mn, and Zn in plant shoots as compared to untreated plants. The findings of El-Sonbaty [24] for Fe-NPs, Abou-Sreea et al. [7] for $FeSO_4$, Erdale [65] for EDTA, and Tavallali [66] for EDDHA were in agreement with these results. In this regard, Gutierrez-Ruelas et al. [18] found that in green bean, the application of Fe sources (Fe-NPs, $FeSO_4$, EDDHA) increased plant Fe concentration. Likewise, 0.2% Fe-EDDHA application amplified Chl a and Chl b and induced a marginal rise in the plant tissue N content [67]. Moreover, El-Sonbaty [24] found that spraying onion plants with Fe-NPs significantly increased N, P, and K content in plant organs over control plants. The role of Fe in increasing nutrient concentration and uptake may be due to increased energy availability and increased deactivated absorption of anions in root cells that increased absorption of cations as potassium [68]. Additionally, the increase in N in plant tissues by Fe sources (Fe-NPs, $FeSO_4$, EDDHA) application may result from the role of Fe in the enhancement of nitrate reductase activity which is increased N uptake and accumulation [18].

Currently, the supplementation of Fe-sources improved phytopharmaceutical accumulation in plant shoots, which was in accord with previous research [25,66]. Numerous phytopharmaceuticales' assembly was documented to be increased by elicitors including Fe [69,70]. The mechanism of elicitation by Fe, was, nonetheless, diverse in different herbs, and in the majority, an 'elicitor–receptor' complex was formed and a massive range of physio-biochemical responses was demonstrated [71]. The existing data have ascertained that Fe encouraged the extra accretion of phenolic in an RSG shoot. This might be because producing signal transduction systems and activating the gene for phenyl aminolyase (PAL), a secondary metabolic pathway, speed up the assimilation of phenols. The most important bioactive molecule with a reliable antioxidant has been determined to be phenolic chemicals. They have received more attention recently since they have been shown to be more effective than ascorbic acid, tocopherol, and carotenoid [72,73]. According to earlier studies [74–76], they have a variety of biological functions, including anti-inflammatory, antioxidant, antiviral, anticarcinogenic, anti-oxidant, antispasmodic, and depressive effects. Epidemiology surveys have discovered that a substantial nutritional intake of flavonoids and phenolics is coupled with lesser rates of cancer incidence [72]. The antioxidant aptitudes of phenolic compounds are mediated by numerous approaches [77]: (1) abolish ROS/reactive nitrogen species (RNS); (2) defeat ROS/RNS assembly by hindering numerous enzymes or chelating ions occupied in ROS; (3) regulate antioxidant capacity. Like total soluble phenolic, flavonoids establish a widespread secondary metabolite with polyphenolic structures and play an imperative function in shielding biological systems alongside oxidation processes [78]. In humans, flavonoids can impede aldose reductase and are occupied in diabetic difficulties i.e., neuropathy, heart disease, and retinopathy as well as attended as antioxidant compounds that lessen the hazard of cancers [79].

5. Conclusions

In the context of sustainable agriculture, prevailing and low-cost, using pioneering nanotechnology in agriculture is considered one of the encouraging attitudes for improving plant productivity. The current outcomes display a solid confirmation of the high effectiveness of nano fertilizer on plant productivity and product quality over conventional Fe-sources. The study recommended that since Fe NPs with humic acid are naturally non-toxic, they have been utilized as Fe-enriching fertilizers to replenish Fe levels in plants, demonstrating the significance of using Fe NPs for commercial purposes.

Supplementary Materials: The following supporting information can be downloaded at: https://www.mdpi.com/article/10.3390/agronomy12092164/s1, Table S1: Mean of monthly climatic data of the experimental site throughout the experimental seasons.

Author Contributions: Conceptualization, A.E.E.-S., S.F. and S.M.A.E.-G.; methodology, A.E.E.-S., S.F., E.F.A., H.M.A.-Y., A.A.S.A.-K. and S.M.A.E.-G.; software, S.F.; validation, A.E.E.-S., S.F., E.F.A., A.A.S.A.-K. and S.M.A.E.-G.; formal analysis, S.F.; investigation, A.E.E.-S., S.F. and S.M.A.E.-G.; resources, A.E.E.-S., S.F., E.F.A., H.M.A.-Y., A.A.S.A.-K. and S.M.A.E.-G.; data curation, A.E.E.-S., S.F. H.M.A.-Y., A.A.S.A.-K. and S.M.A.E.-G.; writing—original draft preparation, A.E.E.-S. and S.M.A.E.-G.; writing—review and editing, S.F., E.F.A. and H.M.A.-Y.; visualization, A.E.E.-S., S.F., E.F.A., H.M.A.-Y., A.A.S.A.-K. and S.M.A.E.-G.; supervision, A.E.E.-S., S.F. and S.M.A.E.-G.; project administration, S.F., E.F.A., H.M.A.-Y. and A.A.S.A.-K.; funding acquisition, A.E.E.-S., S.F., E.F.A., H.M.A.-Y., A.A.S.A.-K. and S.M.A.E.-G. All authors have read and agreed to the published version of the manuscript.

Funding: This research was funded by Taif University Researchers Supporting Project, grant number TURSP-2020/199.

Data Availability Statement: Not applicable.

Acknowledgments: The authors express their gratitude to Mansoura University, and Agricultural Research Center, Egypt, for supporting our manuscript. The authors are also thankful to Taif University Researchers Supporting Project number (TURSP-2020/199), Taif University, Saudi Arabia, for the financial support and research facilities.

Conflicts of Interest: The authors declare no conflict of interest.

References

1. Kumar, D.; Padalia, R.C.; Suryavanshi, P.; Chauhan, A.; Pratap, P.; Verma, S.; Venkatesha, K.T.; Kumar, R.; Singh, S.; Tiwari, A.K. Essential oil yield, composition and quality at different harvesting times in three prevalent cultivars of rose-scented geranium. *J. Appl. Hortic.* **2021**, *23*, 19–23. [CrossRef]

2. Pandey, P.; Upadhyay, R.K.; Singh, V.R.; Padalia, R.C.; Kumar, R.; Venkatesha, K.T.; Tiwari, A.K.; Singh, S.; Tewari, S.K. *Pelargonium graveolens* L. (Rose-scented geranium): New hope for doubling Indian farmers' income. *Environ. Conserv. J.* **2020**, *21*, 141–146. [CrossRef]

3. Blerot, B.; Baudino, S.; Prunier, C.; Demarne, F.; Toulemonde, B.; Caissard, J.C. Botany, agronomy and biotechnology of Pelargonium used for essential oil production. *Phytochem. Rev.* **2015**, *26*, 807–816. [CrossRef]

4. Androutsopoulou, C.; Christopoulou, S.D.; Hahalis, P.; Kotsalou, C.; Lamari, F.N.; Vantarakis, A. Evaluation of essential oils and extracts of rose geranium and rose petals as natural preservatives in terms of toxicity, antimicrobial, and antiviral activity. *Pathogens* **2021**, *10*, 494. [CrossRef]

5. Mosta, N.M.; Soundy, P.; Steyn, J.M.; Learmonth, R.A.; Mojela, N.; Teubes, C. Plant shoot age and temperature effects on essential oil yield and oil composition of rose-scented geranium (*Pelargonium* sp.) grown in South Africa. *J. Essent. Oil Res.* **2006**, *18*, 106–110.

6. Asgarpanah, J.; Ramezanloo, F. An overview on phytopharmacology of *Pelargonium graveolens* L. *Indian J. Tradit. Knowl.* **2015**, *14*, 558–563.

7. Abou-Sreea, A.I.B.; Yassen, A.A.A.; El-Kazzaz, A.A.A. Effects of iron (ii) sulfate and potassium humate on growth and chemical composition of *Coriandrum sativum* L. *Int. J. Agric. Res.* **2017**, *12*, 136–145. [CrossRef]

8. El-Khateeb, M.A.; El-Attar, A.B.; Abo-Bakr, Z.A.M. Effect of nano-microlements on growth, yield and essential oil production of sweet marjoram (*Origanum majorana*) plants. *Plant Arch.* **2020**, *20*, 8315–8324.

9. Filiz, E.; Akbudak, M.A. Investigation of PIC1 (permease in chloroplasts 1) gene's role in iron homeostasis: Bioinformatics and expression analyses in tomato and sorghum. *BioMetals* **2020**, *33*, 29–44. [CrossRef]

10. Gao, F.; Dubos, C. Transcriptional integration of plant responses to iron availability. *J. Exp. Bot.* **2020**, *72*, 2056–2070. [CrossRef]

11. Jeong, J.; Guerinot, M.L. Homing in on iron homeostasis in plants. *Trend Plant Sci.* **2009**, *14*, 280–285. [CrossRef]

12. Wang, Y.; Kang, Y.; Zhong, M.; Zhang, L.; Chai, X.; Jiang, X.; Yang, X. Effects of iron deficiency stress on plant growth and quality in flowering chinese cabbage and its adaptive response. *Agronomy* **2022**, *12*, 875. [CrossRef]

13. Tripathi, D.K.; Singh, S.; Gaur, S.; Singh, S.W.; Yadav, V.; Liu, S.; Singh, V.P.; Sharma, S.; Srivastava, P.; Prasad, S.M.; et al. Acquisition and homeostasis of iron in higher plants and their probable role in abiotic stress tolerance. *Front. Environ. Sci.* **2018**, *5*, 86. [CrossRef]

14. Li, J.; Cao, X.; Jia, X.; Liu, L.; Cao, H.; Qin, W.; Li, M. Iron deficiency leads to chlorosis through impacting chlorophyll synthesis and nitrogen metabolism in *Areca catechu* L. *Front. Plant Sci.* **2021**, *12*. [CrossRef] [PubMed]
15. Bindraban, P.S.; Dimkpa, C.; Nagarajan, L.; Roy, A.; Rabbinge, R. Revisiting fertilisers and fertilisation strategies for improved nutrient uptake by plants. *Biol. Fertil. Soils* **2015**, *51*, 897–911. [CrossRef]
16. Balk, J.; von Wirén, N.; Thomine, S. The iron will of the research community: Advances in iron nutrition and interactions in lockdown times. *J. Exp. Bot.* **2021**, *72*, 2011–2013. [CrossRef]
17. EU Directive. Regulation (EC) No 2003/2003 of the European parliament and of the Council of 13 October 2003 relating to fertilizers. *Off. J. Eur. Union* **2003**.
18. Gutierrez-Ruelas, N.J.; Palacio-Marquez, A.; Sanchez, E.; Munoz-Marquez, E.; Chavez-Mendoza, C.; Ojeda-Barrios, D.L.; Flores-Cordova, M.A. Impact of the foliar application of nanoparticles, sulfate and iron chelate on the growth, yield and nitrogen assimilation in green beans. *Not. Bot. Horti Agrobot.* **2021**, *49*, 12437. [CrossRef]
19. Schonherr, J.; Fernandez, V.; Schreiber, L. Rates of cuticular penetration of chelated FeIII, Role of humidity, concentration, adjuvants, temperature and type of chelate. *J. Agric. Food Chem.* **2005**, *53*, 4484–4492. [CrossRef]
20. Farajzadeh, M.T.; Yarnia, E.M.; Khorshidi, M.B.; Ahmadzadeh, V. Effect of micronutrients and their application method on yield, crop growth rate and net assimilation rate of corn cv. Jeta. *J. Food Agric. Environ.* **2009**, *7*, 611–615.
21. Šramek, F.; Dubsky, M. Occurrence and correction of chlorosis in young petunia plants. *HortScience* **2009**, *36*, 147–153. [CrossRef]
22. Hyvönen, H.; Orama, M.; Saarinen, H.; Aksela, R. Studies on biodegradable chelating agents: Complexation of iminodisuccinic acid (ISA) with Cu(II), Zn(II), Mn(II) and Fe(III) ions in aqueous solution. *Green Chem.* **2003**, *5*, 410–414. [CrossRef]
23. Asadi-Kavan, Z.; Khavari-Nejad, R.A.; Iranbakhsh, A.; Najafi, F. Cooperative effects of iron oxide nanoparticle (α-Fe_2O_3) and citrate on germination and oxidative system of evening primrose (*Oenthera biennis* L.). *J. Plant Interact.* **2020**, *15*, 166–179. [CrossRef]
24. El Conbaty, A.E. Application of potassium humate and magnetite iron nanoparticles to enhance growth and yield quality of onion plants grown on salt affected soil. *Plant Cell Biotechnol. Mol. Biol.* **2021**, *22*, 56–68.
25. Nejad, A.R.; Izadi, Z.; Sepahvand, K.; Mumivand, H.; Mousavi-Far, S. Changes in total phenol and some enzymatic and non-enzymatic antioxidant activities of rose-scented geranium (*Pelargonium graveolens*) in response to exogenous ascorbic acid and iron nutrition. *J. Ornam. Plants* **2020**, *10*, 27–36.
26. Peng, L.; Qin, P.; Lei, M.; Zeng, Q.; Song, H.; Yang, J.; Shao, J.; Liaoa, B.; Gu, J. Modifying Fe_3O_4 nanoparticles with humic acid for removal of Rhodamine B in water. *J. Hazard. Mater.* **2012**, *209*, 193–198. [CrossRef]
27. Black, C.A.; Evans, D.O.; Ensminger, L.E.; White, J.L.; Clark, F.E.; Dinauer, R.C. Methods of soil analysis. Part 2. In *Chemical and Microbiological Properties*, 2nd ed.; Soil Science Society of America, Inc.; American Society of Agronomy, Inc.: Madison, WI, USA, 1982.
28. Abouelatta, A.M.; Keratum, A.Y.; Ahmed, S.I.; El-Zun, H.M. The effect of air-drying and extraction methods on the yield and chemical composition of Geranium (*Pelargonium graveolens* L. 'Hér) essential oils. *Am. J. Appl. Ind. Chem.* **2021**, *5*, 17–21. [CrossRef]
29. Adams, P.R. *Identification of Essential Oil Components by Gas Chromatography/Mass Spectrometry*; Allured Publishing Corporation: Carol Stream, IL, USA, 2007.
30. Motsara, M.R.; Roy, R.N. *Guide to Laboratory Establishment for Plant Nutrient Analysis*; FAO Fertilizer and Plant Nutrition Bulletin: Rome, Italy, 2008.
31. Cooper, T.G. *The Tools of Biochemistry*; A Wiley-Interscience Pub. Wiley: New York, NY, USA, 1977.
32. Bettinelli, M.; Beone, G.M.; Spezia, S.; Baffi, C. Determination of heavy metals in soils and sediments by microwave-assisted digestion and inductively coupled plasma oprical emission spectrometetry analysis. *Anal. Chim. Acta* **2000**, *424*, 289–296. [CrossRef]
33. Lichtenthaler, H.K.; Wellburn, A.R. Determination of total carotenoids and chlorophylls A and B of leaf in different solvents. *Bio. Soc. Trans.* **1985**, *11*, 591–592. [CrossRef]
34. Radojevic, M.; Bashkin, V.N. *Practical Environmental Analysis*, 2nd ed.; RSC Publishing: Cambridge, UK, 2006.
35. Harpaz-Saad, S.; Azoulay, T.; Arazi, T.; Ben-Yaakov, E.; Mett, A.; Shiboleth, Y.M.; Hortensteiner, S.; Gidoni, D.; Gal-On, A.; Goldschmidt, E.E.; et al. Chlorophyllase is a rate-limiting enzyme in chlorophyll catabolism and is post translationally regulated. *Plant Cell* **2007**, *19*, 1007–1022. [CrossRef]
36. Sarropoulou, V.; Dimassi-Theriou, K.; Therios, I.; Koukourikou-Petridou, M. Melatonin enhances root regeneration, photosynthetic pigments, biomass, total carbohydrates and proline content in the cherry rootstock PHL-C (*Prunus avium* × *Prunus cerasus*). *Plant Physiol. Biochem.* **2012**, *61*, 162–168. [CrossRef] [PubMed]
37. Zhang, B.; Zheng, L.P.; Wang, J.W. Nitric oxide elicitation for secondary metabolite production in cultured plant cells. *Appl. Microbiol. Biotechnol.* **2012**, *93*, 455–466. [CrossRef] [PubMed]
38. Meda, A.; Lamien, C.E.; Romito, M.; Millogo, J.; Nacoulma, O.G. Determination of the total phenolic, flavonoid and praline contents in Burkina Fasan honey, as well as their radical scavenging activity. *Food Chem.* **2005**, *91*, 571–577. [CrossRef]
39. Abdel-Aal, E.S.M.; Hucl, P. A rapid method for quantifying total anthocyanin in blue aleurone and purple pericarp wheat. *Cereal Chem.* **1999**, *76*, 350–354. [CrossRef]
40. Kah, M.; Kookana, R.S.; Gogos, A.; Bucheli, T.D. A critical evaluation of nanopesticides and nanofertilizers against their conventional analogues. *Nat. Nanotechnol.* **2018**, *13*, 677–684. [CrossRef] [PubMed]

41. Ghormade, V.; Deshpande, M.V.; Paknikar, K.M. Perspectives for nano-biotechnology enabled protection and nutrition of plants. *Biotechnol. Adv.* **2011**, *29*, 792–803. [CrossRef]

42. Aslani, F.; Bagheri, S.; Julkapli, N.M.; Juraimi, A.; Farahnaz, F.S.; Hashemi, S.; Baghdadi, A. Effects of engineered nanomaterials on plants growth: An Overview. *Sci. World J.* **2014**, *2014*, 1–28. [CrossRef]

43. ElFeky, S.A.; Mohammed, M.A.; Khater, M.S.; Osman, Y.A.H.; Elsherbini, E. Effect of magnetite Nano-Fertilizer on Growth and yield of *Ocimum basilicum* L. *Int. J. Indig. Med. Plants* **2013**, *46*, 1286–1293.

44. Wang, H.; Kou, X.; Pei, Z.; Xiao, J.Q.; Shan, X.; Xing, B. Physiological effects of magnetite (Fe_3O_4) nanoparticles on perennial ryegrass (*Lolium perenne* L.) and pumpkin (*Cucurbita mixta*) plants. *Nanotoxicology* **2011**, *5*, 30–42. [CrossRef]

45. Zahra, Z.; Arshad, M.; Rafique, R.; Mahmood, A.; Habib, A.; Qazi, I.A.; Khan, S.A. Metallic nanoparticle (TiO_2 and Fe_3O_4) application modifies rhizosphere phosphorus availability and uptake by *Lactuca sativa*. *J. Agric. Food Chem.* **2015**, *63*, 6876–6882. [CrossRef]

46. Jia, M.S.H.; Mateen, A.K.; Sohani Das, S.; William, C.M.; Elizabeth, C.T.; Dixie, J.G. Fe^{2+} binds iron responsive element-RNA, selectively changing protein-binding affinities and regulating mRNA repression and activation. *Proc. Natl. Acad. Sci. USA* **2012**, *109*, 8417–8422.

47. Kokina, I.; Plaksenkova, I.; Jermaļonoka, M.; Petrova, A. Impact of iron oxide nanoparticles on yellow medick (*Medicago falcata* L.) plants. *J. Plant Interact.* **2020**, *15*, 1–7. [CrossRef]

48. Markus Lange, B.; Wildung, M.R.; Stauber, E.J.; Christopher, S.; Derek, P.; Rodney, C. Probing essential oil biosynthesis and secretion by functional evaluation of expressed sequence tags from mint glandular trichomes. *Proc. Natl. Acad. Sci. USA* **2000**, *97*, 2934–2939. [CrossRef]

49. Abdel Wahab, M.M.; Taha, S.S. Main sulphur content in essential oil of *Eruca sativa* as affected by nano iron and nano zinc mixed with organic manure. *Agriculture* **2018**, *64*, 65–79. [CrossRef]

50. Sharopov, F.S.; Zhang, H.; Setzer, W.N. Composition of geranium (*Pelargonium graveolens*) essential oil from Tajikistan. *Am. J. Essent. Oils Nat. Prod.* **2014**, *2*, 13–16.

51. Askary, M.; Talebi, S.; Amini, F.; Bangan, A. Effects of iron nano-particles on *Mentha piperita* L. under salinity stress. *Biologia* **2017**, *63*, 65–75.

52. Verma, R.S.; Verma, R.K.; Yadav, A.K.; Chauhan, A. Changes in the essential oil composition of rose-scented geranium (*Pelargonium graveolens* L'Her ex Ait.) due to date of transplanting under hill conditions of Uttarakhand. *Indian J. Nat. Prod. Resour.* **2010**, *1*, 367–370.

53. Weiss, E.A. *Essential Oil Crops*; Centre for Agriculture and Biosciences (CAB) International: New York, NY, USA, 1997.

54. Peterson, A.; Machmudah, S.; Roy, B.C.; Goto, M.; Sasaki, M.; Hirose, T. Extraction of essential oil from geranium (*Pelargonium graveolens*) with supercritical carbon dioxide. *J. Chem. Tech. Biotechnol.* **2006**, *81*, 167–172. [CrossRef]

55. Rao, R.B.R.; Kaul, P.N.; Mallavarapu, G.R.; Ramesh, S. Effect of seasonal climatic changes on biomass yield and terpenoid composition of rose-scented geranium (*Pelargonium species*). *Biochem. Sys. Ecol.* **1996**, *24*, 627–635. [CrossRef]

56. Mohammadi, H.; Hatami, M.; Feghezadeh, K.; Ghorbanpour, M. Mitigating effect of nano-zerovalent iron, iron sulfate and EDTA against oxidative stress induced by chromium in *Helianthus annuus* L. *Acta Physiol. Plant.* **2018**, *40*, 69. [CrossRef]

57. Kroh, G.E.; Pilon, M. Regulation of iron homeostasis and use in chloroplasts. *Int. J. Mol. Sci.* **2020**, *21*, 3395. [CrossRef] [PubMed]

58. Barhoumi, L.; Oukarroum, A.; Taher, L.B.; Smiri, L.S.; Abdelmelek, H.; Dewez, D. Effects of superparamagnetic iron oxide nanoparticles on photosynthesis and growth of the aquatic plant *Lemna gibba*. *Arch. Environ. Contam. Toxicol.* **2015**, *68*, 510–520. [CrossRef]

59. Vleck, L.M.; Gasman, M.L. Reversal of α, α'-dipyridyl-induced porphyrin synthesis in etiolated and greening red kidney bean leaves. *Plant Physiol.* **1979**, *64*, 393–397.

60. Chereskin, B.M.; Castelfranco, P.A. Effects of iron and oxygen on chlorophyll biosynthesis. II Observations on the biosynthetic pathway of isolated etiochloroplasts. *Plant Physiol.* **1982**, *69*, 112–116. [CrossRef]

61. Tombuloglu, H.; Slimani, Y.; Tombuloglu, G.; Korkmaz, A.D.; Baykal, A.; Almessiere, M.; Ercan, I. Impact of superparamagnetic iron oxide nanoparticles (SPIONs) and ionic iron on the physiology of summer squash (*Cucurbita pepo*): A comparative study. *Plant Physiol. Biochem.* **2019**, *139*, 56–65. [CrossRef] [PubMed]

62. Yeshi, K.; Wangchuk, P. Essential oils and their bioactive molecules in healthcare. *Herb. Biomol. Healthc. Appl.* **2022**, 215–237. [CrossRef]

63. Bienfait, H.; Der Mark, F.V. Phytoferritin and its role in iron metabolism. In *Metals and Micronutrients: Uptake and Utilization by Plants*; Robb, D.A., Pierpoint, W.S., Eds.; Academic Press: New York, NY, USA, 1983.

64. Giraldo, J.; Landry, M.; Faltermeier, S.; Nicholas, M.; Iverson, N.; Boghossian, A.; Reuel, N.; Hilmer, A.; Sen, F.; Brew, J.; et al. Plant nanobionics approach to augment photosynthesis and biochemical sensing. *Nat. Mater.* **2014**, *13*, 400–408. [CrossRef] [PubMed]

65. Erdale, I. Effect of foliar iron applications at different growth stages on iron and some nutrient concentrations in strawberry cultivars. *Turk. J. Agric. For.* **2004**, *28*, 421–427.

66. Tavallali, V. Effects of iron nano-complex and Fe-EDDHA on bioactive compounds and nutrient status of purslane plants. *Int. Agrophys.* **2018**, *32*, 411–419. [CrossRef]

67. Sahua, M.P.; Singha, H.G. Effect of sulphur on prevention of iron chlorosis and plant composition of groundnut on alkaline calcareous soils. *J. Agric. Sci.* **1987**, *109*, 73–77. [CrossRef]

68. Moghadam, A.; Vattani, H.; Baghaei, N.; Keshavarz, N. Effect of different levels of fertilizer nano-iron chelates on growth and yield characteristics of two varieties of Spinach (*Spinacia oleracea* L.). *Res. J. Appl. Sci. Eng. Technol.* **2012**, *4*, 4813–4818.

69. Pliankong, P.; Padungsak, S.A.; Wannakrairoj, S. Chitosan elicitation for enhancing of vincristine and vinblastine accumulation in cell culture of *Catharanthus roseus* (L.) G. Don. *J. Agric. Sci.* **2018**, *10*, 287–293. [CrossRef]

70. Gupta, S.; Chaturvedi, P.; Kulkarni, M.G.; Van Staden, J. A critical review on exploiting the pharmaceutical potential of plant endophytic fungi. *Biotechnol. Adv.* **2020**, *39*, 107462. [CrossRef]

71. Bakalova, R.; Zhelev, Z.; Miller, T.; Aoki, I.; Higashi, T. Vitamin C versus cancer: Ascorbic acid radical and impairment of mitochondrial respiration? *Hindawi Oxidative Med. Cell Longev.* **2020**, *2020*, 1504048. [CrossRef]

72. Dai, J.; Mumper, R. Plant phenolics: Extraction, analysis and their antioxidant and anticancer properties. *Molecules* **2010**, *15*, 7313–7352. [CrossRef] [PubMed]

73. Ismail, B.B.; Yusuf, H.L.; Pu, Y.; Zhao, H.; Guo, M.; Liu, D. Ultrasound-assisted adsorption/desorption for the enrichment and purification of flavonoids from baobab (*Adansonia digitata*) fruit pulp. *Ultrason. Sonochem.* **2020**, *65*, 104980. [CrossRef]

74. Matsuda, H.; Morikawa, T.; Ando, S.; Toguchida, I.; Yoshikawa, M. Structural requirements of flavonoids for nitric oxide production inhibitory activity and mechanism of action. *Bioorg. Med. Chem.* **2003**, *11*, 1995–2000. [CrossRef]

75. Ghasemzadeh, A.; Jaafar, H.Z.E. Anticancer and antioxidant activities of Malaysian young ginger (*Zingiber officinale* Roscoe) varieties grown under different CO_2 concentration. *J. Med. Plant Res.* **2011**, *5*, 3247–3255.

76. Basli, A.; Belkacem, N.; Amrani, I. *Health Benefits of Phenolic Compounds against Cancers*; IntechOpen: London, UK, 2017; ISBN 978-953-51-2960-8. [CrossRef]

77. Miguel-Chávez, R.S. *Phenolic Antioxidant Capacity: A Review of the State of the Art, Phenolic Compounds-Biological Activity*; Soto-Hernandez, M., Palma-Tenango, M., Garcia-Mateos, M.R., Eds.; IntechOpen: London, UK, 2017. Available online: https://www.intechopen.com/books/phenolic-compounds-biological-activity/phenolic-antioxidant-capacity-a-review-of-the-state-of-the-art (accessed on 6 March 2022).

78. Halliwell, B.; Gutteridge, J.M. Protection against oxidants in biological system: The superoxide theory of oxygen toxicity. In *Free Radicals in Biology and Medicine*; Halliwell, B., Gutteridge, J.M., Eds.; Clarendon Press: Oxford, UK, 1989; pp. 86–123.

79. Fimognari, C.; Berti, F.; Nusse, M.; Cantelli Forti, G.; Hrelia, P. In vitro antitumor activity of cyanidin-3-O-glucopyranoside. *Chemotherapy* **2005**, *51*, 332–335. [CrossRef]

 *agronomy*

Article

Beneficial Effects of Silicon Fertilizer on Growth and Physiological Responses in Oil Palm

Saowapa Duangpan [1,2,*], Yanipha Tongchu [1], Tajamul Hussain [1], Theera Eksomtramage [2,3] and Jumpen Onthong [3]

[1] Laboratory of Plant Breeding and Climate Resilient Agriculture, Agricultural Innovation and Management Division, Faculty of Natural Resources, Prince of Songkla University, Hat Yai, Songkhla 90110, Thailand; 6010620055@psu.ac.th (Y.T.); 6110630006@psu.ac.th (T.H.)

[2] Oil Palm Agronomical Research Center, Faculty of Natural Resources, Prince of Songkla University, Hat Yai, Songkhla 90110, Thailand; theera.e@psu.ac.th

[3] Agricultural Innovation and Management Division, Faculty of Natural Resources, Prince of Songkla University, Hat Yai, Songkhla 90110, Thailand; jumpen.o@psu.ac.th

* Correspondence: saowapa.d@psu.ac.th; Tel.: +66-74-286-138

Abstract: Vigorous and well-established nursery seedlings are an important component of sustainable oil palm production. We postulated that Si fertilization at the seedling stage could help to achieve improved performance of oil palm seedlings leading to healthy and vigorous nursery establishment. In this study, we evaluated the growth and physiological responses of oil palm *Tenera* hybrid seedlings under three Si fertilization treatments and a control including (i) 0 g Ca_2SiO_4 (T0), (ii) 0.5 g Ca_2SiO_4 (T1), (iii) 3.5 g Ca_2SiO_4 (T2), and (iv) 7.0 g Ca_2SiO_4 (T3) per plant per month. Ca_2SiO_4 was used as the Si fertilizer source and was applied for four consecutive months. Nondestructive data including stem diameter, plant height, leaf length, photosynthetic rate, leaf angle, and leaf thickness and destructive data including leaf, stem, and root fresh weight and dry weight, as well as chlorophyll *a*, Si, and nitrogen contents, were recorded before treatment (0 DAT), as well as 60 (60 DAT) and 120 days after treatment (120 DAT). Results indicated that Si fertilization enhanced Si accumulation in oil palm seedlings, and maximum accumulation was observed in the aerial parts especially the leaves with the highest accumulation of 0.89 % dry weight at T3. Higher Si accumulation stimulated the growth of seedlings; a total fresh weight of 834.28 g and a total dry weight of 194.34 g were observed at T3. Chlorophyll *a* content (0.83 gm^{-2}) and net photosynthetic rate (4.98 µM $CO_2 \cdot m^{-2} \cdot s^{-1}$) were also observed at T3. Leaf morphology was not significantly influenced under Si fertilization, whereas the nitrogen content of seedlings was significantly increased. Correlation analysis revealed a highly significant and positive association among Si accumulation, chlorophyll *a* content, photosynthetic rate, total fresh weight, total dry weight, and nitrogen content of seedlings, indicating that Si fertilization enhanced the performance of these attributes. On the basis of the research evidence, it was concluded that Si fertilization should be considered for improved nutrient management for oil palm seedling and nursery production.

Keywords: silicon fertilization; oil palm; growth; physiological response

Citation: Duangpan, S.; Tongchu, Y.; Hussain, T.; Eksomtramage, T.; Onthong, J. Beneficial Effects of Silicon Fertilizer on Growth and Physiological Responses in Oil Palm. *Agronomy* 2022, 12, 413. https://doi.org/10.3390/agronomy12020413

Academic Editors: Christos Noulas, Shahram Torabian and Ruijun Qin

Received: 31 December 2021
Accepted: 4 February 2022
Published: 7 February 2022

Publisher's Note: MDPI stays neutral with regard to jurisdictional claims in published maps and institutional affiliations.

1. Introduction

Silicon (Si) is a beneficial element for plants and is ranked as the second most abundant element at 28% in the Earth's crust following oxygen [1,2]. The Si content of soil ranges from 1–45% depending on soil type, but Si is usually scarcely soluble, and its availability for plant uptake is limited [3]. A long period of intensive plant cultivation leads to the deprivation of soil Si, subsequently resulting in insufficient Si to sustain productive agriculture [4,5]. Subtropical and tropical agriculture are typically low in available Si, and rational Si fertilization could enhance crop yield [4]. In plants, silicon deficiency affects the development of strong leaves, stems, and roots. Rice with silicon deficiency is susceptible to fungal and bacterial diseases, as well as insect pests. The photosynthetic activity, growth, and grain yield are reduced [6].

The potential of Si in improving growth and yield and in alleviating the negative effects of biotic and abiotic stresses has been studied in multiple crops including rice, tomato, sugarcane, and wheat [4,7–11]. Plants absorb Si from the soil solution in the form of monosilicic acid, also called orthosilicic acid (H_4SiO_4). On average, plants absorb 50–200 kg Si·ha^{-1} [12]. However, the absorption ability of plants differs greatly among species. According to Si concentrations found in the tissues, plants can be classified as low accumulators (<0.1% Si), intermediary accumulators (1% Si), and high accumulators (up to 5% Si) [13]. In general, monocots are classified as either intermediate or high accumulators [4]. Most dicots are unable to accumulate Si and belong to the low accumulator classification. However, some dicots of the Asteraceae, Urticaceae, and Cucurbitaceae families are known to benefit from Si fertilizer, indicating that Si is absorbed and accumulated in the tissues of these species [14–16].

The beneficial effects of Si on plant growth, productivity, photosynthesis, balanced nutrient availability, and the mechanisms for reactive oxygen species scavenging have been demonstrated in numerous studies [17,18]. Thus, the application of Si has been widely implemented on various crops to alleviate the deleterious effects of water, salt, and heavy-metal stresses, as well as protect against pest infestation and disease [1,7,9–11,18–21]. Si has been reported to promote stem strength by increasing lignin accumulation. In rice, stem strengthening helps to reduce lodging, thus preventing mutual shading, maintaining canopy photosynthesis, and consequently improving productivity [20,22]. Ahmad and Haddad [19] demonstrated that Si application positively influenced the antioxidant system in *Triticum aestivum* plants. Song et al. [23] reported that leaf chloroplast was disordered and chlorophyll content was reduced under high-Zn stress, which were counteracted by the addition of Si. An interaction between Si and nitrogen and an increase in the levels of chlorophyll *a* in *Oryza sativa* plants were reported by Ávila et al. [24]. Si has been proven to mediate plant defense against insect and pest infestation. Amorphous Si deposition in plant tissues acts as a physical barrier, contributing to increased rigidity and abrasiveness of plants, thus enabling plants to become less digestible for insects [25–27]. Additionally, in several species, Si appears to increase the level of proline and glycine betaine concentrations under drought and salinity stresses to attenuate their negative impacts [28,29]. However, studies in drought-stressed maize and salt-stressed borage reported a decrease in glycine betaine [30,31].

Oil palm (*Elaeis guineensis* Jacq.) is the highest yielding oil crop with an average annual oil yield of 3.3 t·ha^{-1} [32]. In the beginning, oil palm plantations were centered in tropical areas with the optimal conditions for oil palm to grow [33]. Demand for palm oil has greatly increased over the past few decades because of numerous benefits and its applications in food and nonfood industries. To match the growing demand, there has been an expansion of oil palm plantations, but protected areas and other existing land uses have been major limitations [34]. As a perennial crop which generally produces ongoing fruit for up to 30 years, constant high yield is desired. Factors determining yield, i.e., varieties planted, available rain and fertilization must be taken into account. Sufficient nutrition is necessary during the growth and development stages of the oil palm, since nutrient uptake establishes the plant's production potential [33]. In addition to essential nutrients, other elements such as sodium (Na), silicon (Si), and cobalt (Co) have been applied in some species to promote plant growth and development [35]. According to Munevar and Romeo [36] who assessed oil palm throughout Colombia, Si concentration in oil palm ranges from 0.73% to 1.71% in leaf no. 3 and 1.55% to 4.07% in leaf no. 17, indicating Si accumulation and sensitivity to available Si. However, little is known about the potential effects of Si fertilization on oil palm growth and development, especially without biotic and abiotic stresses. Therefore, in this study, we investigated the beneficial effects of silicon fertilization on the growth and physiological responses of oil palm at the seedling stage under nonstress conditions. The results of this research can be advantageous in oil palm nutrition management for nursery establishment and to sustain high yields.

2. Materials and Methods

2.1. Pre-Plantation Soil Analysis

Topsoil of Hat Yai soil series (clayey, skeletal, kaolinitic, isohyperthermic Typic Paleudults) [37] was obtained from the 0–50 cm layer of an agriculture field in Songkhla Province. Soil was finely prepared, and plant parts and roots were removed using a 2 mm sieve. A uniform and homogeneous soil sample was obtained prior to soil filling in planting bags. Soil was airdried and sent for soil analysis to assess soil properties and nutrient concentrations. The texture of the soil was sandy clay loam, and details of soil properties including texture [38], pH, electrical conductivity, total N [39], available P [40], extractable K^+, extractable Mg, extractable Ca, and available Si [41] are presented in Table 1. Available Si in the planting soil used in this current study was 8.774 $mg \cdot kg^{-1}$, which was considered to be a low level [4].

Table 1. Pre-plantation physicochemical properties of soil used for the experiments.

Properties	Values/Description	Methods	References/Instrument
Texture	Sandy clay loam	Hydrometer	Bouyoucos, 1936 [38]
pH	4.76	pH meter, soil/water = 1:5	Seven Easy (Mettler Toledo)
Electrical conductivity ($ds \cdot m^{-1}$)	0.303	EC meter, soil/water = 1:5	Seven Easy EC Meter (Mettler Toledo)
Total N ($g \cdot kg^{-1}$)	0.262	Kjeldahl method	Kjeldahl, 1883 [39]
Available P ($mg \cdot kg^{-1}$)	1.729	Bray II, molybdenum blue method	Bray, 1945 [40]
Extractable K^+ ($cmol \cdot kg^{-1}$)	0.073	1 M-NH_4OAc (pH 7) atomic absorption spectrophotometry	Spectrophotometer
Extractable Mg ($cmol \cdot kg^{-1}$)	0.020	1 M-NH_4OAc (pH 7) atomic absorption spectrophotometry	Spectrophotometer
Extractable Ca ($cmol \cdot kg^{-1}$)	0.034	1 M-NH_4OAc (pH 7) atomic absorption spectrophotometry	Spectrophotometer
Available Si ($mg \cdot kg^{-1}$)	8.774	Yellow molybdenum blue method	Estefan et al., 2013 [41]

2.2. Seedling Transplantation and Adaptation

Seeds of the oil palm *Tenera* variety, a hybrid between *Dura* and *Pisifera* widely grown in the oil palm industry [33], were germinated in seedling trays. Sandy clay loam soil was filled in black plastic planting bags of 40 × 45 cm size. Four month old seedlings were transplanted in planting bags as single seedlings per bag, and the bags were placed at 30 cm plant-to-plant and row-to-row distance in sheds located at the Faculty of Natural Resources, Prince of Songkla University, Hat Yai, Thailand. Plants were manually irrigated on daily basis, and 1.5 L of water per bag was applied to maintain the water content near to field capacity and avoid water stress. Moreover, 7 g of N–P–K (15–09–15) fertilizer not containing any Si was applied to each planting bag twice a month to help seedling establishment and nursery adaptation. Emerging weeds were manually removed. A net was placed to protect oil palm seedlings from pests, and daily monitoring for disease was performed during the experiment.

2.3. Treatment Application

Planting bags containing 4 month old single seedlings per planting bag were arranged using a completely randomized design (CRD) with 18 replications in the greenhouse. Calcium silicate (Ca_2SiO_4 from Sigma-Aldrich, St. Louis, MO, USA) powder was used as the Si fertilizer source. The water solubility of the Ca_2SiO_4 used was 0.26 g/L at 20 °C. Plants were subjected to four treatments: (i) 0 g Ca_2SiO_4 (T0), (ii) 0.5 g $Ca_2SiO_4 \cdot plant^{-1} \cdot month^{-1}$ (T1), (iii) 3.5 g $Ca_2SiO_4 \cdot plant^{-1} \cdot month^{-1}$ (T2), and (iv) 7.0 g $Ca_2SiO_4 \cdot plant^{-1} \cdot month^{-1}$ (T3). Ca_2SiO_4 was applied at 10 cm soil depth from the base of the plants, every month for four consecutive months.

2.4. Data Collection

Nondestructive data including stem diameter, plant height, and leaf length were recorded on 10 reserved seedlings from each treatment before treatment (0 DAT), as well as 60 (60 DAT) and 120 days after treatment (120 DAT). Stem diameter was recorded by measuring the circumference of the base of the stem near the soil surface in the planting bag. Plant height was recorded as the length of oil palm plants from soil surface to the joint of topmost leaf. Leaf length was recorded from the base of the leaf to the tip. A portable photosynthesis measurement system LCpro-SD (ADC BioScientific Ltd., Hoddesdon, UK) was used to record the photosynthetic rate at the third fully expanded leaf in each treatment between 9:00 and 10:00 a.m. for each recording interval. Leaf angle and leaf thickness were measured using a MultispeQ device at the third fully expanded leaf in each treatment. The relative rate of growth in stem diameter, leaf length, and plant height was calculated using a modification of the formula proposed by Hoffmann and Poorter [42].

$$\text{Relative growth rate (RGR)} = \frac{(\ln A_t - \ln A_i)}{t_2 - t_1}, \tag{1}$$

where "ln" is the natural logarithm, "A_t" is the reference value for specific attributes and units at 60 DAT and 120 DAT, and "A_i" is the reference value for specific attribute and units at 0 DAT; "t_2" refers to 60 DAT and 120 DAT, while "t_1" refers to 0 DAT.

Destructive data including fresh weight and dry weight of four oil palm seedlings from each treatment were recorded at 0 DAT, 60 DAT, and 120 DAT. At 120 DAT, four oil palm seedlings were taken from those reserved for nondestructive data collection. Plant samples were first separated into leaves, roots, and stems to record fresh weight and were then kept in the oven for various time intervals at 75 °C until a constant weight was observed. The foliar content of chlorophyll *a*, the main pigment that participates directly in harvesting light energy for photosynthesis in plants [43], was measured from three randomly selected oil palm seedlings at the third fully expanded leaf in each treatment. Leaves were drilled into circular discs with an area of 0.84 square centimeters. Leaf samples were then placed into a glass tube filled with 4 mL of DMF (*N,N*-dimethylformamide), and tubes were covered and stored in the dark for 24 h at 4 °C to prevent chlorophyll contents from being damaged by light. The DMF solution was used to measure the absorbance at 647 and 664 nm wavelengths with a spectrophotometer using pure DMF solution as a reference. The recorded absorbance was used to calculate the chlorophyll *a* content according to Equation (2).

$$\text{Chlorophyll } a = [-2.99(A_{647}) + 12.64\,(A_{664})] \times \frac{\text{vol}}{X \times \text{Area} \times 100}, \tag{2}$$

where A_{647} is the absorbance at a wavelength of 647 nm, A_{664} is the absorbance at a wavelength of 664 nm, vol is the volume of DMF used to extract chlorophyll (mL), and X is the dilution factor.

2.5. Determination of Silicon and Nitrogen Contents in Oil Palm Seedlings

Plant samples were obtained for silicon and nitrogen content analysis to observe the variations in nutrient absorbance under various treatments. The silicon concentration in plant parts was analyzed using the molybdenum blue method [41] to observe silicon accumulation in plants. The distribution of silicon in different parts of plants was calculated using Equation (3). The Kjeldahl method [39] was used to determine the nitrogen content of oil palm seedlings.

$$\text{Si distribution} = (\text{Si concentration in plant part}/\text{Total Si concentration}) \times 100 \quad (3)$$

2.6. Statistical Analysis

Observed data were statistically analyzed using the Statistix 8.1 package (Analytical software, Tallahassee, FL, USA) to study the impact and the significance of fertilization treatments. Means were compared using Fisher's least significant difference (LSD) method at a 95% confidence level. "Corr" and Corrplot packages [44] of R program were used to compute Pearson's correlation matrices and visuals for various attributes as described by Hussain et al. [45].

3. Results

3.1. Accumulation and Distribution of Si in Oil Palm Seedlings

The concentration of silicon (Si) was highest in leaf followed by root and stem at 120 DAT (Table 2). Si-treated soil resulted in a gradual increase in Si content in stem and leaf, ranging from 0.17% to 0.24% and from 0.63% to 0.89%, respectively. Si concentration in the root was not significantly different in oil palm seedlings grown under nontreated and Si−treated soil. The silicon percentage in root ranged from 0.40% to 0.45%.

Table 2. Silicon concentration in the root, stem, and leaf of oil palm after 4 months of calcium silicate application.

Si Fertilization Treatments	Si Concentration (% Dry Weight)		
	Root	**Stem**	**Leaf**
T0	0.45 ± 0.01	0.17 ± 0.09b	0.66 ± 0.02b
T1	0.44 ± 0.02	0.22 ± 0.01ab	0.63 ± 0.05b
T2	0.45 ± 0.02	0.25 ± 0.01a	0.74 ± 0.02ab
T3	0.40 ± 0.05	0.24 ± 0.01a	0.89 ± 0.04a
F-test	ns	*	*
CV (%)	11.60	7.80	8.54

Data are represented as means ± standard errors; ns indicates non-significant; * indicates significant at $p \leq 0.01$; Different letters in the same column indicate significant differences according to the LSD test at $p \leq 0.01$ ($n = 4$).

Generally, Si accumulation increased significantly with seedling age regardless of Si fertilizer treatment (Figure 1A–D). Leaves accumulated the highest amount of Si ranging from 124–163 g·plant^{-1} and 494–736 g·plant^{-1} at 60 DAT and 120 DAT, respectively (Figure 1C). In stem, Si accumulation of 36–44 g·plant^{-1} and 94–134 g·plant^{-1} was observed at 60 DAT and 120 DAT, respectively (Figure 1B). Si accumulation in root was the lowest compared to accumulation in other tissues at 60 DAT, with the range of 24–43 g·plant^{-1} (Figure 1A). However, at 120 DAT, a range of 141–172 g·plant^{-1} was found in oil palm root, surpassing Si accumulation in the stem. Considering the whole plant, Si accumulation was approximately 121 g·plant^{-1} prior to the start of Si treatment. Si accumulation then increased to 172–249 g·plant^{-1} and 740–1040 g·plant^{-1} at 60 DAT and 120 DAT, respectively (Figure 1D). According to the results, application of Si fertilizer to oil palm seedlings generally enhanced Si accumulation in the shoot, stem, and total plant but not in the root. Si fertilization of 3.5 and 7.0 g·plant^{-1}·month^{-1} significantly increased the stem, leaf, and total Si accumulation in oil palm seedlings starting from 60 DAT.

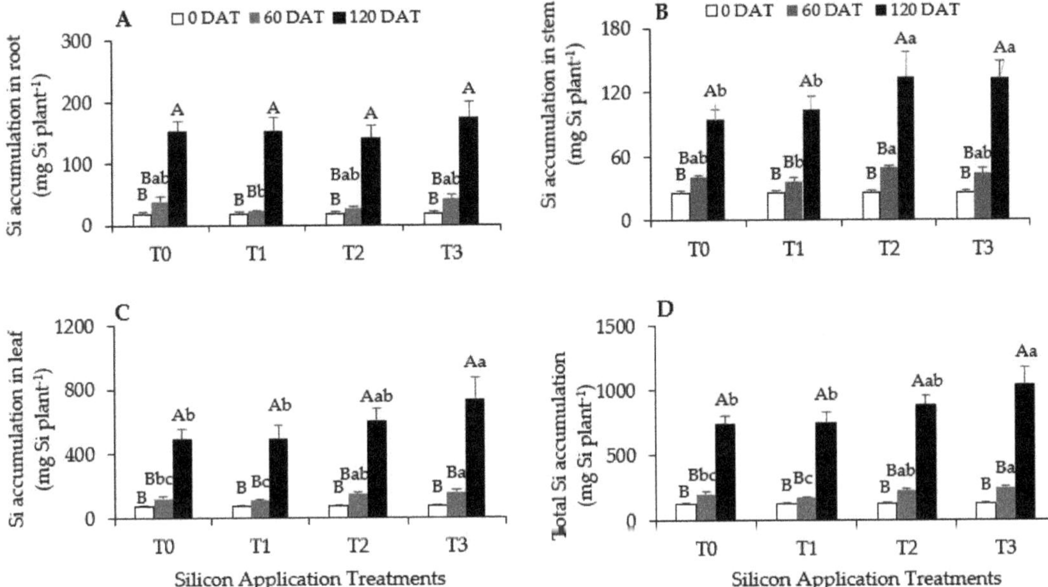

Figure 1. Silicon accumulation in the root (**A**), stem (**B**), leaf (**C**), and total plant (**D**). Data represent the means, and error bars represent the standard errors of the means. Different uppercase letters indicate significant differences in silicon accumulation at 0, 60, and 120 days after treatment (DAT), and different lowercase letters indicate significant differences in silicon accumulation in oil palm seedlings treated with different levels of calcium silicate according to the LSD test at $p \leq 0.01$ (n = 4). T0, T1, T2, and T3 are Si treatments of oil palm seedlings using 0, 0.5, 3.5, and 7.0 g $Ca_2SiO_4 \cdot plant^{-1} \cdot month^{-1}$, respectively.

3.2. Positive Impacts of Silicon on Oil Palm Growth

Application of Si fertilizer stimulated the oil palm growth. The stem, leaf, and total plant biomass of oil palm seedlings was significantly promoted by 3.5 and 7.0 g Si fertilizer·$plant^{-1} \cdot month^{-1}$ treatments (Figures 2 and 3A–D). Root biomass, however, was not affected by Si fertilization. Overall, root dry weight increased 1.56- and 5.30-fold at 60 and 120 DAT, respectively, regardless of Si fertilizer application (Figure 3A). Stem dry weight increased 1.86- and 6.05-fold at 60 and 120 DAT, respectively, without addition of Si fertilizer. Similarly, 1.58- and 5.68-fold increases at 60 and 120 DAT, respectively, were observed in oil palm treated with 0.5 g Si fertilizer·$plant^{-1} \cdot month^{-1}$. A significant increase in stem dry weight with Si supplementation was noted at 60 DAT. Application of Si fertilizer at 3.5 and 7.0 g·$plant^{-1} \cdot month^{-1}$ resulted in 1.95- and 2.07-fold increases, respectively, in stem dry weight at 60 DAT and 7.34- and 7.59-fold increases, respectively, in stem dry weight at 120 DAT (Figure 3B). Similar results were observed for leaf and total plant biomass in oil palm treated with Si fertilizer at 3.5 and 7.0 g·$plant^{-1} \cdot month^{-1}$. At 60 DAT, leaf biomass increased 1.84- and 1.85-fold when silicon fertilizer at 3.5 and 7.0 g·$plant^{-1} \cdot month^{-1}$ was applied, respectively, compared to the 1.58-fold increase in the control treatment. More notable effects of Si fertilizer were demonstrated at 120 DAT. Specifically, 6.38- and 6.39-fold increases in accumulation were observed in leaf biomass of oil palm seedlings with 3.5 and 7.0 g Si·$plant^{-1} \cdot month^{-1}$, respectively, at 120 DAT, whereas oil palm seedlings without Si fertilization and with 0.5 g Si·$plant^{-1} \cdot month^{-1}$ exhibited 5.2- and 4.68-fold increases in leaf dry weight (Figure 3C).

Figure 2. Positive impact of silicon fertilization on growth of oil palm seedlings at 120 days after treatment. T0, T1, T2, and T3 are Si treatments of oil palm seedlings using 0, 0.5, 3.5, and 7.0 g $Ca_2SiO_4 \cdot plant^{-1} \cdot month^{-1}$, respectively.

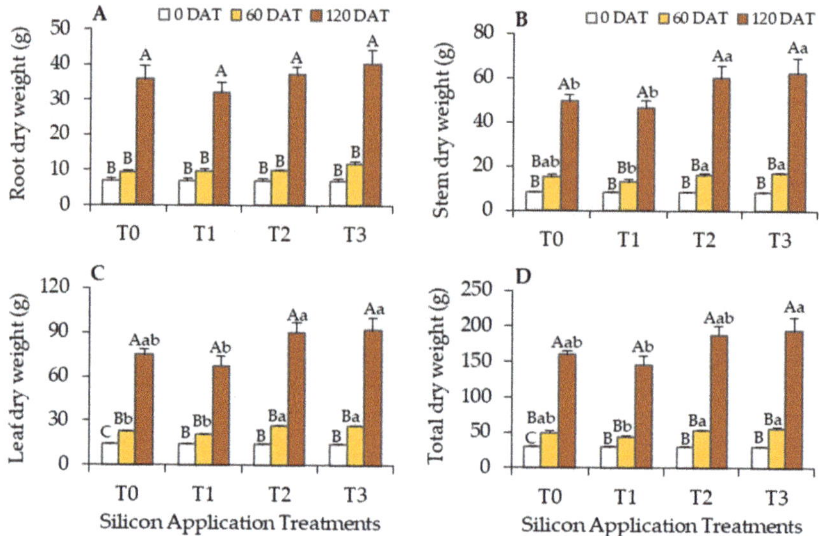

Figure 3. Root (**A**), stem (**B**), leaf (**C**), and total plant (**D**) dry weight. Data represent the means, and error bars represent the standard errors of the means. Different uppercase letters indicate significant differences in silicon accumulation at 0, 60, and 120 days after treatment (DAT), and different lower-case letters indicate significant differences in silicon accumulation in oil palm seedlings treated with different levels of calcium silicate according to the LSD test at $p \leq 0.01$ ($n = 4$). T0, T1, T2, and T3 are Si treatments of oil palm seedlings using 0, 0.5, 3.5, and 7.0 g $Ca_2SiO_4 \cdot plant^{-1} \cdot month^{-1}$, respectively.

Similar patterns in the dry mass accumulation of oil palm seedlings treated with Si fertilizer were noticed for fresh weight (Figure 4). Si fertilization at 0.5 g·plant^{-1}·month^{-1} did not noticeably change the fresh weight accumulation in oil palm seedlings. The stem and leaf but not the root of oil palm seedlings treated with 3.5 and 7.0 g·plant^{-1}·month^{-1} exhibited a significant increase when compared with the control at 60 and 120 DAT (Figure 4). Stem fresh weight increased 1.67-, 1.52-, 2.03-, and 1.76-fold at 60 DAT following control, 0.5, 3.5, and 7.0 g Si·plant^{-1}·month^{-1} treatments, respectively, in contrast to 5.44-, 5.40-, 6.93-, and 6.94-fold, respectively, at 120 DAT (Figure 4B). A significant increase in the leaf fresh weight was noted when at least 3.5 g Si·plant^{-1}·month^{-1} was applied to oil palm seedlings. In the control treatment, leaf fresh weight increased 1.96- and 5.50-fold at 60 and 120 DAT, respectively, whereas Si fertilization rates of 3.5 and 7.0 g·plant^{-1}·month^{-1} significantly stimulated 2.46- and 2.34-fold increases in leaf fresh weight, respectively, at 60 DAT and 6.61- and 6.67-fold increases in leaf fresh weight, respectively, at 120 DAT (Figure 4C).

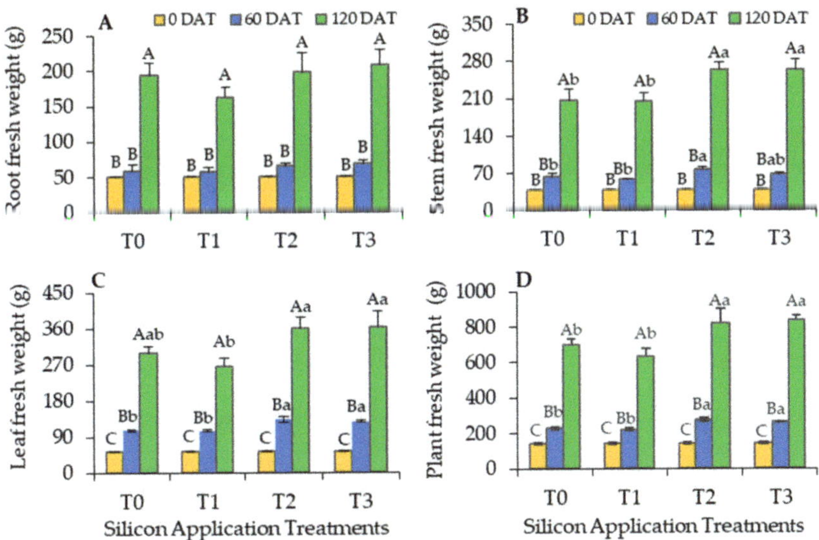

Figure 4. Root (**A**), stem (**B**), leaf (**C**), and total plant (**D**) fresh weight. Data represent the means, and error bars represent the standard errors of the means. Different uppercase letters indicate significant differences in silicon accumulation at 0, 60, and 120 days after treatment (DAT), and different lowercase letters indicate significant differences in silicon accumulation in oil palm seedlings treated with different levels of calcium silicate according to the LSD test at $p \leq 0.01$ ($n = 4$). T0, T1, T2, and T3 are Si treatments of oil palm seedlings using 0, 0.5, 3.5, and 7.0 g Ca$_2$SiO$_4$·plant^{-1}·month^{-1}, respectively.

Si fertilization demonstrated positive effects on the relative growth rate of oil palm stem diameter, leaf length, and plant height (Figure 5). The relative growth rate of stem diameter was generally higher at 60 DAT as compared to 120 DAT with and without Si fertilization. The maximum relative growth rate in stem diameter was observed in oil palm seedlings treated with 7.0 g Ca$_2$SiO$_4$·plant^{-1}·month^{-1} at 60 DAT. The increase in relative growth rate of stem diameter was then abated but still stimulated by the addition of Si fertilizer at 120 DAT (Figure 5A).

Similar trends were demonstrated in the relative growth rate of leaf length and plant height. At 60 DAT, the relative leaf length rate of oil palm seedlings treated with 7.0 g Ca$_2$SiO$_4$·plant^{-1}·month^{-1} was significantly higher as compared to other treatments. At 120 DAT, application of both 3.5 g and 7.0 g Ca$_2$SiO$_4$·plant^{-1}·month^{-1} considerably increased the relative leaf length rate. Under control conditions, the relative leaf length rate at 120 DAT was slightly lower than that at 60 DAT (Figure 5B). The relative plant height rate was also positively affected by Si fertilization. Although the stimulating effects were not observed 60 DAT, a significant increase in the relative plant height rate of oil palm seedlings supplemented with 7.0 g $Ca_2SiO_4 \cdot plant^{-1} \cdot month^{-1}$ was detected 120 DAT (Figure 5C).

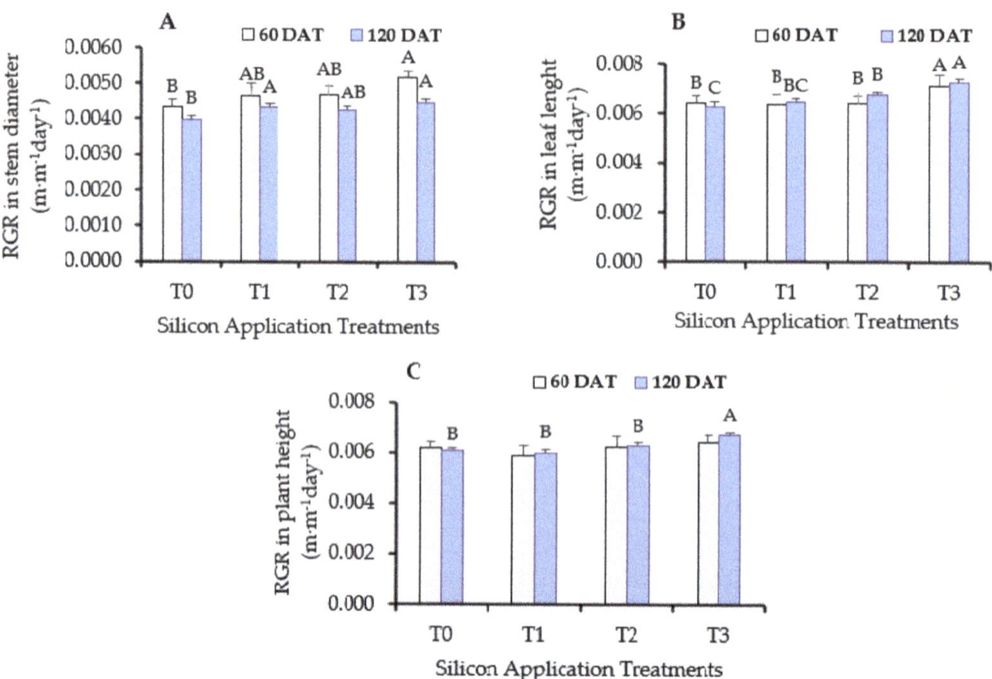

Figure 5. Relative growth rate (RGR) of stem diameter (**A**), leaf length (**B**), and plant height (**C**). Data represent the means, and error bars represent the standard errors of the means. Different uppercase letters indicate significant differences in silicon accumulation at 0, 60, and 120 days after treatment (DAT) in oil palm seedlings treated with different levels of calcium silicate according to the LSD test at $p \leq 0.01$ ($n = 10$). T0, T1, T2, and T3 are Si treatments of oil palm seedlings using 0, 0.5, 3.5, and 7.0 g $Ca_2SiO_4 \cdot plant^{-1} \cdot month^{-1}$, respectively.

3.3. Effects of Silicon Fertilization on Physiological Responses of Oil Palm Seedlings

In the present study, the stimulating effects of Si fertilization on chlorophyll *a* content and photosynthesis were clearly demonstrated (Figure 6). Starting from 60 DAT, the chlorophyll *a* content and photosynthetic rate following treatments with 3.5 and 7.0 g $Ca_2SiO_4 \cdot plant^{-1} \cdot month^{-1}$ were significantly higher than those following control and 0.5 g $Ca_2SiO_4 \cdot plant^{-1} \cdot month^{-1}$ treatments. The chlorophyll *a* content was increased 1.14- and 1.07-fold by the 3.5 and 7.0 g $Ca_2SiO_4 \cdot plant^{-1} \cdot month^{-1}$ treatments, respectively, compared to the control at 60 DAT and increased 1.20- and 1.13-fold, respectively, at 120 DAT (Figure 6A). A significant increase in photosynthetic rate were observed along with an increase in chlorophyll *a* content. At 60 DAT, the photosynthetic rate of oil palm seedlings following 3.5 and 7.0 g $Ca_2SiO_4 \cdot plant^{-1} \cdot month^{-1}$ treatments were increased 1.09- and 1.08-fold, respectively, compared to the control. Similarly, at 120 DAT, 3.5 and 7.0 g $Ca_2SiO_4 \cdot plant^{-1} \cdot month^{-1}$ treatments increased the photosynthetic rate 1.05- and 1.04-fold, respectively, compared to the nontreated seedlings (Figure 6B).

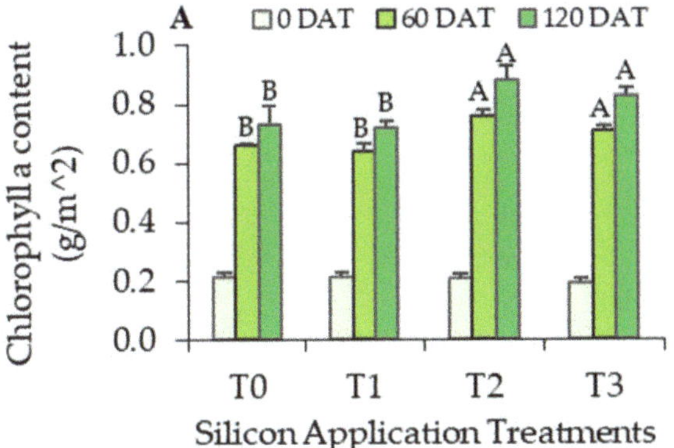
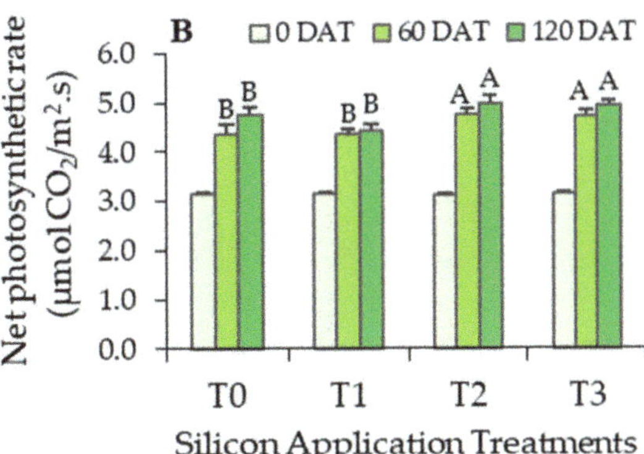

Figure 6. Chlorophyll *a* content (**A**) and net photosynthetic rate (**B**). Data represent the means, and error bars represent the standard errors of the means. Different uppercase letters indicate significant differences in silicon accumulation at 0, 60, and 120 days after treatment (DAT) in oil palm seedlings treated with different level of calcium silicate according to the LSD test at $p \leq 0.01$ ($n = 3$). T0, T1, T2, and T3 are Si treatments of oil palm seedlings using 0, 0.5, 3.5, and 7.0 g $Ca_2SiO_4 \cdot plant^{-1} \cdot month^{-1}$, respectively.

3.4. Effects of Silicon Fertilization on Oil Palm Leaf Morphology

Leaf angle and leaf thickness were measured to elucidate the effects of Si fertilization on leaf morphology (Figure 7). A slight decrease in leaf angle was measured following control and 0.5 g $Ca_2SiO_4 \cdot plant^{-1} \cdot month^{-1}$ treatments, indicating that seedling leaves were less erect, whereas 3.5 and 7.0 g $Ca_2SiO_4 \cdot plant^{-1} \cdot month^{-1}$ treatments led to more upright leaves. At 0 DAT, leaf angle ranged from 44.12° to 44.6° in all treatments. At 120 DAT, leaf angle ranged from 44.89° to 45.08° in 3.5 and 7.0 g $Ca_2SiO_4 \cdot plant^{-1} \cdot month^{-1}$ treatments, respectively, in contrast to 43.53° and 43.45° in control and 0.5 g $Ca_2SiO_4 \cdot plant^{-1} \cdot month^{-1}$ treatments, respectively (Figure 7A). However, leaf thickness was not affected by Si fertilization (Figure 7B). Leaves in all treatments increased in thickness with the age of seedlings, exhibiting a similar pattern across treatments. At 0 DAT, leaf thickness was approximately 0.60 mm, and it increased to the range of 0.64 to 0.68 mm at 60 DAT and to the range of 0.65 to 0.68 mm at 120 DAT (Figure 7B).

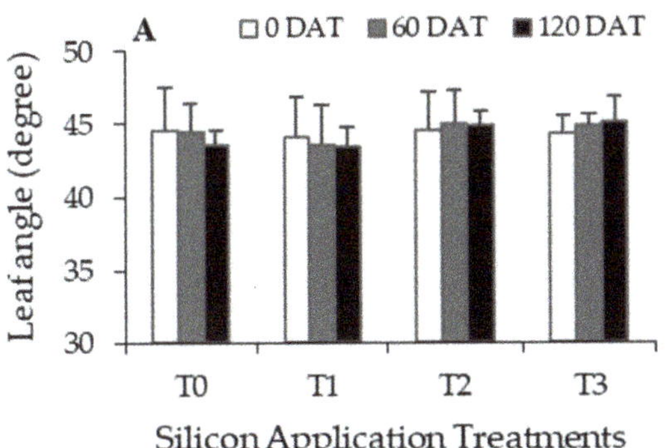
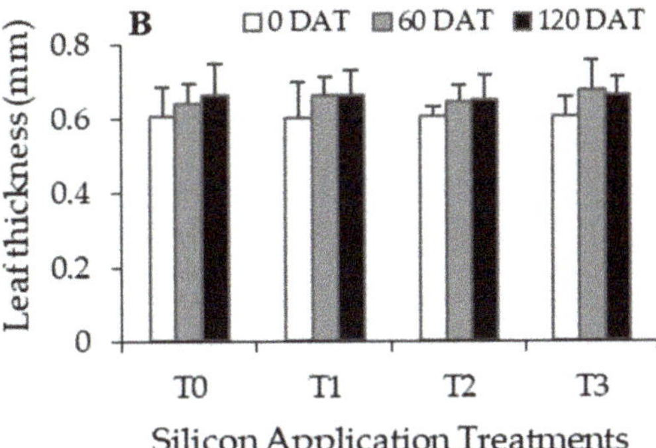

Figure 7. Oil palm leaf angle (**A**) and leaf thickness (**B**). Data represent the means, and error bars represent the standard errors of the means. T0, T1, T2, and T3 are Si treatments of oil palm seedlings using 0, 0.5, 3.5, and 7.0 g $Ca_2SiO_4 \cdot plant^{-1} \cdot month^{-1}$, respectively.

3.5. Effects of Silicon Fertilization on Nitrogen Accumulation in Oil Palm Seedlings

Oil palm seedlings grown in soil treated with 3.5 and 7.0 g $Ca_2SiO_4 \cdot plant^{-1} \cdot month^{-1}$ accumulated considerably higher nitrogen content as compared to the control (Figure 8). Prior to Si treatments, oil palm seedlings contained approximately 0.38 g $N \cdot kg^{-1}$ dry weight. At 60 DAT, oil palm seedlings without Si fertilization accumulated 0.82 g $N \cdot kg^{-1}$ DW, whereas fertilization with 0.5, 3.5, and 7.0 g $Ca_2SiO_4 \cdot plant^{-1} \cdot month^{-1}$ resulted in 0.82, 1.08, and 1.03 g $N \cdot kg^{-1}$ DW, respectively. At 120 DAT, nitrogen accumulation was 2.25, 1.87, 2.43, and 2.58 g $N \cdot kg^{-1}$ DW following treatment with 0, 0.5, 3.5, and 7.0 g $Ca_2SiO_4 \cdot plant^{-1} \cdot month^{-1}$, respectively.

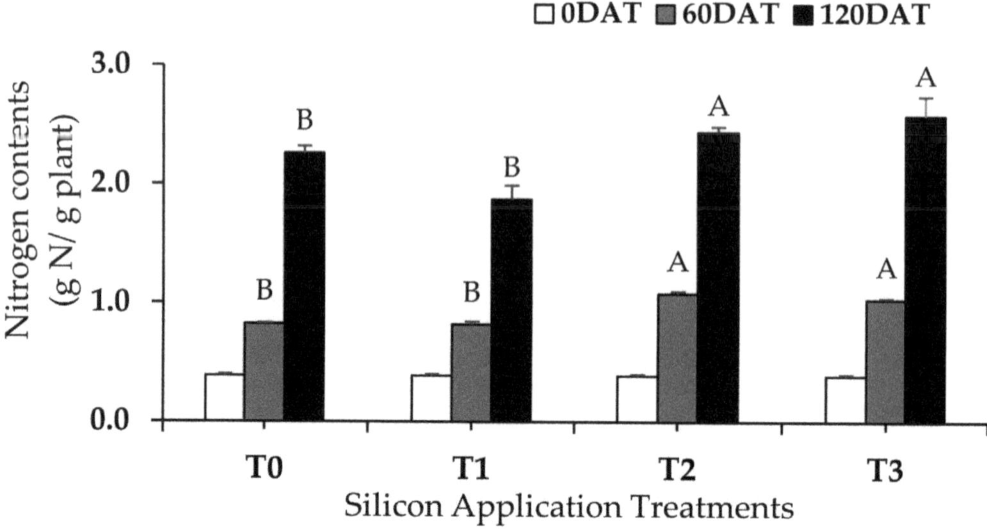

Figure 8. Nitrogen content in oil palm seedlings. Data represent the means, and error bars represent the standard errors of the means. Different uppercase letters indicate significant differences in nitrogen accumulation at 0, 60, and 120 days after treatment (DAT) according to the LSD test at $p \leq 0.01$ ($n = 10$). T0, T1, T2, and T3 are Si treatments of oil palm seedlings using 0, 0.5, 3.5, and 7.0 g $Ca_2SiO_4 \cdot plant^{-1} \cdot month^{-1}$, respectively.

3.6. Correlation Assessment

The correlation assessment (Pearson's) among various attributes (Figure 9) showed high positive associations between Si accumulation and total fresh weight (0.99), total dry weight (0.99), and nitrogen content (0.98). A highly positive correlation was also found between nitrogen content and total fresh weight (0.99) and total dry weight (0.99). A positive relationship was detected between silicon accumulation and chlorophyll *a* content (0.76) and net photosynthesis (0.61). As expected, chlorophyll *a* content and net photosynthesis were highly correlated (0.91). The high associations among these attributes indicated the positive relationship and beneficial impact of Si fertilization. In contrast, negative associations were observed in terms of relative growth rate between stem diameter and total dry weight (−0.60), stem diameter and total fresh weight (−0.59), and stem diameter and nitrogen content (−0.57).

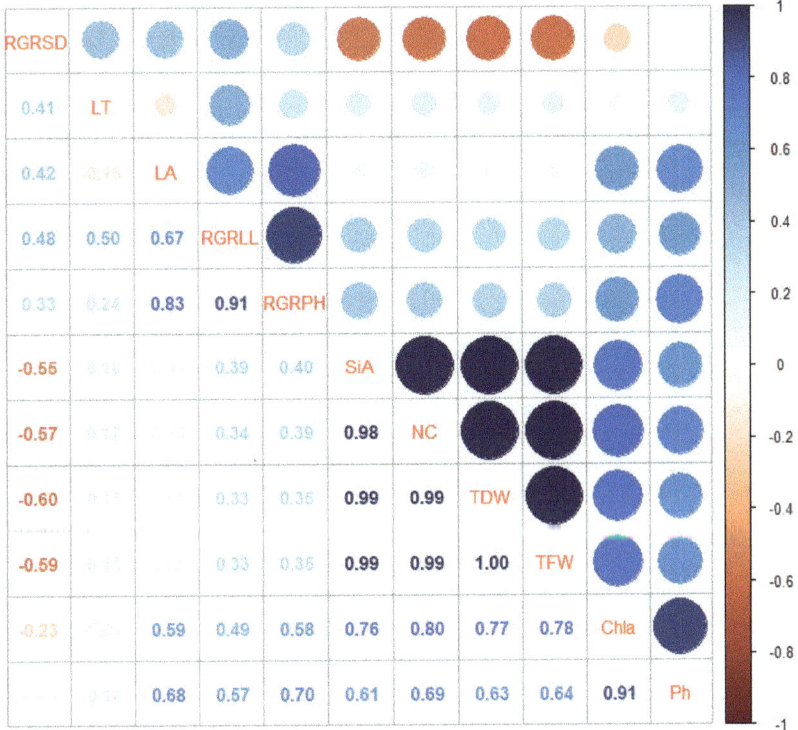

Figure 9. Correlation plot illustrating computed Pearson's correlation coefficients for photosynthesis (Ph), chlorophyll *a* (Chla), total fresh weight (TFW), total dry weight (TDW), nitrogen content (NC), silicon accumulation (SiA), relative growth rate in terms of plant height (RGRPH), relative growth rate in terms of leaf length (RGRLL), leaf angle (LA), leaf thickness (LT), and relative growth rate in terms of stem diameter (RGRSD). Positive and negative associations are indicated by blue and red circles, respectively. Squares with no colored circles represent a nonsignificant association at $p < 0.05$. Computed coefficient values are also listed. The strength of association among different attributes is directly proportional to the color intensity and the size of the circles.

4. Discussion

In the oil palm plantation industry, optimum growth and productivity are not achieved due to various factors, among which nutrient management and well-established nursery seedlings are crucial. Uniform and vigorous seedlings are a key component of improved oil palm production [33,46]. Therefore, improved nutrient management is necessary to achieve healthy and well-adapted vigorous seedlings [33].

The beneficial effects of Si on various crops have been studied intensively under various biotic and abiotic stress conditions. However, the effects of Si on vegetative growth under nonstress conditions have been debated for different plants. Guo et al. [47] reported that alfalfa treated with Si had increased leaf area, height, and forage yield. Costa et al. [48] indicated that Si fertilization at 0.28 to 0.55 g·pot^{-1} provided better growth of passion fruit. In addition, Si application stimulated vegetative growth of rice, sugarcane, strawberry, and soybean [4,12,49]. In contrast, applying Si had no significant effect on the growth of *Spartina anglica* and cowpea [47].

The mechanism via which plants benefit from Si is still unclear. However, for plants to be affected by Si fertilization, Si accumulation needs to be observed [13]. In the present study, oil palm was found to accumulate Si in all tissues, preferably leaf followed by root and stem. According to Tongchu et al. [50], a study on the translocation factor of calcium silicate in oil palm demonstrated that more Si was transferred from the root to the leaf in the presence of calcium silicate compared to the control. Oil palm seedlings accumulated 0.17–0.89% Si (dry weight) across the whole plant, which was, therefore, considered to be an intermediate accumulator according to Ma et al. [13]. A positive relationship between Si accumulation and Si fertilization was noticed in present study. A linear regression relationship between dose of Si application and the concentration of Si in leaf tissues of oil palm seedlings was also reported by Putra et al. [51]. Among plants, Si concentration is found to be higher in monocotyledons than in dicotyledons with an increase in the following order: legumes < fruit crops < vegetables < grasses < grain crops [5]. Si is mainly deposited in plant parts as phytoliths ($SiO_2 \cdot nH_2O$) [52]. This acts as a physical barrier and, thus, improves plant resistance to pathogens and insects [52]. As foliar deposition of Si was evidenced in our results, resistance to leaf spot and leaf blight diseases, commonly found in nursery-stage oil palms, should be investigated.

The present study indicated stimulatory effects of Si on fresh weight, dry weight, and relative vegetative growth rate [17]. Similar results were reported in multiple plants [17,18]. Biomass production is involved in the coordination of different events including the transition from elongation to thickening of stem tissues and synthesis of secondary cell walls impregnated with lignin, many of which are regulated by phytohormones. It has been proposed that the remaining Si in the form of soluble silicic acid ($Si(OH)_4$) may be involved in stimulating biochemical/molecular processes contributing to biomass production under Si supplementation.

Interactions of Si fertilization with other plant nutrients including nitrogen have been well documented [53–55]. Si positively affects almost all aspects of nitrogen nutrition, including nitrogen uptake, assimilation, and remobilization, and it has been reported in many crops [53,56,57]. Beneficial effects of Si on plant growth and production have been reported under low, optimal, and excessive nitrogen supply [58]. In oil palm seedlings, nitrogen content was enhanced with the supplementation of Si. The mechanisms underlying the stimulatory effects of Si have not been reported in any plants, but it has been demonstrated that increased nitrogen concentration in plant parts resulted from enhance nitrogen fixation, as well as upregulation of NO_3^- transporter genes and genes involved in nitrogen uptake.

In the present study, the enhanced growth of oil palm seedlings treated with Si could be attributed to increased chlorophyll *a* content and, consequently, photosynthetic rate. Chlorophyll is an important pigment in photon absorption, transmission, and transportation and is closely related to photosynthesis [59]. Nitrogen is required for the production of chlorophyll, nucleic acids, and enzymes. Therefore, increased nitrogen content can improve the chlorophyll content in crop leaves, thus improving photosynthetic performance [60]. A positive correlation among nitrogen content, chlorophyll *a* content, and photosynthetic rate was also demonstrated in the present study.

Overall, Si fertilization at a certain rate exhibited stimulatory effects on oil palm seedling growth, which could be attributed to increased nitrogen uptake and photosynthesis. An increased growth rate in oil palm seedlings is preferable as it help to shorten the time required for seedling development and establishment. As oil palm plantations are continuously expanded, these beneficial effects of Si on oil palm can facilitate the production of oil palm seedling to be able to keep up with increased demand for seedling materials. As oil palm was demonstrated to be an intermediate Si accumulator, other beneficial effects of Si fertilization under biotic and abiotic stress conditions can be further investigated to contribute to better nutrient management in oil palm.

5. Conclusions

Silicon (Si) fertilization with 3.5 g and 7.0 g $Ca_2SiO_4 \cdot plant^{-1} \cdot month^{-1}$ stimulated growth and physiological processes of oil palm seedlings under nonstress conditions. Oil palm is an intermediate Si accumulator, and it was observed that a higher proportion of Si was deposited in the aerial parts, especially the leaf. Chlorophyll *a* content and photosynthetic rate were positively correlated with Si fertilization and could have contributed to the better growth observed. Nitrogen uptake was demonstrated to be enhanced by Si fertilization. Leaf thickness and leaf angle were not considerably affected by Si fertilization, but slight trends toward more upright leaves with increased Si fertilization were observed and should be further investigated. Overall, Si fertilization provided beneficial effects on growth and physiological responses in oil palm seedlings. Therefore, Si fertilization should be considered for improved nutrient management for healthy and vigorous oil palm seedling and nursery production.

Author Contributions: S.D. conceptualized the study. S.D. and Y.T. conducted the experiments and analyzed data. S.D. and T.H. prepared the first draft. T.E. and J.O. provided technical guidance and edited the manuscript. J.O. proofread the manuscript. All authors contributed to the article and approved the submitted version. All authors have read and agreed to the published version of the manuscript.

Funding: This research was funded by Prince of Songkla University, Hat Yai Campus, Thailand through the grant number NAT601300S.

Institutional Review Board Statement: Not applicable.

Informed Consent Statement: Not applicable.

Data Availability Statement: The data presented in this study are available on request from the corresponding author.

Conflicts of Interest: The authors declare no conflict of interest.

References

1. Hodson, M.J.; Evans, D.E. Aluminium/silicon interactions in higher plants. *J. Exp. Bot.* **1995**, *46*, 161–171. [CrossRef]
2. Hans Wedepohl, K. The composition of the continental crust. *Geochim. Cosmochim. Acta* **1995**, *59*, 1217–1232. [CrossRef]
3. Sommer, M.; Kaczorek, D.; Kuzyakov, Y.; Breuer, J. Silicon pools and fluxes in soils and landscapes—A review. *J. Soil Sci. Plant Nutr.* **2006**, *169*, 310–329.
4. Camargo, M.S.; Keeping, M.G. Silicon in sugarcane: Availability in soil, fertilization, and uptake. *Silicon* **2021**, *13*, 3691–3701. [CrossRef]
5. Meena, V.D. A case for silicon fertilization to improve crop yields in tropical soils. *Proc. Natl. Acad. Sci. India Sect. B Biol. Sci* **2014**, *84*, 505–518. [CrossRef]
6. Dobermann, A.; Fairhurst, T. *Rice: Nutrient Disorder & Nutrient Management*; Potash & Phosphate Institute (PPI), Potash & Phosphate Institute of Canada (PPIC) and International Rice Research Institute: Laguna, Philippines, 2000; p. 191.
7. Bari, M.A.; Prity, S.A.; Das, U.; Akther, M.S.; Sajib, S.A.; Reza, M.A.; Kabir, A.H. Silicon induces phytochelatin and ROS scavengers facilitating cadmium detoxification in rice. *Plant Biol.* **2020**, *22*, 472–479. [CrossRef]
8. Chaiwong, N.; Rekasem, B.; Pusadee, T. Silicon application improves caryopsis development and yield in rice. *J. Sci. Food Agric.* **2021**, *101*, 220–228. [CrossRef]
9. Geng, A.; Wang, X.; Wu, L.; Wang, F.; Wu, Z.; Yang, H.; Chen, Y.; Wen, D.; Liu, X. Silicon improves growth and alleviates oxidative stress in rice seedlings (*Oryza sativa* L.) by strengthening antioxidant defense and enhancing protein metabolism under arsanilic acid exposure. *Ecotoxicol. Environ. Saf.* **2018**, *15*, 266–273.
10. Ahmad, M.; Ahmad, M.; El-Saeid, M.H.; Akram, M.A.; Ahmad, H.R.; Hussain, H.H.A. Silicon fertilization—A tool to boost up drought tolerance in wheat (*Triticum aestivum* L.) crop for better yield. *J. Plant Nutr.* **2016**, *39*, 1283–1291. [CrossRef]
11. Verma, K.K.; Song, X.P.; Zeng, Y.; Guo, D.J.; Singh, M.; Rajput, V.D.; Malviya, M.K.; Wei, K.J.; Sharma, A.; Li, D.P.; et al. Foliar application of silicon boosts growth, photosynthetic leaf gas exchange, antioxidative response and resistance to limited water irrigation in sugarcane (*Saccharum officinarum* L.). *Plant Physiol. Biochem.* **2021**, *166*, 582–592. [CrossRef]
12. Sun, X.; Liu, Q.; Tang, T.; Chen, X.; Luo, X. Silicon fertilizer application promotes phytolith accumulation in rice plants. *Front. Plant Sci.* **2019**, *10*, 425. [CrossRef]
13. Ma, J.F.; Miyake, Y.; Takahashi, E. Silicon as a beneficial element for crop plants. In *Studies in Plant Science*; Datnoff, L.E., Snyder, G.H., Korndörfer, G.H., Eds.; Elsevier: Amsterdam, The Netherlands, 2001; pp. 17–39.
14. Adatia, M.H.; Besford, R. The effects of silicon on cucumber plants grown in recirculating nutrient solution. *Ann. Bot.* **1986**, *58*, 343–351. [CrossRef]
15. Hodson, M. Phylogenetic variation in the silicon composition of plants. *Ann. Bot.* **2005**, *96*, 1027–1046. [CrossRef]
16. Mitani, N. Uptake system of silicon in different plant species. *J. Exp. Bot.* **2005**, *56*, 1255–1261. [CrossRef]
17. Hossain, M.T.; Mori, R.; Soga, K.; Wakabayashi, K.; Kamisaka, S.; Fujii, S.; Yamamoto, R.; Hoson, T. Growth promotion and an increase in cell wall extensibility by silicon in rice and some other *Poaceae* seedlings. *J. Plant Res.* **2002**, *115*, 23–27. [CrossRef]

18. Ahmed, M.; Qadeer, U.; Hassan, F.; Fahad, S.; Naseem, W.; Duangpan, S.; Ahmad, S. Abiotic Stress Tolerance in Wheat, and the Role of Silicon: An Experimental Evidence. In *Agronomic Crops*; Hasanuzzaman, M., Ed.; Springer: Singapore, 2020.
19. Ahmad, T.; Haddad, R. Study of silicon effects on antioxidant enzyme activities and osmotic adjustment of wheat under drought stress. *Czech J. Genet. Plant Breed.* **2011**, *47*, 17–27. [CrossRef]
20. Berahim, Z.; Omar, H.M.; Zakaria, N.; Ismail, M.R.; Rosle, R.; Roslin, N.A.; Che'ya, N.N. Silicon improves yield performance by enhancement in physiological responses, crop imagery, and leaf and culm sheath morphology in new rice line, PadiU Putra. *BioMed Res. Int.* **2021**, *2021*. [CrossRef]
21. Verma, K.K.; Liu, X.H.; Wu, K.C.; Singh, K.R.; Song, Q.; Malviya, M.K.; Song, X.; Singh, P.; Verma, C.L.; Li, Y. The impact of silicon on photosynthetic and biochemical responses of sugarcane under different soil moisture levels. *Silicon* **2020**, *12*, 1355–1367. [CrossRef]
22. Dorairaj, D.; Ismail, M.R.; Sinniah, U.R.; Ban, T.K. Influence of silicon on growth, yield, and lodging resistance of MR219, a lowland rice of Malaysia. *J. Plant Nutr.* **2017**, *40*, 1111–1124. [CrossRef]
23. Song, A.; Li, P.; Li, Z.; Liang, Y. The effect of Silicon on photosynthesis and expression of its relevant genes in rice (*Oryza sativa* L.) under high-zinc stress. *PLoS ONE* **2014**, *9*, e113782.
24. Ávila, F.; Baliza, D.; Faquin, V.; Araujo, J.; Ramos, S. Silicon-nitrogen interaction in rice cultivated under nutrient solution. *Cienc. Agron.* **2010**, *41*, 184–190. [CrossRef]
25. Reynolds, O.L. Silicon: Potential to promote direct and indirect effects on plant defense against arthropod pests in agriculture. *Front. Plant Sci.* **2016**, *7*, 744. [CrossRef] [PubMed]
26. Massey, F.P.; Hartley, S.E. Experimental demonstration of the antiherbivore effects of silica in grasses: Impacts on foliage digestibility and vole growth rates. *Proc. Biol. Sci.* **2006**, *273*, 2299–2304. [CrossRef] [PubMed]
27. Massey, F.P.; Hartley, S.E. Physical defences wear you down: Progressive and irreversible impacts of silica on insect herbivores. *J. Anim. Ecol.* **2009**, *78*, 281–291. [CrossRef]
28. Yang, R.; Howe, J.A.; Golden, B.R. Calcium silicate slag reduces drought stress in rice (*Oryza sativa* L.). *J. Agron. Crop. Sci.* **2019**, *205*, 353–361. [CrossRef]
29. Pei, Z.F.; Ming, D.F.; Liu, D.; Wan, G.L.; Geng, X.X.; Gong, H.J.; Zhou, W.J. Silicon improves the tolerance to water-deficit stress induced by polyethylene glycol in wheat (*Triticum aestivum* L.) seedlings. *J. Plant Growth Regul.* **2010**, *29*, 106–115. [CrossRef]
30. Parveen, A.; Liu, W.; Hussain, S.; Asghar, J.; Perveen, S.; Xiong, Y. Silicon priming regulates morpho-physiological growth and oxidative metabolism in maize under drought stress. *Plants* **2019**, *8*, 431. [CrossRef]
31. Torabi, F.; Majd, A.; Enteshari, S. The effect of silicon on alleviation of salt stress in borage (*Borago officinalis* L.). *Soil Sci. Plant Nutr.* **2015**, *61*, 788–798. [CrossRef]
32. Woittiez, L.; Wijk, M.; Slingerland, M.; van Noordwijk, M.; Giller, K. Yield gaps in oil palm: A quantitative review of contributing factors. *Eur. J. Agron.* **2017**, *83*, 57–77. [CrossRef]
33. Corley, R.H.V.; Tinker, P.B. *The Oil Palm*, 4th ed.; Blackwell Publishing: Oxford, UK, 2003; p. 284.
34. Xin, Y.; Sun, L.; Hansen, M.C. Biophysical and socioeconomic drivers of oil palm expansion in Indonesia. *Environ. Res. Lett.* **2021**, *16*, 034048. [CrossRef]
35. Pilon-Smits, E.A.; Quinn, C.F.; Tapken, W.; Malagoli, M.; Schiavon, M. Physiological functions of beneficial elements. *Curr. Opin. Plant Biol.* **2009**, *12*, 267–274. [CrossRef]
36. Munevar, M.; Romero, F.A. Soil and plant silicon status in oil palm crops in Colombia. *Exp. Agric.* **2015**, *51*, 382–392. [CrossRef]
37. Onthong, J.; Osaki, M.; Nilnond, C.; Tadano, T. Phosphorus status of some highly weathered soils in Peninsular Thailand and availability in Relation to citrate and oxamate application. *Soil Sci. Plant Nutr.* **1999**, *45*, 627–637. [CrossRef]
38. Bouyoucos, G.J. Directios for making mechanical analyses of soils by the hydrometer method. *Soil Sci.* **1936**, *42*, 225–230. [CrossRef]
39. Kjeldahl, J. Neue Methode zur Bestimmung des Stickstoffs in organischen Korpern. *Z. Anal. Chem.* **1883**, *22*, 366–382. [CrossRef]
40. Bray, R.H.; Kurtz, L.T. Determination of toal, organic, and available forms of phosphorus in soils. *Soil Sci.* **1945**, *59*, 39–46. [CrossRef]
41. Estefan, G.; Sommer, R.; Ryan, J. *Methods of Soil, Plant, and Water Analysis: A Manual for the West Asia and North Africa Region*; International Center for Agricultural Research in the Dry Areas: Beirut, Lebanon, 2013.
42. Hoffmann, W.; Poorter, H. Avoiding bias in calculations of relative growth rate. *Ann. Bot.* **2002**, *90*, 37–42. [CrossRef]
43. Lepeduš, H.; Vidaković-Cifrek, Z.; Šebalj, I.; Dunić, J.A.; Cesar, V. Effects of low and high irradiation levels on growth and PSII efficiency in *Lemna minor* L. *Acta Bot. Croat.* **2020**, *79*, 185–192. [CrossRef]
44. Wei, T.; Simko, V. R Package "Corrplot": Visualization of a Correlation Matrix. Available online: https://github.com/taiyun/corrplot (accessed on 10 November 2021).
45. Hussain, T.; Hussain, N.; Ahmed, M.; Nualsri, C.; Duangpan, S. Responses of lowland rice genotypes under terminal water stress and identification of drought tolerance to stabilize rice productivity in southern Thailand. *Plants* **2021**, *10*, 2565. [CrossRef]
46. Duangpan, S.; Buapet, P.; Sujitto, S.; Eksomtramage, T. Early assessment of drought tolerance in oil palm $D \times P$ progenies using growth and physiological characters in seedling stage. *Plant Genet. Resour. C* **2018**, *16*, 544–554. [CrossRef]
47. Guo, Z.G.; Liu, H.X.; Tian, F.P.; Zhang, Z.H.; Wang, S.M. Effect of silicon on the morphology of shoots and roots of alfalfa (*Medicago sativa*). *Anim. Prod. Sci.* **2006**, *46*, 1161–1166. [CrossRef]

48. Costa, B.N.S.; Dias, G.M.G.; Costa, I.J.S.; Assis, F.A.; Silveira, F.A.S.; Pasqual, M. Effects of silicon on the growth and genetic stability of passion fruit. *Acta Sci.* **2016**, *38*, 503–511. [CrossRef]

49. Shamshiripour, M.; Motesharezadeh, B.; Rahmani, H.A.; Alikhani, H.A.; Etesami, H. Optimal concentrations of silicon enhance the growth of soybean (*Glycine Max*, L.) cultivars by improving nodulation, root system architecture, and soil biological properties. *Silicon* **2021**. [CrossRef]

50. Tongchu, C.; Onthong, J.; Duangpan, S. Effects of calcium silicate fertilizer on silicon accumulation of oil palm seedling. *Thai J. Soils Fertil.* **2018**, *41*, 18–25.

51. Putra, E.T.S.; Issukindarsyah, I.; Taryono, T.; Purwanto, B.H.; Indradewa, D. Role of boron and silicon in inducing mechanical resistance of oil palm seedlings to drought stress. *J. Appl. Sci.* **2016**, *16*, 242–251. [CrossRef]

52. Luyckx, M.; Hausman, J.F.; Lutts, S.; Guerriero, G. Impact of silicon in plant biomass production: Focus on bast fibres, hypotheses, and perspectives. *Plants* **2017**, *6*, 37. [CrossRef]

53. Pavlovic, J.; Kostic, L.; Bosnic, P.; Kirkby, E.A.; Nikolic, M. Interactions of silicon with essential and beneficial elements in plants. *Front. Plant Sci.* **2021**, *12*, 697592. [CrossRef]

54. Bokor, B.; Ondoš, S.; Vaculík, M.; Bokorová, S.; Weidinger, M.; Lichtscheidl, I.; Turňa, J.; Lux, A. Expression of genes for Si uptake, accumulation, and correlation of Si with other elements in Ionome of Maize Kernel. *Front. Plant Sci.* **2017**, *8*, 1063. [CrossRef]

55. Greger, M.; Landberg, T.; Vaculík, M. Silicon influences soil availability and accumulation of mineral nutrients in various plant species. *Plants* **2018**, *7*, 41. [CrossRef]

56. Pati, S.; Pal, B.; Badole, S.; Hazra, G.C.; Mandal, B. Effect of silicon fertilization on growth, yield, and nutrient uptake of rice. *Commun. Soil Sci. Plant Anal.* **2016**, *47*, 284–290. [CrossRef]

57. Cuong, T.X.; Ullar, H.; Datta, A.; Hanh, T.C. Effects of silicon-based fertilizer on growth, yield and nutrient uptake of rice in tropical zone of Vietnam. *Rice Sci.* **2017**, *24*, 283–290. [CrossRef]

58. Gou, T.; Yang, L.; Hu, W.; Chen, X.; Zhu, Y.; Guo, J.; Gong, H. Silicon improves the growth of cucumber under excess nitrate stress by enhancing nitrogen assimilation and chlorophyll synthesis. *Plant Physiol. Biochem.* **2020**, *152*, 53–61. [CrossRef] [PubMed]

59. Liu, Z.; Gao, F.; Yang, J.; Zhen, X.; Li, Y.; Zhao, J.; Li, J.; Qian, B.; Yang, D.; Li, X. Photosynthetic characteristics and uptake and translocation of nitrogen in peanut in a wheat-peanut rotation system under different fertilizer management regimes. *Front. Plant Sci.* **2019**, *10*, 86. [CrossRef] [PubMed]

60. Bassi, D.; Menossi, M.; Mattiello, L. Nitrogen supply influences photosynthesis establishment along the sugarcane leaf. *Sci. Rep.* **2018**, *8*, 2327. [CrossRef]

Article

Optimal Fertilization Level for Yield, Biological and Quality Traits of Soybean under Drip Irrigation System in the Arid Region of Northwest China

Jing Li [1], Gengtong Luo [2], Abdulwahab S. Shaibu [1], Bin Li [1], Shengrui Zhang [1] and Junming Sun [1,*]

[1] The National Engineering Research Center of Crop Molecular Breeding, MARA Key Laboratory of Soybean Biology (Beijing), Institute of Crop Sciences, Chinese Academy of Agricultural Sciences, 12 Zhongguancun South Street, Beijing 100081, China; lijing02@caas.cn (J.L.); asshuaibu.agr@buk.edu.ng (A.S.S.); libin02@caas.cn (B.L.); zhangshengrui@caas.cn (S.Z.)

[2] Institute of Crops Research, Xinjiang Academy of Agricultural and Reclamation Sciences, Shihezi 832000, China; sunny1@yeah.net

* Correspondence: sunjunming@caas.cn; Tel.: +86-10-82105805

Abstract: Soybean is one of the most important oilseed crops worldwide. Fertilization severely restricts the yield potential of soybean in the arid regions of Northwest China. A two-year field experiment was conducted to investigate the effects of fertilization on soybean yield in arid areas under a drip irrigation system. The treatment consisted of 14 fertilizer combinations comprising of four rates each of nitrogen (N) (0, 225, 450, and 675 kg ha^{-1}), phosphorus (P) (0, 135, 270, and 405 kg ha^{-1}), and potassium (K) (0, 75, 150, and 225 kg ha^{-1}). The results revealed that grain yield was more sensitive to N fertilizer than to P and K fertilizers. The P and K fertilizers influenced harvest index and biomass, respectively. The optimized combination of fertilizers for high yield, as well as biological and quality traits was obtained by quadratic polynomial regression analysis. The theoretical grain yields based on the performed statistical calculations and plant biomass were greater than 7.21 tons ha^{-1} and 16.38 tons ha^{-1} with 300,000 plants ha^{-1} and were obtained under a fertilization combination of 411.62–418.39 kg ha^{-1} N, 153.97−251.03 kg ha^{-1} P$_2$O$_5$, and 117.77−144.73 kg ha^{-1} K$_2$O. Thus, our findings will serve as a guideline for an effective fertilizer application in order to achieve a balance between grain yield and plant biomass as well as to contribute to the promotion of large-scale cultivation of soybean under drip irrigation.

Keywords: drip irrigation; high-yield; fertilization rate; biomass; soybean (*Glycine max* L. Merrill)

Citation: Li, J.; Luo, G.; Shaibu, A.S.; Li, B.; Zhang, S.; Sun, J. Optimal Fertilization Level for Yield, Biological and Quality Traits of Soybean under Drip Irrigation System in the Arid Region of Northwest China. *Agronomy* **2022**, *12*, 291. https://doi.org/10.3390/agronomy12020291

Academic Editors: Christos Noulas, Shahram Torabian and Ruijun Qin

Received: 27 December 2021
Accepted: 21 January 2022
Published: 24 January 2022

Publisher's Note: MDPI stays neutral with regard to jurisdictional claims in published maps and institutional affiliations.

1. Introduction

Soybean (*Glycine max* [L.] Merrill) is one of the most important oilseed crops with rich protein and oil worldwide. In China, the annual import of soybean reached more than 100 million tons, accounting for more than 83% of China's soybean demand in 2020 according to the General Administration of Customs of the People's Republic of China (GACC, http://www.customs.gov.cn/, accessed on 21 December 2021). A huge potential in yield increase is possible for soybean production in China, however, soybean production in China has decreased in recent years because of lower yield levels and lagging technological progress [1]. Therefore, it is important to increase and sustain the yield of soybean with optimal fertilizer application to ensure food security in China. However, the fertilizer management in the current farmers' practices is not usually in balance with crop demand [2], which limits the soybean yield and results in low nutrient use efficiency [3].

Supply of adequate fertilizer including nitrogen (N), phosphorus (P), and potassium (K) is fundamental in optimizing soybean yield and quality. Grain yield and N relationships have been extensively explored in the scientific literature [4–6], nonetheless, relationships for other nutrients such as P and K have received less attention. Previous research found that treatments with high N rates extended the duration of the seed filling due to its function of biological N fixation [7]. Soybeans require higher nitrogen (8 to 9 kg of N for 100 kg soybeans), but only about 1/3 of it comes from fertilizers. An overdependence on N, P, and K fertilizers may deteriorate soil quality and health and ultimately reduce soil fertility and the size of arable land [8]. Therefore, to achieve a stable planting area, to have grain yield improvement, and to realize industrialization, the balanced requirements of N, P, and K fertilizers and their proper combinations are essential in identifying optimal fertilizer application regimes. For implementation, a robust fertilizer recommendation method must be established to maximize the soybean yield and improve nutrient use efficiency.

Fertilizer along with drip irrigation is a technology that offers precise and accurate irrigation and fertilization and can save more than 30–50% of fertilizer consumption, as well as increasing the nutrients and rainfall use efficiency together with the net-profit [9]. With the strengthening of the concept of sustainable development in people's consciousness, the prominent role of drip irrigation and fertilization in resource utilization and environmental protection has attracted increasing attention [10,11]. Thus, exploring the optimum fertilizer recommendation for soybean under drip irrigation may be one as-yet untried method.

Since 2005, the determination of optimum fertilization has been extensively carried out throughout China. The "3414" fertilizer experiment design [three fertilizer factors (N, P, K), four fertilization levels, and 14 types of proportional fertilizer treatments] developed by the Ministry of Agriculture of China was recommended for soil testing and fertilizer research to develop a fertilization system and to guide farmers on how to apply fertilizer. The design is considered optimal with less regression, high efficiency, and comparability, easy in demonstration and promotion, and satisfying the professional requirements for fertilizer testing and fertilization decisions. Thus, this test scheme has been successfully applied to pumpkin [12], phoebe bournei [13], and adzuki bean [14] to obtain an optimal fertilizer application.

To optimize soybean grain yield, plant biomass is also the main factor that determines biological yield [15]. Previous studies improved grain yield by agronomic measures [16,17], but only a few studies have focused on the increase in grain yield balanced with biomass. Several studies have described the correlation of grain yield with plant biomass and reported contradictory conclusions. Some researchers agreed that a significant positive correlation exists between plant biomass and specific grain yield components, which increases with the process of plant growth and development and reaches the maximum at the seed-filling (stage R5 and R6) [18,19]. Other studies did not observe significant correlations between plant biomass and grain yield [20]. Therefore, the correlation appears to be unclear and the achievement of grain yield and plant biomass harmony are needed to better understand the genetic basis of grain yield and facilitate the pyramiding of the optimal fertilizer amount.

Therefore, the objectives of the current study were to (i) dissect the effect of precision N, P, and K management on soybean under drip irrigation; (ii) determine the relationship among a wide range of agronomic, quality, and biomass traits; and (iii) determine the optimal amount of N, P, K fertilizer and reasonable management balance for grain yield and plant biomass.

2. Materials and Methods

2.1. Experimental Site

A field experiment was conducted in Wulanwusu Agricultural Meteorological Experimental Station in 2014 and 2015, which represents the ecological conditions in northern Xinjiang, China. The annual mean temperatures and cumulative precipitations were 10.18 °C and 158.2 mm and 16.9 °C and 161.4 mm during the soybean growing season in 2014 and 2015, respectively (Figure 1 and Table S1). According to the FAO classification [21], the soil is Haplic Calcisols with sandy loam soil texture and pH of 8.0–8.5. The content of nitrate nitrogen, ammoniacal nitrogen, available phosphorus, and available potassium were 13.85, 12.40, 3.82, and 143.16 mg kg^{-1} DW of soil, respectively.

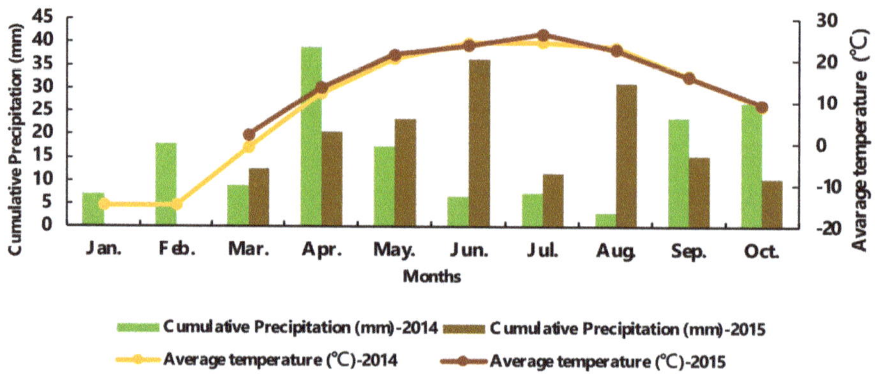

Figure 1. The average temperature and cumulative precipitation for each month in 2014 and 2015.

2.2. Experimental Materials and Design

A high-yielding and drought-sensitive cultivar Zhonghuang 35 was used in our research. The cultivar has high oil, early maturity, broad adaptability [22], and grain yield of 6.32 tons ha^{-1}. The cultivar was evaluated with drip irrigation in the Xinjiang province of China, which showed the advantages of photosynthetic accumulation due to the typical continental climate, the large temperature difference between day and night with an average of 11°C and maximum higher than 20 °C [23].

In 2014 and 2015, the "3414" experiment was conducted with three factors (N, P, and K), each with four levels (0, 1, 2, and 3) giving a combination of 13 treatments (2014) and 14 treatments (2015) (Table 1). For the four levels, 0 indicates no fertilizer, 1 indicates half of the typical fertilizing amount, 2 indicates the typical fertilizer application, and 3 indicates 1.5 times the typical application. All treatments were arranged in a randomized complete block design with three replications.

Table 1. The 3414-fertilization experiment design for soybean.

No.	Treatment	Urea (kg ha^{-1})	Monoammonium Phosphate (kg ha^{-1})	Potassium Chloride (kg ha^{-1})	Total Fertilizer in the Block (kg)		
		(N:46%)	(P$_2$O$_5$:61%)	(K$_2$O:62%)	Urea	Monoammonium Phosphate	Potassium Chloride
1	N0P0K0	0	0	0	0.00	0.00	0.00
2	N0P2K2	0	519	0	0.00	1.05	0.00
3	N1P2K2	374	443	242	0.75	0.89	0.49
4	N2P0K2	978	0	242	1.97	0.00	0.49
5	N2P1K2	921	221	242	1.86	0.45	0.49
6	N2P2K2	863	443	242	1.74	0.89	0.49
7	N2P3K2	805	664	242	1.62	1.34	0.49
8	N2P2K0	863	443	0	1.74	0.89	0.00
9	N2P2K1	863	443	121	1.74	0.89	0.24
10	N3P2K2	1352	443	242	2.73	0.89	0.49
11	N1P1K2	431	221	242	0.87	0.45	0.49
12	N1P2K1	374	443	121	0.75	0.89	0.24
13	N2P1K1	921	221	121	1.86	0.45	0.24
14	N2P2K3	863	443	363	1.74	0.89	0.73

The fertilizers, used for the "3414" experiment, were applied as follows: N fertilizer (urea containing 46% N and diammonium phosphate containing 12% N), P fertilizer (diammonium phosphate containing 61% P$_2$O$_5$ and monopotassium phosphate containing 52% P$_2$O$_5$), and K fertilizer (potassium chloride containing 62% K$_2$O and monopotassium phosphate containing 34% K_2O). To avoid excessive K fertilizer, monopotassium phosphate was used to meet the requirement for P and K fertilizer in N0P2K2 treatment. Detailed fertilization information is presented in Tables S2 and S3.

2.3. Field Management and Cultivation Conditions

The experiment was managed by water-saving drip irrigation under a plastic mulching film. The total irrigation water was 6750 m^3 per ha. Each irrigation pipe was set up between every two soybean rows under the plastic film to facilitate mechanical harvesting. Before planting, the seeds were treated with the appropriate strains of bacteria, plant density was designed with 180,000 seedlings per ha. The sowing date was 14 April in 2014 and 25 April in 2015. The sowing rate was 65 kg ha^{-1}, the depth of sowing was 3 cm and the average of emergence was 80% for all treatments. Independent fertilizer sources were used to apply each treatment to exactly control fertilizer amount. According to the soybean growth stages, fertilizer was applied six times in 2014, including once in June, twice in July, once in August, and twice in September. In 2015, fertilizer was applied seven times, that is once in May, once in June, thrice in July, and twice in August. Detailed information is shown in Table S4.

2.4. Data Observations

The plants were harvested on 14 September 2014 and 13 September 2015. Thirty-one (31) traits relating to grain yield, biomass, and nutrition were measured (Table S5). The identification of 12 yield-related traits, that is from T1 to T12, were measured in the laboratory after harvesting. The biological-related traits (T13 to T16) which include biomass per pod, leaf, stem, and plant were measured as dry matter weight. The nutritional traits (T17 to T31) comprised of N, P, K content in different tissues (seed, pod, leaf, and stem), H_2SO_4–H_2O_2 was used for the combined digestion, N content was determined using standard Kjeldahl by an automatic nitrogen analyzer (ZDDN-III-A, Zhejiang Top Cloud-Agri Technology Co., Zhejiang, China) [24], P content was determined by ultraviolet spectrophotometer (SHIMADZU UV-1800, Shimadzu Corporation, Kyoto, Japan) [24], K content was measured by a flame atomic absorption spectrometer (Model GGX-6, Beijing Haiguang Instrument Co., Beijing, China) after microwave digestion (Discover SP-D Gold, CEM, KampLintfort) using a contrAA 700 high-resolution continuum source atomic absorption spectrometer (Analytik Jena, Jena, Germany)) [24]. In addition, protein, oil, and moisture contents in seeds were determined on 20−30 g seed samples using Fourier-transform near-infrared spectrometry (Bruker Optik GmbH, Ettlingen, Germany) based on a method published previously [25]. The samples were stored at 4 °C in plastic bags in a fridge until measurement. In every case, three replicates were measured.

2.5. Data Analysis

To investigate the relationships among the 31 traits, heatmap for normalized data was conducted using the "pheatmap" package [26], correlation analysis was carried out using the "corrplot" package [27], and principal component analysis using the "cluster" package [28]. To select the optimal model, we compared the adjusted r^2 for the different combinations of the quadratic polynomial regression models using the "leaps" package [29] and the models were fitted by using the "lm" function. All the analyses were conducted using R software. The optimal fertilizer amount and corresponding value were calculated by Matlab software (Mathworks, Inc., Natick, MA, USA) [30].

3. Results

3.1. Principal Component and Correlation Analyses

The PCA result shows that the first two components accounted for 50.5% of the variance; PC1 and PC2 accounted for 29 and 20.6% phenotypic variation, respectively. The yield component traits such as seed number per plant and pod number as well as biomass traits are loaded more in PC1, while nutritional traits especially N and P were loaded more in PC2 (Figure 2A,D). In general, all the traits clustered together as per their grouping into yield, biological, and nutritional traits (Figure 2B). Among the 14 treatments, 11 treatments clustered together while N0P2K2 and N1P2K2 were grouped together and N2P2K3 was alone. This indicated that moderate P and K fertilization is necessary for plant growth and excessive K fertilization affects yield and nutritional traits (Figure 2C,D).

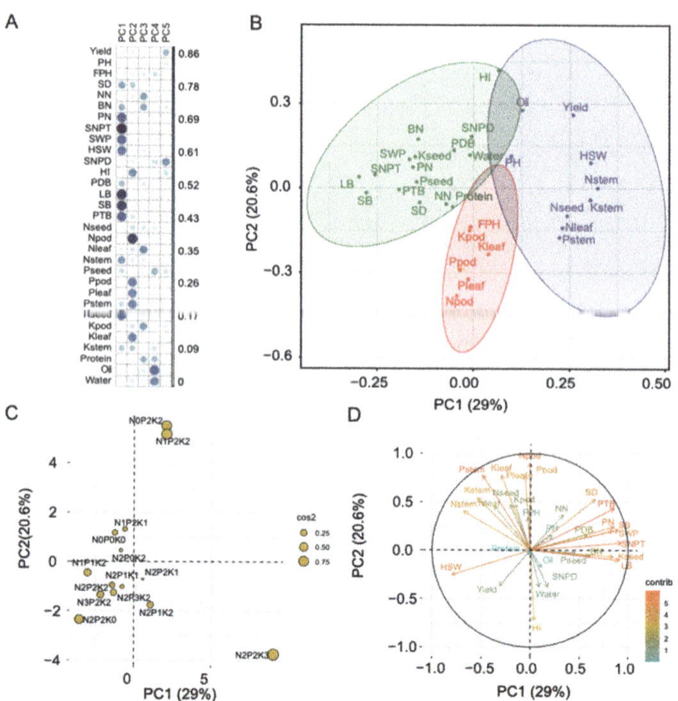

Figure 2. The principal component analysis of 31 traits. (**A**) The heatmap of 31 traits on the first five principal components, the depth of color indicates the level of trait contribution on the corresponding PC, (**B**) PCA plot for 31 traits, ellipses and shapes show clustering of traits, (**C**) principal component analysis for 14 treatments, circle size means the square cosine, (**D**) the contribution for 31 traits on PC1 and PC2, the length indicates the quality of trait. Yield, Grain yield; PH, Plant height; FPH, First pod height; SD, Stem diameter; NN, Number of nodes on main stem; BN, Branch number; PN, Pod number; SNPT, Seed number per plant; SWP, Seed weight per plant; HSW, 100-seed weight; SNPD, Seed number per pod; HI, Harvest index; PDB, Biomass per pod; LB, Biomass per leaf; SB, Biomass per stem; PTB, Biomass per plant; Nseed, N content in seed; Npod, N content in pod; Nleaf, N content in leaf; Nstem, N content in stem; Pseed, P content in seed; Ppod, P content in pod; Pleaf, P content in leaf; Pstem, P content in stem; Kseed, K content in seed; Kpod, K content in pod; Kleaf, K content in leaf; Kstem, K content in stem; Protein, Protein content; Oil, Oil content; Water, Water content.

The correlation analysis showed that biomass per plant was significantly and positively associated with yield component traits such as stem diameter ($r = 0.87$, $p < 0.001$), pod number ($r = 0.86$, $p < 0.001$), seed number per plant ($r = 0.86$, $p < 0.001$), and seed weight per plant ($r = 0.86$, $p < 0.001$), while significantly and negatively associated with hundred seed weight (HSW, $r = -0.72$, $p < 0.01$). There was a negative correlation between protein and oil contents ($r = -0.78$, $p < 0.001$). The K in the seeds was positively correlated with pod number ($r = 0.84$, $p < 0.001$), seed number per plant ($r = 0.91$, $p < 0.001$), seed weight per plant ($r = 0.83$, $p < 0.01$), and biomass per pod ($r = 0.45$, $p < 0.01$). Furthermore, the K in the seeds was positively associated with K in the leaf ($r = 0.73$, $p < 0.01$), stem ($r = 0.67$, $p < 0.01$), and plant ($r = 0.74$, $p < 0.01$). The N in the pod was significantly and negatively associated with harvest index ($r = -0.69$, $p < 0.01$), the N in leaf was significantly and negatively associated with number of branches ($r = -0.72$, $p < 0.01$), while the N in stem was significantly and negatively associated with biomass per leaf ($r = -0.79$, $p < 0.001$) (Figure 3).

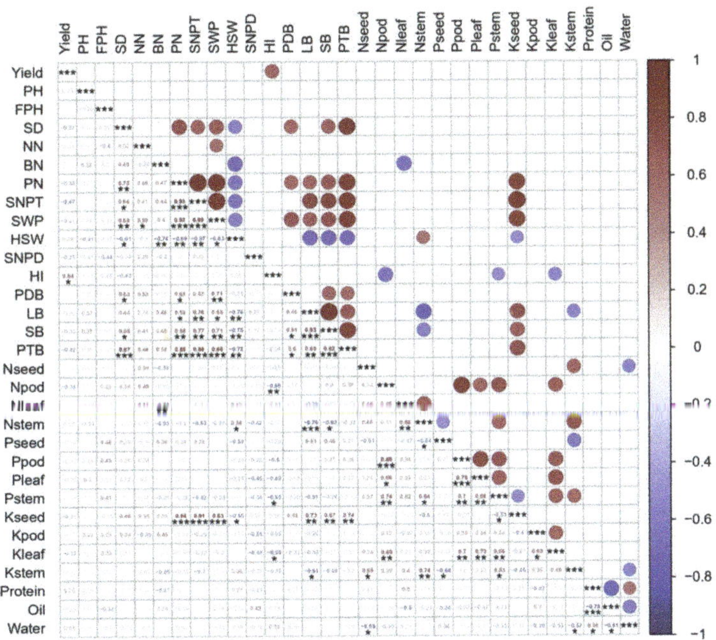

Figure 3. The pairwise correlations (Pearson's *r*) among 31 traits based on the averaged values. The lower diagonal plots show the correlation coefficient, significant differences are indicated by * $p < 0.05$, ** $p < 0.01$, *** $p < 0.001$, values without asterisks were not significant at $p < 0.05$. The upper diagonal only shows the significant correlations which are represented as colored circles, blue indicating negative correlation, and red indicating positive correlation. Yield, Grain yield; PH, Plant height; FPH, First pod height; SD, Stem diameter; NN, Number of nodes on main stem; BN, Branch number; PN, Pod number; SNPT, Seed number per plant; SWP, Seed weight per plant; HSW, 100-seed weight; SNPD, Seed number per pod; HI, Harvest index; PDB, Biomass per pod; LB, Biomass per leaf; SB, Biomass per stem; PTB, Biomass per plant; Nseed, N content in seed; Npod, N content in pod; Nleaf, N content in leaf; Nstem, N content in stem; Pseed, P content in seed; Ppod, P content in pod; Pleaf, P content in leaf; Pstem, P content in stem; Kseed, K content in seed; Kpod, K content in pod; Kleaf, K content in leaf; Kstem, K content in stem; Protein, Protein content; Oil, Oil content; Water, Water content.

3.2. Effects of Different Fertilization Interactions on Grain Yield

Increasing soybean yield is always the core of soybean breeding. To determine the optimal fertilization for grain yield, the 14 treatments were separated into three groups (Table 2) based on grain yield parameters. The first group with a grain yield higher than 4483.75 kg ha^{-1} has harmonious fertilizer and adequate nitrogen, the second group has a grain yield higher than 4112.5 kg ha^{-1}, while the third group consisting of the control has the lowest grain yield of 3726.5 kg ha^{-1}. The percentage grain yield increase of N0P2K2, N2P0K2, N2P2K0, and N2P2K2 over N0P0K0 was 13, 22, 23, and 20%, respectively, and the percentage grain yield difference of N2P2K2 over the other treatments (N0P0K0, N0P2K2, N2P0K2, and N2P2K0) was 19, 9, 1, and 2%, respectively (Figure S1 and Table 3).

Table 2. The average result for 3414 experiment design.

Rep	Treatment	Yield	PH	FPH	SD	NN	BN	PN	SNPT	SWP	HSW	SNPD	HI	PDB	LB	SB	PTB	Nseed	Npod	Nleaf	Nstem	Pseed	Ppod	Pleaf	Pstem	Kseed	Kpod	Kleaf	Kstem	Protein	Oil	Water
	Unit	kg ha^{-1}	cm							g					g								g kg^{-1}								%		
1	N0P0K0	3726.50	79.84	11.33	0.71	17.80	0.80	62.43	149.52	29.60	21.61	2.46	0.51	9.58	29.80	12.19	54.92	48.86	9.57	13.99	9.85	5.22	1.25	1.62	1.42	12.63	17.57	7.79	6.94	39.38	21.87	6.83
2	N0P2K2	4195.88	82.68	12.23	0.77	17.90	1.28	64.17	150.97	29.85	20.30	2.36	0.50	10.92	39.45	17.20	65.22	48.58	11.38	13.69	8.83	5.52	1.85	1.88	1.63	12.49	19.52	9.00	6.01	39.37	21.77	6.87
3	N1P2K2	4190.40	83.31	11.41	0.75	18.72	1.03	69.90	156.10	33.19	20.06	2.45	0.51	12.34	36.86	16.19	60.84	48.99	11.19	15.16	9.83	5.36	1.78	1.89	1.77	12.48	17.69	7.38	6.91	39.15	21.93	6.83
4	N2P0K2	4543.75	82.52	10.83	0.74	18.46	0.82	59.67	133.78	30.17	21.85	2.41	0.55	12.95	34.21	14.39	55.68	49.02	9.80	15.46	11.30	5.08	1.29	1.54	1.38	12.39	17.50	6.64	6.59	39.70	22.10	6.90
5	N2P1K2	4483.75	80.24	10.84	0.74	18.10	1.25	65.02	153.74	31.73	21.34	2.46	0.55	11.48	31.49	12.55	59.19	48.53	8.71	13.02	9.84	5.20	1.14	1.40	1.18	12.48	18.26	6.53	6.20	39.05	22.30	6.93
6	N2P2K2	4590.75	79.10	12.10	0.70	17.73	1.03	57.68	135.95	28.97	21.23	2.45	0.56	10.27	33.38	12.62	52.55	49.13	8.97	13.46	8.85	5.24	1.33	1.52	1.43	12.28	17.52	6.93	6.56	39.00	22.33	6.70
7	N2P3K2	4584.25	82.36	11.96	0.68	17.82	1.13	58.43	132.04	28.51	21.17	2.40	0.57	10.79	34.11	13.09	51.42	47.69	9.30	12.44	9.51	5.37	1.41	1.51	1.49	11.60	17.89	7.00	6.39	39.20	21.98	6.93
8	N2P2K0	4484.00	82.18	10.69	0.66	17.65	1.05	51.89	121.45	25.26	21.86	2.44	0.55	8.79	26.59	10.86	46.80	46.78	8.93	13.61	8.94	5.21	1.45	1.73	1.35	11.64	18.11	7.33	5.70	39.05	22.25	7.03
9	N2P2K1	4500.50	79.38	12.39	0.74	17.60	1.15	66.69	147.95	29.06	21.41	2.36	0.54	12.36	32.25	13.46	54.99	46.84	9.02	14.46	9.69	5.62	1.43	1.56	1.37	12.37	17.39	6.55	5.75	39.82	21.08	7.20
10	N3P2K2	4654.50	82.09	11.26	0.68	17.96	0.89	56.49	131.55	26.01	21.15	2.45	0.53	8.54	34.04	12.13	49.81	48.93	9.23	14.62	9.40	5.56	1.29	1.65	1.49	11.97	16.48	6.74	5.62	39.85	21.23	7.00
11	N1P1K2	4189.25	79.60	11.81	0.70	17.87	0.79	49.95	124.79	25.02	22.05	2.54	0.51	10.51	31.13	13.40	49.62	47.48	10.31	13.88	9.21	5.54	1.47	1.40	1.53	10.78	17.79	6.92	6.30	39.03	22.23	6.87
12	N1P2K1	4219.00	76.49	11.49	0.71	18.52	0.79	63.92	151.40	29.87	21.62	2.47	0.54	10.78	30.27	13.02	57.08	48.04	10.53	15.41	10.73	5.42	1.55	1.68	1.67	12.73	16.34	7.23	6.02	39.17	21.72	7.00
13	N2P1K1	4491.25	81.07	11.18	0.67	18.28	0.66	60.51	146.70	29.63	22.46	2.48	0.55	11.83	35.96	14.17	54.04	48.89	9.82	15.24	9.15	5.21	1.47	1.52	1.24	12.72	16.23	7.07	6.13	40.12	21.47	7.07
14	N2P2K3	4112.50	82.00	11.08	0.74	18.20	1.33	72.65	183.90	34.35	19.57	2.53	0.55	12.33	55.84	20.28	64.77	46.79	8.77	12.08	5.07	5.69	1.22	1.42	0.86	14.30	16.24	5.65	4.50	39.27	22.27	7.00

Yield, Grain yield; PH, Plant height; FPH, First pod height; SD, Stem diameter; NN, Number of nodes on main stem; BN, Branch number; PN, Pod number; SNPT, Seed number per plant; SWP, Seed weight per plant; HSW, 100-seed weight; SNPD, Seed number per pod; HI, Harvest index; PDB, Biomass per pod; LB, Biomass per leaf; SB, Biomass per stem; PTB, Biomass per plant; Nseed, N content in seed; Npod, N content in pod; Nleaf, N content in leaf; Nstem, N content in stem; Pseed, P content in seed; Ppod, P content in pod; Pleaf, P content in leaf; Pstem, P content in stem; Kseed, K content in seed; Kpod, K content in pod; Kleaf, K content in leaf; Kstem, K content in stem; Protein, Protein content; Oil, Oil content; Water, Water 'content.

Table 3. Fertilizer interaction effect and abundance or shortage status of nutrients.

Number	Fertilizer Type	Treatment	Grain Yield (kg ha^{-1})	Percentage Higher than CK	Percentage of N2P2K2 Compared to Other Treatments
1	CK	N0P0K0	3726.50		19
2	P, K	N0P2K2	4195.88	13	9
4	N, K	N2P0K2	4543.75	22	1
6	N, P, K	N2P2K2	4590.75	23	
8	N, P	N2P2K0	4484.00	20	2

From the heatmap, there are wider variations for biological traits and pod/leaf-related nutritional traits than other traits (Figure 4). The N2P2K2 fertilizer treatment ranks second in grain yield (4590.75 kg ha^{-1}) but resulted in the highest nitrogen (49.13 g kg^{-1}) and oil content (22.33%) in the seed. The grain yield of N-deficient treatment (N0P2K2) declined (4195.88 kg ha^{-1}). The N-abundant treatment, N3P2K2, resulted in the highest grain yield (4654.5 kg ha^{-1}) but low biomass per plant (49.81 g), plant height (82.09 cm), and protein content (39.85 %). The grain yield of the P-deficient treatment (N2P0K2) decreased slightly but resulted in increased biomass per pod (12.95 g), N-stem (11.3 g kg^{-1}), and N-leaf (15.46 g kg^{-1}), while the P-abundant treatment, N2P3K2, resulted in the highest HI (0.57). The K deficient treatment, N2P2K0, resulted in low biomass per plant (46.8 g), while for the K-abundant treatment (N2P2K3), the grain yield decreased (4112.5 kg ha^{-1}), but with increased biomass per plant (64.77 g) and seed nutrients.

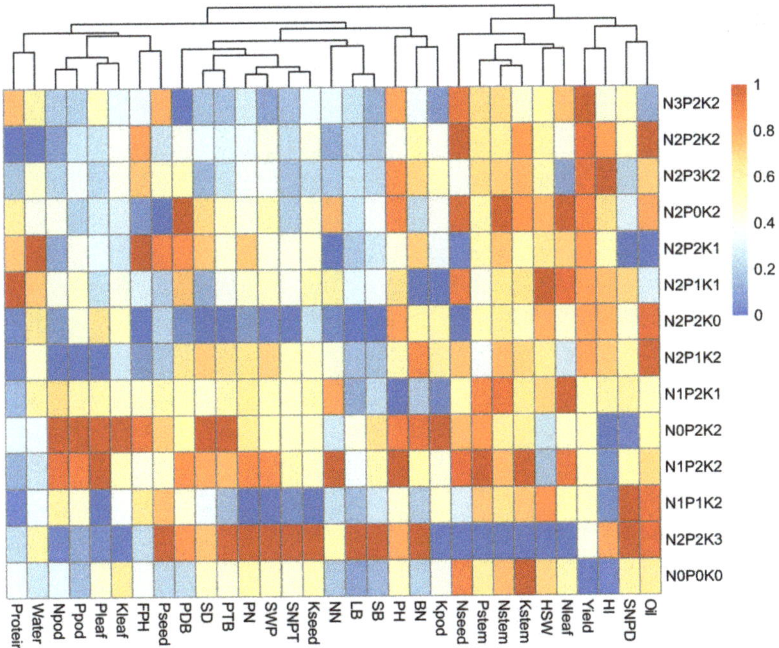

Figure 4. The heatmap of 31 traits for the 3414-experiment design. The color indicates the correlation between treatment and trait.

3.3. Optimal Fertilization Model Development

Based on the adjusted r^2 for the polynomial regression model, the optimal model ($r^2 = 0.89$) for grain yield was selected (Figure S2A). Thus, the polynomial regression equation that governs the effect of grain yield by N, P, and K fertilizers is expressed as $y_1 = 3724.326 + 1.504x_1 - 1.357x_2 + 6.097x_3 + 0.001x_1^2 + 0.004x_2^2 - 0.017x_3^2 - 0.008x_1x_3$ (Table 4). The regression was significant ($p < 0.01$) with N and K fertilizers having significant effects ($p < 0.01$) on grain yield. From the regression coefficient, the effect of different fertilizers on grain yield was in the order K > N > P (Table 4). Grain yield gradually increased with an increase in N and P, while with increasing K fertilizer, grain yield slowly increased and then rapidly decreased (Figure 5A).

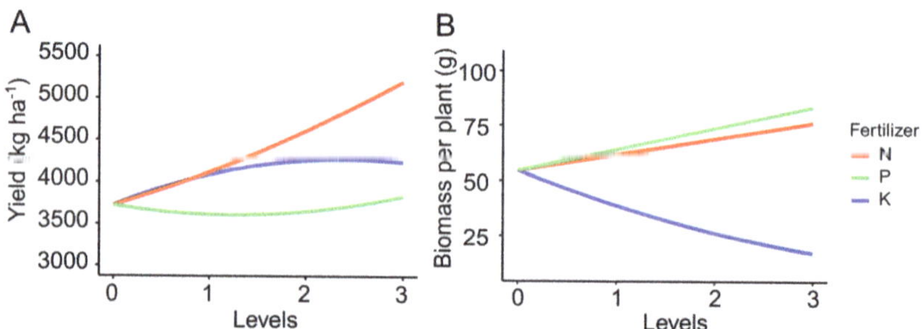

Figure 5. Effect of N, P, and K fertilizer rates on the grain yield (**A**) and biomass per plant (**B**) of soybean.

Table 4. Analysis of variance of the effects of N, P, and K fertilizers on grain yield.

Source of Variation	Estimate	Standard Error	t-Value	p-Value
Intercept	3724.32	82.66	45.06	0.00 ***
N	1.50	0.40	3.72	0.01 **
P	−1.36	0.75	−1.80	0.12
K	6.10	1.21	5.03	0.00 **
N2	0.00	0.00	1.17	0.29
P2	0.00	0.00	2.00	0.09
K2	−0.02	0.01	−2.94	0.03 *
N:K	−0.01	0.00	−2.97	0.03 *

Significant differences are indicated by * $p < 0.05$, ** $p < 0.01$, *** $p < 0.001$.

Besides, there was a significant interaction ($p < 0.05$) between N and K fertilizer in terms of grain yield while no significant interaction between N and P (Table 4, Figure 6A,B). Grain yield slowly increased with the increase in N and K fertilizers (Figure 6B), and the maximum grain yield of 5202.3 kg ha^{-1} was observed at 675 kg ha^{-1} N and 20.5 kg ha^{-1} K$_2$O (Figure 6B).

For biomass per plant, the best fitting model was $y_1 = 54.575 + 0.032x_1 + 0.071x_2 - 0.234x_3 + 0.0003x_3^2 - 0.0008x_1x_2 + 0.0002x_1x_3 + 0.0005x_2x_3$ (Table 5). The regression was significant ($p < 0.05$) and the adjusted r^2 was 0.66 (Figure S2B). The regression result shows that K fertilizer largely affects biomass per plant ($p < 0.05$). The effect of fertilizer on biomass per plant was in the order K > P > N (Table 5). Biomass per plant gradually increased with an increase in N and P fertilizers, while the effect of the increased rate of P on biomass per plant was higher than N. However, with increasing K fertilizer, the biomass per plant sharply declined (Figure 5B).

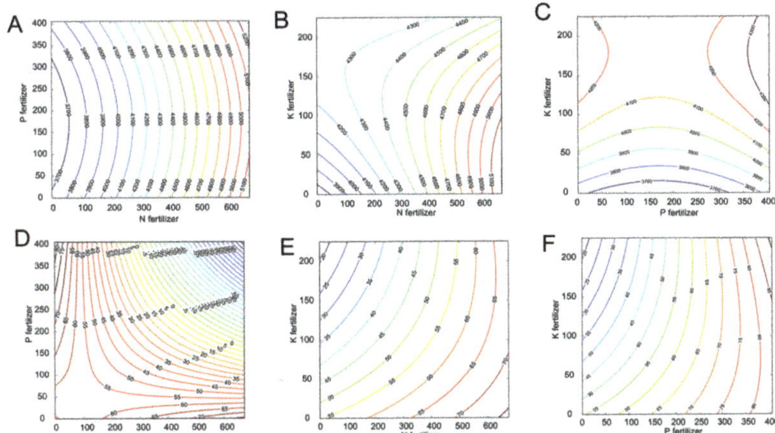

Figure 6. Effects and response surface of interaction among N, P, and K fertilizers on the grain yield (**A–C**) and biomass per plant (**D–F**) of soybean. (**A**) effects and response surface of N fertilizers and P fertilizer on grain yield (fixed factor = 0); (**B**) effects of N fertilizers and K fertilizer on grain yield (fixed factor = 0); (**C**) effects of P fertilizers and K fertilizer on grain yield (fixed factor = 0); (**D**) effects of N fertilizers and P fertilizer on biomass per plant (fixed factor = 0); (**E**) effects of N fertilizers and K fertilizer on biomass per plant (fixed factor = 0); (**F**) effects of P fertilizers and K fertilizer on biomass per plant (fixed factor = 0).

Table 5. Analysis of variance of the effects of N, P, and K fertilizers on biomass per plant.

Source of Variation	Estimate	Standard Error	*t*-Value	*p*-Value
Intercept	54.56	3.18	17.18	0.00 ***
N	0.03	0.03	1.30	0.24
P	0.07	0.04	1.73	0.14
K	−0.23	0.09	−2.74	0.04 *
K2	0.00	0.00	1.51	0.18
N:P	0.00	0.00	−3.46	0.01 *
N:K	0.00	0.00	1.38	0.22
P:K	0.00	0.00	1.62	0.16

Significant differences are indicated by * $p < 0.05$, *** $p < 0.001$

For biomass per plant, a significant interaction between N and P fertilizer ($p < 0.05$) was observed while no interaction effects between P and K (Table 5, Figure 6E,F). Biomass per plant gradually increased and then rapidly decreased with the increase in N and P (Figure 6D). The maximum biomass per plant was 83.33 g with 405 kg ha^{-1} P$_2$O$_5$ and 0 kg ha^{-1} N (Figure 6D).

3.4. Optimal Fertilizer Application

To obtain the optimal fertilizer amount for grain yield, we conducted frequency analysis for the eight treatments with grain yields higher than 4480 kg ha^{-1}. The 95% confidence intervals for N, P$_2$O$_5$, and K$_2$O were 1.83–2.42, 0.86–2.39, and 0.87–2.13, respectively. Thus, the optimal fertilizer amount for high grain yield was 411.62–544.63 kg ha^{-1} N, 115.98–322.77 kg ha^{-1} P$_2$O$_5$, and 65.10−159.90 kg ha^{-1} K$_2$O (Table 6).

Similarly, nine treatments with biomass per plant higher than 54 g were used for the fertilizer frequency analysis. The 95% confidence intervals for N and P$_2$O$_5$ fertilizer were 1.14−1.86 for each and 1.57−1.93 for K$_2$O. Thus, the optimal fertilization requirement for high biomass per plant (>54 g) was 256.61−418.39 kg ha^{-1} N, 153.97−251.03 kg ha^{-1} P$_2$O$_5$, and 117.77−144.73 kg ha^{-1} K$_2$O (Table 7).

Table 6. The frequency distribution and fertilization measures for grain yield greater than 4480 kg ha^{-1}.

Levels	N		P$_2$O$_5$		K$_2$O	
	Times	Frequency	Times	Frequency	Times	Frequency
0	0	0	1	0.125	1	0.125
1	0	0	2	0.25	2	0.25
2	7	0.875	4	0.50	5	0.625
3	1	0.125	1	0.125	0	0
Weight mean	2.13		1.63		1.50	
Standard error	0.35		0.92		0.76	
95% confidence interval	1.83	2.42	0.86	2.39	0.87	2.13
Fertilization measures (kg ha^{-1})	411.62	544.63	115.98	322.77	65.10	159.90

Table 7. The frequency distribution and fertilization measures for biomass per plant > 54 g.

Levels	N		P$_2$O$_5$		K$_2$O	
	Times	Frequency	Times	Frequency	Times	Frequency
0	2.00	0.25	2.00	0.25	1.00	0.13
1	2.00	0.25	2.00	0.25	3.00	0.38
2	5.00	0.625	5.00	0.63	4.00	0.5
3	0.00	0.00	0.00	0.00	1.00	0.13
Weight mean	1.50		1.50		1.75	
Standard error	0.47		0.47		0.23	
95% confidence interval	1.14	1.860	1.14	1.86	1.57	1.93
Fertilization measures (kg ha^{-1})	256.61	418.39	153.97	251.03	117.77	144.73

The average plant density was 178,440 plants per hectare and ranged from 132,270 to 240,195 plants per hectare for plots. The analysis of the relationship between plant number and grain yield in 2014 and 2015 revealed that the best fitting models were $y_i = 440.2033 + 0.022571x_i$ and $y_i = 111.5964 - 0.00019x_i$, respectively, where y_i is the grain yield and biomass per plant for treatment i; and x_i is the plant density for each treatment i. The result shows that the regression was significant for grain yield ($p < 0.001$) and biomass per plant ($p < 0.05$) and the adjusted r^2 is 0.43 for grain yield and 0.13 for biomass per plant. Thus, a plant density of 300,000 plants per hectare gave a grain yield of 7211.5 kg ha^{-1} and plant biomass of 16378.9 kg ha^{-1}.

To determine the optimal fertilizer amount for high grain yield (>7.21 tons ha^{-1}) and high plant biomass (>16.38 tons ha^{-1}), we compared the optimal range for grain yield and plant biomass. The result shows that 411.62–418.39 kg ha^{-1} N, 153.97–251.03 kg ha^{-1} P$_2$O$_5$, and 117.77–144.73 kg ha^{-1} K$_2$O is the optimal combination.

4. Discussion

Mulched drip irrigation is regarded as an effective water-saving irrigation technique that is adopted widely in the arid and semi-arid environments, which can effectively maintain and improve soil health and functionality [31–33]. The combination of film mulching and drip irrigation has been applied to vegetable, corn, and cotton cultivations in Northwest China and provides a potential solution to balance the needs of the rising agricultural production and the sustainability of the oasis agroecosystem [34]. However, little attention has been paid to the optimal fertilization combination for soybean in the agro-ecosystem where mulching is implemented with drip irrigation. There is little potential to further increase current levels of soybean grain yields (2800 up to 4500 kg ha^{-1}) with the exception when plants are grown under some favorable conditions (day length, water availability, etc.) prevailing in the arid and semiarid areas where appropriate irrigation systems have been practiced. Given the unique climate in Xinjiang which is beneficial for nutrient transportation, photosynthetic accumulation, and seed development using drip irrigation [35], it is worth improving the grain yield and plant biomass simultaneously to maintain the high-yield record. Thus, we selected Zhonghuang 35 with different fertilizer treatments in a typical agro-ecosystem to shed new insights into the fertilization effects on soybean grain yield and plant biomass under constant film mulching and drip irrigation.

Nitrogen (N), Phosphorus (P), and potassium (K) are essential to ensure adequate nutrient supply and maximum grain yield, comprising a significant proportion of total fertilizer expenditures, and can be yield limiting in soybean. Determining optimum application of these fertilizers effects has been an ongoing research focus for decades and efforts are continuing to refine recommendations. Studies have shown that excessive or insufficient N, P, and K application suppresses plant growth and dry matter accumulation, as well as the allocation and utilization ratio of nitrogen and phosphate fertilizer [36]. In the present study, fertilization at different degrees promoted the growth of agronomic characters such as plant height and stem number. High variability for grain yield was observed with N fertilization rather than P and K fertilization, thus, nitrogen has a big impact on soybean grain yield. Increasing the nitrogen amount will result in an increase in grain yield, while decreasing the nitrogen amount will promote fertilizer absorption by the nutritious organ. P fertilizer improved the harvest index while K fertilizer largely affected the biomass. Therefore, the effect of K fertilizer on biomass was more as compared to N and P. The result of the findings revealed that soybean performance was largely affected by fertilizer, and it also provides effective fertilization measures for the cultivation and production of soybean.

The "3414" fertilizer experiment design was recommended for the national soil testing and fertilization work to fast build a formula fertilization system and guide farmers in applying fertilizer due to less processing and higher efficiency. The main analysis methods include the fertilizer effect function method, the nutrient balance method, and the soil nutrient abundance index method [37], while the fertilizer effect function method is the most commonly used method in soil measurement. To the best of our knowledge, this is the first time that the recommended fertilizer Zhonghuang 35 in Xinjiang under drip irrigation has been characterized in any research.

For the fertilizer effect function method, the unary quadratic function, binary quadratic function, and quadratic polynomial regression function [38,39] have been used to fit the model and obtain the optimal fertilization application. Studies show that the quadratic polynomial regression function is the best model for the "3414" fertilizer experiment design [40], while there still exists a lot of statistical problems. For instance, there is shown to be a typical fertilizer effect function, the maximum fertilizer for the highest yield loses touch with reality, the optimal fertilizer amount with a detailed number on solving the function is not available for control. Thus, an effective and practical analytical method for the "3414" experimental design needs to be determined. Due to multicollinearity, the quadratic polynomial regression equation is difficult to put into practice and cannot account for the diminishing effects of fertilizer. To overcome this, we set a maximum grain yield target, then obtained a set of fertilization combinations within the 95% confidence interval.

Previous research on fertilizers applied to leguminous species have limited their investigations to short-term data collection with few treatments setting for establishing fertilizer combinations, while little data exploration has occurred [41]. Hence, research on the relationship among a wide range of agronomic, quality, and biomass traits under various fertilizer treatments is necessary. Studies have shown that a decrease in grain yield will also result in a decrease in biomass [19]. Inadequate biomass accumulation caused by environmental changes can lead to grain yield reduction. The correlation analysis in our experiment shows that there is a positive correlation between grain yield components and plant biomass. This is evident from the highest grain yield observed by the treatment N3P2K2 but with low plant biomass, which is consistent with the report of Huang et al. [19]. The high-yielding soybean cultivar, Zhonghuang 35, won a high seed yield record of 6.32 tons ha^{-1} in 2012 for this experiment. Although soybeans require higher nitrogen requirement (100 kg of soybeans require 8 to 9 kg of N), only about 1/3 comes from fertilizers.

To obtain the economically effective fertilizer combinations, the price of the fertilizer was taken into consideration. As of January 2022, the price of urea per ton was \$354, monoammonium phosphate was \$435, and potassium chloride was \$455. The yield difference of 401kg between N2P2K2 (4591 kg) and N1P2K2 (4190 kg), at the latest soybean sales price of \$0.57 a kilogram, will give an increase of \$228 revenue. In this way, \$55 net income per ha will be produced, which will totally cover the \$173 cost of 489 kg urea. Taken together, we concluded that the optimal fertilizer needed to achieve a grain yield greater than 7.21 tons ha^{-1} and plant biomass greater than 16.38 tons ha^{-1} is 411.62–418.39 kg ha^{-1} N, 153.97–251.03 kg ha^{-1} P$_2$O$_5$, and 117.77–144.73 kg ha^{-1} K$_2$O. The recommended fertilizer application would be useful for production and serve as a guideline for efficient fertilization of soybean. This economical fertilizer combination could promote the use of profitable fertilizer in future production of soybean.

5. Conclusions

In this study, we conducted the "3414" experiment under drip irrigation in the arid region of Northwest China based on a super-high yielding soybean cultivar Zhonghuang 35. First, we confirmed that N fertilizer significantly affects grain yield, while P and K fertilizers influence harvest index and biomass, respectively. Second, we clarified the relationship among a wide range of agronomic, quality, and biomass traits under various fertilizer treatments. Third, we offered the optimal fertilizer scheme to obtain a theoretical grain yield and plant biomass of more than 7.21 tons ha^{-1} and 16.38 tons ha^{-1} with 300,000 plants ha^{-1}, respectively. This will serve as a guideline for effective fertilization measures in order to achieve a balance between grain yield and plant biomass as well as to contribute to the promotion of the large-scale cultivation of soybean under drip irrigation, which will increase the efficiency and productivity of farmlands, thereby improving profitability and also helping to minimize environmental risk.

Supplementary Materials: The following supporting information can be downloaded at: https://www.mdpi.com/article/10.3390/agronomy12020291/s1, Figure S1: Grain yield of Zhonghuang 35 based on the different treatment combinations; Figure S2: The adjusted r^2 on the quadratic polynomial regression model with different components for grain yield (A) and plant biomass (B), the number on the left means r^2 of the model with the component colored in "black" for each row; Table S1: Statistics of meteorological data of Wulanwusu agricultural meteorological experimental station in 2014 and 2015; Table S2: The field fertilizer amount of each plot for Zhonghuang 35 in 2014; Table S3: The field fertilizer amount of each plot for Zhonghuang 35 in 2015; Table S4: The growth period and irrigation amount in 2014 and 2015; Table S5: The definition and description information for 31 soybean traits.

Author Contributions: Conceptualization, J.S. and J.L.; methodology, J.L.; software, J.L.; investigation, G.L.; resources, J.S.; writing—original draft preparation, J.L.; writing—review and editing, J.S., A.S.S., B.L., and S.Z.; visualization, J.L.; supervision, J.S.; project administration, J.S.; funding acquisition, J.S. All authors have read and agreed to the published version of the manuscript.

Funding: This research was funded by National Natural Science Foundation of China: 32161143033; National Natural Science Foundation of China: 32001574; and Agricultural Science and Technology Innovation Program of CAAS: 2060203-2.

Institutional Review Board Statement: Not applicable.

Informed Consent Statement: Not applicable.

Data Availability Statement: All data is provided in the manuscript.

Acknowledgments: We thank Jing Zhao, who works in the Institute of Crops Research, Xinjiang Academy of Agricultural and Reclamation Sciences, for helping with the data collection.

Conflicts of Interest: The authors declare no competing financial interest with regard to this manuscript.

References

1. Wang, H.; Ni, C.; Xu, R. Analysis of change and convergence of soybean productivity in China. *Jiangsu J. Agric. Sci.* **2011**, *27*, 199–203.
2. Zhang, F.; Niu, J.; Zhang, W.; Chen, X.; Li, C.; Yuan, L.; Xie, J. Potassium nutrition of crops under varied regimes of nitrogen supply. *Plant Soil* **2010**, *335*, 21–34. [CrossRef]
3. Fu, C.F.; Sun, C.; Dong, Y.M. Effects of nutrient management on NPK uptake and yield of soybean. *Heilongjiang Agr. Sci.* **2011**, *10*, 33–35.
4. Ciampitti, I.A.; Salvagiotti, F. New insights into soybean biological nitrogen fixation. *Agron. J.* **2018**, *10*, 1185–1196. [CrossRef]
5. Salvagiotti, F.; Cassman, K.G.; Specht, J.E.; Walters, D.T.; Weiss, A.; Dobermann, A. Nitrogen uptake, fixation and response to fertilizer N in soybeans: A review. *Field Crop. Res.* **2008**, *108*, 1–13. [CrossRef]
6. Yang, F.; Xu, X.; Wang, W.; Ma, J.; Wei, D.; He, P.; Pampolino, M.F.; Johnston, A.M. Estimating nutrient uptake requirements for soybean using QUEFTS model in China. *PLoS ONE* **2017**, *12*, e0177509. [CrossRef]
7. Zapata, F.; Danso, S.K.A.; Hardarson, G.; Fried, M. Time Course of Nitrogen Fixation in Field-Grown Soybean Using Nitrogen-15 Methodology1. *Agron. J.* **1987**, *79*, 172–176. [CrossRef]
8. Li, L.; Yang, T.; Redden, R.; He, W.F.; Zong, X.X. Soil fertility map for food legumes production areas in China. *Sci. Rep.* **2016**, *6*, 26102. [CrossRef]
9. Fan, J.; Lu, X.; Gu, S.; Guo, X. Improving nutrient and water use efficiencies using water-drip irrigation and fertilization technology in Northeast China. *Agric. Water Manag.* **2020**, *241*, 106352. [CrossRef]
10. Tang, P.; Li, H.; Issaka, Z.; Chen, C. Effect of manifold layout and fertilizer solution concentration on fertilization and flushing times and uniformity of drip irrigation systems. *Agric. Water Manag.* **2018**, *200*, 71–79. [CrossRef]
11. Zou, H.; Fan, J.; Zhang, F.; Xiang, Y.; Wu, L.; Yan, S. Optimization of drip irrigation and fertilization regimes for high grain yield, crop water productivity and economic benefits of spring maize in Northwest China. *Agric. Water Manag.* **2020**, *230*, 105986. [CrossRef]
12. Chen, Y.; Zhou, X.; Lin, Y.; Ma, L. Pumpkin yield affected by soil nutrients and the interactions of Nitrogen, Phosphorus, and Potassium fertilizers. *HortScience* **2019**, *54*, 1831–1835. [CrossRef]
13. Yang, Z.J.; Wu, X.H.; Grossnickle, S.C.; Chen, L.H.; Yu, X.X.; El-Kassaby, Y.A.; Feng, J.L. Formula fertilization promotes Phoebe bournei robust seedling cultivation. *Forests* **2020**, *11*, 781. [CrossRef]
14. Yin, Z.C.; Guo, W.Y.; Liang, J.; Xiao, H.Y.; Hao, X.Y.; Hou, A.F.; Zong, X.X.; Leng, T.R.; Wang, Y.J.; Wang, Q.Y.; et al. Effects of multiple N, P, and K fertilizer combinations on adzuki bean (*Vigna angularis*) yield in a semi-arid region of northeastern China. *Sci. Rep.* **2019**, *9*, 19408. [CrossRef] [PubMed]
15. Parry, M.A.; Reynolds, M.; Salvucci, M.E.; Raines, C.; Andralojc, P.J.; Zhu, X.G.; Price, G.D.; Condon, A.G.; Furbank, R.T. Raising yield potential of wheat. II. Increasing photosynthetic capacity and efficiency. *J. Exp. Bot.* **2010**, *62*, 453–467. [CrossRef]
16. Fischer, R.; Rees, D.; Sayre, K.; Lu, Z.M.; Condon, A.; Saavedra, A.L. Wheat yield progress associated with higher stomatal conductance and photosynthetic rate, and cooler canopies. *Crop Sci.* **1998**, *38*, 1467–1475. [CrossRef]
17. Jin, J.; Liu, X.; Wang, G.; Mi, L.; Shen, Z.; Chen, X.; Herbert, S.J. Agronomic and physiological contributions to the yield improvement of soybean cultivars released from 1950 to 2006 in Northeast China. *Field Crops Res.* **2010**, *115*, 116–123. [CrossRef]
18. Board, J.E.; Modali, H. Dry matter accumulation predictors for optimal yield in soybean. *Crop Sci.* **2005**, *45*, 1790–1799. [CrossRef]
19. Huang, Z.W.; Zhao, T.J.; Gai, J.Y. Dynamic analysis of biomass accumulation and partition in soybean with different yield levels. *Crop Sci.* **2009**, *35*, 1483–1490. [CrossRef]
20. Mehetre, S.S.; Jamadagni, B.M. Biomass partitioning and growth characters in relation to plant architecture in soybean. *Soybean Genet. Newsl.* **1996**, *23*, 92–97.
21. IUSS Working Group WRB. International soil classification system for naming soils and creating legends for soil maps. In *World Reference Base for Soil Resources 2014*; update 2015; Food and Agriculture Organization of the United Nations: Rome, Italy, 2015.
22. Wang, L.; Wang, L.Z.; Zhao, R.J.; Li, Q. Development of new soybean cultivar Zhonghuang 35 with high yielding, high oil, early maturity and broad adaptability. *Soybean Sci.* **2009**, *28*, 360–362.
23. Wang, L.Z.; Luo, G.T.; Wang, L.; Sun, J.M.; Zhang, Y. Development of soybean cultivation technology with the yield over 6 tonnes per hectare for soybean cultivar Zhonghuang 35 in northern Xinjiang province. *Soybean Sci.* **2012**, *31*, 217–223.
24. Bao, S.D. *Soil Agrochemical Analysis*; China Agricultural Press: Beijing, China, 2000; pp. 25–114.
25. Zhang, Y.; He, J.; Wang, H.; Meng, S.; Xing, G.; Li, Y.; Yang, S.; Zhao, J.; Zhao, T.; Gai, J. Detecting the QTL-allele system of seed oil traits using multi-locus genome-wide association analysis for population characterization and optimal cross prediction in soybean. *Front. Plant Sci.* **2018**, *9*, 1793. [CrossRef] [PubMed]
26. Kolde, R. *Pheatmap: Pretty Heatmaps*; R Package Version; CRC Press: Boca Raton, FL, USA, 2012; Volume 61, p. 915.
27. Wei, T.; Simko, V. *Corrplot: Visualization of a Correlation Matrix*; R Package Version; CRC Press: Boca Raton, FL, USA, 2013; p. 230.
28. Maechler, M.; Rousseeuw, P.; Struyf, A.; Hubert, M.; Hornik, K. Package 'cluster'. Dosegljivo na. 2013. Available online: https://cran.r-project.org/web/packages/cluster/cluster.pdf (accessed on 20 January 2022).
29. Lumley, T.; Miller, A. Package 'leaps'. In *Regression Subset Selection*, R package version; CRC Press: Boca Raton, FL, USA, 2009; Volume 2, p. 2366.
30. Grant, M.; Boyd, S.; Ye, Y. *CVX: Matlab Software for Disciplined Convex Programming*; CVX: San Ramon, CA, USA, 2008.

31. Seyfi, K.; Rashidi, M. Effect of drip irrigation and plastic mulch on crop yield and yield components of cantaloupe. *Int. J. Agric. Biol.* **2007**, *9*, 247–249.
32. He, H.; Wang, Z.; Guo, L.; Zheng, X.; Zhang, J.; Li, W.; Fan, B. Distribution characteristics of residual film over a cotton field under long-term film mulching and drip irrigation in an oasis agroecosystem. *Soil Till. Res.* **2018**, *180*, 194–203. [CrossRef]
33. Akhtar, K.; Wang, W.; Ren, G.; Khan, A.; Feng, Y.; Yang, G.; Wang, H. Integrated use of straw mulch with nitrogen fertilizer improves soil functionality and soybean production. *Environ. Int.* **2019**, *132*, 105092. [CrossRef] [PubMed]
34. Li, F.M.; Wang, P.; Wang, J.; Xu, J.Z. Effects of irrigation before sowing and plastic film mulching on yield and water uptake of spring wheat in semiarid Loess Plateau of China. *Agric. Water Manag.* **2004**, *67*, 77–88. [CrossRef]
35. Zhang, D.; Liu, H.B.; Hu, W.L.; Qin, X.H.; Yan, C.R.; Wang, H.Y. The status and distribution characteristics of residual mulching film in Xinjiang, China. *J. Integr. Agric.* **2016**, *15*, 2639–2646. [CrossRef]
36. Galloway, J.N.; Winiwarter, W.; Leip, A.; Leach, A.M.; Bleeker, A.; Erisman, J.W. Nitrogen footprints: Past, present and future. *Environ. Res. Lett.* **2014**, *9*, 115003. [CrossRef]
37. Chen, X.P.; Zhang, F.S. The technical index system of soil testing and fertilization was established through "3414" test. *China Agric. Technol. Ext.* **2006**, *4*, 36–39.
38. Zhao, B.; Wang, Y.; Zhang, Z.Y.; Zhu, X.D.; Yang, R. Identification and optimization of extreme value of effect function of multivariate secondary fertilizer. *Rain Fed Crops* **2001**, *21*, 42–45.
39. Wang, S.R.; Chen, X.P.; Gao, X.Z.; Mao, D.R.; Zhang, F.S. Study on simulation of "3414" fertilizer experiments. *Plant Nutr. Fertil. Sci* **2002**, *8*, 409–413.
40. Peng, S.B.; Cheng, Y.X.; Dong, W.H.; Liu, D.L. "3414" fertilization tests and regression analysis of recommended fertilizer amount of walnut. *Nonwood For. Res.* **2018**, *36*, 27–32.
41. Ma, R.X.; Wang, J.S.; Li, X.Z.; Liu, W.C. Effect of different npk fertilizers cooperating application on yield and quality of high starch maize. In *Applied Mechanics and Materials*; Trans Tech Publications Ltd.: Bäch SZ, Switzerland, 2012; Volume 214, pp. 423–429.

agronomy

Article

Response of Winter Wheat (*Triticum aestivum* L.) to Fertilizers with Nitrogen-Transformation Inhibitors and Timing of Their Application under Field Conditions

Marie Školníková [1], Petr Škarpa [1], Pavel Ryant [1], Zdenka Kozáková [2] and Jiří Antošovský [1,*]

1 Department of Agrochemistry, Soil Science, Microbiology and Plant Nutrition, Faculty of AgriScience, Mendel University in Brno, Zemědělská 1, 61300 Brno, Czech Republic; mar.skolnikova@seznam.cz (M.Š.); petr.skarpa@mendelu.cz (P.Š.); pavel.ryant@mendelu.cz (P.R.)
2 Faculty of Chemistry, Brno University of Technology, Purkyňova 118, 61200 Brno, Czech Republic; kozakova@fch.vutbr.cz
* Correspondence: jiri.antosovsky@mendelu.cz

Abstract: Winter wheat is a widely cultivated crop that requires high inputs of nitrogen (N) fertilization, which is often connected with N losses. The application of fertilizers with nitrification (NI) and urease inhibitors (UI) is an opportunity to eliminate the risk of N losses and improve N availability to plants. The aim of this study is to compare the effect of conventional nitrogen fertilizers with fertilizers containing nitrogen-transformation inhibitors as well as to evaluate the timing of their application on the wheat-grain yield and quality under the conditions of a three-year field experiment. The examined fertilizers with inhibitors were applied in a single dose or in a split application in combination with conventional fertilizers. The single application of urea with NI and/or UI resulted in a relatively average increase in the grain yield, while protein content and the Zeleny-test values were significantly increased compared to the split N application. The more significant effect of urea with NI and UI was found under the moisture-rich conditions compared to the drier conditions. A significant increase in the grain yield (by 6.3%) and in the Zeleny-test value (by 16.5%) was observed after inhibited urea application comparing to the control treatment (without inhibitors).

Keywords: wheat; nitrification and urease inhibitors; split and single application of fertilizer; grain yield; quality of grain

check for
updates

Citation: Školníková, M.; Škarpa, P.; Ryant, P.; Kozáková, Z.; Antošovský, J. Response of Winter Wheat (*Triticum aestivum* L.) to Fertilizers with Nitrogen-Transformation Inhibitors and Timing of Their Application under Field Conditions. *Agronomy* **2022**, *12*, 223. https://doi.org/10.3390/agronomy12010223

Academic Editors: Christos Noulas, Shahram Torabian and Ruijun Qin

Received: 19 December 2021
Accepted: 14 January 2022
Published: 17 January 2022

Publisher's Note: MDPI stays neutral with regard to jurisdictional claims in published maps and institutional affiliations.

1. Introduction

The efficient use of fertilizers is an important factor of sustainable agriculture. It not only influences crop productivity, but it also reduces nutrients losses, which eliminates the detrimental impact on the environment. Nitrogen (N) is an essential nutrient for plant growth and development. Many authors have described its positive effect on the plant biomass production, the grain yield, and the grain protein content [1–3]. The average N efficiency in the field conditions is 32%, and no more than 40% [4]. The efficiency can reach 50−70% in the case of a synchronic effect of N supply and crop demand [5]. Low efficiency of nitrogen fertilizers not only causes economical losses, but it is also environmentally unsafe (it represents leaching losses of NO_3^-, volatilization losses of NH_3, and emission of other N-containing gases) [6]. These losses decrease soil fertility, but also represent a risk to the atmosphere, hydrosphere and human health [7,8]. NH_3 volatilization is mainly involved in surface-applied fertilizers without incorporation into the soil [9]. N losses through NH_3 volatilization can reach up to 60 % of the nitrogen applied by the fertilization [10], and they are responsible for the generation of condensation nuclei, which contribute to the greenhouse effect [11,12].

Wheat is among the most globally cultivated cereal crops [13]. Under conditions of intensive agriculture systems, the traditional wheat cultivation requires high inputs of N fertilizers which is related to the risk of N losses [14]. The nitrogen fertilization is considered crucial for an optimal crop yield. The split application of nitrogen is a common strategy for the optimization of the plants' nutrient uptake and for the reduction of the risk of N losses in conventional agriculture [15]. Positive effects of the split application on the wheat-grain yield, flour quality, and proportions of gliadin and glutenin have been described by several authors [16–18]. Another opportunity to reduce N losses and improve N efficiency is in the use of fertilizers with inhibitors. These effective and environmentally friendly fertilizers contain inhibitors that temporary restrict N changes in the soil (urea hydrolysis or nitrification). The most widely used N fertilizers in the world for the winter wheat production are ureic fertilizers. Urea that is applied to the soil surface is hydrolyzed very quickly, thereby generating CO_2 and high amounts of NH_3 [11]. The urea hydrolysis requires a urease enzyme, the activity of which could be reduced by a urease inhibitor (UI). The application of a UI with the urea helps to delay the conversion of urea into NH_4^+ in the soil [19,20] due to the partial inhibition of the urease activity [21]. The principle of most nitrification inhibitors (NIs) is the influence of the ammonia mono-oxygenase enzyme, which is responsible for the oxidation of ammonium into nitrite in the first step of nitrification (conversion of NH_4^+ into NO_2). NI temporarily binds ammonia mono-oxygenase, which leads to the conservation of immobile NH_4^+ in the soil for a longer period (4−10 weeks). Subsequently, it reduces the amount of NO_3^-, which is very mobile in the soil and is involved in leaching and denitrification [22,23]. Denitrification is also a source of NO and N_2O emissions [24]. NIs are also recommended by The Intergovernmental Panel on Climate Change (IPCC) as an option to reduce N_2O emissions in agriculture [25]. Stabilized fertilizers seem to represent a good opportunity to minimalize the negative impact of N losses on the environment and to improve the agronomic benefits of fertilization since their positive environmental aspects have been reviewed by many authors [4,26–30].

The amount of available N in the soil is influenced not only by fertilization but also by the level of mineralization. Mineralization of N is assumed to supply random and unpredictable amounts of inorganic N from one year to the next [31]. The fluctuation of residual soil N is unpredictable [32] and it is affected by random environmental effects [33]. The mineralization of organic soil N is often accelerated by the application of N fertilizers, which results in interactions of the added N or in a priming effect [34]. N supplied by fertilization can be conserved through its immobilization by micro-organisms (a biotic process) and fixation by soil-clay minerals (an abiotic process). Subsequently, it can be remineralized and further released in order to cover crop demands, thus reducing the N losses [35]. The application of N-transformation inhibitors significantly affects these processes [36].

Another essential nutrient for optimal wheat development is sulfur (S), which plays an important role in the yield formation and protein constitution [37,38]. The interaction between nitrogen and sulfur has an impact not only on the uptake and assimilation of NO_3^- and SO_4^{2-}, but also on N and S metabolism [8]. Sulfur also positively influences the quality of protein in baking [39]. In wheat with S deficiency, asparagine amino acid is accumulated in the grain, which contributes to a higher risk of an unhealthy acrylamide formation during the baking of flour products [40].

This work should contribute to the description of the effect of nitrogen and sulfur fertilizers with inhibitors in combination with single and split applications and emphasize their effect on the winter wheat-grain yield and quality parameters. Based on previous research, the positive effect of NI and UI on the prolongation of nitrogen availability in the soil is expected. Therefore, the basic aim of this work is to confirm if the single or split application of nitrogen fertilizers with inhibitors provides similar or better results in terms of the grain yield and qualitative parameters of winter wheat in comparison with common technology (fertilization without inhibitors split into three doses).

2. Materials and Methods

2.1. Experimental Site and Field Treatments

The 4-year field experiment (2018–2021) was established as a small-plot field observation at the experimental station Žabčice in southern Moravia, the Czech Republic (49°1′18.658″ N and 16°36′56.003″ E). The region is characterized by warm and dry conditions; the annual precipitation ranges from 380−550 mm and the average annual temperature is 10.1 °C. The experiment was established on a silty clay loam soil (clay 38.0%; silt 46.3%; sand 15.7%); the soil type was stagnic fluvisols (FL-st). Each year, the experiment was based on a new plot within the experimental station. The basic physical–chemical parameters of the soil that were determined before the sowing are given in Table 1. The soil nutrient content characterizes the whole area used for the experiment in each growing season. The effect of the soil parameters, including N released by mineralization, was assumed to be the same throughout the experimental area.

Table 1. The physical–chemical properties of the soil.

Soil Parameter	Growing Season					Refs.
	2017/2018	2018/2019	2019/2020	2020/2021	Average	
Cox (%)	1.32	1.43	1.33	1.36	1.36	[41]
CEC (mmol/kg)	219	257	208	234	230	[42]
$pH/CaCl_2$	6.8	6.6	6.4	5.9	6.4	[42]
P (mg/kg)	148	134	152	92	132	[42]
K (mg/kg)	276	247	283	184	248	[42]
Ca (mg/kg)	3644	3321	3641	3934	3635	[42]
Mg (mg/kg)	384	397	411	355	387	[42]
SO_4^{2-} (mg/kg)	14.5	11.3	11.1	10.4	12	[42]
NH_4^+ (mg/kg)	1.84	2.69	1.49	9.73	4	[42]
NO_3^- (mg/kg)	3.86	14.00	4.62	14.30	9	[42]

CEC—Cation exchange capacity; Cox—Soil oxidizable carbon.

The model crop used in this experiment was winter wheat of the Julie variety (Selgen, Prague, Czech Republic). This variety is characterized by a high yield potential and good disease resistance. It belongs to the quality class E, its stated protein content is 13.8%, the density of its grain is 80.4 kg/hL, and its Zeleny-test value is 60 mL [43].

Wheat crops were damaged by pests (voles) in the 2019/2020 growing season. The damage was so severe that it was impossible to assess the impact of the fertilizers' application. Therefore, the results of this season were not included in the evaluation of the 4-year experiment. The experiment was set up using a randomized complete-block design with nine treatments (Table 2); each treatment was repeated four times. The size of each experimental plot for the fertilization was 15 m². All fertilizers were spread by hand separately on each block. The examined fertilizers used in the experiments and their basic characteristics are listed in Table 3.

Table 2. Treatments and doses of fertilizers.

Treatments	Term (T) of Fertilization (Dose of N, S kg/ha)			Total Dose of N, S (kg/ha)
	T1 (BBCH 25)	T2 (BBCH 32)	T3 (BBCH 50)	
control	CAN (55, 0)	CAN (65, 0)	UAN (40, 0)	160, 0
N1	ALZON neo-N (160, 0)			160, 0
N2	CAN (55, 0)	ALZON neo-N (105, 0)		160, 0
N3	UREA^{stabil} (160, 0)			160, 0
N4	CAN (55, 0)	UREA^{stabil} (105, 0)		160, 0
NS1	CAN (55, 0)	ASN (105, 52)		160, 52
NS2	ASN (120, 60)		UAN (40, 0)	160, 60
NS3	ENSIN (160, 80)			160, 80
NS4	ENSIN (120, 60)		UAN (40, 0)	160, 60

BBCH—phase of growing according to Lancashire [44], BBCH 25–tillering; BBCH 32—stem elongation, BBCH 50–beginning of heading. UAN—Urea ammonium nitrate; CAN—Calcium ammonium nitrate; ASN—Ammonium sulphate nitrate; ENSIN—Ammonium sulphate nitrate with NIs; UREA^{stabil}—urea with UI; ALZON neo-N—urea with UI and NI.

Table 3. Type of used fertilizers.

Fertilizers	Nutrients Content (%)		Inhibitors	Producer
	N	S		
ALZON neo-N	46	0	nitrification (NI) and urease (UI)	(SKW Piesteritz, Wittenberg, Germany)
UREA^{stabil}	46	0	urease (UI)	(AGRA GROUP a.s., Střelské Hoštice, the Czech Republic)
ENSIN	26	13	nitrification (NI)	(Duslo, a.s., Šaľa, the Slovak Republic)
UAN	30	0	none	(ADW AGRO, a.s., Okříšky, the Czech Republic)
CAN	27	0	none	(Duslo, a.s., Šaľa, the Slovak Republic)
ASN	26	13	none	(Duslo, a.s., Šaľa, the Slovak Republic)

UAN—Urea ammonium nitrate; CAN—Calcium ammonium nitrate; ASN—Ammonium sulphate nitrate; ENSIN—Ammonium sulphate nitrate with NIs; UREA^{stabil}—urea with UI; ALZON neo-N—urea with UI and NI. ALZON neo-N contains NI: (MPA−N-[3(5)-methyl-1H-pyrazol-1-yl] methyl] acetamide) and UI: (2-NPT−N-(2-nitrophenyl) Phosphoric Triamide); UREA^{stabil} contains UI: NBPT−N-(butyl) Thiophosphoric Triamide; ENSIN contains NIs: DCD—dicyandiamide, TZ—triazol.

Table 4 presents the terms of sowing, fertilizer applications and the date of the harvest. In all experimental years, *winter wheat* was grown after winter wheat (a pre-crop). Winter wheat, grown as the pre-crop, was cultivated identically throughout the experimental area in each year. It was fertilized with a N fertilizer at the same rate (120 kg/ha N). The harvest was performed at the stage of fully ripe (BBCH 89). The winter wheat was harvested by the harvester Sampo Rosenlew 2035 (Sampo Rosenlew Ltd., Pori, Finland).

Table 4. Terms of sowing, fertilization, and harvest.

Growing Season (GS)	Sowing	T1 (BBCH 25)	T2 (BBCH 32)	T3 (BBCH 50)	Harvest (BBCH 89)
GS1: 2017–2018	6 October 2017	5 March 2018	9 April 2018	2 May 2018	4 July 2018
GS2: 2018–2019	9 October 2018	28 February 2019	29 April 2019	10 May 2019	11 July 2019
GS3: 2020–2021	8 October 2020	3 March 2021	20 April 2021	29 May 2021	24 July 2021

BBCH—phase of growing according to Lancashire [44], BBCH 25–tillering; BBCH 32—stem elongation, BBCH 50–beginning of heading, BBCH 89–fully ripe.

The distribution of precipitation was not regular during the experimental years, as presented in Figure 1. The highest amount of precipitation was noticed in the GS3 (total precipitation of 377.20 mm/GS3). The GS1 and GS2 both had less rainfall and almost the same amount of precipitation (the total precipitation of 286.06 mm/GS1 and 282.40 mm/GS2). In the GS2, the lowest temperature was recorded in January while the other terms had their minimum temperatures in February. The GS2 and GS3 showed a similar temperature and precipitation development in the periods of May and June. The May period was colder and rainier while the temperature sharply rose in June. May in the GS3 was a little bit richer in precipitation than May in the GS2.

2.2. Yield and Grain Quality Measurement

The parameters observed over all experimental years were the grain yield and qualitative parameters such as the hectoliter weight of the grain, the content of protein and gluten in the grain, and the Zeleny-test (ZT) value. Four repetitions from all variants were analyzed. The weight of the harvested grain was determined using the digital scale Kern ECE 20K-2N (KERN and Sohn GmbH, Balingen, Germany). The test weight scale Wile 241 (Farmcomp OY, Tuusula, Finland) was used for determination of the hectoliter weight. The content of protein in the grain was determined by the Kjeltec 2300 device (Foss, Hillerød, Denmark) followed by the multiplication by a 6.25 coefficient (the Kjeldal method). The gluten content and the Zeleny-test value were estimated by the NIR (Near Infrared Spectroscopy) method on the Inframatic 9500 NIR grain analyzer (Perten Instruments, Hägersten, Sweden). The principle of the NIR method is the transmittance or reflectance measurement of radiation within the wavelength range of 800 to 2500 nm (12,500–4000 cm^{-1}) which is related to the different chemical groups contained in the sample [45,46].

2.3. Statistical Analysis

The program Statistica 12 CZ [47] was used for the statistical evaluation of monitored parameters. The Shapiro–Wilk and the Levene tests (at $p \leq 0.05$) were performed for the verification of normality and homogeneity of variances. The values of these parameters were subsequently evaluated by the analysis of variance (ANOVA) and by the follow-up tests according to Fisher (LSD test) at the 95% level ($p \leq 0.05$) of significance. The results were expressed as the arithmetical mean $\pm$ standard error (SE).

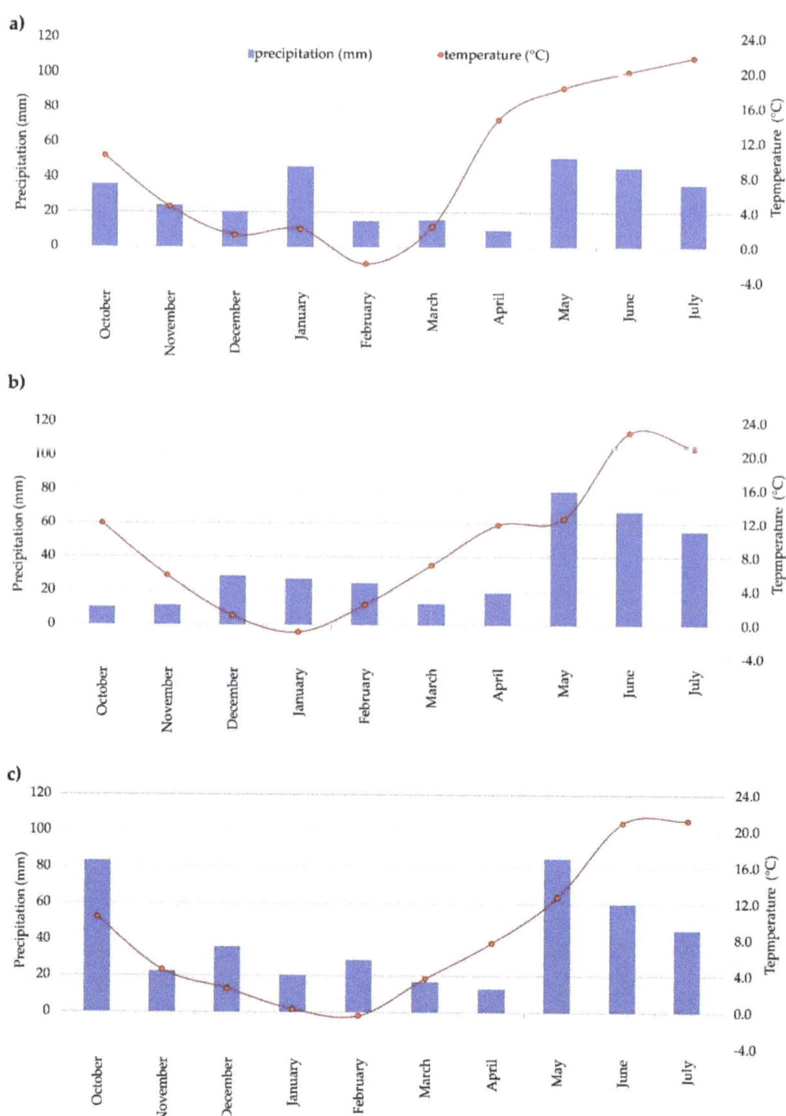

Figure 1. Weather conditions of the field experiments. (**a**) growing season 1 (2017−2018), (**b**) growing season 2 (2018−2019), (**c**) growing season 3 (2020−2021).

3. Results

3.1. The Grain Yield

The average grain yields over three growing seasons as well as the average grain yields in each year of the experiment are given in Figure 2. Almost no differences between the grain yield in the GS1 and GS2 were found. The yield of the N1−N4 treatments and the NS1−NS4 treatments were slightly increased in comparison to the control treatment, but the differences were statistically insignificant. The statistical differences were observed in the GS3. The significantly highest yield was observed in the N1 treatment fertilized by ALZON neo-N (urea with NI and UI), which was 6.3% higher than the control without the application of inhibitors. The N1 treatment was also significantly higher than the NS1 treatment in the GS3.

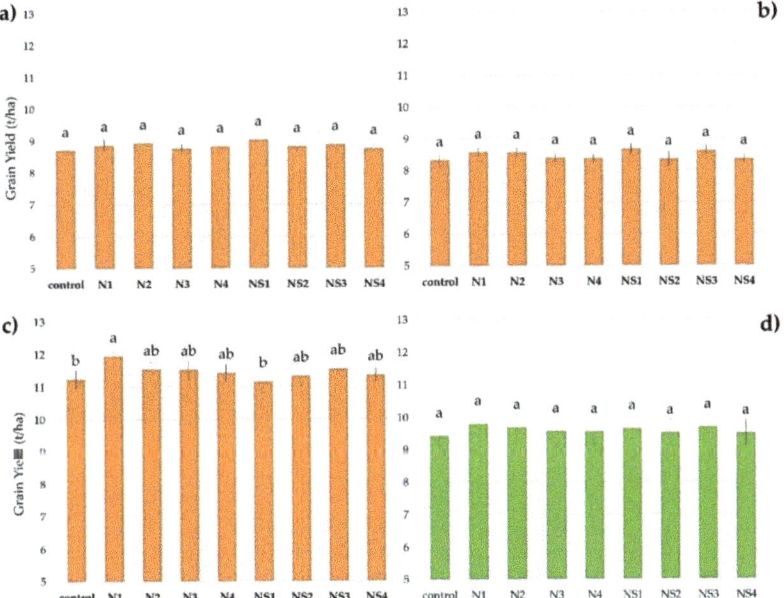

Figure 2. The effect of fertilizer treatments on grain yield (t/ha), (**a**) growing season 1 (2017−2018), (**b**) growing season 2 (2018−2019), (**c**) growing season 3 (2020−2021), (**d**) average yield. The same letters above the columns describe no statistically significant differences between treatments (LSD test). Each growing season was evaluated separately. Standard error (SE) is expressed by error bars.

According to the average values, the split fertilization did not result in an increased yield or a yield reduction. The treatments with the single application of N fertilizers with inhibitors provided a slightly increased yield in comparison to the treatments with the split application of N. The combination of nitrogen and sulfur fertilization showed a similar trend with the average increase of 2.1% in the NS3 treatment (the single application) in comparison with the NS4 (the split application).

3.2. Qualitative Parameters of Wheat

According to the standard of the Commission Regulation (EC) No. 824/2000, the required minimum value of the hectoliter weight of the wheat grain is 73 kg/hL. All of the treatments complied with this standard. The differences among the hectoliter weights of the individual fertilizer treatments were minimal in the GS1 and the GS2. Significant differences were observed in the GS3 (Table 5). On average, the hectoliter weight on the N2 treatment (the split application of CAN and ALZON neo-N) was significantly higher compared to the NS3 and NS4 treatments (with the sulfur application).

Table 5. The effect of fertilizer treatments on the hectoliter weight of the wheat grain (kg/hL).

Treatments	GS1	GS2	GS3	Average
control	80.43 [a] ± 0.09	80.85 [a] ± 0.26	80.50 [abc] ± 0.11	80.59 [ab] ± 0.11
N1	80.68 [a] ± 0.22	80.93 [a] ± 0.22	80.58 [ab] ± 0.23	80.73 [ab] ± 0.12
N2	80.98 [a] ± 0.37	81.03 [a] ± 0.45	80.83 [a] ± 0.15	80.94 [a] ± 0.18
N3	80.33 [a] ± 0.49	81.20 [a] ± 0.07	80.20 [bc] ± 0.07	80.58 [ab] ± 0.20
N4	80.73 [a] ± 0.18	80.95 [a] ± 0.19	80.50 [abc] ± 0.22	80.73 [ab] ± 0.12
NS1	80.88 [a] ± 0.11	81.13 [a] ± 0.16	80.30 [bc] ± 0.15	80.77 [ab] ± 0.13
NS2	80.43 [a] ± 0.19	80.95 [a] ± 0.27	80.33 [bc] ± 0.21	80.57 [ab] ± 0.14
NS3	80.45 [a] ± 0.17	81.05 [a] ± 0.31	80.08 [c] ± 0.10	80.53 [b] ± 0.16
NS4	80.60 [a] ± 0.16	80.65 [a] ± 0.33	80.23 [bc] ± 0.17	80.59 [b] ± 0.13

Results are expressed as the mean ± standard error. The mean values with different letters are significantly different ($p < 0.05$) according to the LSD test (each growing season was evaluated separately). GS1—growing season 1 (2017–2018), GS2—growing season 2 (2018–2019), GS3—growing season 3 (2020–2021).

The results presented in Figure 3 clearly demonstrate an inconsistent protein content in the individual growing seasons. All treatments accomplished the minimal 10.5% value of the protein content set by the standard of the Commission Regulation (EC) No. 824/2000. The protein contents from the GS2 were higher compared to other growing seasons. Further, they were very uniform with no significant differences among the observed fertilization treatments. The protein contents observed in the GS1 and the GS3 showed more inconsistent values. In the GS1, the protein content in the N1 treatment was significantly lower compared with the N4 and NS2 treatments. These treatments were also significantly increased compared to the control; both were higher by 1.8%. The opposite trend was observed in the GS3. The protein content in the wheat grain of the N1 treatment was higher compared with the N4 and NS2 treatments. A significant difference was observed only in comparison with the NS2 treatment, which was lower by than N1 10.5%. The N1 treatment was distinguished by the highest protein content in the GS3 and in the average values. The protein content of the N1 treatment was higher by 5.4% than the control in the GS3 and higher by 2.3% than the control in the total average values.

The single application (N1, N3, and NS3 treatments) of fertilizers proved to have a significant effect on the protein content, which was increased in comparison with the split application (N2, N4, NS1, NS2, and NS4 treatments) in the GS3.

The gluten content is not commonly used in the EU countries as a technological quality criterion for the wheat grain exported to the food industry. Nevertheless, the gluten content is an important indicator of baking quality, which influences the properties of dough and bakery products. It is obvious from Figure 4 that the N2, N4, NS1, NS2, and NS4 treatments had a significantly higher gluten content compared to the control treatment in the GS1. The highest increase in the gluten content was found in the N4 treatment (higher by 3.4% than the control). The GS2 did not induce any significant differences in the gluten content among the treatments. Such results are in contrast with the GS3, which showed a slight decrease in the gluten content in comparison with the other terms. A significant difference was observed between the N1 treatment (the single application) and the N4, NS2, and NS4 treatments (with two applications). The N4 treatment was lower by 9.5%, the NS2 treatment by 10.1% and the NS4 treatment by 8.9% than the N1 treatment. The average values indicated the enhancement in the N1 (by 3.6%), N2 (by 2.5%), NS1 (by 1.9%), and NS3 treatments (by 1.8%) compared to the control.

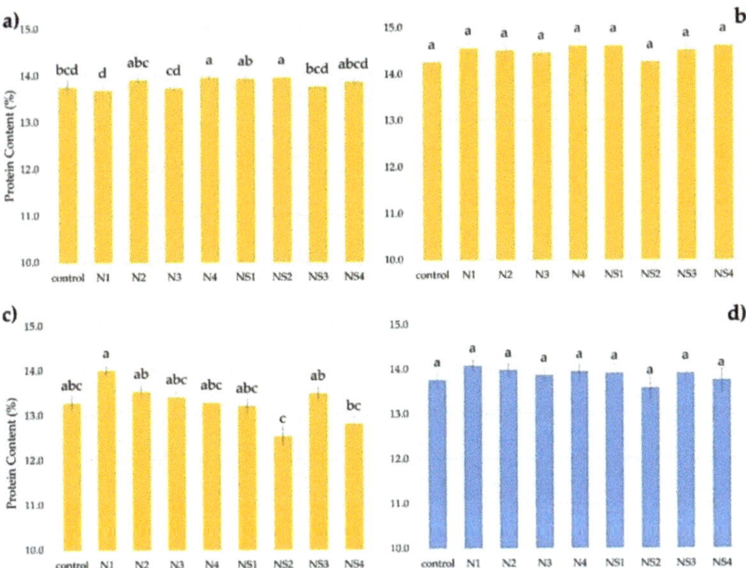

Figure 3. The effect of fertilizer treatments on protein content (%) in the grain, (**a**) growing season 1 (2017–2018), (**b**) growing season 2 (2018–2019), (**c**) growing season 3 (2020–2021), (**d**) average yield. The same letters above the columns describe no statistically significant differences between treatments (LSD test). Each growing season was evaluated separately. Standard error (SE) is expressed by error bars.

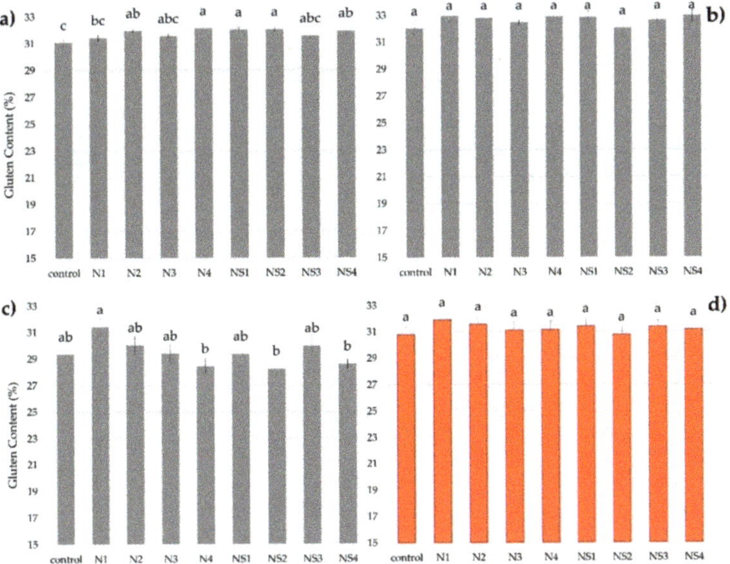

Figure 4. The effect of fertilizer treatments on gluten content (%) in the grain, (**a**) growing season 1 (2017–2018), (**b**) growing season 2 (2018–2019), (**c**) growing season 3 (2020–2021), (**d**) average yield. The same letters above the columns describe no statistically significant differences between treatments (LSD test). Each growing season was evaluated separately. Standard error (SE) is expressed by error bars.

The gluten content was significantly influenced by the application of two doses of fertilizer in the GS1 when it was higher than the gluten content in the control treatment (the application of three doses). A significant increase in the gluten content was observed in the treatments with the single application compared to the treatments with the split fertilization in the GS3 (two and three applications).

The values of the Zeleny test, which are displayed in Table 6, showed significant differences between the treatments in the GS1 and the GS3. The minimal value of ZT is 22 mL, according to the standard of the Commission Regulation (EC) No. 824/2000, and all the determined values were above this standard. Significant differences were observed in the N2, N4, NS1, and NS2 treatments in comparison with the control treatment in the GS1. The treatments without the sulfur fertilization (N2 and N4) were increased by 8%, the treatments fertilized with S (NS1 and NS2) showed ZT values that were higher by 7.3% compared to the control in the GS1. The highest ZT value was found in the N1 treatment in the GS3, which was significantly higher than the control (by 16.5%). In the N3, N4, and NS2 treatments, the ZT values were even higher by 17.1%, and by 14.1% in the NS4 treatment. A significant difference was also found between the NS3 (the single application) and the NS4 treatments (two applications of fertilizer) whereas the NS3 had a higher ZT value by 8.9%.

Table 6. The effect of fertilizer treatments on the Zeleny-test values (mL).

Treatments	GS1	GS2	GS3	Average
control	37.8 ^{c} ± 0.3	44.0 ^{a} ± 1.8	44.1 ^{de} ± 1.6	41.9 ^{b} ± 1.2
N1	39.0 ^{bc} ± 0.6	47.0 ^{a} ± 1.2	51.3 ^{a} ± 0.6	45.8 ^{a} ± 1.6
N2	40.8 ^{a} ± 0.3	46.5 ^{a} ± 1.7	50.0 ^{ab} ± 0.7	45.8 ^{a} ± 1.3
N3	39.0 ^{bc} ± 0.4	45.8 ^{a} ± 1.0	47.3 ^{bcd} ± 1.3	44.0 ^{ab} ± 1.2
N4	40.8 ^{a} ± 0.3	46.8 ^{a} ± 1.3	47.7 ^{bc} ± 1.3	45.1 ^{ab} ± 1.1
NS1	40.5 ^{a} ± 0.5	46.5 ^{a} ± 1.3	48.1 ^{abc} ± 0.9	45.0 ^{ab} ± 1.1
NS2	40.5 ^{a} ± 0.5	44.0 ^{a} ± 2.2	43.9 ^{e} ± 0.5	42.8 ^{ab} ± 0.9
NS3	39.0 ^{bc} ± 0.4	45.8 ^{a} ± 1.7	49.0 ^{ab} ± 1.9	44.6 ^{ab} ± 1.5
NS4	40.0 ^{ab} ± 0.7	47.0 ^{a} ± 1.8	45.0 ^{cde} ± 0.7	44.0 ^{ab} ± 1.0

Results are expressed as the mean ± standard error. The mean values with different letters are significantly different ($p < 0.05$) according to the LSD test (each growing season was evaluated separately). GS1—growing season 1 (2017–2018), GS2—growing season 2 (2018–2019), GS3—growing season 3 (2020–2021).

Significant differences were observed between each type of fertilization (single, split and control) in the GS1. The highest values of ZT were observed in the treatments with the split application of fertilizers. Differences between effects of three and one or two applications on the ZT values were significant in the GS3. The average values showed a significant difference between the control (three applications of fertilizers) and the treatments with one application (N1, N3, and NS3).

4. Discussions

The application of fertilizers with the NIs leads to partial ammonium nutrition in plants which could affect the crop yield and quality. Many studies have observed this effect, but the results diverge. Some studies have not found any changes in the wheat yield after the use of NI [48–50]. On the other hand, other studies describe a slight increase in the wheat yield after the NI application [51,52]. Our examined treatment that was fertilized by ammonium sulphate nitrate (ASN) with the NIs of dicyandiamide (DCD) in combination with triazole (TZ) (the NS3 treatment, fertilizer ENSIN) did not provide any significant increase in the grain yield, although the yield was slightly higher compared to the control treatment. A significant increase in the grain yield was observed after the fertilization with ALZON neo-N (the N1 treatment, urea with NI and UI) compared to the control and the NS1 treatment in the moisture-certain GS3. Other studies in the literature [53–55] report that many environmental factors including rainfalls and soil moisture have an effect on NIs as reducers of N_2O emissions, which subsequently influences the amount of N available to plants. The increase in the wheat yield after the split application of conventional N fertilizers was described in [16], which is contrary to our results in the N1 treatment with the single application in the GS3. A relative yield increase in the N1 treatment was observed in the GS1 and the GS2. The other treatments with the single application of fertilizers with NI and/or UI (N3 and NS3) also showed a relative increase in the yield compared to the treatments with the split application in each growing season. The simplification of the split application from two or three doses into a single application is an evident benefit of the use of NI and UI [56].

Grain quality is assessed by physical and chemical characteristics, which are influenced by genetic potential and environmental conditions [57]. The hectoliter weight of the wheat grain is a parameter of milling quality, and it is also connected with the wheat milling yield. It depends on agricultural inputs, variety, and weather conditions [58]. The average hectoliter weights of the wheat grain were slightly above 80.4 kg/hL in our experiment, which was the value declared by the plant breeder [43]. According to the average values, the S application did not result in an increase in this parameter. In the N2 treatment (without sulfur application), the value was significantly higher than in the NS2 and NS4 treatments (with the sulfur application). This result is similar to that reported by Hoel [59], who found a significant decrease in the hectoliter weight of the wheat grain after the S fertilization.

A study by Abalos et al. [60] explained that the application of NIs could positively affect the grain quality due to the inhibitors' induction followed by a subsequent increase in the NH_4^+ nutrition and a decrease in the N losses by the leaching of N_2O emissions. Some studies [48,51,61] have not found any changes in the grain quality after different NI applications, while Guardia et al. [62] observed slight differences after the application of the DMPSA (3,4-Dimethylpyrazole-succinic acid) inhibitor. Regarding the protein content, the N1 treatment (urea with NI and UI) showed a relative enhancement of this parameter. The increase in the grain protein content after the fertilization by the ammonium-nitrate fertilizer with NI dicyandiamide (DCD) was reported by Peltonen and Virtanen [63]. Our results are in contrast with this study because a relatively slight increase in the grain protein content was observed in the NS3 and NS4 treatments with DCD and TZ inhibitors (fertilizer ENSIN). The references on the impact of the sulfur fertilization on the grain protein content are inconsistent. Some studies [64–66] described no effect of S on the protein content, while another [67] reported the increase in grain protein after the S application. A significant increase in the protein content was observed in the NS2 treatment compared with the control and the N1, N3, NS3, and NS4 treatments, but the average values showed neither a decrease nor an increase in the protein content after the S fertilization. Järvan et al. [68] claimed that weather conditions as well as the sulfur application affect whether the correlation between the grain yield and the protein content is positive or negative. Other studies [69,70] suggested that the grain protein content is influenced by the late N fertilization. The significant effect of the split application on the increased protein content in the grain was observed in the N4 and NS2 treatments (the split application) compared to the treatments with the single application in the GS1 (the season with the lower level of precipitation). On the contrary, a stronger impact of the single application on the level of grain protein was observed in the more moisture-certain GS3.

The gluten content in the grain is affected by the amount of N supplied [71]. When comparing the gluten content of the N1 treatment (urea with MPA and 2-NPT inhibitors) and the NS4 treatment (ASN with DCD and TZ inhibitors) in the rainfall-rich GS3, we noticed a significantly higher gluten content in the N1 treatment. The studies conducted by Cahalan et al. [72] and by Shepherd et al. [73] found a correlation between the increased levels of precipitation and the higher leaching of the DCD inhibitor. The leaching of DCD can cause a lower amount of available N as well as a subsequent decrease in the gluten content in the GS3, in contrast to the drier GS1 and GS2. McGeough et al. [74] even stated that DCD showed a higher inhibition effect on arable soils. It is known that sulfur plays an important role in the amount of gluten [75] and in the composition of gluten proteins [65,76]. Nevertheless, the positive effect of the S fertilization on the gluten content was observed only in the GS1, when the NS1, NS3, and NS4 treatments provided a significantly higher content of gluten than the control. The significant impact of the split N application on the increased gluten content was found in the N2, N4, NS1, NS2, and NS4 treatments in the GS1, which is consistent with studies [18,77] describing the increase in gluten concentration after the split N application. On the contrary, a significant difference was observed between the N1 treatment with the single application and the treatments with two applications (N4, NS2, and NS4) in the GS3.

The Zeleny test represents viscoelastic characteristics of gluten proteins, which determine baking quality [78]. Meteorological factors such as temperature and rainfall during the wheat growth strongly influence this parameter [79]. The significantly highest values of the ZT in the GS3 as well as on average were observed in the N1 treatment (compared to the control). Further, the application of fertilizers with NI and UI positively influenced this quality parameter under the conditions with higher precipitation. Although other studies [66,80] reported that the S application could positively influence the value of ZT, the effect of S fertilization on the ZT values was variable in this experiment. The NS1 and NS2 treatments (fertilized with N and S) had significantly increased ZT values compared to the control in the GS1. Nevertheless, the same effect of S was not found in the other growing seasons. The single application of fertilizers brought a significant impact on the ZT, whereas the values in these treatments (N1, N3, and NS3) were higher in comparison with the control treatment (three doses of fertilizer).

Urease activity is dependent on soil moisture because the rate of urea hydrolysis is low in dry conditions [81]. The response of wheat to the inhibited fertilizer UREA^{stabil} (the N3 treatment) was influenced by weather conditions in this experiment. The relatively long period without rainfall, which was recorded after the first fertilization date (T1) in the seasons GS1 (12 days) and GS2 (11 days), increased the effectiveness of the UI. Nitrogen contained in this fertilizer (the N3 treatment) was largely retained in the soil for later use, which had a positive effect on the grain quality (protein and gluten content). The GS3 was characterized by different rain distribution. A rainfall of 12.4 mm occurred the second day after the first term of the N application (T1), and the values of the protein and gluten content in the N3 treatment (UREA^{stabil}) were lower in comparison with the drier conditions of the GS1 and the GS2. The conditions for the UI activity were not suitable in the GS3 because the rain could mitigate the gaseous loss. It could also increase the risk of leaching, which probably occurred in this case. Subsequently, the rain negatively influenced the protein and gluten content in this treatment. A highly critical factor in the application of inhibitors is the timing of their application. Their effect on the grain yield and the N efficiency is strongly affected by environmental conditions during the period of their probable inhibition activity (a period of 1–8 weeks) [82]. Chien et al. [56] stated that it is convenient if a rainfall comes 5–7 days after the fertilization by urea with inhibitors, i.e., at the time of a potentially high inhibitor activity. It is also reported that UI is suitable for arable soils on which they are capable of reducing NH_3 emission, and thus they create a better opportunity for plants to use N at later growth phases [83,84]. In contrast, the treatments with ALZON neo-N fertilizer containing both types of inhibitors showed higher values of quality parameters in the rainfall-rich GS3 compared to the treatments fertilized with the fertilizer containing only UI. These results indicate a greater effect of fertilization with NI and UI on the better N availability for plants in more humid areas.

5. Conclusions

The application of N fertilizers with NI and/or UI has great potential in today's agriculture. The cultivation of winter wheat requires high inputs of N fertilization and the application of inhibited fertilizers can reduce the risk of N losses, which negatively affect the environment as well as the efficiency of N fertilization. Neither the grain yield nor the wheat quality was reduced after the single application of fertilizers with inhibitors. In addition, a relatively average increase in the observed parameters was noticed. The protein content and ZT value were even significantly higher after the single application in comparison with the split application. Differences in the effects of the applied fertilizers with inhibitors on wheat in relation to the rainfall in each growing season were observed. The fertilizer with NI and UI ALZON neo-N (the N1 treatment) seemed to be more effective in the moisture-rich conditions of the GS3 due to the significant increase in the grain yield and ZT value in the GS3 were observed, as well as to the other observed parameters that were also relatively increased in contrast to the application under drier conditions. The use of fertilizers containing N-transformation inhibitors in wheat nutrition provides the benefit of the possibility of combining split N rates. Reducing the number of land entries has an impact on the economics of wheat cultivation, reduces soil compaction and contributes to environmental protection.

Author Contributions: Conceptualization, M.Š. and P.Š.; methodology, M.Š. and P.Š.; validation, M.Š., P.Š. and J.A.; formal analysis, M.Š. and P.Š.; investigation, M.Š., P.Š. and J.A.; resources, M.Š.; data curation, P.Š; writing—original draft preparation, M.Š.; writing—review and editing, P.Š., P.R., Z.K. and J.A.; visualization, M.Š. and J.A.; supervision, P.Š.; project administration, M.Š., P.Š. and J.A. All authors have read and agreed to the published version of the manuscript.

Funding: This research received no external funding.

Institutional Review Board Statement: Not applicable.

Informed Consent Statement: Not applicable.

Data Availability Statement: Presented data in this study are available on request from the corresponding author.

Conflicts of Interest: The authors declare no conflict of interest.

References

1. Raun, W.R.; Johnson, G.V. Improving nitrogen use efficiency for cereal production. *Agron. J.* **1999**, *91*, 357–363. [CrossRef]
2. Chen, D.; Suter, H.; Islam, A.; Edis, R.; Freney, J.; Walker, C. Prospects of improving efficiency of fertiliser nitrogen in Australian agriculture: A review of enhanced efficiency fertilisers. *Soil Res.* **2008**, *46*, 289–301. [CrossRef]
3. Mohammed, Y.A.; Kelly, J.; Chim, B.K.; Rutto, E.; Waldschmidt, K.; Mullock, J.; Torres, G.; Desta, K.G.; Raun, W. Nitrogen fertilizer management for improved grain quality and yield in winter wheat in Oklahoma. *J. Plant Nutr.* **2013**, *36*, 749–761. [CrossRef]
4. Subbarao, G.V.; Ito, O.; Sahrawat, K.L.; Berry, W.L.; Nakahara, K.; Ishikawa, T.; Watanabe, T.; Suenaga, K.; Rondon, M.; Rao, I.M. Scope and strategies for regulation of nitrification in agricultural systems—Challenges and opportunities. *Crit. Rev. Plant Sci.* **2006**, *25*, 303–335. [CrossRef]
5. Wiesler, F. Comparative assessment of the efficacy of various nitrogen fertilizers. *J. Crop Prod.* **1998**, *1*, 81–114. [CrossRef]
6. Engel, R.; Jones, C.; Wallander, R. Ammonia volatilization from urea and mitigation by NBPT following surface application to cold soils. *Soil Sci. Soc. Am. J.* **2011**, *75*, 2348–2357. [CrossRef]
7. Cameron, K.C.; Di, H.J.; Moir, J.L. Nitrogen losses from the soil/plant system: A review. *Ann. Appl. Biol* **2013**, *162*, 145–173. [CrossRef]
8. Pilbeam, D.J. Nitrogen. In *Handbook of Plant Nutrition*, 2nd ed.; Barker, A.V., Pilbeam, D.J., Eds.; CRC Press Taylor and Francis Group: Boca Raton, FL, USA, 2015; pp. 17–63.
9. Cantarella, H.; Otto, R.; Soares, J.R.; de Brito Silva, A.G. Agronomic efficiency of NBPT as a urease inhibitor: A review. *J. Adv. Res.* **2018**, *13*, 19–27. [CrossRef] [PubMed]
10. Schlesinger, W.H.; Hartley, A.E. A global budget for atmospheric NH_3. *Biogeochemistry* **1992**, *15*, 191–211. [CrossRef]
11. Harrison, R.; Webb, J. A review of the effect of N fertilizer type on gaseous emissions. *Adv. Agron.* **2001**, *73*, 65–108.
12. Bowles, T.M.; Atallah, S.S.; Campbell, E.E.; Gaudin, A.C.M.; Wieder, W.R.; Grandy, A.S. Addressing agricultural nitrogen losses in a changing climate. *Nat. Sustain.* **2018**, *1*, 399–408. [CrossRef]
13. Reyer, C.P.O.; Leuzinger, S.; Rammig, A.; Wolf, A.; Bartholomeus, R.P.; Bonfante, A.; Lorenzi, F.D.; Dury, M.; Gloning, P.; Jaoudé, R.A.; et al. Plant's perspective of extremes: Terrestrial plant responses to changing climatic variability. *Glob. Chang. Biol.* **2013**, *19*, 75–89. [CrossRef] [PubMed]
14. Erisman, J.; Bleeker, A.; Galloway, J.; Sutton, M. Reduced nitrogen in ecology and the environment. *Environ. Pollut.* **2007**, *150*, 140–149. [CrossRef]
15. Gaudin, R. The kinetics of ammonia disappearance from deep-placed urea supergranules (USG) in transplanted rice: The effects of split USG application and PK fertilizer. *Paddy Water Environ.* **2012**, *10*, 1–5. [CrossRef]

16. Fisher, R.A.; Howe, G.N.; Ibrahim, Z. Irrigated spring wheat and timing and amount of nitrogen fertilizer. I. Grain yield and protein content. *Field Crop. Res.* **1993**, *33*, 37–56. [CrossRef]

17. Sowers, K.E.; Miller, B.C.; Pan, W.L. Optimizing yield and grain protein in soft white winter wheat with split nitrogen applications. *Agron. J.* **1994**, *86*, 1020–1025. [CrossRef]

18. Xue, C.; Erley, G.S.A.; Rossmann, A.; Schuster, R.; Koehler, P.; Mühling, K.H. Split Nitrogen Application Improves Wheat Baking Quality by Influencing Protein Composition Rather Than Concentration. *Front. Plant Sci.* **2016**, *7*, 738. [CrossRef]

19. Manunza, B.; Deiana, S.; Pintore, M.; Gessa, C. The binding mechanism of urea, hydroxamic acid and N-(N-butyl)-phosphoric triamide to the urease active site. A comparative molecular dynamics study. *Soil Biol. Biochem.* **1999**, *31*, 789–796. [CrossRef]

20. Turner, D.; Edis, R.; Chen, D.; Freney, J.; Denmead, O.; Christie, R. Determination and mitigation of ammonia loss from urea applied to winter wheat with N-(n-butyl) thiophosphorictriamide. *Agr. Ecosyst Environ.* **2010**, *137*, 261–266. [CrossRef]

21. Trenkel, M.E. *Slow- and Controlled-Release and Stabilized Fertilizers: An Option for Enhancing Nutrient Efficiency in Agriculture*, 2nd ed.; IFA: Paris, France, 2010; p. 160.

22. Cui, L.; Li, D.; Wu, Z.; Xue, Y.; Xiao, F.; Zhang, L.; Song, Y.; Li, Y.; Zheng, Y.; Zhang, J.; et al. Effects of Nitrification Inhibitors on Soil Nitrification and Ammonia Volatilization in Three Soils with Different pH. *Agronomy* **2021**, *11*, 1674. [CrossRef]

23. Ruser, R.; Schulz, R. The effect of nitrification inhibitors on the nitrous oxide (N_2O) release from agricultural soils—A review. *J. Plant Nutr. Soil Sci.* **2015**, *178*, 171–188. [CrossRef]

24. Braker, G.; Conrad, R. Diversity, Structure, and Size of N2O-Producing Microbial Communities in Soils—What Matters for Their Functioning? *Adv Appl Microbiol.* **2011**, *75*, 33–70.

25. IPCC. *Climate Change 2014: Mitigation of Climate Change. Contribution of Working Group III to the Fifth Assessment Report of the Intergovernmental Panel on Climate Change*; Cambridge University Press: Cambridge, UK; New York, NY, USA, 2014.

26. Slangen, J.H.G.; Kerkhoff, P. Nitrification inhibitors in agriculture and horticulture: A literature review. *Fertil. Res.* **1984**, *5*, 1–76. [CrossRef]

27. Prasad, R.; Power, J.F. Nitrification inhibitors for agriculture, health, and the environment. *Adv. Agron.* **1995**, *54*, 233–281.

28. Singh, S.N.; Verma, A. The potential of nitrification inhibitors to manage the pollution effect of nitrogen fertilizers in agricultural and other soils: A review. *Env. Pr.* **2007**, *9*, 266–279. [CrossRef]

29. Kim, D.G.; Saggar, S.; Roudier, P. The effect of nitrification inhibitors on soil ammonia emissions in nitrogen managed soils: A meta-analysis. *Nutr. Cycl. Agroecosyst.* **2012**, *93*, 51–64. [CrossRef]

30. Rose, T.J.; Wood, R.H.; Rose, M.T.; Van Zwieten, L. A re-evaluation of the agronomic effectiveness of the nitrification inhibitors DCD and DMPP and the urease inhibitor NBPT. *Agric. Ecosyst. Env.* **2018**, *252*, 69–73. [CrossRef]

31. Stevens, W.B.; Hoeft, R.G.; Mulvaney, R.L. Fate of nitrogen-15 in a long-term nitrogen rate study. *Agron. J.* **2005**, *97*, 1046–1053. [CrossRef]

32. Onken, A.B.; Matheson, R.L.; Nesmith, D.M. Fertilizer nitrogen and residual nitrate-nitrogen effects on irrigated corn yield. *Soil Sci. Soc. Am. J.* **1985**, *49*, 134–139. [CrossRef]

33. Jokela, W.E.; Randall, G.W. Corn yield and residual soil nitrate as affected by time and rate of nitrogen application. *Agron. J.* **1989**, *81*, 720–726. [CrossRef]

34. Kuzyakov, Y. Priming effects: Interactions between living and dead organic matter. *Soil Biol. Biochem.* **2010**, *42*, 1363–1371. [CrossRef]

35. Nieder, R.; Benbi, D.K.; Scherer, H.W. Fixation and defixation of ammonium in soils: A review. *Biol. Fertil Soils* **2011**, *47*, 1–14. [CrossRef]

36. Sugihara, S.; Funakawa, S.; Kilasara, M.; Kosaki, T. Dynamics of microbial biomass nitrogen in relation to plant nitrogen uptake during the crop growth period in a dry tropical cropland in Tanzania. *Soil Sci. Plant Nutr.* **2010**, *56*, 105–114. [CrossRef]

37. Klikocka, H.; Cybulska, M.; Barczak, B.; Narolski, B.; Szostak, B.; Kobiałka, A.; Nowak, A.; Wójcik, E. The effect of sulphur and nitrogen fertilization on grain yield and technological quality of spring wheat. *Plant Soil Environ.* **2016**, *62*, 230–236. [CrossRef]

38. Dostálová, Y.; Hřivna, L.; Kotková, B.; Burešová, I.; Janečková, M.; Šottníková, V. Effect of nitrogen and sulphur fertilization on the quality of barley protein. *Plant Soil Environ.* **2015**, *61*, 399–404. [CrossRef]

39. Zhao, F.J.; Salmon, S.E.; Withers, P.J.A.; Evans, E.J.; Monaghan, J.M.; Shewry, P.R.; McGrath, S.P. Responses of breadmaking quality to sulphur in three wheat varieties. *J. Sci. Food Agric.* **1999**, *79*, 1865–1874. [CrossRef]

40. Halford, N.G.; Curtis, T.Y.; Muttucumaru, N.; Postles, J.; Elmore, J.S.; Mottram, D.S. The acrylamide problem: A plant and agronomic science issue. *J. Exp. Bot.* **2012**, *63*, 2841–2851. [CrossRef]

41. Schumacher, B.A. *Methods for the Determination of Total Organic Carbon (TOC) in Soils and Sediments*; United States Environmental Protection Agency, Environmental Sciences Division National, Exposure Research Laboratory: Las Vegas, NV, USA, 2002.

42. Zbíral, J.; Malý, S.; Váňa, M. *Soil Analysis*, 3rd ed.; Central Institute for Supervising and Testing in Agriculture: Brno, Czech Republic, 2011; pp. 18–52. (In Czech)

43. Selgen. Available online: https://selgen.cz/psenice-ozima/julie/ (accessed on 25 September 2021).

44. Lancashire, P.D.; Bleiholder, H.; Boom, T.V.D.; Langelüddeke, P.; Stauss, R.; Weber, E.; Witzenberger, A. A uniform decimal code for growth stages of crops and weeds. *Ann. Appl. Biol.* **1991**, *119*, 561–601. [CrossRef]

45. Rodriguez-Saona, L.E.; Fry, F.S.; Calvey, E.M. Use of Fourier Transform Near-Infrared Reflectance Spectroscopy for Rapid Quantification of Castor Bean Meal in a Selection of Flour-Based Products. *J. Agric. Food Chem.* **2000**, *48*, 5169–5177. [CrossRef]

46. Azizian, H.; Kramer, J.K.G.; Winsborough, S. Factors influencing the fatty acid determination in fats and oils using Fourier transform near-infrared spectroscopy. *Eur. J. Lipid Sci. Technol.* **2007**, *109*, 960–968. [CrossRef]

47. StatSoft, Inc. STATISTICA (Data Analysis Software System), Version 12. 2013. Available online: www.statsoft.com (accessed on 5 November 2021).

48. Arregui, L.M.; Quemada, M. Strategies to improve nitrogen use efficiency in winter cereal crops under rainfed conditions. *Agron. J.* **2008**, *100*, 277–284. [CrossRef]

49. Carrasco, I.; Villar, J.M. Field evaluation of DMPP as a nitrification inhibitor in the area irrigated by the Canal d'Urgell (Northeast Spain). *J. Plant Nutr.* **2001**, *92*, 764–765.

50. Polychronaki, E.; Douma, C.; Giourga, C.; Loumou, A. Assessing nitrogen fertilization strategies in winter wheat and cotton crops in northern Greece. *Pedosphere* **2002**, *22*, 689–697. [CrossRef]

51. Villar, J.M.; Guillaumes, E. Use of nitrification inhibitor DMPP to improve nitrogen recovery in irrigated wheat on a calcareous soil. *Span. J. Agric. Res.* **2010**, *8*, 1218. [CrossRef]

52. Pasda, G.; Hähndel, R.; Zerulla, W. Effect of fertilizers with the new nitrification inhibitor DMPP (3,4-Dimethylpyrazole phosphate) on yield and quality of agricultural and horticultural crops. *Biol. Fertil. Soils.* **2002**, *34*, 85–97. [CrossRef]

53. Barth, G.; von Tucher, S.; Schmidhalter, U. Influence of soil parameters on the effect of 3,4-dimethylpyrazole-phosphate as a nitrification inhibitor. *Biol. Fertil. Soils* **2001**, *34*, 98–102.

54. Weiske, A.; Benckiser, G.; Herbert, T.; Ottow, J.C.G. Influence of the nitrification inhibitor 3,4-dimethylpyrazole phosphate (DMPP) in comparison to dicyandiamide (DCD) on nitrous oxide emissions, carbon dioxide fluxes and methane oxidation during 3 years or repeated application in field experiments. *Biol. Fertil. Soils* **2001**, *34*, 109–117.

55. Di, H.J.; Cameron, K.C.; Podolyan, A.; Robinson, A. Effect of soil moisture status and a nitrification inhibitor, dicyandiamide, on ammonia oxidizer and denitrifier growth and nitrous oxide emissions in a grassland soil. *Soil Biol. Biochem.* **2014**, *73*, 59–68. [CrossRef]

56. Chien, S.H.; Prochnow, L.I.; Cantarella, H. Chapter 8 Recent Developments of Fertilizer Production and Use to Improve Nutrient Efficiency and Minimize Environmental Impacts. *Adv. Agron.* **2009**, *102*, 267–322.

57. Johansson, E.; Prieto-Linde, M.L.; Svensson, G. Influence of nitrogen application rate and timing on grain protein composition and gluten strength in Swedish wheat cultivars. *J. Plant Nutr. Soil Sci.* **2004**, *167*, 345–350. [CrossRef]

58. Zimolka, J. *Pšenice, Pěstování, Hodnocení a Užití Zrna*; Profi Press: Prague, Czech Republic, 2005; p. 181. (In Czech)

59. Hoel, B.O. Effects of sulphur application on grain yield and quality, and assessment of sulphur status in winter wheat (*Triticum aestivum* L.). *Acta Agric. Scand. B Soil Plant Sci.* **2011**, *61*, 499–507.

60. Abalos, D.; Jeffery, S.; Sanz-Cobena, A.; Guardia, G.; Vallejo, A. Meta-analysis of the effect of urease and nitrification inhibitors on crop productivity and nitrogen use efficiency. *Agric. Ecosyst. Environ.* **2014**, *189*, 136–144. [CrossRef]

61. Liu, C.; Wang, K.; Zheng, X. Effects of nitrification inhibitors (DCD and DMPP) on nitrous oxide emission, crop yield and nitrogen uptake in a wheat-maize cropping system. *Biogeosciences* **2013**, *10*, 2427–2437. [CrossRef]

62. Guardia, G.; Sanz-Cobena, A.; Sanchez-Martín, L.; Fuertes-Mendizábal, T.; González-Murua, C.; Álvarez, J.M.; Chadwick, D.; Vallejo, A. Urea-based fertilization strategies to reduce yield-scaled N oxides and enhance bread-making quality in a rainfed Mediterranean wheat crop. *Agric. Ecosyst. Environ.* **2018**, *265*, 421–431. [CrossRef]

63. Peltonen, J.; Virtanen, A. Effect of nitrogen fertilizers differing in release characteristics on the quantity of storage proteins in wheat. *Cereal Chem.* **1994**, *71*, 1–5.

64. Steinfurth, D.; Zörb, C.; Braukmann, F.; Mühling, K.H. Time dependent distribution of sulphur, sulphate and glutathione in wheat tissues and grain as affected by three sulphur fertilization levels and late S fertilization. *J. Plant Physiol.* **2012**, *169*, 72–77. [CrossRef]

65. Zhao, F.J.; Salmon, S.E.; Withers, P.J.; Monaghan, J.M.; Evans, E.J.; Shewry, P.R.; McGrath, S.P. Variation in the bread-making quality and rheological properties of wheat in relation to sulphur nutrition under field conditions. *J. Cereal Sci.* **1999**, *30*, 19–31. [CrossRef]

66. Erekul, O.; Götz, K.P.; Koca, Y.O. Effect of sulphur and nitrogen fertilization on bread-making quality of wheat (*Triticum aestivum* L.) varieties under Mediterranean climate conditions. *J. Appl. Bot. Food Qual.* **2012**, *85*, 17–22.

67. Shahsavani, S.; Gholami, A. Effect of sulphur fertilization on breadmaking quality of three winter wheat varieties. *Pak. J. Biol. Sci.* **2008**, *11*, 2134–2138. [CrossRef]

68. Järvan, M.; Edesi, L.; Adamson, A. The effect of sulphur fertilization on yield, quality of protein and baking properties of winter wheat. *J. Agric. Sci.* **2009**, *20*, 8–15.

69. Wieser, H.; Seilmeier, W. The influence of nitrogen fertilisation on quantities and proportions of different protein types in wheat flour. *J. Sci. Food Agric.* **1998**, *76*, 49–55. [CrossRef]

70. Blandino, M.; Vaccino, P.; Reyneri, A. Late-season nitrogen increases improver common and durum wheat quality. *Agron. J.* **2015**, *107*, 680–690. [CrossRef]

71. Kozlovský, O.; Balík, J.; Černý, J.; Kulhánek, M.; Kos, M.; Prášilová, M. Influence of nitrogen fertilizer injection (CULTAN) on yield, yield components formation and quality of winter wheat grain. *Plant Soil Environ.* **2018**, *55*, 536–543. [CrossRef]

72. Cahalan, E.; Ernfors, M.; Müller, C.; Devaney, D.; Laughlin, R.J.; Watson, C.J.; Hennessy, D.; Khalil, M.I.; McGeough, K.L.; Richards, K.G. 2014. The effect of the nitrification inhibitor dicyandiamide (DCD) on nitrous oxide and methane emissions after cattle slurry application to Irish grassland. *Agric. Ecosyst. Environ.* **2014**, *199*, 339–349. [CrossRef]

73. Shepherd, M.; Wyatt, J.; Welten, B. Effect of soil type and rainfall on dicyandiamide concentrations in drainage from lysimeters. *Soil Res.* **2012**, *50*, 67–75. [CrossRef]

74. McGeough, K.L.; Watson, C.J.; Müller, C.; Laughlin, R.J.; Chadwick, D.R. Evidence that the efficacy of the nitrification inhibitor dicyandiamide (DCD) is affected by soil properties in UK soils. *Soil Biol. Biochem.* **2016**, *94*, 222–232. [CrossRef]

75. Wieser, H.; Gutser, R.; von Tucher, S. Influence of sulphur fertilisation on quantities and proportions of gluten protein types in wheat flour. *J. Cereal Sci.* **2004**, *40*, 239–244. [CrossRef]

76. Zörb, C.; Steinfurth, D.; Seling, S.; Langenkämper, G.; Koehler, P.; Wieser, H.; Lindhauer, M.G.; Mühling, K.H. Quantitative protein composition and baking quality of winter wheat as affected by late sulfur fertilization. *J. Agric. Food Chem.* **2009**, *57*, 3877–3885. [CrossRef] [PubMed]

77. Martre, P.; Jamieson, P.D.; Semenov, M.A.; Zyskowski, R.F.; Porter, J.R.; Triboi, E. Modelling protein content and composition in relation to crop nitrogen dynamics for wheat. *Eur. J. Agron.* **2006**, *25*, 138–154. [CrossRef]

78. Moot, D.J.; Every, D. A comparison of bread baking, falling number, amylase essay and visual methods for assessment of pre-harvest sprouting in wheat. *J. Cereal Sci.* **1990**, *11*, 225–234. [CrossRef]

79. Koga, S.; Böcker, U.; Moldestad, A.; Tosi, P.; Shewry, P.R.; Mosleth, E.F.; Uhlen, A.K. Influence of temperature during grain filling on gluten viscoelastic properties and gluten protein composition. *J. Sci. Food Agric.* **2016**, *96*, 122–130. [CrossRef] [PubMed]

80. Schäfer, T.; Honermeier, B. Einfluss unterschiedlicher N- und SDüngung auf Kornertrag und Backqualitaet von Win-terweizen (*Triticum aestivum* L.). *Mitt. Ges. Pflanzenbauwiss* **2007**, *19*, 8–9.

81. Volk, G.M. Efficiency of fertilizer urea as affected by method of application, soil moisture, and lime. *Agron. J.* **1966**, *58*, 249–252. [CrossRef]

82. Quemada, M.; Baranski, M.; Nobel-de Lange, M.N.J.; Vallejo, A.; Cooper, J.M. Meta-analysis of strategies to control nitrate leaching in irrigated agricultural systems and their effects on crop yield. *Agric. Ecosyst. Environ.* **2013**, *174*, 1–10. [CrossRef]

83. Abalos, D.; Sanz-Cobena, A.; Misselbrook, T.; Vallejo, A. Effectiveness of urease inhibition on the abatement of ammonia, nitrous oxide and nitric oxide emissions in a non-irrigated Mediterranean barley field. *Chemosphere* **2012**, *89*, 310–318. [CrossRef] [PubMed]

84. Francisco, S.S.; Urrutia, O.; Martin, V.; Peristeropoulos, A.; Garcia-Mina, J.M. Efficiency of urease and nitrification inhibitors in reducing ammonia volatilization from diverse nitrogen fertilizers applied to different soil type sand wheat straw mulching. *J. Sci. Food Agric.* **2011**, *91*, 1569–1575. [CrossRef] [PubMed]

Article

Yield of Winter Oilseed Rape (*Brassica napus* L. var. *napus*) in a Short-Term Monoculture and the Macronutrient Accumulation in Relation to the Dose and Method of Sulphur Application

Mariusz Stepaniuk[1] **and Aleksandra Głowacka**[2,*]

1. STEPPOL AGRO, Łykoszyn 61, 22-652 Telatyn, Poland; stepaniuk.m@op.pl
2. Department of Plant Cultivation Technology and Commodity Science, University of Life Sciences in Lublin, 15 Akademicka Street, 20-950 Lublin, Poland
* Correspondence: aleksandra.glowacka@up.lublin.pl; Tel.: +48-81-445-66-23

Abstract: The objective of this study was to assess the yield efficiency of sulphur-enhanced fertilisers, depending on the dose and application method, in a short-lived (three-year) monoculture of winter oilseed rape under the climate and soil conditions of south-eastern Poland. The experiment was carried out between 2010 and 2013 on winter oilseed rape (*Brassica napus* L. var. *napus*) of the Orlando variety, fertilised with different sulphur doses—0, 20, 40 or 60 kg S ha^{-1} applied in different method—soil application sowing, foliar application in the spring, and soil application sowing + foliar application in the spring (combined application). Following the harvest, seed and straw yields and the content of macroelements (N, S, P, K, Ca and Mg) in the seed and straw samples were determined. The harvest indices were also established for each of these elements. The impact of sulphur on winter oilseed rape yield depended significantly on both the dose and the application method. Even at the lowest dose (20 kg·ha^{-1}), sulphur materially increased seed yield, regardless of the application method. With autumn soil application and foliar application, differences between the lowest dose and the higher doses (40 and 60 kg·ha^{-1}) were not significant. However, with combined application, the highest dose (60 kg·ha^{-1}) significantly increased yield compared to the lower doses. In general, all the fertilisation approaches significantly increased the N, P, K, Ca and Mg contents compared to the control sample, but the differences between them were not substantial. Each of the sulphur application approaches decreased the harvest index for sulphur. The foliar application of each of the doses decreased the harvest indices for N, P, K and Ca. The soil application of 20 kg·ha^{-1}, and the mixed application of 40 and 60 kg·ha^{-1}, all increased the harvest indices for P, K and Ca.

Keywords: oilseed rape; short-lived monoculture; sulphur; yield; seeds; straw; macroelements; harvest index

Citation: Stepaniuk, M.; Głowacka, A. Yield of Winter Oilseed Rape (*Brassica napus* L. var. *napus*) in a Short-Term Monoculture and the Macronutrient Accumulation in Relation to the Dose and Method of Sulphur Application. *Agronomy* **2022**, *12*, 68. https://doi.org/10.3390/agronomy12010068

Academic Editors: Christos Noulas, Shahram Torabian and Ruijun Qin

Received: 1 December 2021
Accepted: 24 December 2021
Published: 28 December 2021

Publisher's Note: MDPI stays neutral with regard to jurisdictional claims in published maps and institutional affiliations.

1. Introduction

In terms of cultivation area and production volume, oilseed rape is the second largest oilseed crop in the world, after soybean. Like other species of the family *Brassicaceae*, oilseed rape has a high demand for sulphur—twice as high as legumes and four times as high as cereals, grasses, maize, or potatoes [1–3]. This is because sulphur performs a number of important physiological functions and is essential for the normal growth and development of these plants [4]. The first information about sulphur deficiencies on oilseed rape plantations appeared in the late 1990s, and since that time, sulphur fertilisation of crop plants, particularly oilseed rape and other species of the family *Brassicaceae*, has been the subject of numerous studies [5–8]. Sulphur deficiency decreases yield, limits nitrogen uptake, and reduces the content of sulphur-rich metabolites responsible for plants' resistance to biotic and abiotic stress [9]. Many recent experiments have confirmed the beneficial effect of sulphur not only on the yield of crops, but on their quality as well. However, the effect of sulphur application on yield depends on a number of factors, including the amount applied and time of application, the form of sulphur used, the sulphur content in the soil (which largely determines its availability for plants), and climate conditions [10–18].

Fertilisation of crop fields with sulphur can raise its content in soil and increase the accumulation of certain minerals in plants due to the acidifying effect of sulphur on soil. Excessive content of these elements can be harmful both for the plant itself and for consumers. According to Podleśna [19], sulphur used in fertilisers can upset the balance in the rhizosphere and thereby have a strong effect on the uptake of other nutrients. Many other authors point out that fertilisation with sulphur can modify pH and the microbial activity of soil, thus increasing the concentrations of some elements in the plant tissue, particularly heavy metals [20–22]. Furthermore, some elements increase the growth rate of plant mass, causing a decrease or increase in the concentrations of other elements, and thus in their total accumulation. This is important because the mineral concentrations in plant tissues are associated not only with plant growth and development dynamics, but also with resistance to disease, freezing, or other stress conditions during the growing period, and therefore with crop quality [19]. According to Jankowski et al. [7,22], sulphur fertilisation of oilseed rape, by modifying the content of minerals in the plants, can significantly influence the fertiliser value of crop residues (roots or straw) and the quality of seeds used for oil extraction.

The literature contains information confirming the effect of sulphur application on the content and accumulation of minerals in crop plants. Podleśna [19] found that sulphur application at a rate of 80–100 kg S·ha^{-1} increased the concentrations of sulphur and nitrogen in the seeds and straw of oilseed rape and of calcium only in the straw. Changes in the content of other macroelements, i.e., phosphorus, potassium and magnesium, were not significant. The simultaneous increase in the dry matter of rape significantly increased the total accumulation of the elements which tested relative to the control plants. Jankowski et al. [7] reported that sulphur applied at a rate of 60 kg S·ha^{-1} (in the form of ammonium sulphate) increased the content of potassium, calcium, magnesium and sulphur in winter rapeseed cakes and in the straw (except for Mg). The nitrogen content decreased in both the seed cakes and straw, while phosphorus remained unchanged. Lošák et al. [23] showed that fertilisation with sulphur had no effect on nitrogen, phosphorus, or potassium concentrations, but significantly increased the content of sulphur and calcium. Similarly, McGrath and Zhao [24] found that sulphur had no significant effect on the nitrogen concentration in oilseed rape stems.

In the last two decades, sulphur fertilisation of crops, particularly oilseed rape and other species of the family *Brassicaceae*, has been the subject of many studies. However, there have been no field studies on sulphur fertilisation of winter oilseed rape grown in monoculture. Due to the significant increase in winter oilseed rape production, it may be grown on the same field more often. The aim of the study was to assess the effect of fertilisation with sulphur (Na_2SO_4) on yield, depending on the rate and means of application, in a short-term (three-year) monoculture of the Orlando cultivar of winter oilseed rape in the climate and soil conditions of south-eastern Poland. A second objective was to assess the effect of sulphur application on the content of macroelements in the seeds and straw of rapeseed.

2. Materials and Methods

2.1. Site Description and Experimental Design

A field experiment was carried out in the years 2010–2013 in the village of Kolonia Franusin (50°32′40″ N 23°51′30″ E) in Telatyn Commune, Tomaszów County, Lubelskie Voivodeship. The experiment was set up on mineral brown soil of the group Cambisols (WRB. IUSS Working Group 2014) rich in organic carbon, with the granulometric composition of clayey silt [25], and a slightly acidic reaction (pH$_{KCl}$ 6.2). The most important physicochemical properties of the soil are presented in Table 1. The soil pH was determined by potentiometric method (in the suspensions of soil and 1 M solution of KCl 1:2.5) according to PN-ISO 10390:1997 [26]; Corg. was determined by the Tiurin method (the oxidation of organic carbon to CO_2 using potassium dichromate $K_2Cr_2O_7$, in the presence of concentrated sulphuric acid H_2SO_4 and the silver sulphate—Ag_2SO_4), according to KQ/PB-34 [27]; the content of available P and K was determined by the Egner–Riehm method, according to PN-R-04023:1996 [28] and PN-R-04022:1996 [29], after extraction with calcium lactate; the Mg content was determined by the Schachtschabel method after extracting from the soil with $CaCl_2$ solution, according to PN-R-04020:1996 + Az1:2004 [30]; sulphate sulphur (S-SO$_4$) content was determined by nephelometric method, according to KQ/PB-44 [27]; the granulometric composition was determined by Casagrande's areometric method as modified by Prószyński, according to KQ/PB-33 [27].

Table 1. Selected soil properties (0–30 cm depth) before setting up the experiment (2010).

Parameters	Unit	Value
Granulometric composition		Clayey silt
Sand (2.0–0.05 mm)		12.0
Dust (0.05–0.02 mm)		39.0
(0.02–0.002 mm)	%	35.0
Loam (<0.002 mm)		14.0
pH_{KCl}		6.2
$C_{org.}$		18.1
$N_{tot.}$	g kg^{-1}	1.6
Content of available nutrients		
P		34.9
K		93.8
Mg	mg kg^{-1}	64.5
S-SO$_4$		8.23

The subject of the study was the Orlando cultivar of winter oilseed rape fertilised with varied amounts of sulphur, applied in different ways, for a total of 10 sulphur fertilisation treatments. The experiment was set up in a single-factor, completely randomized design in four replicates. The design of the field experiment is presented in Table 2. The phonological stages (BBCH) of winter oilseed rape were encoding according to [31].

Table 2. Applied variants of sulphur fertilisation.

(I) Sulphur Dose	(II) Method of Sulphur Application		
	Pd Soil Application Sowing	D Foliar	PdD Mixed = $\frac{1}{2}$ Pd + $\frac{1}{2}$ D
S0		S0	
S1 20 kg S ha^{-1}	S1Pd	S1D One-time application—before budding (BBCH 35)	S1PdD Foliar application: I. before budding (BBCH 35)—10 kg S ha^{-1}.
S2 40 kg S ha^{-1}	S2Pd	S2D (1/2 DI + 1/2 DII) I. before budding (BBCH 35), II. beginning of flowering (BBCH 57)	S2PdD ($\frac{1}{2}$Pd + $\frac{1}{2}$D) Foliar application: I. before budding (BBCH 35)—20 kg S ha^{-1}
S3 60 kg S ha^{-1}	S3Pd	S3D (1/3 DI + 1/3 DII + 1/3 DIII) I. before budding (BBCH 35), II. beginning of flowering (BBCH 57), III. full flowering (BBCH 65)	S3PdD ($\frac{1}{2}$Pd + $\frac{1}{2}$D) Foliar application: I. before budding (BBCH 35)—15 kg S ha^{-1}, II. beginning of flowering (BBCH 57)—15 kg S ha^{-1}

The area of a single plot was 100 m^2. Winter oilseed rape was grown in a three-year monoculture on a field where the previous crop had been ryegrass (*Festuca perennis*) grown for seed. After harvesting the seeds, the straw of the forecrop was plowed in. The cultivation technology was in accordance with current recommendations. After harvesting the forecrop, a set of post-harvest crops was made with a cultivator (Konskilde Delta, depth 6–8 cm), and then, at seven-day intervals, harrowing was performed twice with a heavy harrow. Then, plowing was carried out with a reversible plow to a depth of 22 cm. A pre-sowing cultivator was used to prepare the soil for sowing and to mix fertilisers. Sowing was performed with a pneumatic seed drill. Prior to sowing, mineral fertiliser was applied as follows (in kg·ha^{-1}): P_2O_5—40 kg·ha^{-1} (Super Fos Dar), K_2O—60 kg·ha^{-1} (potassium salt) and the first rate of nitrogen at N—31 kg·ha^{-1} (ammonium nitrate). The second nitrogen rate of 96 kg ha^{-1} was applied in the spring before the start of growth (BBCH 25), using ammonium nitrate. The third nitrogen rate of 48 kg ha^{-1} was applied at the beginning of budding (BBCH 34) also using ammonium nitrate. Sulphur was applied in the form of anhydrous sodium sulphate (Na_2SO_4). The rates, means, and times (BBCH stages) of application are shown in Table 2. For foliar application, sodium sulphate was dissolved in water in the amount of 200 dm^3·ha^{-1}.

Seeds of the Orlando cultivar of winter oilseed rape (F_1 hybrid) were sown at a rate of 3.8 kg·ha^{-1}, for a density of 50 plants per m^2. Seeds were sown between 24 and 28 August, depending on the year, so that during overwintering, rapeseed entered the rosette stage with 8 true leaves and a root neck 7 mm thick. Plant protection against pathogens, weeds and pests was carried out according to the recommendations of the Institute of Plant Protection in Poznań [32]. Rapeseed was harvested with a combine (Deyth Fahr 4554H) during the fully ripe stage (BBCH 89). Each year of the experiment, the rape straw was plowed in.

2.2. Meteorological Conditions

The weather conditions during the experiment are presented in Figures 1 and 2 and in Tables 3 and 4. Data pertaining to precipitation and air temperature were used to calculate Selyaninov's hydrothermal coefficient for the months in which the average daily air temperature exceeded 8 °C, because in this case, reliable results are obtained [33]. The scale developed by Skowera et al. [34] to define ranges of values for the coefficient was adopted. Precipitation totals in the first season of 2010/2011 (August–July) amounted to 791.7 mm, which was 82 mm higher than the long-term average. Particularly high precipitation in this season was noted in August and September, substantially exceeding the long-term averages for these months, by 80.8 and 39.5 mm, respectively. In the other two seasons, the precipitation totals during the period from August to July were lower than the long-term average (Figure 1). Particularly low precipitation was noted in the second season of the study (2011/2012). Precipitation in the autumn growing period (August–November) of 2011 was only 38% of the long-term average (Table 3). In all three seasons of the experiment, precipitation during winter dormancy was much lower than the long-term average (1996–2013). The average temperature during the spring growing season (April–July) in all years of the study was higher than the long-term average (Figure 2).

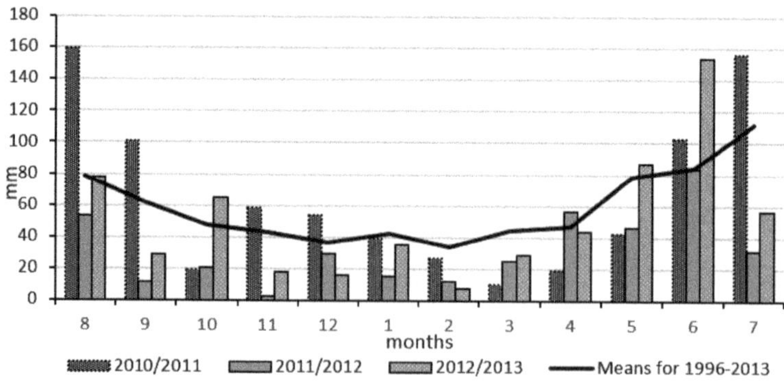

Figure 1. Rainfall in months (according to the COBORU Meteorological Station in Ulhówek) as compared to the long-term means (according to the meteorological station in Tomaszów Lubelski, Institute of Meteorology and Water Management—National Research Institute in Warsaw Meteorological Station).

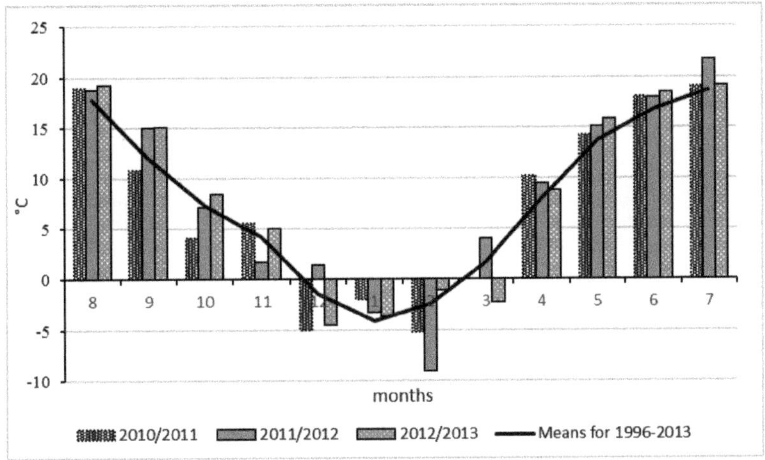

Figure 2. Air temperature in months (according to the COBORU Meteorological Station in Ulhówek) as compared to the long-term means (according to the meteorological station in Tomaszów Lubelski, Institute of Meteorology and Water Management—National Research Institute in Warsaw Meteorological Station).

Table 3. Precipitation and air temperature in months (according to the COBORU Meteorological Station in Ulhówek) compared to the long-term average (1996–2013 according to the meteorological station in Tomaszów Lubelski, Institute of Meteorology and Water Management—National Research Institute in Warsaw).

The Growing Season	Temperature (°C)				Rainfall			
	Deviation from the Long-term Average			Long-Term Average	% of the Long-Term Average			Long-Term Average (mm)
	2010/2011	2011/2012	2012/2013		2010/2011	2011/2012	2012/2013	
Autumn August–November	−0.4	+0.4	+1.7	10.3	146.7	38.2	82.2	230.8
Winter dormancy December–March	−1.5	−0.2	−1.3	−1.6	83.6	52.7	55.8	157.4
Spring April–July	+1.1	+1.7	+1.3	14.3	100	67.9	106.3	321.5

Table 4. Selyaninov's hydrothermal coefficient (K) in 2010–2013 compared to the long-term average (1996–2013).

Year	Months					
	April	May	June	July	August	September
2010	–	–	–	–	2.72	3.12
2011	0.64	0.92	1.90	2.66	0.91	0.25
2012	2.01	0.99	1.55	0.47	1.30	0.64
2013	1.77	1.78	2.78	0.96	–	–
The long-term average for 1996–2013	1.97	1.84	1.68	1.93	1.43	1.71

K—Selyaninov's coefficient [k = (p × 10)/Σt], where: p—sum of monthly precipitation (mm), Σt—sum of average daily air temperatures for a month (°C).

Based on Selyaninov's hydrothermal coefficient, the periods from April to September 1995–2013 can be described as optimal or fairly wet (Table 4). In contrast, during the experiment, there were dry or very dry periods (April, May, August and September 2011 and July and September 2012) as well as very wet periods (August and September 2010 and June 2013).

2.3. Measurements

Each year after harvest, the seed yield from each plot was determined and, following correction to 13% moisture, calculated per ha. The straw yield from each plot was determined as well.

After harvest, samples of plant material (seeds and straw) were taken from each plot for chemical analysis. The following were determined in the seed and straw samples: total sulphur by nephelometry (KQ/PB-31 version 04 of 02/07/12)—the method consists in the oxidation of organic and inorganic sulphur to SO_4 in an electric muffle furnace at 550 °C in the presence of sodium bicarbonate and oxygen from the air, followed by nephelometric determination in the form of $BaSO_4$; total N by the Kjeldahl method (KQ/PB-70 version 02 of 01/12/10)—the method consists in converting the amide form of nitrogen into ammonia by mineralisation with concentrated sulphuric acid, then distillation of ammonia and absorption in a sulphuric acid solution and titration of the excess acid with a standard sodium hydroxide solution; phosphorus by the vanadomolybdate method (KQ/PB-24 version 03 of 01/12/10)—the method consists in measuring by spectrophotometry the intensity of the yellow color of the phosphor-vanadium-molybdic acid complex, which is formed by orthophosphate and vanadium ions in the presence of molybdate in an acidic environment, the measurement is made at a wavelength of 470 nm; potassium and calcium (KQ/PB-25 version 03 of 01/12/10) by flame photometry—after mineralisation in concentrated nitric acid, the nebulized solution is mixed with the flammable gas, and next, the mixture burns at the exit of the burner. The burner flame, in which the elements are excited, is a radiation source located in an optical system containing appropriate filters and a detector. The K-77J filter is used to determine the potassium content, and the Ca-63J filter for the determination of calcium; magnesium (KQ/PB-26 version 03 of 01/12/10) by atomic absorption spectrometry—the method is based on measuring the absorption of radiation emitted by magnesium atoms released when the solution is sprayed into an acetylene-air flame.

The analyses of the content of N, S, P, K, Ca and Mg were performed in the accredited laboratory of the Regional Chemical and Agricultural Testing Station in Lublin according to the procedures developed on the basis of the methodology of Institute of Soil Science and Plant Cultivation—State Research Institute in Puławy [27,35].

Then the harvest index for accumulation of N, S, P, K, Ca and Mg was calculated, i.e., the amount of the element accumulated in the seeds as a percentage of its total amount accumulated in the plant.

2.4. Statistical Analysis

Statistical analysis of the results was performed using Excel 2016 and STATISTICA 13 PL software (Tulsa, USA). Two-way analysis of variance (ANOVA) was performed, and Tukey's test was used to verify the significance between means at $\alpha = 0.05$.

3. Results and Discussion

3.1. Yield of Seeds and Straw

Oilseed rape yield depends on the genetically determined yield potential of a given cultivar and its response to environmental and agrotechnical conditions. Weather conditions were varied in the years of the study (2010–2013), and weather was found to significantly influence both the seed yield and the straw yield of rapeseed. Extreme moisture conditions were noted in the 2011/2012 season. In September 2011, only 2.5 mm of rain fell, and Selyaninov's hydrothermal coefficient (k = 0.25) confirms that this was a dry period. The

Agronomy 2022, 12, 68

precipitation total for the period from August 2011 to July 2012 was only 55% of the long-term average for this period. Moreover, in that season, there were very low temperatures in February—6.6 °C lower than the long-term average. The unfavourable weather conditions clearly affected the seed and straw yield of rapeseed (Table 5). Many previous studies confirm that weather conditions strongly affect seed yields of rape [36–39]. Wielebski [40] obtained the highest rapeseed yield in the season when moderate temperatures in the spring were accompanied by a beneficial rainfall distribution. These favourable moisture and temperature parameters ensured good pod formation and seed formation in the pod. In the present study, the rainfall and temperature distribution in the spring growing period was favourable in 2011 and in 2013, when the highest yields were obtained. The autumn growing period in 2010 (August–November) was distinguished by heavy rainfall, amounting to 147% of the long-term average, and Selyaninov's hydrothermal coefficient was 2.27 (very wet) for August and 3.12 (extremely wet) for September. The following April (2011) was a very dry period (k = 0.64), and May was dry (k = 0.92) [27]. This may have been the cause of the poorer yields obtained in 2011 in comparison to 2013.

Table 5. The influence of variants of sulphur fertilisation on the yield of seeds and straw of winter rape.

Variants of Sulphur Fertilisation [1]	Seeds Yield (dt ha^{-1})			Straw Yield (dt ha^{-1})		
	2011	2012	2013	2011	2012	2013
S0	39.16 [de] ± 0.75	30.60 [a] ± 0.47	38.36 [d] ± 1.30	83.10 [cd] ± 2.91	78.40 [bc] ± 1.21	94.00 [fgh] ± 3.18
S1Pd	39.79 [def] ± 0.42	32.88 [ab] ± 0.40	41.55 [fg] ± 1.20	94.78 [ghi] ± 0.90	74.20 [ab] ± 0.89	107.40 [no] ± 3.10
S2Pd	40.95 [efg] ± 0.73	33.12 [bc] ± 0.54	41.33 [efg] ± 0.77	106.5 [l-m] ± 2.54	69.50 [a] ± 1.13	115.60 [pq] ± 2.16
S3Pd	41.84 [fgh] ± 0.75	33.72 [c] ± 0.71	41.54 [fg] ± 0.95	90.2 [efg] ± 1.95	94.70 [ghi] ± 1.99	101.70 [j-n] ± 2.34
S1D	41.76 [fg] ± 0.74	30.96 [abc] ± 0.99	44.19 [hij] ± 0.79	117.8 [n] ± 2.07	100.70 [i-l] ± 3.22	108.60 [o] ± 1.94
S2D	40.76 [efg] ± 0.80	32.76 [abc] ± 0.71	42.74 [g-j] ± 0.88	104.6k [l-m] ± 2.06	102.10 [k-l] ± 2.21	100.10 [h-l] ± 2.06
S3D	42.28 [g-j] ± 0.75	32.64 [abc] ± 1.03	44.42 [j] ± 0.53	100.2 [h-l] ± 1.78	96.00 [g-j] ± 3.02	101.80 [j-n] ± 1.21
S1PdD	40.76 [efg] ± 0.56	32.52 [abc] ± 0.71	41.29 [efg] ± 0.86	87.9 [def] ± 1.43	104.40 [l-o] ± 2.28	93.40 [fg] ± 1.94
S2PdD	41.64 [fg] ± 0.92	32.28 [abc] ± 0.86	41.20 [efg] ± 0.59	106.3 [k-m] ± 1.98	77.70 [bc] ± 2.06	109.70 [op] ± 1.58
S3PdD	42.10 [f-i] ± 0.80	33.60 [c] ± 0.59	44.50 [j] ± 0.89	86.80 [de] ± 1.66	71.30 [a] ± 1.25	109.40 [op] ± 2.20
Mean	41.10 [B] ± 1.21	32.51 [A] ± 1.21	42.11 [C] ± 1.97	97.91 [B] ± 10.69	86.88 [A] ± 13.34	104.17 [C] ± 7.10

[1] See Table 2; Means followed by the same letter are not statistically different at the α = 0.05 level. Small letters (a-p) for variants of sulphur fertilisation × years, capital letters (A, B, C) for years. ±—standard deviation.

For three seasons, oilseed rape was grown on the same site. The highest yield was obtained in the third year, which indicates that cultivation of winter oilseed rape in a short-term monoculture did not significantly affect yield (Table 5). According to many authors, the yield of oilseed rape decreases when it is grown for a second consecutive season on the same site [41]. An insufficient interval between oilseed rape crops on the same site increases the occurrence of pathogens, pests, and weeds, including pathogenic fungi, which can pose a threat to yield and to crop quality [37,42]. Hegewald et al. [43] also draw attention to the importance of crop rotation in maintaining seed and oil yield of rape. According to many authors, however, oilseed rape occasionally grown for a second consecutive season or in a short monoculture can produce similar yields to crops preceded by cereals [44], or even slightly higher [45]. In a study by Jaskulska et al. [46], the yield of rapeseed grown for a second consecutive season was 4.6% higher than when preceded by winter wheat and 8.5%, significantly, higher than when preceded by spring barley.

Analysis of our results showed that sulphur application from 20 to 60 kg S·ha^{-1} significantly increased the seed and straw yield of rape, on average by 8.02% and 14.2% (Figure 3a,b). Many studies in various parts of the world have confirmed the positive effect of sulphur application on oilseed rape yield in conditions of sulphur deficiency in the soil [13–15,17,19,47,48]. According to many authors [2,8,23,24], the effect of sulphur on yield mainly results from its strong effect on nitrogen metabolism. Sulphur plays an important role in nitrogen metabolism, increasing the rate at which nitrogen taken up by the plant is transformed into protein. Nitrogen is the nutrient with the greatest effect on yield, and therefore by influencing nitrogen metabolism, sulphur directly affects seed yield.

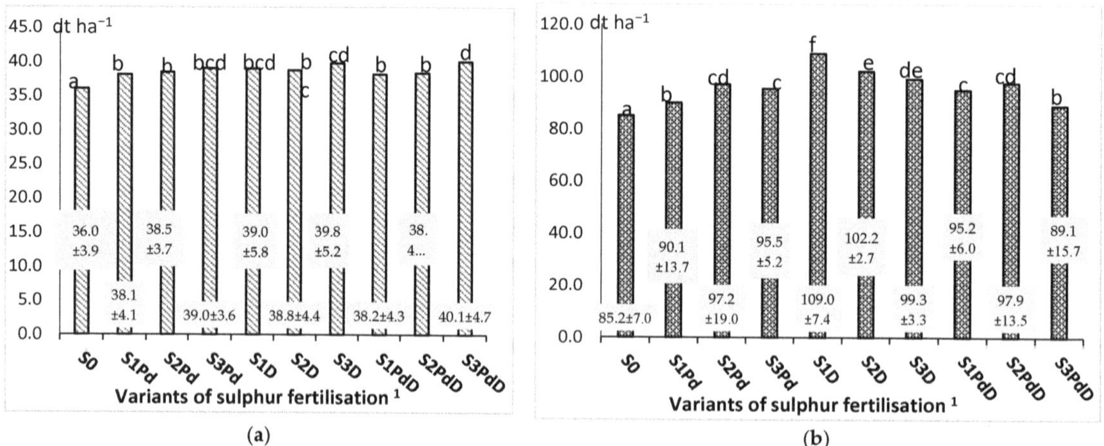

(a)

(b)

Figure 3. (**a**) Average yield of rape seeds for variants of sulphur fertilisation; (**b**) Average yield of rape straw for variants of sulphur fertilisation. [1] See Table 2; Means followed by the same letter are not statistically different at the $\alpha = 0.05$ level. ±—standard deviation.

However, the effect of sulphur on yield is not always favourable. There is no question that the positive effect of sulphur on yield is particularly pronounced in conditions of sulphur deficiency. This is confirmed by research by Wielebski and Wójtowicz [49], in which sulphur application in conditions of optimal and very high sulphur did not significantly affect seed yield. Similar findings were reported by Jakubus and Toboła [48] in an experiment conducted on soil with average total sulphur content. The increase in seed yield observed by the authors as a result of sulphur application proved to be statistically non-significant. Many studies have shown no effect or a minor effect on sulphur fertilisation on yield even in conditions of poor soil content of this element [10,19,50].

The effect of sulphur on yield depends on many factors, including the application rate and when it is applied. The present study only partially confirmed this dependency. On average in the experiment, even the lowest level of sulphur application (20 kg S·ha^{-1}), for each means of application, significantly increased seed yield. Increasing the amount applied did not significantly affect seed yield in the case of application of the entire amount of fertiliser to the soil before sowing or to the leaves. Significant differences in the effect of application rates on yield were only observed when sulphur was applied in a mixed manner, i.e., half to the soil before sowing and half to the leaves. In this case, application in the amount of 60 kg S·ha^{-1} had the most beneficial effect on seed yield (Figure 3a).

Sienkiewicz-Cholewa and Kieloch [16] report that sulphur application before sowing in the amount of 20 kg S·ha^{-1} had no effect on rapeseed yield. Only higher application rates (40 or 60 kg S·ha^{-1}) significantly increased seed yield, by 11–12% relative to the control (3.5–6.1 dt·ha^{-1}). In a study by Wielebski [51], application of sulphur in early spring at levels from 15 to 60 kg S·ha^{-1} increased seed yield by 4.6% on average. Application of 15 kg S·ha^{-1} was sufficient to achieve the highest yield, and higher application of sulphur (30 and 60 kg S·ha^{-1}) did not significantly differentiate yield compared to 15 kg S·ha^{-1}. Dash et al. [15], on the other hand, obtained the most beneficial effect on oilseed rape yield by applying 40 kg S·ha^{-1}. In the present study, fertilisation with 60 kg·ha^{-1} resulted in the highest yield for every means of application. However, when the entire amount of fertiliser was applied to the soil before sowing or to the leaves, the differences in yield (56–102 kg·ha^{-1}) between this level of sulphur and lower application rates were not significant

(Figure 3a). In the case of mixed application of 60 kg, the second portion of sulphur for foliar application (15 kg S·ha^{-1}) was applied at the start of the flowering stage (BBCH 57). According to many authors, oilseed rape has high sulphur requirements, especially from the budding stage to the pod forming stage [52,53]. Availability of sulphur during this period ensures the proper growth and development of rape. Janzen and Bettany [54] also claim that sulphur fertilisers fundamentally affect crop yield. The authors observed particularly favourable effects of fertiliser applied at the budding stage. According to Grant et al. [55], if sulphur deficiencies are observed during the growing season, application of sulphate sulphur during the period from the start of flowering may be beneficial, although yields will generally be lower than if sulphur is available from the start of plant growth.

On average in our experiment, the method of application did not significantly influence the effect of individual levels of sulphur application on yield (Figure 3a). Contrasting results were obtained in a pot experiment by Podleśna [56], in which foliar feeding of rape plants with magnesium sulphate during flower bud formation resulted in a seed yield that was only 73% of that obtained following soil application of sulphur. Oilseed rape with access to mineral sulphur from the start of the growing period produced a significantly higher seed yield and greater weight of stems, leaves, pods, and roots. The effect obtained from soil application may be explained by the fact that the sulphur compounds contained in the solution applied to the leaves may cause leaf burn, which interferes with photosynthesis and transpiration and negatively affects crop yields [57]. Zhao et al. [58] observed that sulphur was much less effective when applied in autumn than in spring, and that poor utilisation of sulphur was due to losses via leaching. According to Podleśna [56], only limited amounts of sulphur can be applied to the leaves, because it is not easily transported from the leaves to other organs. Booth et al. [59] claim that sulphur uptake efficiency by leaves is only 2%. The rest ends up in the soil, where it is gradually released and taken up by the root system.

The statistical analysis confirmed a significant interaction of the sulphur fertilisation treatments with the years of the study (Table 5). In the case of soil application in the 2010/2011 season, only the highest application rate of 60 kg S·ha^{-1} was effective, but this means of application was more effective in the 2011/2012 season. Overall, sulphur fertilisation was least effective in 2011/2012. In August and September 2012, precipitation was very heavy, far exceeding the long-term average. According to Podleśna [19], high levels of precipitation are conducive to leaching of sulphates supplied to the soil in autumn. In our study, sulphur was applied in the form of sodium sulphate. SO$_4^{2-}$ ions are highly mobile, and their retention in the upper layers of the soil in these conditions is very low. According to Wielebski [40], in these conditions, the amount of sulphur introduced with fertiliser before sowing significantly decreases, and in spring, when oilseed rape has the greatest need for sulphur, the plants have a smaller pool of it at their disposal.

It should be noted that sulphur application was most effective in the third year of cultivation (Table 5). All sulphur application rates (20, 40, 60 kg S·ha^{-1}) significantly increased seed yield in comparison to the control. In the case of soil or foliar application, differences between application rates were not significant, but in conditions of mixed application (half soil, half foliar), seed yield following application of 60 kg S·ha^{-1} was significantly higher in comparison to the other levels.

This can be linked to the effect of sulphur in reducing the occurrence of pathogens. Growing oilseed rape in a monoculture leads to accumulation of pests and pathogens, which can negatively affect the development and yield of the plants [42]. According to Heneklaus et al. [60], the plant's natural resistance to stress caused by diseases and pests decreases in conditions of sulphur deficiency. Increased natural resistance to infections by pathogenic fungi is an effect of greater sulphur availability for plants. The high toxicity of glucosinolate breakdown products for fungal pathogens is significant here. Numerous literature sources confirm that sulphur fertilisation increases the content of glucosinolates in plants [1,61]. The beneficial effect of sulphur application in reducing fungal diseases in rape plants has been confirmed by many authors [62,63].

3.2. The Macronutrient Content and the Harvest Index of the Accumulation of Elements

3.2.1. Nitrogen and Sulphur

The content of macroelements in the rape seeds and straw as well as their accumulation were significantly influenced by the years of the study and the sulphur fertilisation treatments, i.e., the amount and time of application. On average for the experiment, the nitrogen content in the seeds was highest following application of 60 kg S·ha^{-1} entirely to the leaves or in a mixed manner (Figure 4a). The nitrogen content in the straw also increased in conditions of sulphur fertilisation (Figure 4b), but the extent of these changes was highly varied in different years of the study. A significant increase in nitrogen in the straw under the influence of various sulphur application rates and times was observed in the third year of the study (Table S1). Plants well supplied with sulphur take up more nitrogen and utilise it better from fertilisers, especially from large amounts of the nutrient. According to many authors, the use of sulphur in fertiliser influences the nitrogen balance of winter oilseed rape, primarily biosynthesis of protein nitrogen compounds [64–66]. This explains the higher nitrogen content in rape following inclusion of sulphur in fertilisation in other studies as well [17]. The effect of sulphur fertilisation on nitrogen accumulation in plants of the family *Brassicaceae*, in both the vegetative parts and the seeds, is confirmed by Jan et al. [67] and Barczak et al. [68].

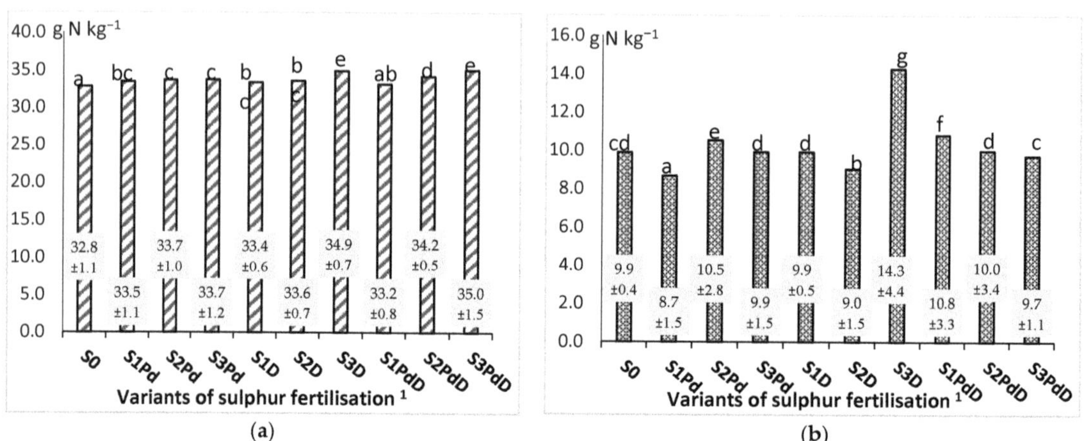

Figure 4. (**a**) Nitrogen content in the rape seeds on average for variants of sulphur fertilisation; (**b**) Nitrogen content in the rape straw on average for variants of sulphur fertilisation. [1] See Table 2; Means followed by the same letter are not statistically different at the α = 0.05 level. ±—standard deviation.

Sulphur content in the seeds ranged from 2.31 to 3.14 g·kg^{-1} (Table S2) and was much lower than that reported by Jakubus and Toboła [48] and by Podleśna [19], but similar to the content determined by Stępień et al. [69] The sulphur content in plants is determined by a genetic factor, but also by the content of its available forms in the soil, moisture, temperature, and the vicinity of industrial areas [70]. Sulphur fertilisation significantly increased the sulphur content in the rapeseeds, on average by 0.18–0.29 g relative to the control (Figure 5a). In the case of soil application, the rate of 20 kg S·ha^{-1} was the most effective. In the case of foliar or mixed application (half soil + half foliar), higher application rates had significantly better effects. The sulphur content in the straw ranged from 1.71 to 4.35 g·kg^{-1}. Even the lowest level (20 kg S·ha^{-1}) was effective, and as the level of application increased, there was a further increase in the content of sulphur (Figure 5b).

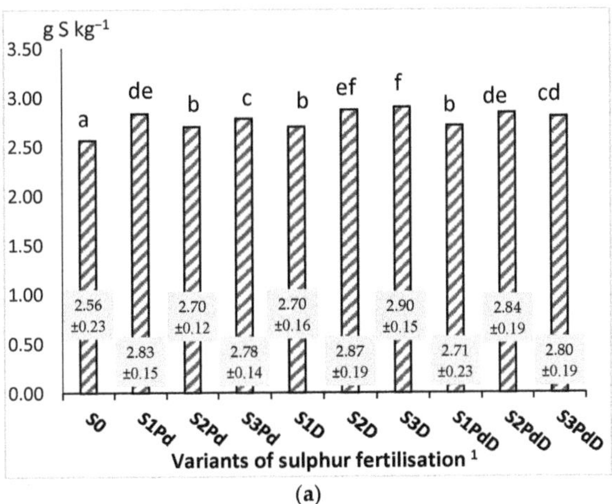

(a)

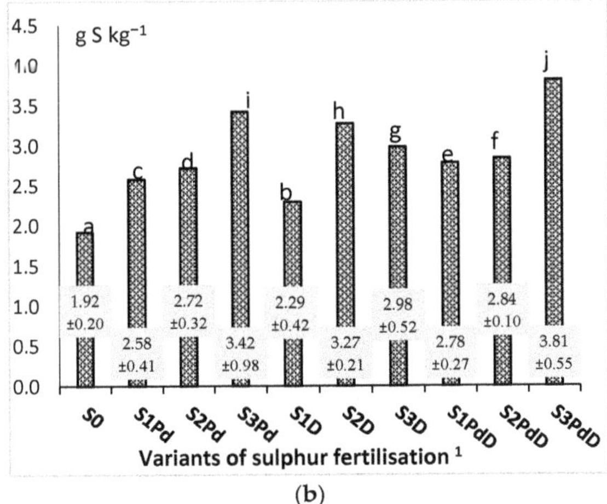

(**b**)

Figure 5. (**a**) Sulphur content in the rape seeds on average for variants of sulphur fertilisation; (**b**) Sulphur content in the rape straw on average for variants of sulphur fertilisation. [1] See Table 2; Means followed by the same letter are not statistically different at the $\alpha = 0.05$ level. $\pm$—standard deviation.

Podleśna [19] reported that autumn sulphur fertilisation of oilseed rape at 80–100 kg·ha⁻¹ significantly increased the content of this macroelement in both the seeds and the vegetative parts. Higher sulphur content in oilseed rape following sulphur fertilisation was also noted by McGrath and Zhao [24], Lošák i in. [23] and Podleśna [56]. In contrast, Jakubus and Toboła [48] did not observe an increase in either the content or the uptake of sulphur in seeds following sulphur application; its content in the seeds showed little variation, ranging from 4.2 to 4.7 g·kg⁻¹, depending on the time and form of fertilisation. A favourable effect was obtained following soil application of ammonium sulphate in autumn and soil and foliar application in spring. Janzen and Bettany [54] report that the time of application of sulphur fertilisers fundamentally affects uptake of sulphur and crop yield. The authors found that fertilisation during the budding stage had a particularly beneficial effect on both parameters.

The nitrogen harvest index, i.e., the amount of the nitrogen accumulated in the seeds as a percentage of its total accumulation in the plant canopy, was 58.4% for rape that was not fertilised with sulphur. Sulphur application had varied effects. Foliar application decreased the harvest index. An increase in the index was noted following soil application or mixed application of 60 kg S·ha^{-1} (Figure 6a, Table S3). In a study by Podleśna [19], more than 70% of the nitrogen taken up by mature oilseed rape was found in the seeds, and applied sulphur did not affect its distribution among individual organs. According to the author, the high value of the harvest index confirms reports of intensive redistribution of this element from the vegetative organs to the reproductive organs [71]. In another study, Podleśna [55] reported that the amount of nitrogen accumulated in seeds relative to the whole pool of accumulated nitrogen was 66% and did not differ significantly between crops with soil application vs. foliar spraying of sulphur fertiliser. Grzebisz [71], on the other hand, reported a nitrogen harvest index for winter oilseed rape in a wide range from 45% to even 70%.

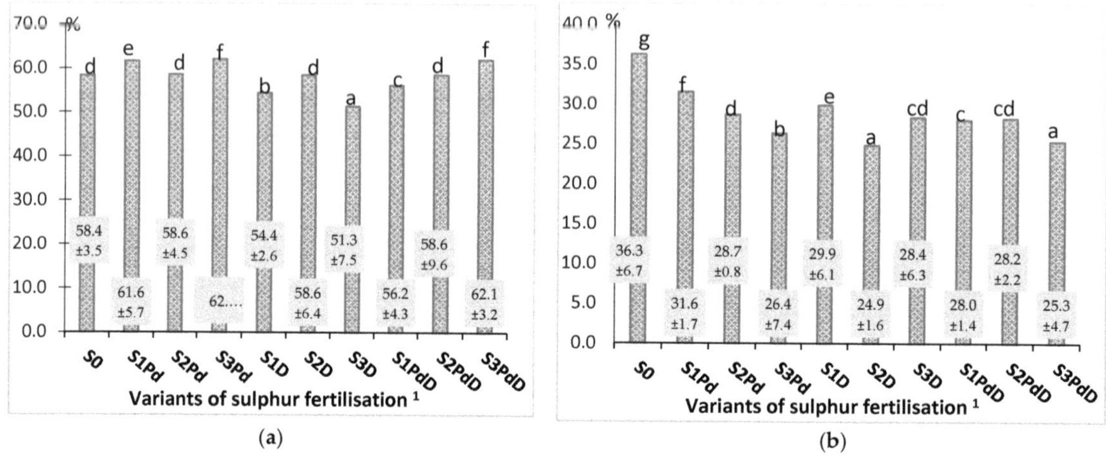

(a)

(b)

Figure 6. (a) Nitrogen harvest index on average for variants of sulphur fertilisation; (b) Sulphur harvest index on average for variants of sulphur fertilisation. [1] See Table 2; Means followed by the same letter are not statistically different at the $\alpha = 0.05$ level. ±—standard deviation.

In the present study, the sulphur harvest index was 36.3% for the control treatment. It was significantly decreased by sulphur application, especially foliar application of 40 kg S·ha^{-1} and mixed application of 60 kg S·ha^{-1}; for these treatments, it ranged from 24.9% to 25.3% (Figure 6b). According to Grzebisz [71], in currently cultivated oilseed rape cultivars, i.e., 00 varieties, sulphur accumulates mainly in the vegetative organs—slightly more in the pods than in the shoots. On the other hand, a study by Podleśna [19] found a sulphur harvest index of about 50%, and it did not depend on sulphur fertilisation.

3.2.2. Phosphorus and Potassium

According to Skwierawska et al. [70], the sulphur content in the soil can indirectly affect the rate and level of uptake of other nutrients. This is an important problem in agriculture because it ultimately affects crop quality. The phosphorus content in the seeds in the experiment showed little variation, amounting to 5.87 g P·kg^{-1} in the control and ranging from 6.21 to 6.55 g P·kg^{-1} in the sulphur fertilisation treatments (Table S4). Sulphur fertilisation significantly increased the phosphorus content in the seeds, with a better effect obtained in the case of foliar application. Application of 60 kg S·ha^{-1} reduced the content of phosphorus in comparison to 40 kg S·ha^{-1} (Figure 7a). A similar reaction to fertilisation with various amounts of sulphur was observed by Barczak et al. [72] in an experiment on

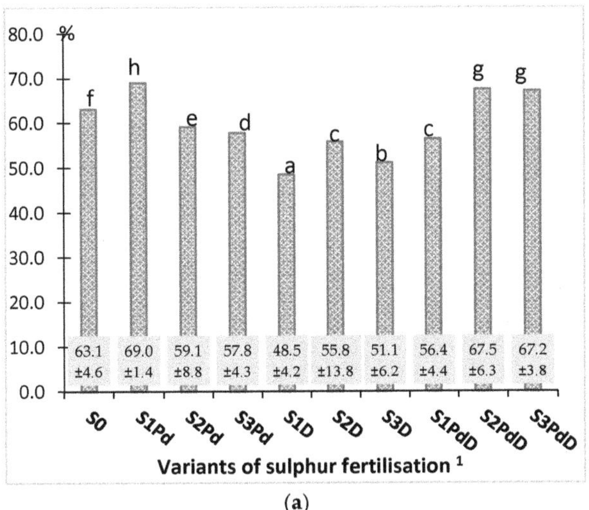

(a)

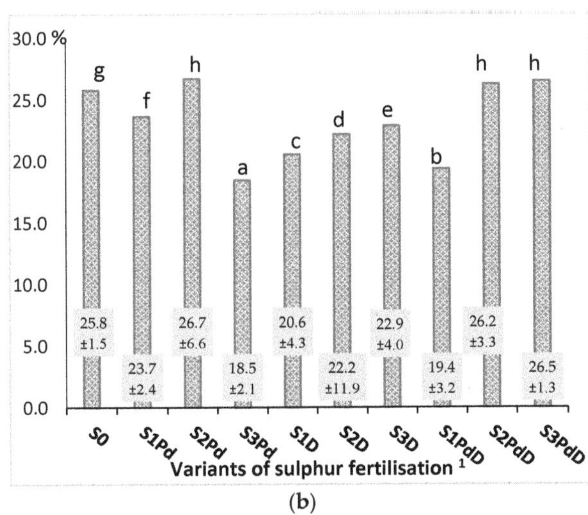

(b)

Figure 9. (a) Phosphorus harvest index on average for variants of sulphur fertilisation; (b) Potassium harvest index on average for variants of sulphur fertilisation. [1] See Table 2; Means followed by the same letter are not statistically different at the $\alpha = 0.05$ level. $\pm$—standard deviation.

The harvest index for potassium accumulation ranged from 13.26% to 30.36%. Soil application of sulphur (40 kg S·ha^{-1}) and mixed application (40 and 60 kg S·ha^{-1}) significantly increased the potassium harvest index from 25.8% to 26.5% (Figure 9b). These results are similar to those obtained by Podleśna [19], who reported that more than 80% of potassium remained in the straw of oilseed rape, due to its concentration in the stems and leaves and physiological functions. According to Grzebisz [71], plants accumulate potassium mainly in the vegetative organs, which is confirmed in research by other authors [48].

3.2.3. Calcium and Magnesium

The calcium content in the seeds during the study showed little variation, ranging from 3.44 to 4.56 g Ca·kg^{-1} (Table S7). On average in the experiment, all sulphur fertilisation treatments increased calcium content in comparison with the control (from 3.51 g Ca·kg^{-1}

to 3.64–3.88 g Ca·kg^{-1}) (Figure 10a). Its content in the straw was much higher and varied during the study period, from 10.43 to 32.09 g Ca·kg^{-1} (Figure 10b). Skwierawska et al. [73] and Brodowska and Kaczor [76] noted an increase in calcium content in the grain of cereals following sulphur application. It should be noted that in 2013, the content of this macroelement in the straw was nearly twice as high as in 2011 and markedly higher than in 2012 (Table S7).

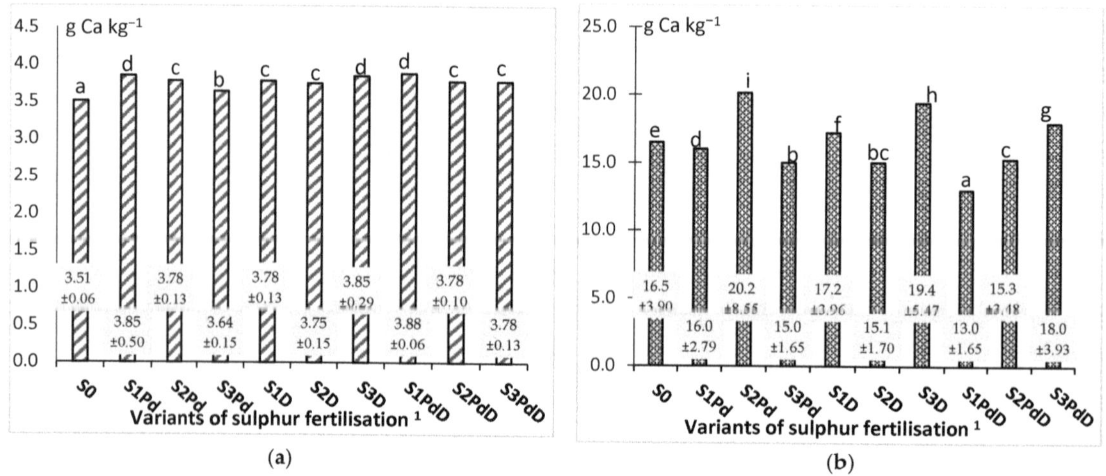

Figure 10. (a) Calcium content in the rape seeds on average for variants of sulphur fertilisation; (b) Calcium content in the rape straw on average for variants of sulphur fertilisation. [1] See Table 2; Means followed by the same letter are not statistically different at the α = 0.05 level. ±—standard deviation.

The magnesium content in the seeds was also varied (Table S8). Sulphur application generally increased the content of this macroelement, but in a narrow range from 2.94 g kg^{-1} for the control to 3.04–3.17 g kg^{-1} for the sulphur fertilisation treatments (Figure 11a). In a study by Stępień et al. [69], the magnesium content in rapeseeds ranged from 2.90 to 3.15 g kg^{-1}, and differences in the intensity of cultivation technology, including sulphur fertilisation, did not significantly affect magnesium content in the seeds. Jarecki [77] also reported that foliar fertilisation of oilseed rape did not affect the magnesium content in the seeds. Magnesium content in the straw was 1.42 g kg^{-1} and was significantly increased only by foliar application of 60 kg S·ha^{-1} (Figure 11b). Podleśna [56] reported that foliar feeding with magnesium sulphate resulted in higher content and uptake of magnesium in comparison to soil fertilisation of oilseed rape. The highest magnesium content, however, was noted in the plants that were not fertilised with sulphur, which had very high concentrations of this macroelement in the leaves, stems, and roots. According to the author, the excessive content of magnesium in conditions of sulphur deficiency indicates that it was taken up from the environment but could not be transported or distributed in the plants, because the lack of sulphur inhibited the growth and development of their organs. Thus, a phenomenon opposite to 'dilution' occurred, i.e., the nutrients were 'concentrated' in the reduced mass of the plant. In an experiment by Brodowska and Kaczor [76], the most pronounced increase in magnesium uptake was noted following application of sulphur in the form of sodium sulphate, while elemental sulphur and sodium thiosulphate were not effective. According to the authors, the presence of sulphate ions in the soil was probably conducive to uptake of magnesium ions.

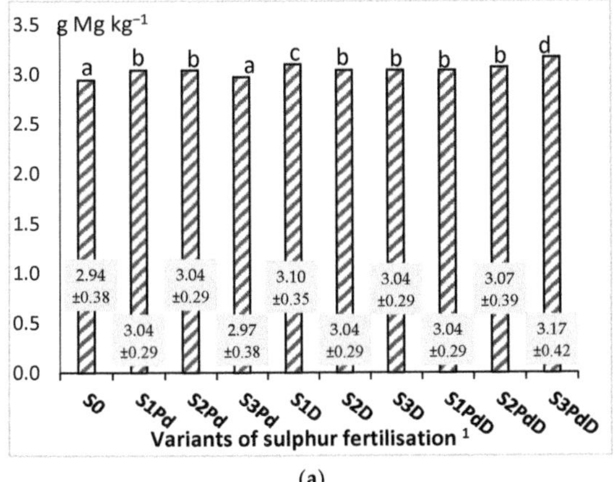

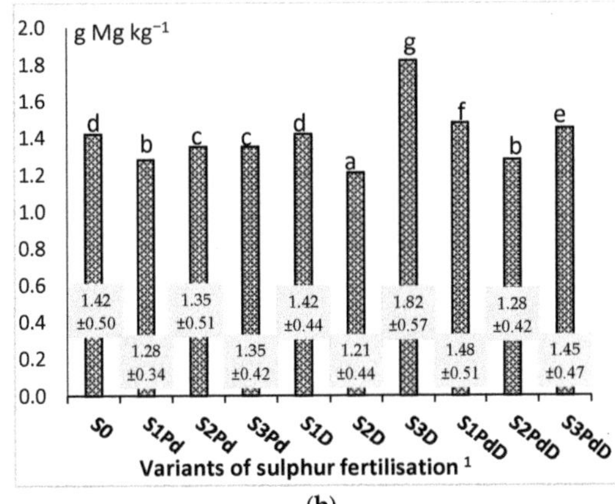

Figure 11. (**a**) Magnesium content in the rape seeds on average for variants of sulphur fertilisation; (**b**) Magnesium content in the rape straw on average for variants of sulphur fertilisation. [1] See Table 2; Means followed by the same letter are not statistically different at the α = 0.05 level. ± standard deviation.

The harvest index of calcium accumulation ranged from 4.07% to 13.40% (Table S9). Foliar application of sulphur decreased the value of the index, while mixed application significantly increased the share of calcium accumulated in the seeds in its total accumulation (Figure 12a). According to Grzebisz [71], calcium accumulates mainly in the pods, and only marginally in the seeds. In a study by Podleśna [19], sulphur fertilisation significantly increased the content and accumulation of calcium in the straw of oilseed rape. The author suggests that this may be linked to greater resistance to certain fungal diseases [78]. A higher concentration of calcium in the cell walls increases the resistance of plants, preventing infection by pathogens by making it more difficult for the enzymes they secrete to macerate the cell wall. It is possible that the positive effect of sulphur fertiliser on the natural immunity of oilseed rape to stress factors, including fungal diseases, may result not only from increased production of sulphur compounds such as glucosinolates or glutathione, but also from morphological changes in the cell wall of the leaves and stems [19].

The harvest index of magnesium accumulation ranged from 33.02% to 61.33% (Table S9). Soil application or mixed application of sulphur generally increased its value (Figure 12b). According to Grzebisz [71], distribution of magnesium in the organs of oilseed rape (shoots, pods and seeds) is roughly balanced, with slightly greater accumulation in the seeds, and the magnesium harvest index is 30–40%. Podleśna [19] reported that sulphur fertilisation did not affect magnesium content in the seeds or straw of rape, but significantly increased its uptake in the biomass of straw. According to the author, 55–57% of magnesium remained in the straw of rape.

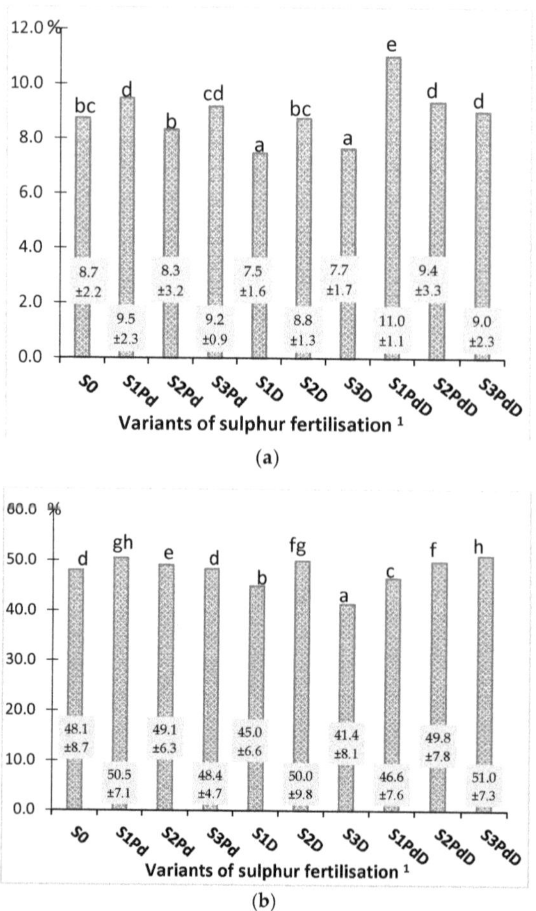

Figure 12. (a) Calcium harvest index on average for variants of sulphur fertilisation; (b) Magnesium harvest index on average for variants of sulphur fertilisation. [1] See Table 2; Means followed by the same letter are not statistically different at the $\alpha = 0.05$ level. ±—standard deviation.

4. Conclusions

The effect of sulphur on oilseed rape yield significantly depended on the amount and means of sulphur application. Even the lowest level of 20 kg S·ha^{-1}, irrespective of the means of application, significantly increased the seed yield. In the case of autumn soil application and foliar application alone, the differences between the lowest rate and higher ones (40 and 60 kg S·ha^{-1}) were not significant. In the case of mixed application, 60 kg S·ha^{-1} significantly increased yield in comparison with lower application. Sulphur fertilisation also significantly increased the straw yield, with foliar application found to be the most favourable to the development of the vegetative parts.

Sulphur fertilisation affected the mineral composition of rapeseeds. Overall, all sulphur application treatments significantly increased the content of N, P, K, Ca and Mg in the seeds compared with the control treatment, but the differences were not great. The content of macroelements in the straw was more varied than in the seeds. Sulphur fertilisation increased the content of S, P and K in the vegetative parts, but in the case of nitrogen, calcium, and magnesium, the effect of sulphur application was much smaller.

Each of the sulphur fertilisation treatments reduced the harvest index of sulphur. Each level of foliar application decreased the harvest index of nitrogen, phosphorus, potassium, and calcium, whereas soil application of 20 kg·ha^{-1} and mixed application of 40 and 60 kg·ha^{-1} increased the harvest index of phosphorus, potassium and calcium.

Supplementary Materials: The following are available online at https://www.mdpi.com/article/10.3390/agronomy12010068/s1, Table S1. Nitrogen content in rape seeds and straw, mean values for interaction variants of sulphur fertilisation × years, Table S2. Sulphur content in rape seeds and straw, mean values for interaction variants of sulphur fertilisation × years, Table S3. Nitrogen and sulphur harvest index, mean values for interaction variants of sulphur fertilisation × years, Table S4. Phosphorus content in rape seeds and straw, mean values for interaction variants of sulphur fertilisation × years, Table S5. Potassium content in rape seeds and straw, mean values for interaction variants of sulphur fertilisation × years, Table S6. Phosphorus and potassium harvest index, mean values for interaction variants of sulphur fertilisation × years. Table S7. Calcium content in rape seeds and straw, mean values for interaction variants of sulphur fertilisation × years, Table S8. Magnesium content in rape seeds and straw, mean values for interaction variants of sulphur fertilisation × years, Table S9. Calcium and magnesium harvest index, mean values for interaction variants of sulphur fertilisation × years.

Author Contributions: Conceptualization, M.S. and A.G.; methodology, M.S. and A.G., Formal analysis, M.S. and A.G.; writing—original draft preparation, M.S. and A.G. All authors have read and agreed to the published version of the manuscript.

Funding: This research received no external funding.

Data Availability Statement: The data presented in this study are available on request from the corresponding author.

Conflicts of Interest: The authors declare no conflict of interest.

References

1. Kozłowska-Strawska, J.; Badora, A. Selected problems of sulfur management in crops. *Pol. J. Nat. Sci.* **2013**, *28*, 309–316.
2. Barczak, B.; Skinder, Z.; Piotrowski, R. Sulphur as a factor that affects nitrogen effectiveness in spring rapeseed agrotechnics. Part III. Agronomic use efficiency of nitrogen. *Acta Sci. Pol. Agric.* **2017**, *16*, 179–189. [CrossRef]
3. Groth, D.A.; Sokólski, M.; Jankowski, K.J. A Multi-criteria evaluation of the effectiveness of nitrogen and sulfur fertilization in different cultivars of winter rapeseed-productivity, economic and energy balance. *Energies* **2020**, *13*, 4654. [CrossRef]
4. Marazzi, C.; Städler, E. Influence of plant sulphur nutrition on oviposition and larval performance of the diamondback moth. *Entomol. Exp. Appl.* **2004**, *111*, 225–232. [CrossRef]
5. Ahmad, A.; Khan, I.; Anjum, N.A.M.; Diva, I.; Abdin, M.Z.; Iqbal, M. Effect of timing of sulfur fertilizer application on growth and yield of rapeseed. *J. Plant Nutr.* **2005**, *28*, 1049–1059. [CrossRef]
6. Barłóg, P.; Grzebisz, W.; Diatta, J. Effect of timing and nitrogen fertilizers on nutrients content and uptake by winter oilseed rape. Part II. Dynamics of nutrients uptake. In *Chemistry for Agriculture*; Górecki, H., Dobrzański, Z., Kafarski, P., Eds.; Czech-Pol Trade: Prague, Czech Republic; Brussels, Belgium, 2005; Volume 6, pp. 113–123.
7. Jankowski, K.J.; Kijewski, Ł.; Groth, D.; Skwierawska, M.; Budzyński, W.S. The effect of sulfur fertilization on macronutrient concentrations in the post-harvest biomass of rapeseed (*Brassica napus* L. ssp. *oleifera* Metzg). *J. Elem.* **2015**, *20*, 585–597. [CrossRef]
8. Ahmad, A.; Abdin, M.Z. Interactive effect of sulphur and nitrogen on the oil and protein contents and on the fatty acid profiles of oil in the seeds of rapeseed (*Brassica campestris* L.) and mustard (*Brassica juncea* L. Czern. and Coss.). *J Agron. Crop Sci.* **2000**, *185*, 49–54. [CrossRef]
9. Rausch, T.; Wachter, A. Sulfur metabolism: A versatile platform for launching defense operations. *Trends Plant Sci.* **2005**, *10*, 503–509. [CrossRef]
10. Farahbakhsh, H.; Pakgohar, N.; Karimi, A. Effects of nitrogen and sulphur fertilizers on yield, yield components and oil content of oilseed rape (*Brassica napus* L.). *Asian J. Plant Sci.* **2006**, *5*, 112–115.
11. Jankowski, K.J.; Budzyński, W.; Szymanowski, A. Influence of the rate and timing of sulphur fertilization on winter oilseed rape yield. *Rośliny Oleiste Oilseed Crops* **2008**, *XXIX*, 76–90.
12. Egesel, C.Ö.; Gül, M.K.; Kahrıman, F. Changes in yield and seed quality traits in rapeseed genotypes by sulphur fertilization. *Eur. Food Res. Technol.* **2009**, *229*, 505–513. [CrossRef]
13. Brennan, R.F.; Bell, R.W.; Raphael, C.; Eslick, H. Sources of sulfur for dry matter, seed yield, and oil concentration of canola grown in sulfur deficient soils of south-western Australia. *J. Plant Nutr.* **2010**, *33*, 1180–1194. [CrossRef]
14. Begum, F.; Hossain, F.; Mondal, M.d.R.I. Influence of sulphur on morpho-physiological and yield parameters of rapeseed (*Brassica campestris* L.). *Bangladesh J. Agric. Res.* **2012**, *37*, 645–652. [CrossRef]

15. Dash, N.R.; Ghosh, G.K. Efficacy of gypsum and magnesium sulfate as sources of sulfur to rapeseed in lateritic soils. *J. Plant Nutr.* **2012**, *35*, 2156–2166. [CrossRef]
16. Sienkiewicz-Cholewa, U.; Kieloch, R. Effect of sulphur and micronutrients fertilization on yield and fat content in winter rape seeds (*Brassica napus* L.). *Plant Soil Environ.* **2015**, *61*, 164–170. [CrossRef]
17. Chwil., S. The effect of foliar feeding under different soil fertilization conditions on the yield structure and quality of winter oilseed rape (*Brassica napus* L.). *Electron. J. Pol. Agric. Univ.* **2016**, *19*, 2.
18. Tuncturk, R.; Tuncturk, M. The effect of different sulphur doses on the yield and quality of rapeseed (*Brassica napus* L.). *Fresenius Environ. Bull.* **2017**, *26*, 6952–6957.
19. Podleśna, A. The effect of sulfur fertilization on concentration and uptake of nutrients by winter oilseed rape. (Wpływ nawożenia siarką na zawartość i pobieranie składników pokarmowych przez rzepak ozimy). *Rośliny Oleiste Oilseed Crops* **2004**, *XXV*, 627–636.
20. Kaczor, A.; Brodowska, M. Effect of liming and sulphur fertilization on the growth and yielding of spring forms of wheat and rape. Part II. Rape. (Wpływ wapnowania i nawożenia siarką na wzrost, rozwój i plonowanie form jarych pszenicy i rzepaku. Cz. II. Rzepak). *Acta Agrophys.* **2003**, *1*, 661–666.
21. Cui, Y.; Wang, Q.; Dong, Y.; Li, H.; Christie, P. Enhanced uptake of soil Pb and Zn by Indian mustard and winter wheat following combined soil application of elemental sulphur and EDTA. *Plant Soil* **2004**, *261*, 181–188. [CrossRef]
22. Jankowski, K.; Kijewski, Ł.; Skwierawska, M.; Krzebietke, S.; Mackiewicz-Walec, E. Effect of sulfur fertilization on the concentrations of copper, zinc and manganese in the roots, straw and oil cake of rapeseed (*Brassica napus* L. ssp *oleifera* Metzg). *J. Elem.* **2014**, *19*, 433–446. [CrossRef]
23. Lošák, T.; Hrivna, L.; Richter, R. Effect of increasing doses of nitrogen and sulphur on yields, quality and chemical composition of winter rape. *Zesz. Probl. Post. Nauk. Rol.* **2000**, *472*, 481–487.
24. McGrath, S.P.; Zhao, F.J. Sulphur uptake, yield responses and the interactions between nitrogen and sulphur in winter oilseed rape (*Brassica napus*). *J. Agric. Sci.* **1996**, *126*, 53–62. [CrossRef]
25. WRB. IUSS Working Group. World Reference Base for Soil Resources 2014, update 2015 International soil classification system for naming soils and creating legends for soil maps. In *World Soil Resources Reports*; FAO: Rome, Italy, 2014; No. 106. 201.
26. The Polish Committee for Standarization. PKN Polish Standard PN-ISO 10390. In *Soil Quality-Determination of pH. PKN*; The Polish Committee for Standarization: Warszawa, Poland, 1997.
27. Anonymous. *Catalog of Research Methods at Chemical and Agricultural Stations*; Regional Chemical and Agricultural Station in Lublin: Lublin, Poland, 2010.
28. The Polish Committee for Standarization. PKN Polish Standard PN-R-04023. In *Chemical and Agricultural Analysis of Soil–Determination of Available Phosphorus in Mineral Soils*; PKN: Warszawa, Poland, 1996.
29. The Polish Committee for Standarization. PKN Polish Standard PN-R-04022. In *Chemical and Agricultural Analysis of Soil–Determination of Available Potassium in Mineral Soils*; PKN: Warszawa, Poland, 1996.
30. The Polish Committee for Standarization. PKN Polish Standard PN-R-04020:1994/Az1. In *Chemical and Agricultural Analysis of soil–Determination of Available Magnesium Content*; PKN: Warszawa, Poland, 2004.
31. Lancashire, P.D.; Bleiholder, H.; Van den Boom, T.; Langelüddeke, P.; Strauss, R.; Weber, E.; Witzenberger, A. A uniform decimal code for growth stages of crops and weeds. *Ann. Appl. Biol.* **1991**, *119*, 561–601. [CrossRef]
32. Available online: Databaseofplantprotectionproducts.www.ior.poznan.pl (accessed on 30 November 2021).
33. Kapuściński, J.; Nowak, R. The frequency of the occurrence of droughts and post-droughts periods in mid-west Poland on the example of Poznań, Wałcz and Wieluń. In *Kształtowanie i ochrona środowiska leśnego*; Miler, A., Ed.; Klimat a las. Wyd. AR: Poznań, Poland, 2003; pp. 76–88.
34. Skowera, B.; Jędrszczyk, E.; Kopcińska, J.; Ambroszczyk, A.M.; Kołtun, A. The effects of hydrothermal conditions during vegetation period on fruit quality of processing tomatoes. *Pol. J. Environ. Stud.* **2014**, *23*, 195–202.
35. Kamińska, W.; Kardasz, T.; Strahl, A. Metody badań laboratoryjnych w stacjach chemiczno-rolniczych. Część 2. Badanie materiału roślinnego. In *Laboratory Test Methods in Chemical and Agricultural Stations. Part 2. Examination of Plant Material. Wyd.*; IUNG: Puławy, Poland, 1981.
36. Zając, T.; Klimek-Kopyra, A.; Oleksy, A.; Lorenc-Kozik, A.; Ratajczak, K. Analysis of yield and plant traits of oilseed rape (*Brassica napus* L.) cultivated in temperate region in light of the possibilities of sowing in arid areas. *Acta Agrobot.* **2016**, *69*, 1–13. [CrossRef]
37. Brachaczek, A.; Kaczmarek, J.; Jędryczka, M. Warm and wet autumns favour yield losses of oilseed rape caused by phoma stem canker. *Agronomy* **2021**, *11*, 1171. [CrossRef]
38. Marjanović-Jeromela, A.; Terzić, S.; Jankulovska, M.; Zorić, M.; Kondić-Špika, A.; Jocković, M.; Hristov, N.; Crnobarac, J.; Nagl, N. Dissection of year related climatic variables and their effect on winter rapeseed (*Brassica napus* L.) development and yield. *Agronomy* **2019**, *9*, 517. [CrossRef]
39. Jarecki, W. The size and quality of winter rape seed yield depending on the cultivar type. *Agron. Sci.* **2021**, *LXXVI*, 5–14. [CrossRef]
40. Wielebski, F. Share of yield components in the creation of yield of winter oilseed rape hybrids. (Udział elementów struktury plonu w kształtowaniu plonu nasion mieszańcowych odmian rzepaku ozimego). *Rośliny Oleiste Oilseed Crops* **2005**, *XXVI*, 87–98.
41. Stępień, A.; Wojtkowiak, K.; Pietrzak-Fiećko, R. Nutrient content, fat yield and fatty acid profile of winter rapeseed (*Brassica napus* L.) grown under different agricultural production systems. *Chil. J. Agric. Res.* **2017**, *77*, 266–277. [CrossRef]

42. Cwalina-Ambroziak, B.; Stępień, A.; Kurowski, T.P.; Głosek-Sobieraj, M.; Wiktorski, A. The health status and yield of winter rapeseed (*Brassica napus* L.) grown in monoculture and in crop rotation under different agricultural production systems. *Arch. Agron. Soil Sci.* **2016**, *62*, 1722–1732. [CrossRef]
43. Hegewald, H.; Koblenz, B.; Wensch-Dorendorf, M.; Christen, O. Impacts of high intensity crop rotation and N management on oilseed rape productivity in Germany. *Crop Pasture Sci.* **2016**, *67*, 439–449. [CrossRef]
44. Różyło, K.; Pałys, E. Influence of crop rotation and row spacing on weed infestation of winter rape grown on rendzina soil. *Acta Sci. Pol. Agric.* **2011**, *10*, 57–64.
45. Sieling, K.; Christen, O.; Nemati, B.; Hanus, H. Effects of previous cropping on seed yield and yield components of oil-seed rape (*Brassica napus* L.). *Eur. J. Agron.* **1997**, *6*, 215–223. [CrossRef]
46. Jaskulska, I.; Jaskulski, D.; Kotwica, K.; Piekarczyk, M.; Wasilewski, P. Yielding of winter rapeseed depending on the forecrops and soil tillage methods. (Plonowanie rzepaku ozimego w zależności od przedplonów i sposobów uprawy roli). *Ann. UMSC Sec. Agric.* **2014**, *69*, 30–38.
47. Jankowski, K.J.; Budzyński, W.; Szymanowski, A. Effect of sulfur on the quality of winter rape seeds. *J. Elem.* **2008**, *13*, 521–534.
48. Jakubus, M.; Toboła, P. Content of total and sulphate sulphur in winter oilseed rape depending on fertilization. (Zawartość siarki ogólnej i siarczanowej w rzepaku ozimym w zależności od nawożenia). *Rośliny Oleiste Oilseed Crops* **2005**, *XXIV*, 149–162.
49. Wielebski, F.; Wójtowicz, M. Effect of spring sulphur fertilization on yield and glucosinolate content in seeds of winter oilseed rape composite hybrids. (Wpływ wiosennego nawożenia siarką na plon i zawartość glukozynolanów w nasionach odmian mieszańcowych złożonych rzepaku ozimego). *Rośliny Oleiste Oilseed Crops* **2003**, *XXIV*, 109–119.
50. Gallejones, P.; Castellón, A.; del Prado, A.; Unamunzaga, O.; Aizpurua, A. Nitrogen and sulphur fertilization effect on leaching losses, nutrient balance and plant quality in a wheat-rapeseed rotation under a humid Mediterranean climate. *Nutr. Cycl. Agroecosyst.* **2012**, *93*, 337–355. [CrossRef]
51. Wielebski, F. The effect of sulphur fertilization on the yield of different breeding forms of winter oilseed rape in the conditions of diverse nitrogen rates. (Wpływ nawożenia siarką w warunkach stosowania zróżnicowanych dawek azotu na plonowanie różnych typów odmian rzepaku ozimego). *Rośliny Oleiste–Oilseed Crops* **2011**, *XXXII*, 61–78.
52. Zhao, F.J.; Evans, E.J.; Bilsborrow, P.E. Varietal differences in sulphur uptake and utilization in relation to glucosinolate accumulation in oilseed rape. In Proceedings of the 9th International Rapeseed Congress, Cambridge, UK, 4–7 July 1995; Volume 1, pp. 271–273.
53. Withers, P.J.A.; Zhao, F.J.; McGrath, S.P.; Evans, E.J.; Sinclair, A.H. Sulphur inputs for optimum yields of cereals. *Asp. Appl. Biol.* **1997**, *50*, 191–197.
54. Janzen, H.H.; Bettany, J.R. Sulfur nutrition of rapeseed. II. Effect of time of sulfur application. *Soil Sci. Soc. Am. J.* **1984**, *48*, 107–112. [CrossRef]
55. Grant, C.A.; Mahli, S.S.; Karamanos, R.E. Sulfur management for rapeseed. *Field Crops Res.* **2012**, *128*, 119–128. [CrossRef]
56. Podleśna, A. The effect of soil and foliar application of sulfur on the yield and mineral composition of winter oilseed rape plants. (Wpływ doglebowego i dolistnego stosowania siarki na plon i skład mineralny roślin rzepaku ozimego). *Ann. UMCS Sec. E* **2009**, *LXIV*, 68–75.
57. Phillips, S.B.; Mullins, G.L. Foliar burn and wheat grain yield responses following topdress-applied nitrogen and sulphur fertilizers. *J. Plant Nutr.* **2004**, *27*, 921–930. [CrossRef]
58. Zhao, F.J.; McGrath, S.P.; Blake-Kalff, M.M.A.; Link, A.; Tucker, M. Crop response to sulphur fertilization in Europe. *Nawozy I Nawożenie–Fertil. Fertil.* **2003**, *3*, 26–51.
59. Booth, E.J.; Batchelor, S.E.; Walker, K.C. The effect of foliar applied sulphur on individual glucosinolates in oilseed rape seed. *J. Plant. Nutr. Soil Sci.* **1995**, *158*, 87–88. [CrossRef]
60. Heneklaus, S.; Bloem, E.; Schnug, E. Sulphur in agroecosystems. *Folia Univ. Agric. Stetin.* **2000**, *204*, 17–32.
61. Fismes, J.; Vong, P.C.; Guckert, A.; Frossard, E. Influence of sulfur on apparent N-use efficiency, yield and quality of oilseed rape (*Brassica napus* L.) grown on a calcareous soil. *Eur. J. Agron.* **2000**, *12*, 127–141. [CrossRef]
62. Dłużniewska, J.; Nadolnik, M.; Kulig, B. Fungal diseases of winter oilseed rape under the different level of nitrogen and sulphur fertilization. (Choroby rzepaku ozimego w zależności od poziomu zaopatrzenia roślin w azot i siarkę). *Prog. Plant Protect. Post. Ochr. Roś.* **2011**, *51*, 1811–1815.
63. Kurowski, T.; Majchrzak, B.; Jankowski, K. Effect of sulfur fertilization on the sanitary state of plants of the family *Brassicaceae*. *Acta Agrobot.* **2010**, *63*, 171–178. [CrossRef]
64. De Kok, L.J.; Castro, A.; Durenkamp, M.; Stuiver, C.E.; Westerman, S.; Yang, L.; Stulen, I. Sulphur in plant phisiology. *Nawozy I Nawożenie–Fertil. Fertil.* **2003**, *2*, 55–80.
65. Anjum, N.A.; Gill, S.S.; Umar, S.; Ahmad, I.; Duarte, A.C.; Pereira, E. Improving growth and productivity of oleiferous Brassicas under changing environment: Significance of nitrogen and sulphur nutrition, and underlying mechanisms. *Sci. World J.* **2012**, 657808. [CrossRef]
66. Eriksen, J.; Nielsen, M.; Mortensen, J.V.; Schjorring, J.K. Redistribution of sulphur during generative growth of barley plants with different sulphur and nitrogen status. *Plant Soil* **2001**, *230*, 239–246. [CrossRef]
67. Jan, A.; Ahmad, G.; Arif, M.; Jan, M.T.; Marwat, K.B. Quality parameters of canola as affected by nitrogen and sulfur fertilization. *J. Plant Nutr.* **2010**, *33*, 381–390. [CrossRef]

68. Barczak, B.; Barczak, T.; Skinder, Z.; Piotrowski, R. Proportions of nitrogen and sulphur in spring rapeseeds depending on fertilization with these elements. *J. Elem.* **2020**, *25*, 1385–1398. [CrossRef]
69. Stępień, A.; Wojtkowiak, K.; Pietrzak-Fiećko, R. Influence of a crop rotation system and agrotechnology level on the yielding and seed quality of winter rapeseed (*Brassica napus* L.) varieties Castille and Nelson. *J. Elem.* **2018**, *23*, 1281–1293. [CrossRef]
70. Skwierawska, M.; Benedycka, Z.; Jankowski, K.; Skwierawski, A. Sulphur as a fertiliser component determining crop yield and quality. *J. Elem.* **2016**, *21*, 609–623. [CrossRef]
71. Grzebisz, W. Technologie nawożenia roślin uprawnych–fizjologia plonowania. In *Tom 1. Oleiste, Okopowe i Strączkowe*; PWRiL: Poznań, Poland, 2011; p. 413.
72. Barczak, B.; Knapowski, T.; Kozera, W.; Ralcewicz, M. Effects of sulphur fertilisation on the content and uptake of macroelements in narrow-leaf lupin. *Rom. Agric. Res.* **2014**, *31*, 245–251.
73. Skwierawska, M.; Zawartka, L.; Zawadzki, B. The effect of different rates and forms of applied sulphur on nutrient composition of planted crops. *Plant Soil Environ.* **2008**, *54*, 179–189. [CrossRef]
74. Majumder, S.; Halder, T.K.; Saha, D. Integrated nutrient management of rapeseed (*Brassica campestris* L. var. *yellow sarson*) grown in a typic haplaquept soil. *J. Nat. Appl. Sci.* **2017**, *9*, 1151–1156. [CrossRef]
75. Szczepanek, M.; Siwik-Ziomek, A. P and K accumulation by rapeseed as affected by biostimulant under different NPK and S fertilization doses. *Agronomy* **2019**, *9*, 477. [CrossRef]
76. Brodowska, M.; Kaczor, A. The effect of various forms of sulphur and nitrogen on calcium and magnesium content and uptake in spring wheat (*Triticum aestivum* L.) and cocksfoot (*Dactylis glomerata* L.) *J. Elem.* **2009**, *14*, 641–647. [CrossRef]
77. Jarecki, W. The reaction of winter oilseed rape to different foliar fertilization with macro- and micronutrients. *Agriculture* **2021**, *11*, 515. [CrossRef]
78. Sadowski, Cz.; Baturo, A.; Lenc, L.; Trzciński, J. Downy mildew (*P. parasitica*) and powdery mildew (*E. cruciferarum*) occurrence on spring oilseed rape cv. Star depending on differentiated fertilisation with nitrogen and sulphur. (Występowanie mączniaka rzekomego (*Perono-spora parasitica*/Pers.ex Fr./Fr.) i mączniaka prawdziwego (*Erysiphe crucifererum* Opiz ex L. Junell) na rzepaku jarym odmiany Star przy zróżnicowanym nawożeniu azotem i siarką). *Rośliny Oleiste Oilseed Crops* **2002**, *XXIII*, 391–408.

 *agronomy*

Article

Improving Nitrogen Status Estimation in Malting Barley Based on Hyperspectral Reflectance and Artificial Neural Networks

Karel Klem [1,2,*], Jan Křen [2], Ján Šimor [2], Daniel Kováč [1], Petr Holub [1], Petr Míša [3], Ilona Svobodová [3], Vojtěch Lukas [2], Petr Lukeš [1], Hana Findurová [1,2] and Otmar Urban [1]

1 Global Change Research Institute CAS, Bělidla 986/4a, 603 00 Brno, Czech Republic; kovac.d@czechglobe.cz (D.K.); holub.p@czechglobe.cz (P.H.); lukes.p@czechglobe.cz (P.L.); findurova.h@czechglobe.cz (H.F.); urban.o@czechglobe.cz (O.U.)
2 Faculty of AgriSciences, Mendel University in Brno, Zemědělská 1/1665, 613 00 Brno, Czech Republic; jan.kren@mendelu.cz (J.K.); jan.simor@gmail.com (J.Š.); vojtech.lukas@mendelu.cz (V.L.)
3 Agrotest Fyto, Ltd., Havlíčkova 2787/121, 767 01 Kromeriz, Czech Republic; misa@vukrom.cz (P.M.); svobodova@vukrom.cz (I.S.)
* Correspondence: klem.k@czechglobe.cz; Tel.: +420-724-285-737

Abstract: Malting barley requires sensitive methods for N status estimation during the vegetation period, as inadequate N nutrition can significantly limit yield formation, while overfertilization often leads to an increase in grain protein content above the limit for malting barley and also to excessive lodging. We hypothesized that the use of N nutrition index and N uptake combined with red-edge or green reflectance would provide extended linearity and higher accuracy in estimating N status across different years, genotypes, and densities, and the accuracy of N status estimation will be further improved by using artificial neural network based on multiple spectral reflectance wavelengths. Multifactorial field experiments on interactive effects of N nutrition, sowing density, and genotype were conducted in 2011–2013 to develop methods for estimation of N status and to reduce dependency on changing environmental conditions, genotype, or barley management. N nutrition index (NNI) and total N uptake were used to correct the effect of biomass accumulation and N dilution during plant development. We employed an artificial neural network to integrate data from multiple reflectance wavelengths and thereby eliminate the effects of such interfering factors as genotype, sowing density, and year. NNI and N uptake significantly reduced the interannual variation in relationships to vegetation indices documented for N content. The vegetation indices showing the best performance across years were mainly based on red-edge and carotenoid absorption bands. The use of an artificial neural network also significantly improved the estimation of all N status indicators, including N content. The critical reflectance wavelengths for neural network training were in spectral bands 400–490, 530–570, and 710–720 nm. In summary, combining NNI or N uptake and neural network increased the accuracy of N status estimation to up 94%, compared to less than 60% for N concentration.

Keywords: artificial neural network; grain yield; *Hordeum vulgare*; nitrogen status; hyperspectral reflectance

Citation: Klem, K.; Křen, J.; Šimor, J.; Kováč, D.; Holub, P.; Míša, P.; Svobodová, I.; Lukas, V.; Lukeš, P.; Findurová, H.; et al. Improving Nitrogen Status Estimation in Malting Barley Based on Hyperspectral Reflectance and Artificial Neural Networks. *Agronomy* **2021**, *11*, 2592. https://doi.org/10.3390/agronomy11122592

Academic Editors: Christos Noulas, Shahram Torabian and Ruijun Qin

Received: 12 November 2021
Accepted: 17 December 2021
Published: 20 December 2021

Publisher's Note: MDPI stays neutral with regard to jurisdictional claims in published maps and institutional affiliations.

1. Introduction

Malting barley ranks among the most challenging crops, concerning its nitrogen (N) nutrition. This is mainly due to the very narrow N optima of barley, which is constrained by grain yield on the one hand, and by grain protein content on the other, extensively affecting malting quality [1,2]. Insufficient N nutrition, particularly at the beginning of vegetation, leads to weak tillering, reduced grain size, and low grain yield [3]. Conversely, excessive N nutrition and, in particular, greater N availability at later growth stages leads to an immediate increase of protein content in grain, also negatively influencing other malting quality parameters [4]. The grain protein content is closely related to canopy N due to its extensive translocation to grain and the transformation into grain protein during grain filling. However, these processes are also significantly modulated by water availability and temperatures, limiting the starch synthesis and accumulation in grain, resulting in an altered relative proportion of protein and starch in grain [3,5].

Moreover, N overfertilization has adverse effects through increasing crop density and subsequent lodging, which can lead to reduced yields and other indirect negative impacts on grain quality, such as contamination by mycotoxins [6,7]. The difference between suboptimal and excessive N nutrition status may, in malting barley, be less than 30 kg N ha^{-1}, and its demands change considerably through different stages of the crop's development. Precise optimization of N nutrition is thus essential to ensure high production and top quality of malting barley in parallel.

Methods presently used for optimizing N status based on soil and plant analyses are relatively expensive, labor-intensive, and scarcely allow for evaluating spatial and temporal variability in N availability on large scales [8]. Therefore, the application of remote sensing, and in particular of reflectance spectroscopy, has recently received considerable attention for estimating N status [9,10]. Diagnostics of N status in plants using spectral reflectance is usually based on a very close correlation between the concentrations of N and chlorophyll, and thus on the absorption properties of chlorophyll a + b [11]. Although specific absorption coefficients of chlorophylls are high for the critical absorption bands (red and blue), the depth of light penetration into the leaf is low, and the changes of reflectance in these bands with changing chlorophyll content are saturated already at medium chlorophyll concentrations per area unit [12]. As a result, widely applied indices based on red and blue reflectance have been found to be insufficiently sensitive to changes at medium and high chlorophyll or N content [13]. One of the essential prospects for improving the estimation of N status thus lies in using green and, particularly, red-edge reflectance regions [14], which provide greater sensitivity to chlorophyll and N at higher contents with simultaneous increase of linearity [15]. Despite these improvements in N status estimation, many authors still indicate considerable effects of genotype, year, growth stage, and crop density on estimation accuracy [16–19].

Besides the use of sensitive vegetation indices, the way to improve the estimation of N status under such variable conditions leads through estimating indicators of N status, which are less dependent on growth stage, biomass, variety, or density. The relative N nutrition index (NNI) was introduced in order to allow expression of N status under the rapid dilution of N content during growth [20]. NNI provides the opportunity to express the relative N status independently of the growth stage based upon the so-called critical or dilution N curve, which reflects the minimum N concentration required to achieve maximum growth or yield [21]. NNI or integrated N uptake per area unit (N uptake) also seem to be more easily estimated by remote sensing methods and are more stable in time than the estimation of N content in plant biomass [10].

Neither the use of more sensitive reflectance indices based on green and red-edge wavebands nor NNI or N uptake can provide complete decoupling of the relationship between spectral reflectance and N status from the effects of such other factors as canopy structure, genotype, or other environmental conditions. However, it is possible to eliminate uncertainties given by the impact of different factors on spectral reflectance signature by using sophisticated multifactorial methods such as PLS regression or artificial neural networks if trained on sufficiently complex data [22].

We hypothesized that the combination of vegetation indices based on red-edge and green reflectance with NNI and N uptake as indicators of N status in malting barley would provide extended linearity and higher accuracy in estimating N status across different years, genotypes, and densities, as compared to indices based on red or blue reflectance, and also compared to N content in dry matter. We also assumed that an artificial neural network based on multiple reflectance wavelengths would reduce the uncertainties of N status estimation caused mainly by interannual variation, genotype, and canopy structure. The main objective of this study was to improve prediction of N status in malting barley using spectral reflectance, and particularly to increase the versatility of its use across variable years, genotypes, and management practices, by using three approaches: (i) selecting vegetation indices providing improved linearity of response to N status and, at the same time, low interannual variability of response; (ii) using the N status indicators which are less sensitive to changing biomass, growth stage, and rapid dilution of N; (iii) employing the artificial neural networks trained on a wide range of factors that may affect the direct estimation of N status in malting barley by vegetation indices.

2. Materials and Methods

Evaluation of the N status, canopy reflectance, and yield of spring barley was performed in small-plot field experiments established at the edge of the city of Kroměříž in Central Moravia (Czech Republic) with coordinates 49°17′5″ N, 17°21′35″ E within the period of three consecutive years (2011–2013). The location is characterized by a warm, slightly humid climate with a mean annual temperature of 9.1 °C and precipitation of 567.7 mm. Weather data during the vegetation period in individual years were collected by a permanent meteorological station located within 500 m of the experimental field (Table 1). The soil type at the site is Luvi-Haplic Chernozem, and the texture is clay loam. The previous crop was maize for grain in all three years. Standard plant protection measures were used during the growing season to avoid negative interactions with weeds, diseases, and pests. Each experimental treatment represented a combination of three factors, ensuring contrasting differences in the canopy density, structure, and nutritional status: (i) malting barley varieties with different tillering intensity (Bojos—middle, Prestige—low, and Sebastian—high); (ii) sowing density (2.5, 4, and 5.5 million germinating seeds ha^{-1}); (iii) N nutrition (0, 45, and 90 kg N ha^{-1}). N was applied before sowing in the form of ammonium nitrate. Each treatment was established in five replications. The plot size was 10 m^2, and the five replications were arranged in randomized block design. Three replications were harvested for grain yield, and two were used for sampling during vegetation, which enabled the analyses of aboveground biomass and N content. In sampling plots, squares of size 0.25 m^2 (0.5 × 0.5 m) were marked out to obtain plants for analyses of canopy structure in two developmental stages: end of tillering to the beginning of stem elongation (DC 29–31) and end of stem elongation (DC 39). Measurements of canopy reflectance in the range 350–2500 nm were made using a FieldSpec 4 HiRes spectroradiometer (ASD, Boulder, CO, USA) and fiber optic cable fixed in pistol grip (25° field of view). The reflectance measurements were conducted from a distance of ca. 0.8 m perpendicular to the canopy surface, which ensured the collection of spectra from a circle with a diameter of about 0.36 m. Smooth movement during spectra collection in the distance ca. 2 m along the plot allowed measurement of reflectance for an area of ca. 0.2 m^2. Two reflectance spectra were taken from each plot/replication. Before each new plot, the reference spectrum was measured using the reflectance standard (Spectralon; Labsphere, North Sutton, NH, USA). The reflected radiances were directly converted to spectral reflectance within the RS^3 Spectral Acquisition Software (ASD). PCA analysis and preliminary correlation with N nutrition indicators showed that outside the reflectance range 380–850 nm was the contribution to the explanation of N status as marginal; therefore, only the average reflectance spectrum in this range from all replications of each treatment was used to calculate vegetation indices (Supplementary Table S1), perform correlation analysis for simple reflectance ratios and normalized difference indices, and to train neural networks.

Analyses of canopy structure and biomass production were performed manually, which involved assessing the numbers of plants and tillers (data not shown) and the quantity of aboveground dry biomass. The aboveground biomass was dried in an oven at 80 °C until constant weight. The dried plant samples were then analyzed for N content using a LECO elemental analyzer (LECO, St. Joseph, MI, USA). The harvest was carried out using a Sampo 2010 small-plot harvester equipped with an automatic weighing and sampling system (Sampo Rosenlew, Pori, Finland). The correlation analysis (coefficients of determination, R^2) and neural network training were performed using Statistica 12 software (StatSoft Inc., Tulsa, USA). Prior to neural network training, the number of input reflectance wavelengths was reduced on the basis of PCA analysis, when reflectance wavelength with the highest PCA scores and lowest (or close to 180°) angle in relation to N status indicators were selected. For the neural network training, each dataset was randomly divided into training (70%), test (15%), and validation (15%) sub-datasets. The training was conducted on 10,000 networks with the maximum number of hidden units as 20 and identity, logistic, exponential, and hyperbolic tangents used for both hidden and output neurons. This allowed achieving high diversity of neural networks from which a set of 10 networks was chosen for each training set on the basis of the lowest training and validation error and the highest R^2 for the relationship between predicted and observed values. These networks were subjected to another 50 training cycles with new random dividing on training, test, and validation datasets for each. The network showing the lowest variation between training cycles was then selected as the best network. If the variation of performance during 50 training cycles was higher than 25%, the process was repeated. The correlation matrix for normalized difference indices combining all wavelengths of the selected range was analyzed in the software R 3.1.1 (R Foundation for Statistical Computing, Vienna, Austria).

Table 1. Mean daily temperatures and monthly precipitation sums for vegetation period in 2011–2013 and comparison to long-term average values for the period 1971–2010.

Characteristic	Year	March	April	May	June	July
Mean temperature (°C)	2011	5.3	11.8	14.6	18.4	18.0
	2012	6.6	10.6	16.3	19.1	20.8
	2013	1.2	10.1	14.2	17.5	21.2
	Long-term average (1971–2010)	4.3	9.4	14.5	17.3	19.2
Precipitation sum (mm)	2011	35.9	45.5	84.2	72.0	119.7
	2012	3.1	29.2	23.8	137.2	35.3
	2013	51.0	33.3	87.2	129.1	2.7
	Long-term average (1971–2010)	32.8	40.7	66.1	80.6	73.6

3. Results

3.1. Correlations between Spectral Reflectance and Indicators of N Status

Evaluation of the index of determination (R^2) was made for the normalized difference indices $NDI_{xy} = (R_x - R_y)/(R_x + R_y)$ at 1 nm step, wherein R_x and R_y represent reflectance in wavelengths x and y. These normalized indices were correlated to the relative N content in dry aboveground biomass (% mass), N nutrition index (NNI; dimensionless), and N uptake by aboveground biomass per area unit (N uptake; kg ha^{-1}) separately in the growth stages DC 29–31 and DC 39. Correlations were made across all experimental years, barley genotypes, sowing densities, and N doses. The results of the correlation analysis are shown in Figure 1, in which the R^2 are expressed using a color scale for each combination of wavelengths. These results generally show low values of R^2 for N content in the aboveground biomass, while for NNI, and particularly for N uptake, R^2 increased significantly. This analysis identified several combinations of reflectance wavelengths with the potential for evaluating N status. Higher R^2 values were achieved for the combination of reflectance wavelengths in the range of 400–500 nm with 650–690 nm. Another area with relatively high R^2 values is delineated on both axes by wavelengths in the range 450–490 nm. This narrowly defined area has a particularly closer correlation to the N status in the growth stage DC 29–31. Similarly, a very narrowly delineated area exhibiting high R^2 in relation to N status is defined by reflectance wavelengths in the range 530–550 nm on both axes. In this case, however, the higher R^2 values were obtained only at the later growth stage (DC 39). Similar results were obtained in the area delineated by reflectance wavelengths 560–590 nm. Reliable estimation of N status was also provided by NDI_{xy} when combining reflectance in the range 710–730 nm with band 730–800 nm. The combination of these bands achieves similar results in both the DC 29–31 and DC 39 growth stages. Comparable results were also performed by a combination of bands 550–650 nm and 730–800 nm. Generally, although these areas overlap for the N content in plants, NNI, and N uptake, the highest values of R^2 were obtained for N uptake and the lowest R^2 for N content.

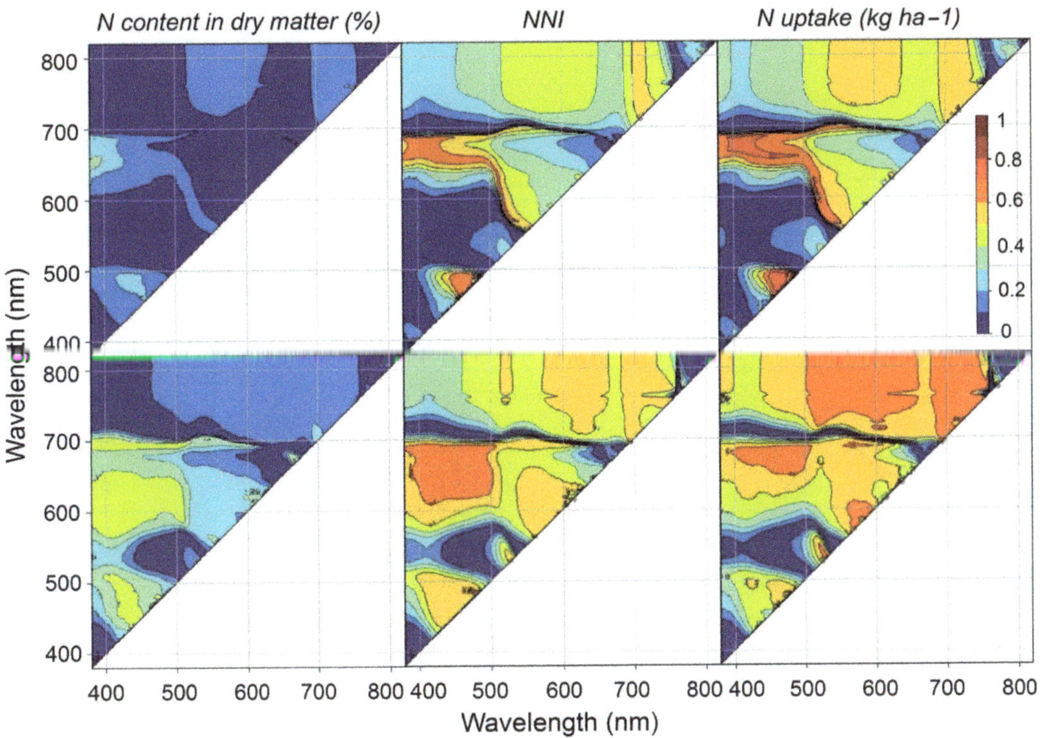

Figure 1. Matrix plots of coefficients of determination (R^2) for the relation of all combinations of wavelengths in the range 380–820 nm used for a linear regression analysis of the normalized difference index $NDI_{xy} = (R_x - R_y)/(R_x + R_y)$, wherein R_x and R_y represent reflectance in wavelengths x and y, against N content in dry matter (**left**), nitrogen nutrition index (NNI, **middle**), and N uptake by aboveground biomass (**right**). The correlations were calculated separately at stage DC 29–31 (**upper**) and DC 39 (**bottom**).

3.2. Evaluation of N Status Using Vegetation Indices

From the results of correlation analysis across all experimental years (Table 2), it is evident that in both sampling dates, generally higher R^2 values were achieved for estimating NNI and N uptake than for the N content in plants. At the same time, the closest correlation to N status was demonstrated for indices based on a combination of reflectances in bands 430 and 680 nm (SRPI, NPCI) and also for indices based on reflectance in the red-edge region (700–730 nm), e.g., NRERI, ZM, or $ANMB_{650-725}$. For such widely used vegetation indices as NDVI, RDVI, and OSAVI, which exploit the differences in reflectance between red and near-infrared bands, the increase of R^2 with later growth stage was evident. Even in the growth stage DC 39, however, these indices do not outperform the indices based on reflectance in regions around 430 nm or red-edge band in estimating N status. As the performances of indices within the abovementioned groups were very similar, we used for further analyses the representatives showing the highest average performance across both growth stages and all N status indicators: SRPI for the region around 430 nm and NRERI for the red-edge region. The vegetation index NDVI was selected as representative of the group of indices based on reflectance in the red region due to its common use. However, its performance was slightly lower compared to some others within this group.

Table 2. Coefficients of determination (R^2) and root mean square errors (RMSE) for linear relationships between reflectance indices and nitrogen status parameters (N content in dry matter, NNI and N uptake by aboveground biomass) at individual growth stages (DC 29–31 and DC 39). Coefficients of determination significant at $p < 0.05$ are indicated in bold.

| Reflectance Indices | N Content in Dry Matter (%) | | | | NNI | | | | N Uptake (kg ha^{-1}) | | | |
| | DC 29–31 | | DC 39 | | DC 29–31 | | DC 39 | | DC 29–31 | | DC 39 | |
	R^2	RMSE	R^2	RMSE	R^2	RMSE	R^2	RMSE	R^2	RMSE	R^2	RMSE
ANMB$_{650-750}$	**0.19**	0.437	**0.21**	0.378	**0.61**	0.137	**0.52**	0.126	**0.68**	17.00	**0.60**	19.68
NDVI	**0.08**	0.464	**0.18**	0.385	**0.40**	0.169	**0.50**	0.128	**0.47**	22.00	**0.60**	19.55
NDGI	**0.13**	0.452	**0.16**	0.390	**0.47**	0.159	**0.51**	0.128	**0.54**	20.54	**0.64**	18.70
NRERI	**0.23**	0.424	**0.17**	0.386	**0.62**	0.135	**0.53**	0.125	**0.68**	17.02	**0.66**	17.98
RDVI	**0.09**	0.461	**0.23**	0.372	**0.42**	0.167	**0.58**	0.119	**0.50**	21.38	**0.65**	18.43
MSR	**0.11**	0.457	**0.13**	0.397	**0.37**	0.174	**0.47**	0.133	**0.44**	22.63	**0.62**	19.20
MT VI1	**0.08**	0.463	**0.25**	0.368	**0.39**	0.171	**0.56**	0.121	**0.47**	21.97	**0.60**	19.52
TCARI	0.01	0.482	**0.11**	0.401	0.03	0.215	**0.08**	0.175	0.07	29.17	0.03	30.50
OSAVI	**0.09**	0.461	**0.21**	0.377	**0.42**	0.166	**0.56**	0.121	**0.50**	21.32	**0.65**	18.35
TCARI/OSAVI	**0.32**	0.399	0.01	0.422	**0.44**	0.164	**0.15**	0.169	**0.41**	23.18	**0.26**	26.70
G	0.07	0.468	**0.18**	0.385	**0.29**	0.184	**0.48**	0.139	**0.36**	24.08	**0.56**	20.68
TVI	**0.08**	0.465	**0.25**	0.367	**0.38**	0.172	**0.56**	0.121	**0.47**	22.07	**0.60**	19.75
ZM	**0.17**	0.440	**0.13**	0.397	**0.50**	0.155	**0.47**	0.133	**0.57**	19.85	**0.63**	18.98
SRPI	**0.20**	0.433	**0.36**	0.339	**0.65**	0.129	**0.67**	0.105	**0.73**	15.60	**0.66**	18.20
NPQI	**0.11**	0.457	**0.26**	0.365	0.04	0.215	**0.36**	0.146	0.01	29.98	**0.31**	25.74
PRI	**0.18**	0.439	**0.25**	0.368	**0.61**	0.136	**0.29**	0.154	**0.66**	17.61	**0.23**	27.23
NPCI	**0.19**	0.437	**0.34**	0.344	**0.65**	0.129	**0.64**	0.109	**0.73**	15.65	**0.63**	18.82
SIPI	**0.09**	0.461	**0.15**	0.391	**0.40**	0.169	**0.48**	0.131	**0.47**	22.06	**0.60**	19.68
VOG3	**0.18**	0.439	**0.12**	0.399	**0.47**	0.159	**0.46**	0.134	**0.54**	20.57	**0.63**	18.87
VOG2	**0.18**	0.439	**0.12**	0.398	**0.48**	0.158	**0.47**	0.133	**0.54**	20.36	**0.64**	18.67
GM1	**0.15**	0.447	0.08	0.408	**0.42**	0.167	**0.40**	0.141	**0.48**	21.81	**0.59**	19.93
GM2	**0.16**	0.443	**0.14**	0.394	**0.47**	0.159	**0.48**	0.131	**0.54**	20.40	**0.63**	18.85

Because the effect of year was the main source of variability in the relationships between vegetation indices and indicators of N status, detailed regression analysis for selected vegetation indices was carried out separately for individual years. The results are summarized in Table 3 and Figures 2–4. It is evident from these analyses that splitting the relationships by individual year led, in major cases, to higher R^2 values and lower root mean square errors (RMSE), and this was particularly true for indices NDVI and NRERI and for N status indicators NNI and N uptake. If the relationships are analyzed separately in each year, the R^2 values are very similar for all three indices (NDVI, NRERI, and SRPI), but especially for estimation of NNI and N uptake. This means that, in particular, the index NDVI, and partly also NRERI, shows a significant effect of the experimental year on relationships between vegetation index and N status indicators. In the case of NDVI, the between-year variation is evident in both the intercept and slope of linear relationships. In NDVI, moreover, an evident change of the index value range was observed for the later growth stage. The NDVI values increased with the later growth stage while the range simultaneously narrowed. This resulted in a rising slope with the later growth stage. In the case of NRERI, a shift of intercept was particularly evident. The slope of linear relationships varied far less, even when comparing the two growth stages. The smallest between-year variation in relationships to N status was found for the index SRPI. Generally, the highest between-year variation in relationships is evident for the N content in aboveground dry matter. Similarly, the largest differences in relationships between growth stages were observed for this indicator of N status. The smallest differences in the relationships among growth stages were recorded for N uptake.

Table 3. Regression parameters of linear relationships between selected reflectance indices (y = y0 + a*x, where [illegible] is the nitrogen status parameter, a is slope of linear relationship, and y0 is the intercept), coefficients of determination (R^2), and root mean square errors (RMSE) analyzed separately for individual year and all years together. Coefficients of determination significant at $p < 0.05$ are indicated in bold.

Growth Stage DC	Nitrogen Status Parameter	Reflectance Index	a				y0				R^2				RMSE			
			2011	2012	2013	2011–2013	2011	2012	2013	2011–2013	2011	2012	2013	2011–2013	2011	2012	2013	2011–2013
29–31	N content in dry matter (%)	NDVI	4.72	7.95	0.56	1.38	−1.34	−3.45	2.54	1.77	**0.29**	**0.46**	0.04	0.08	0.375	0.458	0.317	0.464
		NRERI	3.29	9.59	2.29	3.14	0.97	−2.19	1.84	1.21	**0.39**	**0.6**	0.13	**0.23**	0.346	0.393	0.302	0.424
		SRPI	1.82	4.38	0.51	1.55	1.35	−0.63	2.57	1.64	**0.35**	**0.46**	0.06	**0.2**	0.359	0.457	0.314	0.433
	NNI	NDVI	4.07	3.37	1.34	1.38	−2.89	−1.89	−0.29	−0.3[illegible]	**0.72**	**0.64**	**0.73**	**0.4**	0.128	0.133	0.096	0.169
		NRERI	2.53	3.81	3.39	2.31	−0.7	−1.22	−0.95	−0.4[illegible]	**0.78**	**0.74**	**0.86**	**0.62**	0.113	0.113	0.068	0.135
		SRPI	1.45	1.87	1.02	1.27	−0.45	−0.7	−0.06	−0.2[illegible]	**0.74**	**0.66**	**0.74**	**0.65**	0.124	0.13	0.095	0.129
	N uptake (kg ha^{-1})	NDVI	609.5	408.1	205	206.9	−478.9	−253.9	−97.8	−10[illegible]	**0.76**	**0.68**	**0.8**	**0.47**	17.409	14.777	12.1	21.997
		NRERI	375.4	448.1	508.9	335.2	−150.4	−166	−192.9	−11[illegible]	**0.81**	**0.74**	**0.91**	**0.68**	15.587	13.272	8.279	17.017
		SRPI	216.3	229.2	157.5	185.7	−114.9	−112.2	−62.9	−83[illegible]	**0.77**	**0.71**	**0.83**	**0.73**	17.01	14.041	11.326	15.602
39	N content in dry matter (%)	NDVI	0.23	6.69	4.81	3.41	1.52	−3.75	−2.07	−1	0.01	**0.55**	0.16	**0.18**	0.182	0.302	0.334	0.385
		NRERI	0.37	5.13	3.48	2.23	1.5	−0.89	0.11	−0.6[illegible]	0.04	**0.59**	**0.25**	**0.17**	0.179	0.287	0.315	0.386
		SRPI	0.24	2.37	2.08	1.67	1.54	0.06	0.4	0.6[illegible]	0.05	**0.58**	**0.29**	**0.36**	0.178	0.29	0.306	0.339
	NNI	NDVI	1.83	3.14	5.02	2.47	−1.01	−2.03	−3.84	−1.[illegible]	**0.66**	**0.69**	**0.64**	**0.5**	0.081	0.104	0.113	0.128
		NRERI	1.24	2.39	3.04	1.68	−0.15	−0.67	−1.19	−0.[illegible]	**0.75**	**0.73**	**0.71**	**0.53**	0.07	0.097	0.1	0.125
		SRPI	0.68	1.09	1.73	0.97	0.08	−0.22	−0.85	−0.1[illegible]	**0.72**	**0.71**	**0.75**	**0.67**	0.074	0.101	0.093	0.105
	N uptake (kg ha^{-1})	NDVI	452.3	477	949.2	460.4	−326	−333.1	−784.7	−33[illegible]3	**0.76**	**0.74**	**0.75**	**0.6**	15.745	13.976	16.425	19.546
		NRERI	300.5	360.4	556.3	320	−108.2	−125.6	−271.8	−11[illegible]3	**0.83**	**0.78**	**0.79**	**0.66**	13.388	12.918	15.095	17.979
		SRPI	164.5	164.6	312	149.9	−52	−57.7	−206.1	14.[illegible]	**0.78**	**0.75**	**0.8**	**0.66**	15.01	13.721	14.499	18.203

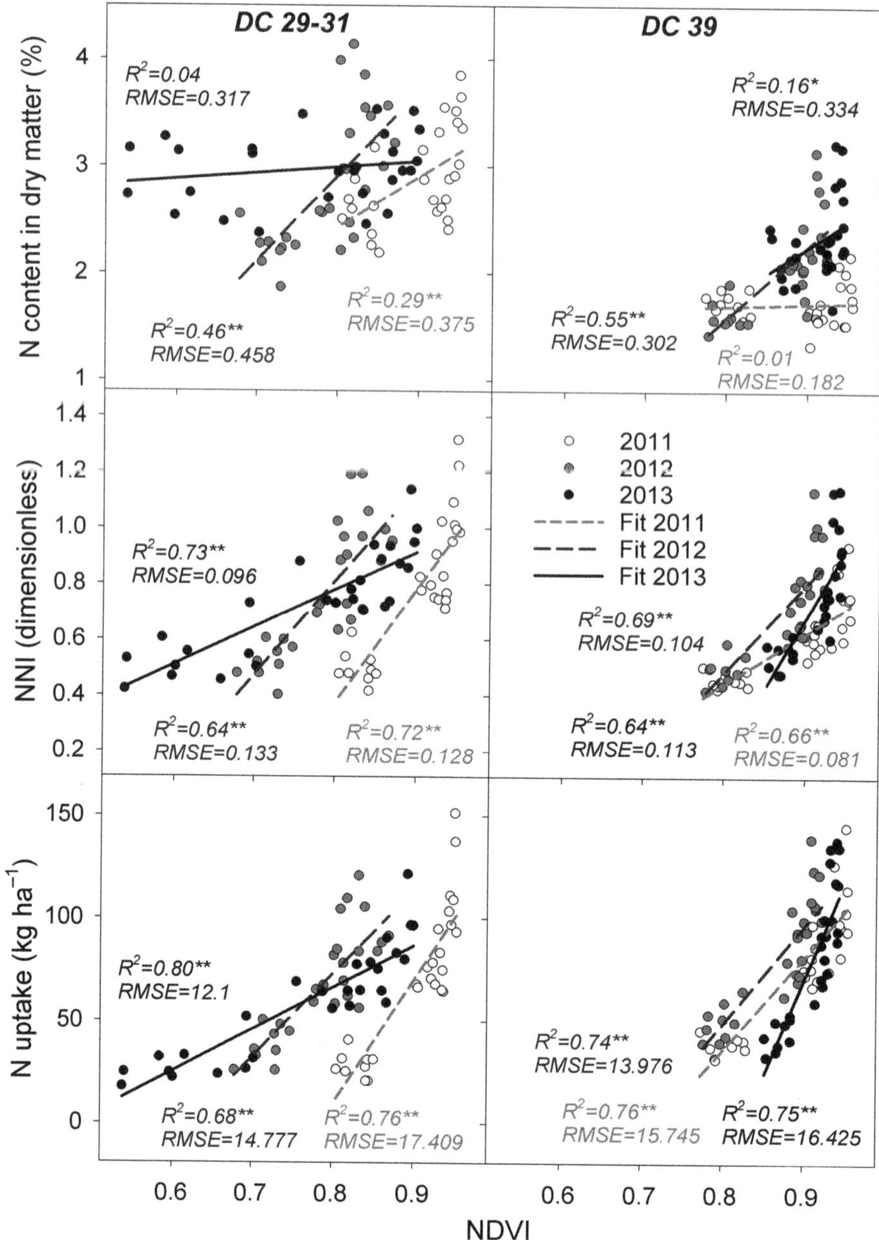

Figure 2. Linear relationships between vegetation index NDVI and nitrogen content in aboveground dry mass (**upper**), nitrogen nutrition index (NNI, **middle**), and nitrogen uptake by aboveground biomass per area unit (**lower**), analyzed separately for growth stage beginning of stem elongation (DC 29–31, **left**) and end of stem elongation (DC 39, **right**), and for individual years (2011—white points and light grey line, 2012—grey points and dark grey line, and 2013—black points and black line). Coefficients of determination (R^2), root mean square errors (RMSE), and significant correlations (* at $p < 0.05$ and ** at $p < 0.01$) are indicated for each relationship.

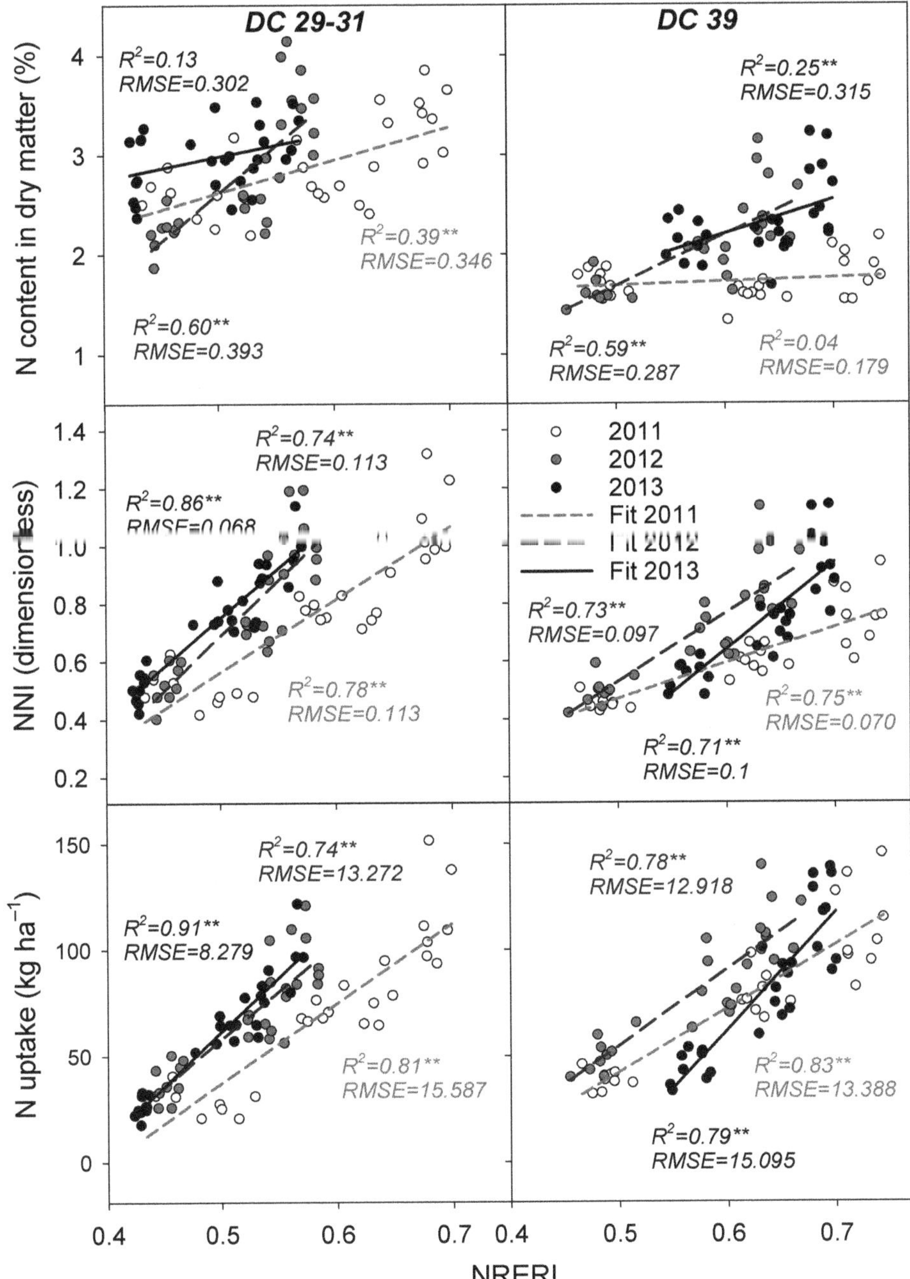

Figure 3. Linear relationships between reflectance index NRERI and nitrogen content in aboveground dry mass (**upper**), nitrogen nutrition index (NNI, **middle**), and nitrogen uptake by aboveground biomass per area unit (**lower**), analyzed separately for growth stage beginning of stem elongation (DC 29–31, **left**) and end of stem elongation (DC 39, **right**), and for individual years (2011—white points and light grey line, 2012—grey points and dark grey line, and 2013—black points and black line). Coefficients of determination (R^2), root mean square errors (RMSE), and significant correlations (* at $p < 0.05$ and ** at $p < 0.01$) are indicated for each relationship.

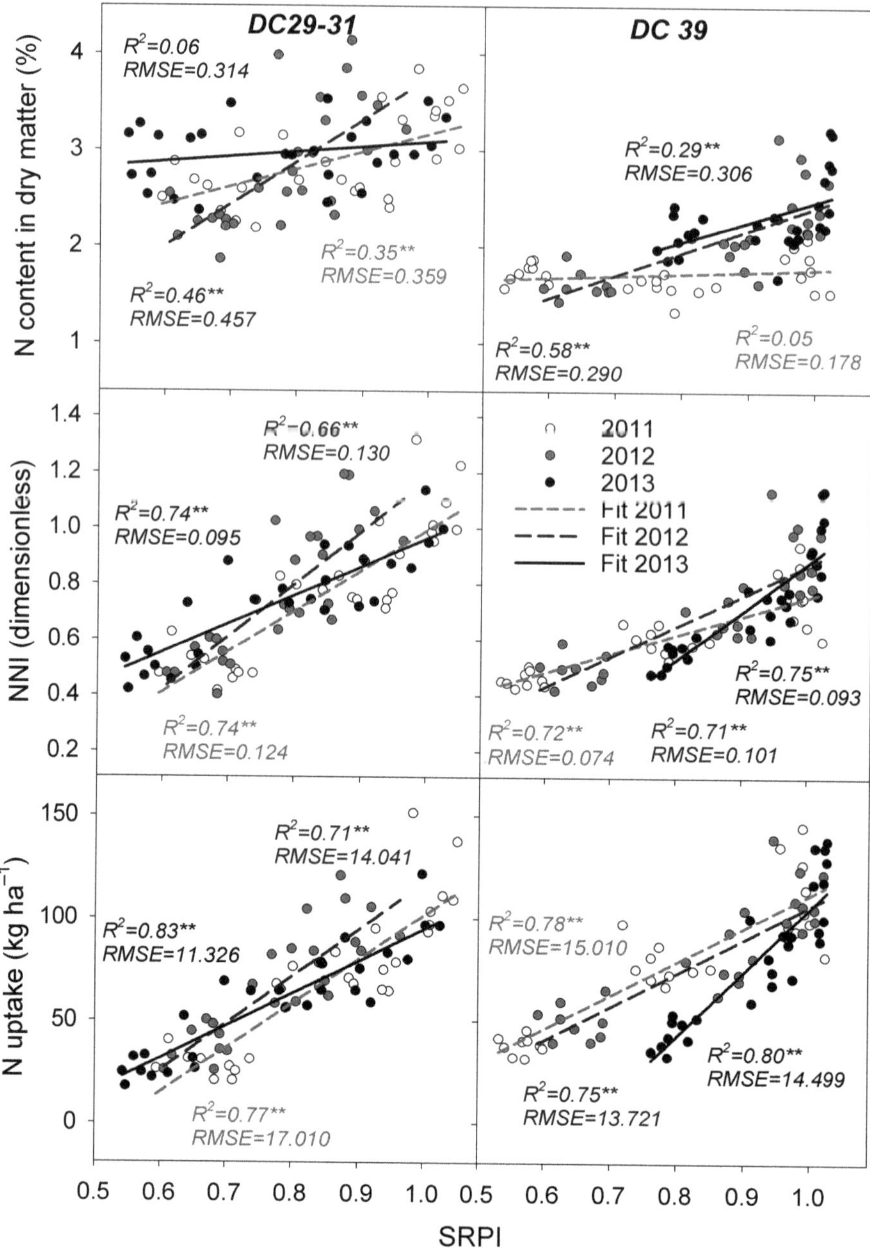

Figure 4. Linear relationships between reflectance index SRPI and nitrogen content in aboveground dry mass (**upper**), nitrogen nutrition index (NNI, **middle**), and nitrogen uptake by aboveground biomass per area unit (**lower**), analyzed separately for growth stage beginning of stem elongation (DC 29–31, **left**) and end of stem elongation (DC 39, **right**), and for individual years (2011—white points and light grey line, 2012—grey points and dark grey line, and 2013—black points and black line). Coefficients of determination (R^2), root mean square errors (RMSE), and significant correlations (* at $p < 0.05$ and ** at $p < 0.01$) are indicated for each relationship.

3.3. Relationships between Indicators of N Status and Grain Yield

The relationships between indicators of nutritional status (i.e., N content in aboveground dry matter, NNI index, and N uptake per area unit) and grain yield achieved were evaluated separately for each year (Figure 5). Generally, these relationships exhibit a rectangular hyperbola shape, although it is evident that in some cases, the upper asymptote was not reached. The greatest variation in relationships was observed for the N content in dry matter. While in 2013, there was almost no correlation between N content and yield, in 2011, a very steep and significant relationship without reaching the upper asymptote was found. In 2012, the typical rectangular hyperbola relationship was observed. Conversely, the relationships between NNI and N uptake per area unit showed the rectangular hyperbola course in virtually all combinations of year and growth stage. At the same time, the shape of the relationship changed only a little, but the upper asymptote changed significantly between years. This means that the maximum yield is generally obtained in all years at the same level of nutritional status, whether it is defined as NNI or N uptake. For NNI, the maximum yield was achieved at values just above 1.0 in both growth stages, whereas for N uptake, a slight shift of the maximum was found. The yield maximum was reached at N uptake of about 100 and 120 kg N ha^{-1} in growth stage DC 29–31 and DC 39, respectively.

3.4. Artificial Neural Networks Based on Hyperspectral Data for Estimating N Status

For purposes of training neural networks, it was first necessary to reduce the initial reflectance dataset to optimize the range of the input variables. Reflectance data were first reduced by using the 10 nm step and in the range 380–850 nm. Subsequently, principal component analysis (PCA) was used to select only the reflectance wavelengths with major explanatory importance for indicators of N status. In total, 17 reflectance wavelengths with the highest PCA scores (components 1 and 2), and also minimum differential angle (or close to 180° for negative association) to the PCA loadings for N status indicators, were selected for subsequent neural network training. The set of 162 observations was randomly divided into training (114 observations), test, and validation (each with 24 observations) datasets. From 10,000 trained networks for each combination of N status indicator and growth stage, a set of 10 networks was chosen for each training set (N status indicator and growth stage combination), having the lowest training and validation error and highest R^2 for the relationship between predicted and observed values. These networks were subjected to 50 training cycles with new random dividing on training, test, and validation datasets for each. The network showing the lowest variation between training cycles was then selected as the best network. If the variation of performance during 50 training cycles was higher than 25%, the process was repeated. The scatterplots of predicted and observed data for best networks and their comparison with the 1:1 line are shown in Figure 6. It is evident from these results that neural networks substantially enable elimination of interannual variability, and therefore significantly improve the accuracy of N status estimation. This is especially the case for N content in the dry aboveground biomass. While for the best of vegetation indices—SRPI—R^2 values of 0.2 and 0.36 had been achieved in the growth stages DC 29–31 and 39, respectively, through the use of neural networks, R^2 was increased to 0.75 and 0.8, respectively. For NNI index and N uptake, the R^2 value increased from the range 0.58–0.73 for the best vegetation indices to 0.83–0.87 when using neural networks. An overview of the best neural networks, their performance in the training, test, and validation dataset, and also the activation functions used for each estimated parameter of N status and growth stage are shown in Table 4. A sensitivity analysis describing the explanatory value of reflectance in each individual wavelength for the neural network model is shown in Supplementary Table S2. It is noteworthy that the reflectance wavelength with the highest sensitivity ranking differs substantially between estimated N status parameters and also between growth stages.

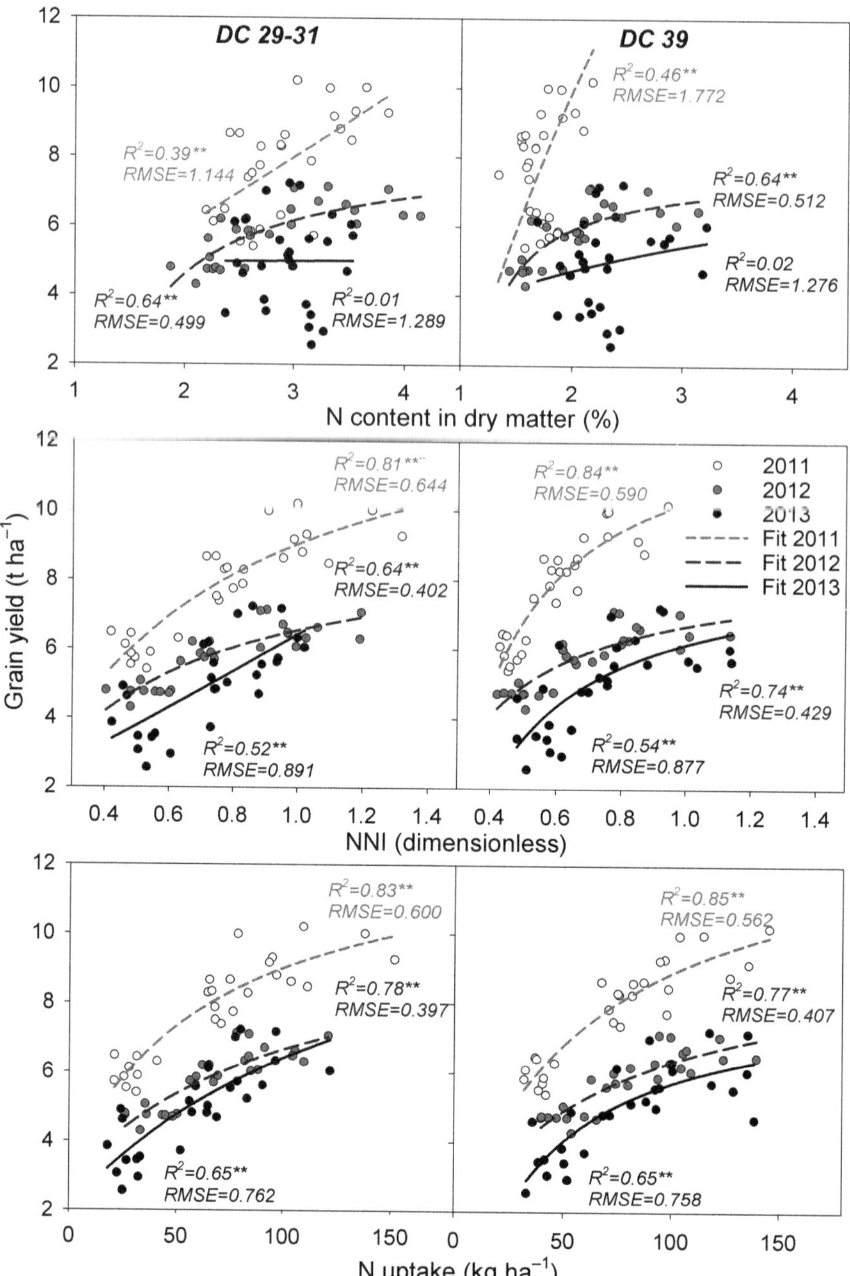

Figure 5. Rectangular hyperbola relationships between nitrogen content in aboveground dry mass (**upper**), nitrogen nutrition index (NNI, **middle**), nitrogen uptake by aboveground biomass per area unit (**lower**) and grain yield, analyzed separately for growth stage beginning of stem elongation (DC 29–31, **left**) and end of stem elongation (DC 39, **right**), and for individual years (2011—white points and light grey line, 2012—grey points and dark grey line, and 2013—black points and black line). Coefficients of determination (R^2), root mean square error (RMSE), and significant correlations (* at $p < 0.05$ and ** at $p < 0.01$) are indicated for each relationship.

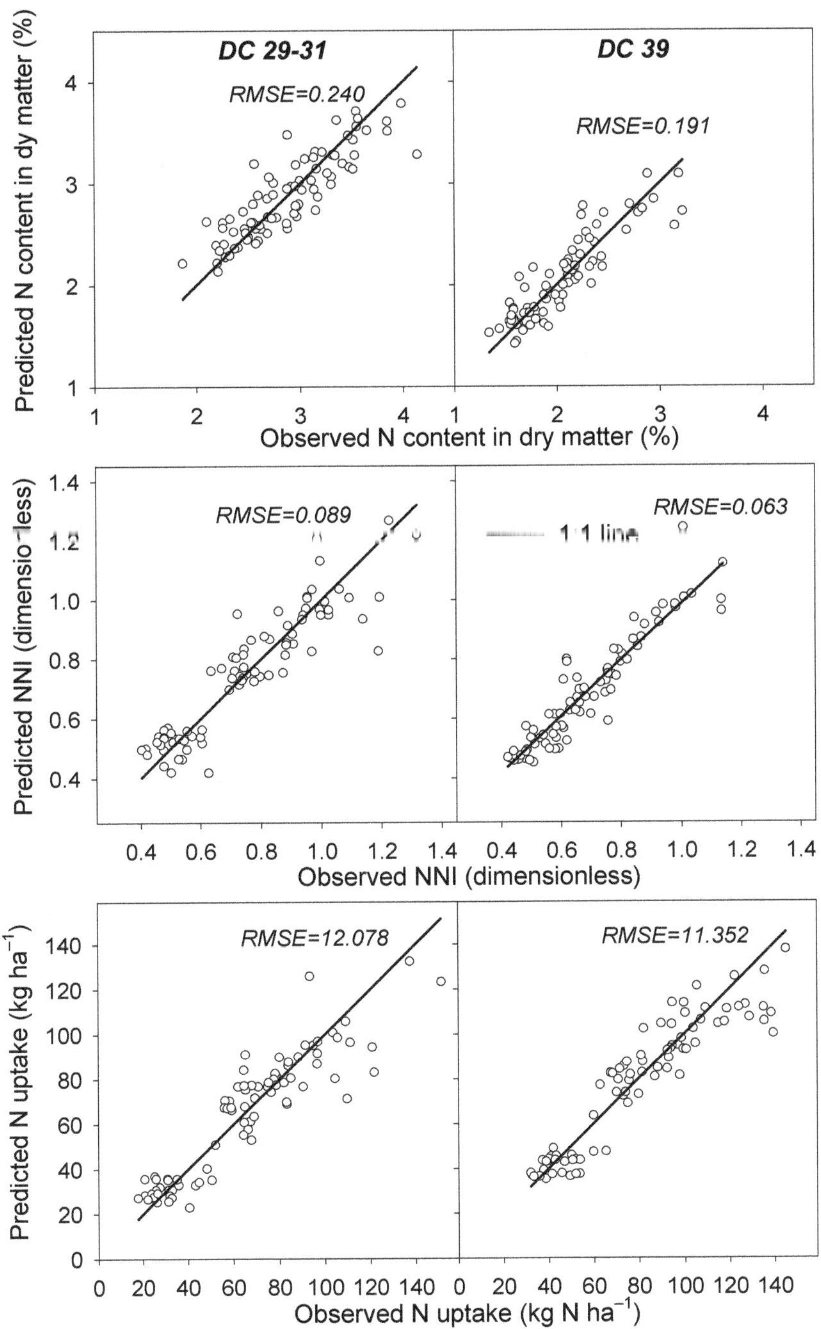

Figure 6. Comparison of observed nitrogen content in aboveground dry mass (**upper**), nitrogen nutrition index (NNI, **middle**), and nitrogen uptake by aboveground biomass per area unit (**lower**) with predicted values by winning artificial neural networks. The 1:1 lines are plotted to evaluate deviations from observed data. Root mean square errors (RMSE) are indicated for each predicted set of data.

Table 4. Characteristics and performance of winning neural networks (MLP—multilayer perceptron) for estimation of individual nitrogen status parameters at growth stage DC 29–31 and DC 39. Network performance is given separately for training, test and validation datasets. Exponential, logistic, and hyperbolic tangent activation functions (TanH) are indicated for the hidden and output layer of the neural network.

Growth Stage	Estimated Parameter	Network Name	Network Performance			Activation Function	
			Training	Test	Validation	Hidden	Output
DC 29–31	N content in dry matter (%)	MLP 17-15-1	0.897	0.753	0.923	TanH	TanH
	NNI	MLP 17-8-1	0.927	0.856	0.938	Exponential	TanH
	N uptake (kg ha^{-1})	MLP 17-11-1	0.920	0.877	0.958	Exponential	Logistic
DC 39	N content in dry matter (%)	MLP 17-13-1	0.924	0.887	0.883	TanH	Logistic
	NNI	MLP 17-6-1	0.955	0.950	0.950	TanH	Exponential
	N uptake (kg ha^{-1})	MLP 17-12-1	0.922	0.975	0.965	TanH	Exponential

4. Discussion

The N status of malting barley requires relatively precise optimization because nitrogen deficiency leads, in particular, to reduced tillering and impaired yield formation. On the other hand, the N excess causes an increase in grain protein content, aggravating the malting quality, but it also leads to increased lodging and associated adverse effects such as yield reduction or mycotoxin contamination [6,7]. From the economic point of view, the impact of N nutrition on grain protein content is crucial. Although it also depends on a number of other factors, such as sufficient water availability and optimum temperature, it significantly impacts the price and economics of malting barley production. Water stress and excessive temperatures limit starch accumulation more than nitrogen translocation, resulting in higher grain protein content, while under sufficient water availability is the grain protein content, transformed from translocated N, diluted in higher starch accumulation [23]. Although environmental conditions can significantly modulate the translation of N status into grain protein content and should always be carefully considered when making decisions about N nutrition in malting barley, N status is still one of the most critical parameters for the final grain protein content that can be affected by agronomic practices. Therefore, accurate and reliable methods of N status monitoring during vegetation are required to achieve the malting quality of grain needed.

The main objective of this study was to improve the estimation of N status in malting barley using hyperspectral data, and in particular, to eliminate the sources of variability given by the effect of year, canopy structure, and genotype, doing so through the use of more robust reflectance indices, N status indicators, and, finally, by using artificial neural networks.

4.1. Hyperspectral Estimation of NNI

The N nutrition index (NNI) was introduced to assess the relative N status with respect to rapidly changing N content in plant biomass during plant development, which is characterized by the so-called critical or dilution curve [20]. N dilution is caused by a gradual change in the ratio between metabolic and structural tissues in favor of structural tissues, which contain much less N [24]. NNI represents the ratio between actual and critical N content and indicates the relative value of actual N status compared to minimum N content for maximum biomass or yield production [20]. The main advantage of using NNI to assess N nutrition status is that it provides an objective parameter that is comparable throughout the entire vegetation season, irrespective of growth stage and biomass accumulation. An NNI value greater than 1 indicates N surplus, and an NNI value less than 1 points to N deficiency. Although NNI is an effective tool in terms of increasing the accuracy and simplifying the assessment of N status, it has rarely been used for estimating N status from hyperspectral data. One of the first studies within which NNI was estimated in winter wheat using the red edge inflexion point (REIP) spectral reflectance index was published by Mistele and Schmidhalter [25]. They also showed the effects of year and growth stage on the relationship between REIP and NNI. Especially at early growth stages, the relationship differs from the typical relationship found after canopy closure. Houles et al. [26] also showed that NNI corresponds to the N status of the whole canopy rather than to actual N or chlorophyll content at the leaf level, which can explain some differences due to canopy development. We achieved similar results within the present study. While the relationships between vegetation indices and N content in the aboveground dry matter were generally weak and highly variable between years, for NNI, close relationships were found that varied only a little between years. The R^2 values for relationships between NNI and vegetation indices were approximately three times higher than those determined for relationships between the indices and N content in the aboveground dry matter (Table 3, Figures 2–4). However, the relationships to NNI showed an effect of growth stage on both intercept and slope of relationship, indicating that NNI cannot completely decouple the effect of biomass and leaf area even for the best vegetation indices.

Later, Erdle et al. [27] demonstrated that significantly more accurate estimation of the N nutrition status by spectral reflectance is achieved using NNI in comparison with the detection of N content in plants. Generally, the best estimate of NNI was achieved using indices based on the red-edge band reflectance, and in particular, using the REIP index. Chen et al. [28] introduced a new spectral index based on red-edge, known as the double-peak canopy N index (DCNI), for improved estimation of N content as input for NNI calculation. Such indirect estimation of NNI requires, however, the knowledge of biomass that may increase the estimation error because it is necessary to estimate two parameters showing interactions in their effects on spectral reflectance. Our results, too, showed a reliable estimation of NNI using spectral indices based on the red-edge band (NRERI, ANMB$_{650-725}$, ZM; Table 2, Figure 3). Even better results, however, were achieved for indices employing reflectance at 430 nm (R$_{430}$; SRPI and NPCI; Table 2, Figure 4). These indices are mostly related to the ratio between total carotenoids and chlorophylls [29]. Although the correlation matrix for NNI (Figure 1) shows that the sensitive area in the blue reflectance region narrows with the growth stage, the correlation for reflectance indices based on R$_{430}$ is almost not changed with the growth stage (Table 1). The sensitivity to N status of reflectance indices based on R$_{430}$ had been shown already by Filella et al. [30]. N deficiency generally increases the total carotenoids and chlorophylls ratio because carotenoids persist longer than do chlorophylls in senescing leaves [31], and N deficit induces leaf senescence [32].

Based on our results, direct estimation of NNI using spectral reflectance appears to be more accurate than the separate estimation of N content and plant biomass, even though the relationship is slightly modified by the growth stage. On the contrary, the relationships for N content in aboveground dry biomass show variation in slope and R^2 between both growth stages and years, probably due to the rapid dilution of N in biomass. This leads to a significant decrease of R^2 if the data are analyzed across years, and even more so if performed across growth stages. Conversely, only small changes in R^2 were observed when such data were analyzed for NNI. The use of calibration, even in the case of NNI, however, may be beneficial for improving estimation accuracy.

4.2. Hyperspectral Estimation of N Uptake by Aboveground Biomass

Optimization of N nutrition based upon total N uptake is, in principle, a simpler method than evaluating NNI, and it also has some practical advantages. These lie particularly in the possibility to optimize N nutrition on the basis of total N balance for expected yield and grain protein content, the latter of which is one of the key grain quality parameters for malting barley. N uptake can also be relatively easily estimated by spectral reflectance with the best estimation performance of the normalized red-edge index [33]. Similarly, Erdle et al. [27] demonstrated the best performance for the simple ratio index based on red-edge reflectance for estimation of both NNI and N uptake. This is in line with our results showing a good estimation of N uptake by red-edge indices NRERI, ZM, or $ANMB_{650-725}$. On the other hand, N uptake does not enable direct estimation of relative N status, which requires comparison with optimum N uptake for a given growth stage or biomass. This can be derived from the critical N curve as a critical N uptake [34], but it requires knowledge of biomass production. In our study, estimation of N uptake by aboveground biomass using spectral reflectance generally had significantly higher accuracy than did the estimation of N content, and it also slightly outperformed the estimation of N status by NNI. This is due to the fact that N uptake per unit area integrates the effects both of biomass and of N content in dry matter. Higher N uptake may be given by higher biomass production at the same N content or by higher N content at the same biomass production. However, actual N uptake at the given stage can be compared with optimum uptake for required yield without any information about biomass production.

In earlier studies in winter wheat, it was demonstrated that N optimization could be successfully carried out on the basis of remote sensing estimation of early-season N uptake and estimation of grain yield potential [35,36]. These optimization algorithms were based on NDVI estimation between the end of tillering and the beginning of stem elongation, and the yield potential was estimated as NDVI divided by growing degree days. Such estimation of yield potential can also be useful in spring barley, as the NDVI show particularly higher values at the high yielding year 2011 when evaluated at the early growth stage (DC 29–31) even for the same N uptake levels. On the other hand, SRPI can be used to estimate the N uptake without the effect of yield potential. Optimum N dose can thus be derived from yield potential (and optimal N uptake for such yield) estimated from NDVI and actual N uptake estimated from SRPI. The maximum yield for a given N response curve is limited mainly by genotype and environment/year [37]. As demonstrated in our recent work [3], the yield potential of spring barley is determined in the early growth stages up to the beginning of stem elongation (DC 31), which supports the idea of early estimation of yield potential from NDVI for N dose optimization.

4.3. Artificial Neural Network for Estimating N Status Using Hyperspectral Data

The application of artificial neural networks has, in recent decades, become rather popular in the analysis of remotely sensed data [38]. The most common applications are in image processing, particularly as classification algorithms (e.g., for land-cover classification). Artificial neural networks also have been applied to retrieve biophysical parameters of vegetation. For example, Bacour et al. [39] developed neural network algorithms for estimating leaf area index, canopy chlorophyll content, and absorbed photosynthetically active radiation, Liu et al. [40] used neural networks to discriminate fungal infection in rice panicles from hyperspectral reflectance, and Christensen et al. [41] adopted neural network for classification of barley growth stage from hyperspectral data.

In our study, we demonstrated the considerable potential of artificial neural networks for improving the direct estimation of N content in barley aboveground biomass from hyperspectral data in comparison with best spectral reflectance indices. The neural network model reduces not only the variability within one year given by genotype or sowing density but also the interannual variation, which is a much more important source of inaccuracy. The effect of year changes both intercepts and slopes of relationships between spectral reflectance indices and N content. A similar improvement of N content estimation by the artificial neural network, compared to the linear regression model, was reported for rice by Yi et al. [42]. They also demonstrated that both multiple linear regression and principal component analysis were suitable to select input variables for neural network training. Wang et al. [43] demonstrated the strength of neural network predictions in comparison with stepwise regression for N content in rape. Recently, Sun et al. [44] showed that the neural network provided the best performance in estimating N content in rice from both active and passive reflectance sensors.

The employment of artificial neural networks in remote sensing provides a number of advantages in comparison to regression models [45]. First, the neural network allows the development of complex models with multiple input parameters and by employing diverse nonlinear response functions. Secondly, neural networks allow combining different types of input data, including categorical ones and thus improving the model by introducing the data, e.g., regarding the preceding crop, soil, genotype, weather, etc. Finally, high adaptivity, given by the possibility to retrain the model on the new dataset, is of significant advantage if the model is used in new conditions. Neural networks also allow joint estimation of multiple biophysical and biochemical variables [39].

5. Conclusions

We demonstrated that the use of indirect N status indicators, such as NNI or N uptake per ground area unit, could significantly improve the estimation accuracy from spectral reflectance data, compared to N content in dry aboveground biomass. From spectral reflectance indices tested, the red-edge indices (NRERI, ZM, or $ANMB_{650-725}$) and carotenoid indices (SRPI and NPCI) showed the lowest interannual variability. Although the maximum yield varied significantly between years, the NNI and N uptake estimated by spectral reflectance may be used in both growth stages as accurate estimators of N status in malting barley. The use of the artificial neural network for N status estimation from hyperspectral data reduced both intra- and interannual variability significantly and provided estimation accuracy above 90%. A more than the twofold increase of estimation accuracy with the use of a neural network model was found for N content in aboveground dry mass. Although the significance of individual wavelength varied with the N status indicator and growth stage, the key reflectance bands for most neural network models were around 430, 530, and 710 nm.

Supplementary Materials: The following are available online at https://www.mdpi.com/article/10.3390/agronomy11122592/s1, Table S1: List of vegetation indices used in correlation analyses with nitrogen status parameters, equations for their calculation from reflectances in given wavelength (R_x is the reflectance in wavelength x nm), and the references for each index. Table S2: Sensitivity analysis for individual input reflectance wavelength in winning neural networks for estimation of nitrogen status parameters.

Author Contributions: Conceptualization, J.K., K.K. and P.M.; methodology, P.M., J.K., K.K., I.S. and V.L.; formal analysis, K.K., D.K., J.Š. and H.F.; investigation, K.K., I.S., P.M., J.K. and V.L.; data curation, K.K., P.M., P.H., D.K., H.F. and J.Š.; writing—original draft preparation, K.K., J.K., P.M., P.L., P.H. and O.U.; writing—review and editing, O.U., P.H., K.K., P.L., J.Š., P.M. and J.K.; visualization, D.K., K.K., P.H. and P.L.; supervision, J.K. All authors have read and agreed to the published version of the manuscript.

Funding: This research was funded by the grant agency of the Ministry of Agriculture of the Czech Republic (NAZV) project number QK1910197. J.Š. was funded by the internal grant agency of Mendel University in Brno project number AF-IGA2019-IP043. K.K., D.K., P.H., P.L., H.F. and O.U. were funded by the project SustES "Adaptation strategies for sustainable ecosystem services and food security under adverse environmental conditions" (CZ.02.1.01/0.0/0.0/16_019/0000797).

Data Availability Statement: Data is contained within the article.

Conflicts of Interest: The authors declare no conflict of interest.

References

1. Weston, D.T.; Horsley, R.D.; Schwarz, P.B.; Goos, R.J. Nitrogen and Planting Date Effects on Low-Protein Spring Barley. *Agron. J.* **1993**, *85*, 1170–1174. [CrossRef]
2. Baethgen, W.E.; Christianson, C.B.; Lamothe, A.G. Nitrogen fertiliser effects on growth, grain yield, and yield components of malting barley. *Field Crop. Res.* **1995**, *43*, 87–99. [CrossRef]
3. Křen, J.; Klem, K.; Svobodová, I.; Míša, P.; Neudert, L. Yield and grain quality of spring barley as affected by biomass formation at early growth stages. *Plant Soil Environ.* **2014**, *60*, 221–227. [CrossRef]
4. Edney, M.J.; O'Donovan, J.T.; Turkington, T.K.; Clayton, G.W.; McKenzie, R.; Juskiw, P.; Lafond, G.P.; Brandt, S.; Grant, C.A.; Harker, K.N.; et al. Effects of seeding rate, nitrogen rate and cultivar on barley malt quality. *J. Sci. Food Agric.* **2012**, *92*, 2672–2678. [CrossRef] [PubMed]

5. Hansen, P.M.; Jørgensen, J.R.; Thomsen, A. Predicting grain yield and protein content in winter wheat and spring barley using repeated canopy reflectance measurements and partial least squares regression. *J. Agric. Sci.* **2002**, *139*, 307–318. [CrossRef]

6. Langseth, W.; Stabbetorp, H. The effect of lodging and time of harvest on deoxynivalenol contamination in barley and oats. *J. Phytopathol.* **1996**, *144*, 241–245. [CrossRef]

7. Nakajima, T.; Yoshida, M.; Tomimura, K. Effect of lodging on the level of mycotoxins in wheat, barley, and rice infected with the *Fusarium graminearum* species complex. *J. Gen. Plant Pathol.* **2008**, *74*, 289. [CrossRef]

8. Olfs, H.W.; Blankenau, K.; Brentrup, F.; Jasper, J.; Link, A.; Lammel, J. Soil- and plant-based nitrogen-fertiliser recommendations in arable farming. *J. Plant Nutr. Soil Sci.* **2005**, *168*, 414–431. [CrossRef]

9. Hatfield, J.L.; Gitelson, A.A.; Schepers, J.S.; Walthall, C.L. Application of spectral remote sensing for agronomic decisions. *Agron. J.* **2008**, *100*, S-117–S-131. [CrossRef]

10. Chen, P. A Comparison of Two Approaches for Estimating the Wheat Nitrogen Nutrition Index Using Remote Sensing. *Remote Sens.* **2015**, *7*, 4527–4548. [CrossRef]

11. Yoder, B.J.; Pettigrew-Crosby, R.E. Predicting nitrogen and chlorophyll content and concentrations from reflectance spectra (400–2500 nm) at leaf and canopy scales. *Remote Sens. Environ.* **1995**, *53*, 199–211. [CrossRef]

12. Merzlyak, M.N.; Gitelson, A.A. Why and what for the leaves are yellow in autumn? On the interpretation of optical spectra of senescing leaves (*Acerplatanoides* L.). *J. Plant Physiol.* **1995**, *145*, 315–320. [CrossRef]

13. Buschmann, C.; Nagel, E. In vivo spectroscopy and internal optics of leaves as basis for remote sensing of vegetation. *Int. J. Remote Sens.* **1993**, *14*, 711–722. [CrossRef]

14. Clevers, J.G.P.W.; Gitelson, A.A. Remote estimation of crop and grass chlorophyll and nitrogen content using red-edge bands on Sentinel-2 and -3. *Int. J. Appl. Earth Obs.* **2013**, *23*, 344–351. [CrossRef]

15. Schlemmer, M.; Gitelson, A.A.; Schepers, J.; Ferguson, R.; Peng, Y.; Shanahan, J.; Rundquist, D. Remote estimation of nitrogen and chlorophyll contents in maise at leaf and canopy levels. *Int. J. Appl. Earth Obs.* **2013**, *25*, 47–54. [CrossRef]

16. Sembiring, H.; Lees, H.L.; Raun, W.R.; Johnson, G.V.; Solie, J.B.; Stone, M.L.; DeLeon, M.J.; Lukina, E.V.; Cossey, D.A.; LaRuffa, J.M.; et al. Effect of growth stage and variety on spectral radiance in winter wheat. *J. Plant Nutr.* **2000**, *23*, 141–149. [CrossRef]

17. Aparicio, N.; Villegas, D.; Araus, J.L.; Casadesús, J.; Royo, C. Relationship between growth traits and spectral vegetation indices in durum wheat. *Crop Sci.* **2002**, *42*, 1547–1555. [CrossRef]

18. Hansen, P.M.; Schjoerring, J.K. Reflectance measurement of canopy biomass and nitrogen status in wheat crops using normalised difference vegetation indices and partial least squares regression. *Remote Sens. Environ.* **2003**, *86*, 542–553. [CrossRef]

19. Li, F.; Miao, Y.; Hennig, S.D.; Gnyp, M.L.; Chen, X.; Jia, L.; Bareth, G. Evaluating hyperspectral vegetation indices for estimating nitrogen concentration of winter wheat at different growth stages. *Precis. Agric.* **2010**, *11*, 335–357. [CrossRef]

20. Lemaire, G.; Jeuffroy, M.H.; Gastal, F. Diagnosis tool for plant and crop N status in vegetative stage: Theory and practices for crop N management. *Eur. J. Agron.* **2008**, *28*, 614–624. [CrossRef]

21. Gastal, F.; Lemaire, G. N uptake and distribution in crops: An agronomical and ecophysiological perspective. *J. Exp. Bot.* **2002**, *53*, 789–799. [CrossRef]

22. Mouazen, A.M.; Kuang, B.; De Baerdemaeker, J.; Ramon, H. Comparison among principal component, partial least squares and back propagation neural network analyses for accuracy of measurement of selected soil properties with visible and near infrared spectroscopy. *Geoderma* **2010**, *158*, 23–31. [CrossRef]

23. Savin, R.; Nicolas, M.E. Effects of short periods of drought and high temperature on grain growth and starch accumulation of two malting barley cultivars. *Funct. Plant Biol.* **1996**, *23*, 201–210. [CrossRef]

24. Lemaire, G.; Gastal, F. N uptake and distribution in plant canopies. In *Diagnosis of the Nitrogen Status in Crops*; Springer: Berlin/Heidelberg, Germany, 1997; pp. 3–43.

25. Mistele, B.; Schmidhalter, U. Estimating the nitrogen nutrition index using spectral canopy reflectance measurements. *Eur. J. Agron.* **2008**, *29*, 184–190. [CrossRef]

26. Houlès, V.; Guérif, M.; Mary, B. Elaboration of a nitrogen nutrition indicator for winter wheat based on leaf area index and chlorophyll content for making nitrogen recommendations. *Eur. J. Agron.* **2007**, *27*, 1–11. [CrossRef]

27. Erdle, K.; Mistele, B.; Schmidhalter, U. Comparison of active and passive spectral sensors in discriminating biomass parameters and nitrogen status in wheat cultivars. *Field Crop. Res.* **2011**, *124*, 74–84. [CrossRef]

28. Chen, P.; Haboudane, D.; Tremblay, N.; Wang, J.; Vigneault, P.; Li, B. New spectral indicator assessing the efficiency of crop nitrogen treatment in corn and wheat. *Remote Sens. Environ.* **2010**, *114*, 1987–1997. [CrossRef]

29. Peñuelas, J.; Baret, F.; Filella, I. Semi-empirical indices to assess carotenoids/chlorophyll a ratio from leaf spectral reflectance. *Photosynthetica* **1995**, *31*, 221–230.

30. Filella, I.; Serrano, L.; Serra, J.; Peñuelas, J. Evaluating wheat nitrogen status with canopy reflectance indices and discriminant analysis. *Crop Sci.* **1995**, *35*, 1400–1405. [CrossRef]

31. Hörtensteiner, S. Chlorophyll degradation during senescence. *Annu. Rev. Plant Biol.* **2006**, *57*, 55–77. [CrossRef] [PubMed]

32. Agüera, E.; Cabello, P.; De La Haba, P. Induction of leaf senescence by low nitrogen nutrition in sunflower (*Helianthus annuus*) plants. *Physiol. Plant.* **2010**, *138*, 256–267. [CrossRef] [PubMed]

33. Magney, T.S.; Eitel, J.U.H.; Vierling, L.A. Mapping wheat nitrogen uptake from RapidEye vegetation indices. *Precis. Agric.* **2017**, *18*, 429–451. [CrossRef]

34. Sadras, V.O.; Lemaire, G. Quantifying crop nitrogen status for comparisons of agronomic practices and genotypes. *Field Crop. Res.* **2014**, *164*, 54–64. [CrossRef]
35. Lukina, E.V.; Freeman, K.W.; Wynn, K.J.; Thomason, W.E.; Mullen, R.W.; Stone, M.L.; Solie, J.B.; Klatt, A.R.; Johnson, G.V.; Elliott, R.L.; et al. Nitrogen fertilisation optimisation algorithm based on in-season estimates of yield and plant nitrogen uptake. *J. Plant Nutr.* **2001**, *24*, 885–898. [CrossRef]
36. Raun, W.R.; Solie, J.B.; Johnson, G.V.; Stone, M.L.; Mullen, R.W.; Freeman, K.W.; Thomason, W.E.; Lukina, E.V. Improving Nitrogen Use Efficiency in Cereal Grain Production with Optical Sensing and Variable Rate Application. *Agron. J.* **2002**, *94*, 815–820. [CrossRef]
37. Lawlor, D.W. Carbon and nitrogen assimilation in relation to yield: Mechanisms are the key to understanding production systems. *J. Exp. Bot.* **2002**, *53*, 773–787. [CrossRef]
38. Mas, J.F.; Flores, J.J. The application of artificial neural networks to the analysis of remotely sensed data. *Int. J. Remote Sens.* **2008**, *29*, 617–663. [CrossRef]
39. Bacour, C.; Baret, F.; Béal, D.; Weiss, M.; Pavageau, K. Neural network estimation of LAI, fAPAR, fCover and LAI×Cab, from top of canopy MERIS reflectance data: Principles and validation. *Remote Sens. Environ.* **2006**, *105*, 313–325. [CrossRef]
40. Liu, Z.Y.; Wu, H.F.; Huang, J.F. Application of neural networks to discriminate fungal infection levels in rice panicles using hyperspectral reflectance and principal components analysis. *Comput. Electron. Agric.* **2010**, *72*, 99–106. [CrossRef]
41. Christensen, L.K.; Bennedsen, B.S.; Jørgensen, R.N.; Nielsen, H. Modelling nitrogen and phosphorus content at early growth stages in spring barley using hyperspectral line scanning. *Biosyst. Eng.* **2004**, *88*, 19–24. [CrossRef]
42. Yi, Q.X.; Huang, J.F.; Wang, F.M.; Wang, X.Z.; Liu, Z.Y. Monitoring Rice Nitrogen Status Using Hyperspectral Reflectance and Artificial Neural Network. *Environ. Sci. Technol.* **2007**, *41*, 6770–6775. [CrossRef] [PubMed]
43. Wang, Y.; Wang, F.; Huang, J.; Wang, X.; Liu, Z. Validation of artificial neural network techniques in the estimation of nitrogen concentration in rape using canopy hyperspectral reflectance data. *Int. J. Remote Sens.* **2009**, *30*, 4493–4505. [CrossRef]
44. Sun, J.; Yang, J.; Shi, S.; Chen, B.; Du, L.; Gong, W.; Song, S. Estimating Rice Leaf Nitrogen Concentration: Influence of Regression Algorithms Based on Passive and Active Leaf Reflectance. *Remote Sens.* **2017**, *9*, 951. [CrossRef]
45. Panda, S.S.; Ames, D.P.; Panigrahi, S. Application of Vegetation Indices for Agricultural Crop Yield Prediction Using Neural Network Techniques. *Remote Sens.* **2010**, *2*, 673–696. [CrossRef]

Article

Effect of Nitrogen Supply on Growth and Nitrogen Utilization in Hemp (*Cannabis sativa* L.)

Yang Yang, Wenxin Zha, Kailei Tang, Gang Deng, Guanghui Du and Feihu Liu *

Laboratory of Plant Improvement and Utilization, Yunnan University, Kunming 650500, China;
yjy@ynu.edu.cn (Y.Y.); wsyjy198797@163.com (W.Z.); kailei.tang@ynu.edu.cn (K.T.);
denggang1986@ynu.edu.cn (G.D.); dgh2012@ynu.edu.cn (G.D.)
* Correspondence: dmzpynu@126.com

Abstract: Hemp is a multipurpose crop that is cultivated worldwide for fiber, oil, and cannabinoids. Nitrogen (N) is a key factor for getting a higher production of hemp, but its application is often excessive and results in considerable losses in the soil–plant–water continuum. Therefore, a rational N supply is important for increasing N efficiency and crop productivity. The main objective of this paper was to determine the responses of four hemp cultivars to different levels of exogenous-N supply as nutrient solution during the vegetative growing period. The experiment was conducted at Yunnan University in Kunming, China. Yunma 1, Yunma 7, Bamahuoma, and Wanma 1 were used as the experimental materials, and five N supplying levels (1.5, 3.0, 6.0, 12.0, and 24.0 mmol/L NO_3-N in the nutrient solution) were set by using pot culture and adding nutrient solution. The root, stem, and leaf of the plant were sampled for the determination of growth indexes, dry matter and N accumulation and distribution, and physiological indicators. The plant height, stem diameter, plant dry weight, and plant N accumulation of four hemp cultivars were significantly increased with the increase in exogenous-N supply. Root/shoot dry weight ratios, stem mass density, and N use efficiency decreased significantly with the increase in exogenous-N supply. Nitrogen accumulation, chlorophyll content, soluble protein content, and nitrate reductase activity in leaves were increased with the increase in exogenous-N supply. Among the four indexes, the increase in N accumulation was more than the increase in NR activity. The activities of superoxide dismutase and peroxidase in leaves were increased first and then decreased with the increase in exogenous-N supply, with the maximum value at N 6.0 mmol/L, while the content of malondialdehyde in leaves increased significantly when the level of exogenous-N supply exceeded 6.0 mmol/L. These results revealed that increasing the exogenous-N supply could improve the plant growth, dry matter accumulation, and N accumulation in hemp during the vegetative growth period, but N supply should not exceed 6.0 mmol/L. Among four hemp cultivars, Wanma 1 performed well at 6.0 mmol/L N application.

Keywords: hemp (*Cannabis sativa* L.); nitrogen nutrition; nitrogen utilization efficiency; vegetative growth

Citation: Yang, Y.; Zha, W.; Tang, K.; Deng, G.; Du, G.; Liu, F. Effect of Nitrogen Supply on Growth and Nitrogen Utilization in Hemp (*Cannabis sativa* L.). *Agronomy* **2021**, *11*, 2310. https://doi.org/10.3390/agronomy11112310

Academic Editors: Christos Noulas, Shahram Torabian and Ruijun Qin

Received: 15 October 2021
Accepted: 13 November 2021
Published: 15 November 2021

Publisher's Note: MDPI stays neutral with regard to jurisdictional claims in published maps and institutional affiliations.

1. Introduction

As an important component of macromolecular substances such as proteins, nucleic acids, phospholipids, hormones, and chlorophyll in plants, nitrogen (N) is one of the essential macronutrients for plant growth and development. Sufficient N supply is an important factor to improve crop yield and quality [1]; however, an overdose of N fertilizer will lead to the decrease in fertilizer efficiency in the plant and cause a large amount of N loss that gives rise to a series of environmental issues [2,3]. Therefore, the rational application of N fertilizer is one of the key measures for the high yield and effective cultivation of crops.

Hemp (*Cannabis sativa* L.) is a multipurpose crop. Cannabinoids and other secondary metabolites extracted from hemp inflorescence and leaves are widely used in medical treatment, beauty, and health care. The multipurpose utilization of hemp significantly improves its economic value and planting [4–6]. The high-yield and high-efficiency cultivation mode of hemp has also become one of the hot spots in hemp production [7–9]. N is the macronutrient absorbed by hemp, and sufficient N supply can greatly improve the stem weight and leaf biomass in hemp [10,11]. Papastylianou et al. [12] reported that hemp biomass yield, stem dry weight, and inflorescence weight increased by 37.3%, 48.2%, and 16%, respectively, with the application of 240 kg N ha^{-1} when compared with the unfertilized control. It is easy to diagnose and mitigate N deficiencies during the hemp-growing season, while excessive N is difficult to diagnose. Therefore, farmers prefer to apply N fertilizer in excess in attempts to increase the yield. Nevertheless, the excessive N fertilizer not only impedes increase hemp productivity but also reduces hemp yield caused by plant lodging and serious pests and diseases. According to studies conducted in the Latvia and Western Canada, N fertilization rates ranged between 50 and 200 kg N ha^{-1}, while the highest bast fiber yields were obtained at N rates between 50 and 150 kg N ha^{-1} [13,14]. Amaducci et al. [15] reported that 100 kg ha^{-1} of N was the natural availability in the soil, and each additional kg of N supplied via fertilization increased hemp stem dry matter production by 20 kg but increased plant mortality. Therefore, it is very important to understand the utilization capacity of exogenous-N for carrying out the high-yield and high-efficiency cultivation of hemp.

A plethora of research reported the utilization of exogenous-N supply in hemp under field conditions. However, due to the difficulty of precise regulation of the N nutrient status in the field, there is still a lack of deep understanding in this research area. Previous work has shown that hemp is sensitive to N fertilizer during the vegetative phase by absorbing 88.2% of the total amount N in the whole growth period, which directly affects the accumulation of plant biomass [16]. Thus, the present study was conducted on hemp using pot culture and irrigation of nutrient solution with two specific objectives i.e., (i) to assess the utilization capacity of exogenous-N during the vegetative growing period in hemp; and (ii) to investigate the effects of excessive N on the growth and physiological attributes of hemp. The results will contribute to the understanding of hemp response to an environment with excessive N supply.

2. Materials and Methods

In this study, four commercial domestic hemp cultivars were selected as the experimental materials. Of them, Yunma 1 (YM1, fiber-type) and Yunma 7 (YM7, fiber-type) were provided by the Yunnan Academy of Agricultural Sciences, Bamahuoma (BM, seed-type) was provided by the Guangxi Academy of Agricultural Sciences, and Wanma 1 (WM1, fiber-type) was provided by Lu'an Agricultural Science Research Institute of Anhui Province.

The pot experiment was conducted in a greenhouse of Yunnan University, Kunming, China. The seeds were sown in 30 cm × 20 cm (diameter × depth) plastic pots filled with 1.2 kg peat (dry weight). In this kind of peat, total N, total P, total K, hydrolyzable N, Olsen phosphorus, and available potassium were 5.46, 0.37, 1.04, 0.14, 0.14, and 0.42 g kg^{-1}, respectively, which were measured by using standard methods. Before sowing, peat was irrigated to 30% peat water content. Three weeks after sowing, ten uniform hemp plants were left per pot after thinning. A half liter of nutrient solution was applied every three days in each pot and lasted seven weeks. In the nutrient solution, N concentration was set at five levels as 1.5, 3.0, 6.0, 12.0, and 24.0 mmol/L NO$_3$-N. The nutrient solution was based on Hoagland formula with modifications as KCl replacing KNO$_3$, and CaCl$_2$ supplementing Ca^{2+} for the low N level solutions. Three replicates for each N level were implemented in the experiment.

Seven weeks after treatment with different N levels, ten plants were sampled from each pot. Five plants were used to determine the growth-related indexes and N accumulation (NA), and the remaining five plants were used to measure the contents of chlorophyll (Chl), soluble protein (SP), and malondialdehyde (MDA) as well as the activity of superoxide dismutase (SOD), peroxidase (POD), and nitrate reductase (NR).

After measuring plant height and stem diameter, plants were divided into root (flushed and washed with running water), leaf, and stem; these parts were placed in an oven, first at 108 °C for 20 min and then at 80 °C until drying out. The dry weights (DW) of the root, leaf, and stem were recorded, and then, the average weight per plant was calculated. The dried root, stem, and leaf were pulverized and sieved; then, they were digested with H_2O_2-H_2SO_4. The total nitrogen content (TN) was determined by the Kjeldahl method [17]. The indicators were calculated according to the formulas listed below:

Total plant DW (DWP, g/plant) = root DW + stem DW + leaf DW;
Ratio of root to shoot = root DW/(stem DW + leaf DW);
Stem mass density (SMD, g/cm^3) = stem DW/stem volume;
NA of plant part (g/plant) = plant part DW $\times$ TN;
NA per plant (NAP, g/plant) = root NA + stem NA + leaf NA;
Nitrogen use efficiency (NUE, g/g) = DWP/NAP.

Fully expanded leaves were separated from the upper part of the other five plants. Leaf laminae without midrib were mixed for determining physiological indexes (Chl, SP, MDA, SOD, and POD) using the methods described by Wang et al. [18]. NR activity was measured following the protocol described by Silveira et al. [19].

The data were handled fundamentally by Excel 2016 software. Standard deviation was calculated for each treatment in each cultivar. Effects of N levels on the growth indexes, N accumulation, and physiological indices were tested with one-way ANOVA followed by Duncan's test at $p < 0.05$ (SPSS 23.0).

The 'R' was used to determine the Pearson's correlation coefficients (https://cran.r-project.org/web/packages/Hmisc/index.html, 12 November 2021) and for a heatmap (https://www.r-graph-gallery.com/heatmap/, 12 November 2021).

3. Results

3.1. Effect of N Supply on Plant Growth

The plant height and stem diameter of hemp cultivars were increased with the increase in exogenous-N level, reaching the maximum value at 24.0 mmol/L, while the root/shoot dry weight ratio and SMD showed a contrary pattern. Plant height and root/shoot weight ratio did not change significantly when the N concentration was 12.0 mmol/L or more; the plant height of YM1 and SMD of WM1 did not change significantly when the N concentration was 6.0 mmol/L or more; the SMD of YM7 showed a significant difference among N concentrations (Table 1).

Table 1. Effect of N treatments on the hemp growth parameters.

Cultivar	Nitrogen (mmol/L)	Plant Height (cm)	Stem Diameter (mm)	Root/Shoot Ratio	Stem Mass Density (g/cm^3)
YM1	1.5	62 ± 8 c	3.3 ± 0.2 c	0.37 ± 0.03 a	0.90 ± 0.09 a
	3.0	75 ± 5 b	3.5 ± 0.3 c	0.36 ± 0.03 ab	0.78 ± 0.09 b
	6.0	89 ± 2 a	4.6 ± 0.4 b	0.33 ± 0.02 b	0.40 ± 0.01 c
	12.0	94 ± 5 a	5.0 ± 0.5 b	0.28 ± 0.01 c	0.36 ± 0.02 cd
	24.0	95 ± 1 a	6.1 ± 0.5 a	0.28 ± 0.02 c	0.29 ± 0.01 d
BM	1.5	45 ± 2 d	3.9 ± 0.2 b	0.43 ± 0.03 a	0.82 ± 0.04 a
	3.0	54 ± 2 c	4.2 ± 0.5 b	0.39 ± 0.06 ab	0.73 ± 0.23 b
	6.0	68 ± 1 b	4.7 ± 0.5 ab	0.35 ± 0.07 abc	0.50 ± 0.21 b
	12.0	70 ± 4 ab	4.7 ± 0.5 ab	0.31 ± 0.02 bc	0.49 ± 0.16 b
	24.0	72 ± 1 a	5.2 ± 0.7 a	0.30 ± 0.06 c	0.43 ± 0.18 b
YM7	1.5	58 ± 4 d	3.1 ± 0.4 d	0.36 ± 0.02 a	1.14 ± 0.09 a
	3.0	71 ± 7 c	4.0 ± 0.5 c	0.32 ± 0.02 b	0.61 ± 0.05 b
	6.0	81 ± 3 b	4.4 ± 0.4 c	0.29 ± 0.01 c	0.51 ± 0.04 c
	12.0	87 ± 4 ab	5.2 ± 0.4 b	0.28 ± 0.01 c	0.40 ± 0.01 d
	24.0	92 ± 1 a	6.1 ± 0.4 a	0.28 ± 0.01 c	0.30 ± 0.03 e

Table 1. *Cont.*

Cultivar	Nitrogen (mmol/L)	Plant Height (cm)	Stem Diameter (mm)	Root/Shoot Ratio	Stem Mass Density (g/cm^3)
WM1	1.5	69 ± 6 c	3.2 ± 0.2 d	0.27 ± 0.02 a	1.42 ± 0.04 a
	3.0	76 ± 2 c	3.5 ± 0.2 c	0.26 ± 0.01 ab	0.96 ± 0.17 b
	6.0	90 ± 4 b	4.6 ± 0.2 b	0.24 ± 0.02 b	0.55 ± 0.05 c
	12.0	98 ± 6 ab	4.8 ± 0.3 ab	0.21 ± 0.02 c	0.50 ± 0.06 c
	24.0	104 ± 4 a	5.0 ± 0.1 a	0.21 ± 0.01 c	0.47 ± 0.02 c
YM1	1.5	62 ± 8 c	3.3 ± 0.2 c	0.37 ± 0.03 a	0.90 ± 0.09 a
	3.0	75 ± 5 b	3.5 ± 0.3 c	0.36 ± 0.03 ab	0.78 ± 0.09 b
	6.0	89 ± 2 a	4.6 ± 0.4 b	0.33 ± 0.02 b	0.40 ± 0.01 c
	12.0	94 ± 5 a	5.0 ± 0.5 b	0.28 ± 0.01 c	0.36 ± 0.02 cd
	24.0	95 ± 1 a	6.1 ± 0.5 a	0.28 ± 0.02 c	0.29 ± 0.01 d

Different letters following the numbers within columns represent significant difference at p = 0.05 within a cultivar. Hemp cultivars: Yunma 1 (YM1), Bamahuoma (BM), Yunma 7 (YM7), Wanma 1 (WM1).

3.2. Effect of N Supply on Hemp Biomass

The biomass (total plant dry weight) of YM7 was increased first and then decreased with the increase in N concentration. A maximum value of dry weight was found at 12.0 mmol/L, while in other cultivars, it was increased along with the N supply levels, reaching the maximum value at 24.0 mmol/L; the biomass of YM7 and BM did not change significantly when the N concentration was 6.0 mmol/L or more. Among plant parts, with the increase in N concentration, the dry weights of stem and leaf were increased greatly, but this increase in root dry weight increased was very little, and the root/total plant dry weight ratio decreased continuously (Figure 1). Among four hemp cultivars, WM1 performed well at 6.0 mmol/L N application.

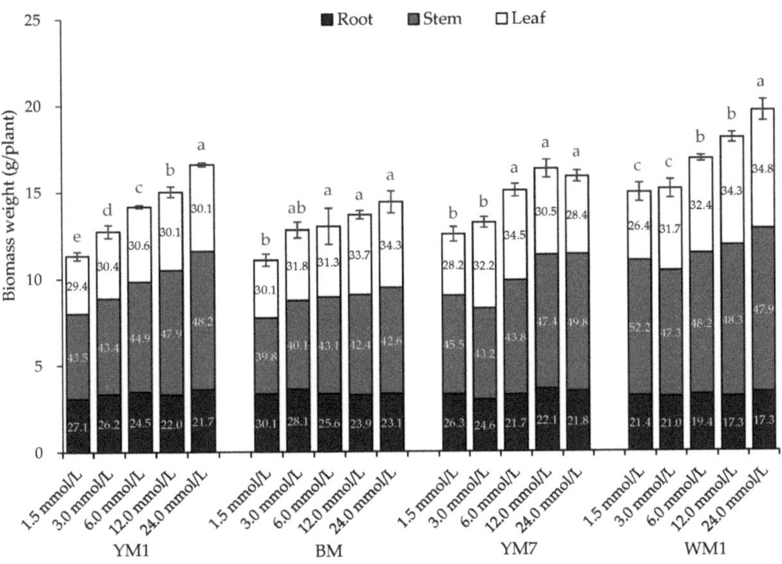

Figure 1. Effect of N treatments on hemp biomass weight. Different letters above the columns within a cultivar represent significant differences in total biomass at p = 0.05. Numbers within a column represent the percentage of root, stem, or leaf weight over the total biomass (%).

3.3. Effect of N Supply on NA and NUE in Hemp Plant

The nitrogen accumulation (NA) of hemp cultivars increased evidently with the increase in exogenous-N supply, reaching the maximum value at 24.0 mmol/L N level, which was about three times that at the 1.5 mmol/L N level. Under all the N concentrations, NA in different plant parts was found as: leaf > stem > root, with the only exception for hemp cultivar BM under 1.5 and 3.0 mmol/L N levels. The ratio of NA in the leaves of BM and YM7 was increased with the increase in N concentration, while that of YM1 and WM1 increased first and then decreased, reaching the highest point at 12.0 mmol/L N level. The ratio of NA in the roots decreased with the increase in N concentration, and the ratio was reduced (about 50%) at 24.0 mmol/L N in comparison with that at 1.5 mmol/L N (Figure 2). However, the NUE of hemp cultivars was decreased significantly with the increase in exogenous-N supply and the lowest NUE value was observed at 24.0 mmol/L N that was about half as that at 1.5 mmol/L N (Figure 3). Among four hemp cultivars, WM1 performed well at 6.0 mmol/L N application.

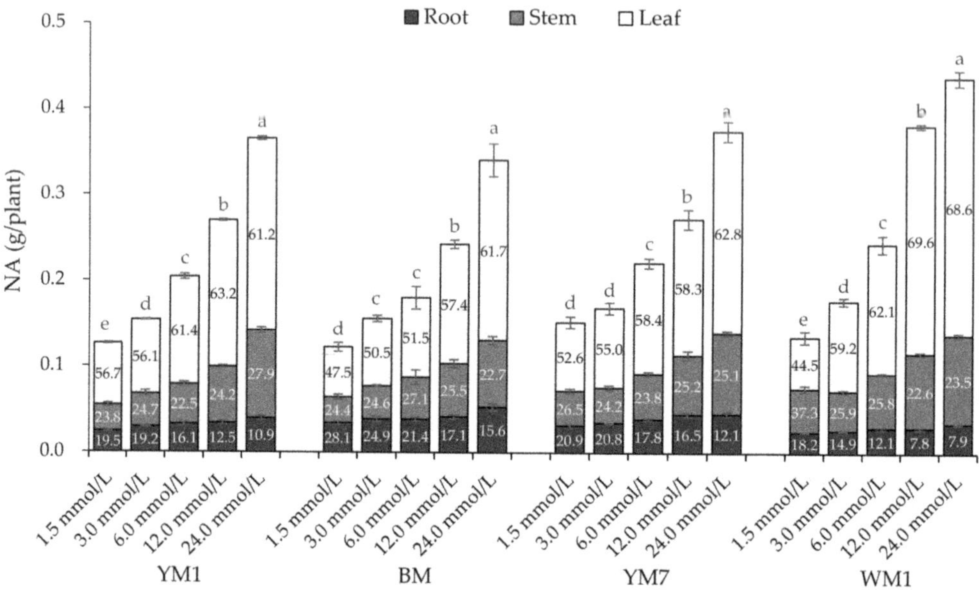

Figure 2. Effect of N treatments on N accumulation (NA) in hemp plant. Different letters above the columns within a cultivar represent significant differences in total N accumulation at $p = 0.05$. Numbers within the column represent the percentage of root, stem, or leaf N accumulation over the total N accumulation (%).

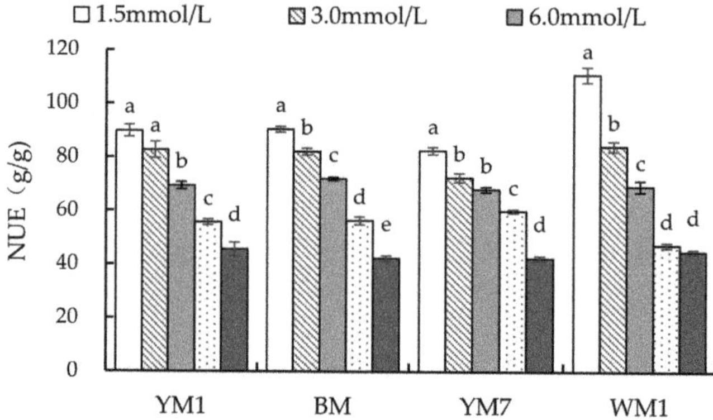

Figure 3. Effect of N treatments on N utilization efficiencies in hemp. Different letters above the columns within a cultivar represent significant differences in N utilization efficiencies at $p = 0.05$.

3.4. Effect of N Supply on Chl, SP, and NR in Hemp Leaf

Chlorophyll (Chl) content in hemp leaves were increased with the increase in exogenous-N supply, reaching maximum value at 24.0 mmol/L N, although not showing a linear correlation between Chl content and N level (Figure 4a).

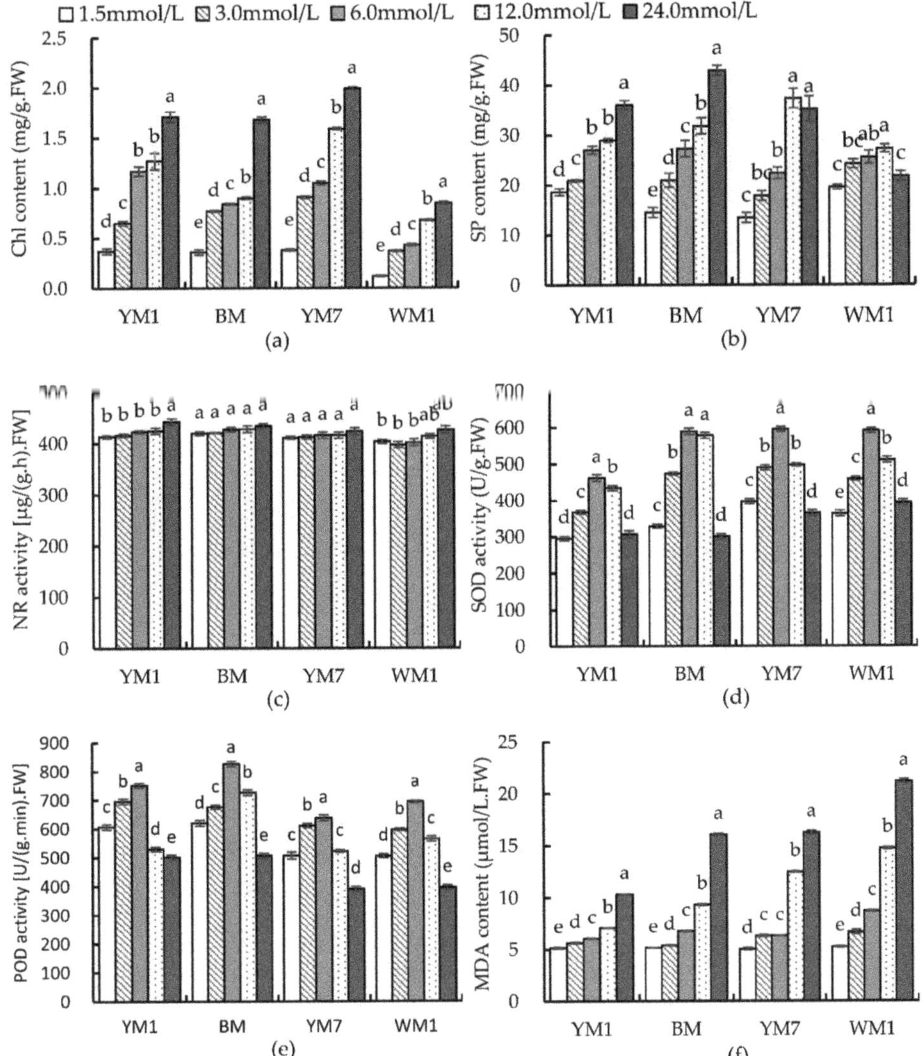

Figure 4. Effect of N treatments on the physiological indicators of hemp leaf. Different letters above the columns within a cultivar represent significant differences in physiological indicators of leaf at $p = 0.05$. (**a**) The content of Chl of four hemp cultivars under different levels of N. (**b**) The content of SP of four hemp cultivars under different levels of N. (**c**) The activity of NR of four hemp cultivars under different levels of N. (**d**) The activity of SOD of four hemp cultivars under different levels of N. (**e**) The activity of POD of four hemp cultivars under different levels of N. (**f**) The content of MNA of four hemp cultivars under different levels of N.

The soluble protein (SP) contents of YM1 and BM were increased with the increase in exogenous-N supply, reaching a maximum value at 24.0 mmol/L N and significantly surpassing those under other N concentrations; while the SP contents of YM7 and WM1 were increased first and then decreased, reaching the maximum value at 12.0 mmol/L (Figure 4b).

With the increase in exogenous-N supply, nitrate reductase (NR) activity in hemp leaves showed an increasing trend. Among the cultivars, NR activity in BM and YM7 did not change significantly under different N concentrations (Figure 4c).

3.5. Effect of N Supply on SOD, POD, and MDA in Hemp Leaf

According to the results presented in Figure 4d,e, the activities of SOD and POD increased first and then decreased with the increase in exogenous-N supply, and they reached maximum value at 6.0 mmol/L N, which was significantly higher than those under other N concentrations (except for SOD in BM from 6.0 to 12.0 mmol/L N levels). In contrast, the malondialdehyde (MDA) content in hemp cultivars increased significantly with the increase in exogenous-N supply, while it showed a smaller increase from N levels 1.5 to 6.0 mmol/L and a larger increase from 6.0 to 24.0 mmol/L N level. From 1.5 to 6.0 mmol/L N level, the average MDA content of the four cultivars increased by 35% only, but it increased by 127% from 6.0 to 24.0 mmol/L N level (Figure 4f).

3.6. Relationships

In order to study the relationship between the studies' parameters, Pearson correlation was carried out (Figure 5). The results showed a significant positive relationship between plant height, stem diameter, leaf weight, stem weight, root weight, biomass weight, chlorophyll content, and N accumulation. Inversely, these attributes showed a negative relationship with SOD and POD activity. Furthermore, this relationship showed a close link between N accumulation and the growth of hemp plants. In addition, hierarchical clustering analysis showed a relation between interactive treatments and studied parameters (Figure 6).

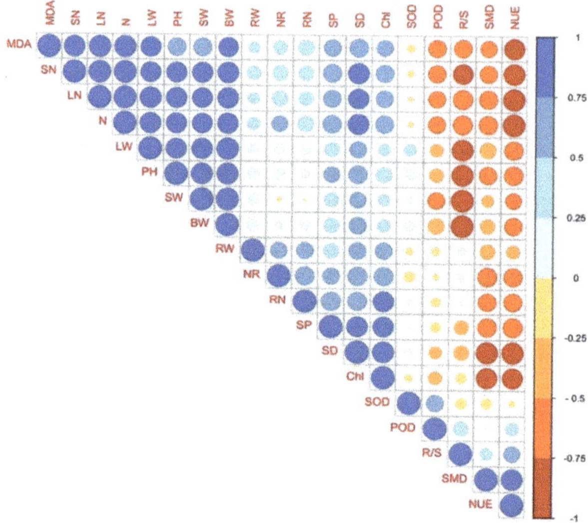

Figure 5. Correlation between studied parameters of growth, antioxidant capacity, and N accumulation in hemp. PH, plant height; SD, stem diameter; LW, leaf weight; SW, stem weight; RW, root weight; BW, biomass weight; R/S, root to shoot ratio; SMD, stem mass density; Chl, chlorophyll content; SP, soluble protein; NR, nitrate reductase activity; SOD, superoxide dismutase; POD, peroxidase; MDA, malondialdehyde; RN, root nitrogen; SN, stem nitrogen; LN, leaf nitrogen; N, total nitrogen; NUE, nitrogen use efficiency.

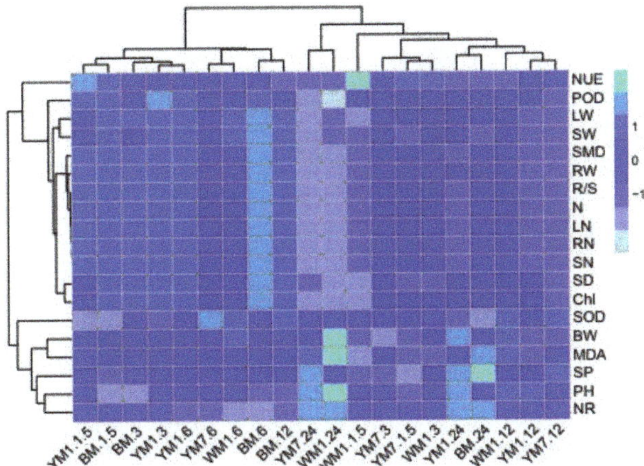

Figure 6. A heatmap showing the relationship of treatments with the studied parameters of growth, antioxidant capacity, and N accumulation in hemp. The colors represent variations in the data. PH, plant height; SD, stem diameter; LW, leaf weight; SW, stem weight; RW, root weight; BW, biomass weight; R/S, root to shoot ratio; SMD, stem mass density; Chl, chlorophyll content; SP, soluble protein; NR, nitrate reductase activity; SOD, superoxide dismutase; POD, peroxidase; MDA, malondialdehyde; RN, root nitrogen; SN, stem nitrogen; LN, leaf nitrogen; N, total nitrogen; NUE, nitrogen use efficiency. YM1.1.5, cultivar YN1 + 1.5 mmol/L N; YM1.3, cultivar YN1 + 3 mmol/L N; YM1.6, cultivar YN1 + 6 mmol/L N; YM1.12, cultivar YN1 + 12 mmol/L N; YM1.24, cultivar YN1 + 24 mmol/L N; BM.1.5, cultivar BM + 1.5 mmol/L N; BM.3, cultivar BM + 3 mmol/L N; BM.6, cultivar BM + 6 mmol/L N; BM.12, cultivar BM + 12 mmol/L N; BM.24, cultivar BM + 24 mmol/L N; YM7.1.5, cultivar YM7 + 1.5 mmol/L N; YM7.3, cultivar YM7 + 3 mmol/L N; YM7.6, cultivar YM7 + 6 mmol/L N; YM7.12, cultivar YM7 + 12 mmol/L N; YM7.24, cultivar YM7 + 24 mmol/L N; WM1.1.5, cultivar WM1 + 1.5 mmol/L N; WM1.3, cultivar WM1 + 3 mmol/L N; WM1.6, cultivar WM1 + 6 mmol/L N; WM1.12, cultivar WM1 + 12 mmol/L N; WM1.24, cultivar WM1 + 24 mmol/L N.

4. Discussion

Hemp is a short-day plant with a stronger stem and well-developed root system to avoid lodging and getting high yield [14,15]. Islam et al. [20] revealed that the dry weight per unit length of basal internodes was the main factor determining the mechanical strength of stems. However, Sperling et al. [21] found that high nitrogen (N) conditions limit photosynthetic productivity in almond trees, which is not good for dry matter accumulation. In the current study, the plant height and stem diameter of hemp were increased significantly with the increase in exogenous-N supply, but SMD was decreased significantly. Present results are consistent with the previous studies performed on rice [22,23]. There might be a reason that the plant height and stem diameter are sensitive to N, which lead toward insufficient dry matter accumulation during the rapid growth of hemp. We found that the average plant height and stem diameter of four cultivars were increased by 59.3% and 67.8% respectively, while the dry matter was increased by 32.8% only, when the exogenous-N supply was increased from 1.5 to 24.0 mmol/L. Our results are consistent with the previous study performed on wheat, where an excessive application of N fertilizer did not significantly increase the rate of dry matter accumulation [24].

In response to N deficiency in crops, assimilates are preferentially used for root development rather than shoot development [25]. Meanwhile, under a higher supply of N, the development of the aerial parts increased and the development of the roots decreased [26,27]. In the present study, the root/shoot dry weight ratio in hemp decreased significantly with the increase in exogenous-N supply, suggesting that excessive N application maximized the shoot development and minimized the root growth. These results are consistent with the previous studies on sweet pepper and maize crops [25,28]. Thus, we observed that the weak plant stem and underdeveloped root were caused by excessive N application, which is prone to lodging and affects the yield in hemp.

Nitrogen application can improve the N content in soil, which is conducive to the absorption by crops [29]. After uptake, a large quantity of N is transferred to leaves, where it is assimilated into amino acids, proteins, and other nitrogenous compounds [30,31]. In the present study, increasing the supply of exogenous-N greatly increases the N accumulation in hemp plants. The accumulated N is mainly concentrated in leaves, and it corresponds to the Chl and SP of leaves significantly increasing, which is consistent with the research results of Li et al. on cotton [32]. Thus, a higher level of exogenous-N was beneficial to N accumulation and the synthesis of nitrogenous compounds in hemp. However, we found that the efficiency of accumulated N converted into dry matter in hemp was reduced at a higher level of exogenous-N, suggesting that the accumulated N was not effectively metabolized. Our results are consistent with a previous study on maize [33].

The activity of related metabolic enzymes is the key to improving the N utilization capacity of plants, such as NR; its activity is affected by the nitrate concentration in the plants [34,35]. The primary assimilation of N is accompanied by various enzymatic activities in higher plants. Nitrate reductase plays a central role in the transformation of NO^{3-}; thus, it regulates the NO^{3-} and amino acids level in the plant cells. However, some investigations revealed that a small amount of nitrate is sufficient for enzymes induction; namely, the activity of NR is not induced by nitrate, when nitrate concentration was higher than a certain level [36]. The result of our experiment is consistent with the above report, because when the level of exogenous-N supply was 1.5 mmol/L, the accumulation of nitrate in leaves of hemp could induce NR to maintain a high activity, and further increasing the level of N supply had little effect on it. Thus, that the NR activity did not increase significantly is one of the major factors indicating that the accumulated N was not effectively metabolized under a high level of exogenous-N in hemp.

Reactive oxygen species (ROS) was considered to be toxic by-products of normal metabolism in plant, such as photosynthesis and respiration [37]. To control the level of ROS in cells, plants developed numerous strategies for the detoxification of ROS. Among antioxidative enzymes, SOD and POD play key roles in the ROS detoxification in cells [38,39]. In the present study, we observed that the levels of two antioxidant enzymes (SOD and POD) were decreased in the leaves of hemp when the level of exogenous-N supply was 12.0 mmol/L or more. These results are consistent with a previous study on wheat [40]. The application of N fertilizer beyond a tolerable limit may have adverse effects on plant growth; thus, it inhibited the activities of ROS scavenging enzymes, which resulted in increased oxidative stress.

Some evidence suggests that the excessive accumulation of ROS can cause a series of oxidative damages to proteins, lipids, and DNA, resulting in lipid peroxidation, cellular damage, and cell death [41–43]. The level of MDA is used normally to indicate the extent of lipid peroxidation in leaves. In the present study, the damage degree of membrane lipid peroxidation was increased significantly, when the level of exogenous-N supply was 12.0 mmol/L or more. Thus, our results indicated that hemp plants suffered from a greater degree of oxidative damage when the exogenous-N supply level exceeded 6.0 mmol/L, which might affect the normal physiological metabolism in hemp. The SP content of YM7 and WM1 decreased at 24 mmol/L N supplying level. These results have also been found in macrophytes, where excess N supply led to the reduction of osmotic regulation substance content, such as SP, reducing its stress resistance [44]. This might be linked to the oxidative stress in hemp, but further research is needed.

In addition, it is also worth paying attention to the accumulation of N far beyond its utilization capacity during the vegetative growth period when N is sufficient. One of the possible explanations might be that excess N is a reserve for later reproductive growth, which is a peak period of nutrient consumption in hemp [10,12]. It is important for hemp to adapt to a changeable variable environment. When N is sufficient, hemp can accumulate N as much as possible, and the stored N can be used for reproductive growth to alleviate the adverse effects of late N deficiency in the environment. However, these contents need to be further studied in hemp, especially the situation of N absorption and utilization during the reproductive growth period.

5. Conclusions

In this study, an increase in the exogenous-N supplying level was good for hemp growth, but it was not to exceed 6.0 mmol/L during vegetative growth periods. Under excessive N supply, plants grow rapidly, but the dry matter accumulation was evidently insufficient, which weakened the hemp plants. Moreover, the large amount of N accumulated in plants could not be effectively assimilated and utilized, and it caused oxidative stress to hemp plants. Thus, the present study suggests that N application up to 6.0 mmol/L is sufficient to regulate the morpho-physiological attributes, antioxidant capacity, and N accumulation to achieve the optimal growth of hemp. Among four hemp cultivars, "Wanma 1" performed well at 6.0 mmol/L N application.

Author Contributions: Conceptualization, F.L.; methodology, W.Z.; formal analysis, Y.Y.; investigation, W.Z.; resources, F.L.; data curation, Y.Y. and G.D. (Gang Deng); writing—original draft preparation, Y.Y.; writing—review and editing, K.T. and G.D. (Guanghui Du); funding acquisition, F.L. All authors have read and agreed to the published version of the manuscript.

Funding: The study was supported by China Agriculture Research System of MOF and MARA for Bast and Leaf Fiber Plants (CARS-16-E15).

Data Availability Statement: The data sets generated for this study are available on request to the corresponding author.

Acknowledgments: The authors gratefully acknowledge all staff members and students involved in the study.

Conflicts of Interest: The authors declare no conflict of interest.

References

1. Liu, Q.; Wu, K.; Fu, X.; Li, S.; Liu, X.; Gao, X. Sustainable crop yields from the coordinated modulation of plant growth and nitrogen metabolism. *Chin. Sci. Bull.* **2019**, *64*, 2633–2640. [CrossRef]
2. Asif, I.; Dong, Q.; Wang, Z.; Wang, X.G.; Gui, H.P.; Zhang, H.H.; Pang, N.C.; Zhang, X.L.; Song, M.Z. Growth and nitrogen metabolism are associated with nitrogen-use efficiency in cotton genotypes. *Plant Physiol. Biochem.* **2020**, *149*, 61–74. [CrossRef]
3. Ahmed, M.; Rauf, M.; Mukhtar, Z.; Saeed, N.A. Excessive use of nitrogenous fertilizers: An unawareness causing serious threats to environment and human health. *Environ. Sci. Pollut. Res.* **2017**, *24*, 26983–26987. [CrossRef]
4. Harm, V.B.; Jake, M.S.; Atina, G.C.; Carling, M.T.; Andrew, G.S.; Timothy, R.H.; Jonathan, E.P. The Draft Genome and Transcriptome of *Cannabis sativa*. *Genome Biol.* **2011**, *12*, R102. Available online: http://genomebiology.com/2011/12/10/R102 (accessed on 12 November 2021).
5. Farinon, B.; Molinari, R.; Costantini, L.; Merendino, N. The seed of industrial hemp (*Cannabis sativa* L.): Nutritional Quality and Potential Functionality for Human Health and Nutrition. *Nutrients* **2020**, *12*, 1935. [CrossRef]
6. Ranalli, P.; Venturi, G. Hemp as a raw material for industrial applications. *Euphytica* **2004**, *140*, 1–6. [CrossRef]
7. Salentijn, E.M.J.; Petit, J.; Trindade, L.M. The Complex Interactions Between Flowering Behavior and Fiber Quality in Hemp. *Front. Plant Sci.* **2019**, *10*, 614. [CrossRef]
8. Liu, F.H.; Du, G.H.; Yang, Y.; Deng, G.; Tang, K.L. *Green and Efficient Cultivation Techniques for Industrial Hemp of Flower Heads and Leaf Uses*, 1st ed.; Yunnan University Press: Kunming, China, 2020; pp. 15–18.
9. Deng, G.; Du, G.; Yang, Y.; Bao, Y.; Liu, F. Planting Density and Fertilization Evidently Influence the Fiber Yield of Hemp (*Cannabis sativa* L.). *Agronomy* **2019**, *9*, 368. [CrossRef]
10. Kakabouki, I.; Kousta, A.; Folina, A.; Karydogianni, S.; Zisi, C.; Kouneli, V.; Papastylianou, P. Effect of Fertilization with Urea and Inhibitors on Growth, Yield and CBD Concentration of Hemp (*Cannabis sativa* L.). *Sustainability* **2021**, *13*, 2157. [CrossRef]
11. Forrest, C.; Young, J.P. The Effects of Organic and Inorganic Nitrogen Fertilizer on the Morphology and Anatomy of Cannabis sativa "Fédrina" (Industrial Fibre Hemp) Grown in Northern British Columbia, Canada. *J. Ind. Hemp* **2006**, *11*, 3–24. [CrossRef]
12. Papastylianou, P.; Kakabouki, I.; Travlos, I. Effect of Nitrogen Fertilization on Growth and Yield of Industrial Hemp (*Cannabis sativa* L.). *Not. Bot. Hort. Agrobot. Cluj-Napoca* **2017**, *46*, 197–201. [CrossRef]
13. Aubin, M.; Seguin, P.; Vanasse, A.; Tremblay, G.F.; Mustafa, A.F.; Charron, J. Industrial Hemp Response to Nitrogen, Phosphorus, and Potassium Fertilization. *CFTM* **2015**, *1*, 1–10. [CrossRef]

14. Sausserde, R.; Adamovics, A. Effect of nitrogen fertilizer rates on industrial hemp (*Cannabis sativa* L.) biomass production. In Proceedings of the International Multidisciplinary Scientific Geo Conference, Varna, Bulgaria, 16–22 June 2013; Volume 1, pp. 339–346.

15. Amaducci, S.; Errani, M.; Venturi, G. Response of hemp to plant population and nitrogen fertilization. *Ital. J. Agron.* **2002**, *6*, 103–111.

16. Wylie, S.E.; Ristvey, A.G.; Fiorellino, N.M. Fertility management for industrial hemp production: Current knowledge and future research needs. *GCB Bioenergy* **2021**, *13*, 517–524. [CrossRef]

17. Li, B.; Xin, W.; Sun, S.; Shen, Q.; Xu, G. Physiological and Molecular Responses of Nitrogen-starved Rice Plants to Re-supply of Different Nitrogen Sources. *Plant Soil* **2006**, *287*, 145–159. [CrossRef]

18. Wang, X.K. *Experimental Principle and Technology of Plant Physiology and Biochemistry*, 3rd ed.; Higher Education Press: Beijing, China, 2015; pp. 50–105.

19. Silveira, J.; Matos, J.; Cecatto, V.; Viegas, R.; Oliveira, J. Nitrate reductase activity, distribution, and response to nitrate in two contrasting Phaseolus species inoculated with Rhizobium spp. *Environ. Exp. Bot.* **2001**, *46*, 37–46. [CrossRef]

20. Islam, M.S.; Peng, S.; Visperas, R.M.; Ereful, N.; Bhuiya, M.S.U.; Julfiquar, A. Lodging-related morphological traits of hybrid rice in a tropical irrigated ecosystem. *Field Crop. Res.* **2007**, *101*, 240–248. [CrossRef]

21. Sperling, O.; Karunakaran, R.; Erel, R.; Yasuor, H.; Klipcan, L.; Yermiyahu, U. Excessive nitrogen impairs hydraulics, limits photosynthesis, and alters the metabolic composition of almond trees. *Plant Physiol. Biochem.* **2019**, *143*, 265–274. [CrossRef]

22. Wu, X.R.; Zhang, W.J.; Wu, L.M.; Wang, F.; Li, G.H.; Liu, Z.H.; Tang, S.; Ding, C.G.; Wang, S.H.; Ding, Y.F. Characteristics of lodging resistance of super-Hybrid indica rice and its response to nitrogen. *Sci. Agric. Sin.* **2015**, *48*, 2705–2717. [CrossRef]

23. Zhang, J.; Li, G.-H.; Song, Y.-P.; Zhang, W.-J.; Yang, C.-D.; Wang, S.-H.; Ding, Y.-F. Lodging Resistance of Super-Hybrid Rice Y Liangyou 2 in Two Ecological Regions. *Acta Agron. Sin.* **2013**, *39*, 682. [CrossRef]

24. Song, M.D.; Li, Z.P.; Feng, H. Effects of irrigation and nitrogen regimes on dry matter dynamic accumulation and yield of winter wheat. *Trans. Chin. Soc. Agric. Eng.* **2016**, *32*, 119–126. [CrossRef]

25. Grasso, R.; de Souza, R.; Peña-Fleitas, M.T.; Gallardo, M.; Thompson, R.B.; Padilla, F.M. Root and crop responses of sweet pepper (*Capsicum annuum*) to increasing N fertilization. *Sci. Hortic.* **2020**, *273*, 109645. [CrossRef]

26. Garnett, T.; Conn, V.; Kaiser, B.N. Root based approaches to improving nitrogen use efficiency in plants. *Plant Cell Environ.* **2009**, *32*, 1272–1283. [CrossRef]

27. Drew, M.C.; Saker, L.R.; Ashley, T.W. Nutrient Supply and the Growth of the Seminal Root System in Barley. *J. Exp. Bot.* **1973**, *24*, 1189–1202. [CrossRef]

28. Anderson, E.L. Tillage and N fertilization effects on maize root growth and root:shoot ratio. *Plant Soil* **1988**, *108*, 245–251. [CrossRef]

29. Ju, X.T.; Gu, B.J. Status-quo, problem and trend of nitrogen fertilization in China. *J. Plant Nutr. Fertil.* **2014**, *20*, 783–795. [CrossRef]

30. Qu, L.B.; Gu, H.Y.; Liu, J.J.; Qing, G.J. *Plant Biology*, 1st ed; The Science Press: Beijing, China, 2012; pp. 303–320.

31. Ren, B.; Dong, S.; Zhao, B.; Liu, P.; Zhang, J. Responses of Nitrogen Metabolism, Uptake and Translocation of Maize to Waterlogging at Different Growth Stages. *Front. Plant Sci.* **2017**, *8*, 1216. [CrossRef] [PubMed]

32. Li, C.P.; Dong, H.L.; Liu, A.Z.; Liu, J.R.; Li, R.Y.; Sun, M.; Li, Y.B.; Mao, S.C. Effects of nitrogen application rates on physiological characteristics of functional leaves, nitrogen use efficiency and yield of cotton. *J. Plant Nutr. Fertil.* **2015**, *21*, 81–91. [CrossRef]

33. Jian, L.L.; Han, L.S.; Han, X.R.; Zhan, X.M.; Zuo, R.H.; Wu, Z.C.; Yuan, C. Effects of nitrogen on growth, root morphological traits, nitrogen uptake and utilization efficiency of maize seedlings. *J. Plant Nutr. Fertil.* **2011**, *17*, 247–253. [CrossRef]

34. Iqbal, A.; Dong, Q.; Wang, X.; Gui, H.; Zhang, H.; Zhang, X.; Song, M. Variations in Nitrogen Metabolism are Closely Linked with Nitrogen Uptake and Utilization Efficiency in Cotton Genotypes under Various Nitrogen Supplies. *Plants* **2020**, *9*, 250. [CrossRef] [PubMed]

35. Krapp, A.; Berthomé, R.; Orsel, M.; Mercey-Boutet, S.; Yu, A.; Castaings, L.; Elftieh, S.; Major, H.; Renou, J.-P.; Daniel-Vedele, F. Arabidopsis Roots and Shoots Show Distinct Temporal Adaptation Patterns toward Nitrogen Starvation. *Plant Physiol.* **2011**, *157*, 1255–1282. [CrossRef] [PubMed]

36. Chen, B.-M.; Wang, Z.-H.; Li, S.-X.; Wang, G.-X.; Song, H.-X.; Wang, X.-N. Effects of nitrate supply on plant growth, nitrate accumulation, metabolic nitrate concentration and nitrate reductase activity in three leafy vegetables. *Plant Sci.* **2004**, *167*, 635–643. [CrossRef]

37. Mittler, R. Oxidative stress, antioxidants and stress tolerance. *Trends Plant Sci.* **2002**, *7*, 405–410. [CrossRef]

38. Bowler, C.; Montagu, M.V.; Inze, D. Superoxide Dismutase and Stress Tolerance. *Annu. Rev. Plant Biol.* **1992**, *43*, 83–116. [CrossRef]

39. Willekens, H.; Chamnongpol, S.; Davey, M.; Schraudner, M.; Langebartels, C.; Van Montagu, M.; Inzé, D.; Van Camp, W. Catalase is a sink for H2O2 and is indispensable for stress defence in C3 plants. *EMBO J.* **1997**, *16*, 4806–4816. [CrossRef] [PubMed]

40. Kong, L.; Xie, Y.; Hu, L.; Si, J.; Wang, Z. Excessive nitrogen application dampens antioxidant capacity and grain filling in wheat as revealed by metabolic and physiological analyses. *Sci. Rep.* **2017**, *7*, 43363. [CrossRef]

41. Mitsuhara, I.; Malik, K.A.; Miura, M.; Ohashi, Y. Animal cell-death suppressors Bcl-xL and Ced-9 inhibit cell death in tobacco plants. *Curr. Biol.* **1999**, *9*, 775–778. [CrossRef]

42. Ishida, H.; Anzawa, D.; Kokubun, N.; Makino, A.; Mae, T. Direct evidence for non-enzymatic fragmentation of chloroplastic glutamine synthetase by a reactive oxygen species. *Plant Cell Environ.* **2002**, *25*, 625–631. [CrossRef]

43. Domínguez-Valdivia, M.D.; Aparicio-Tejo, P.M.; Lamsfus, C.; Cruz, C.; Martins-Loução, M.A.; Moran, J.F. Nitrogen nutrition and antioxidant metabolism in ammonium-tolerant and -sensitive plants. *Physiol. Plant.* **2008**, *132*, 359–369. [CrossRef]
44. Liu, H.W.; Zhang, R.R.; Liu, Y.L.; Xie, C.H.; Lin, H.; Lin, Z.X.; Ye, W.H. Effect of Nitrogen on Pennisetum sp. Seeding Stage Growth and Photosynthetic Physiological Characteristic. *Northern Hortic.* 2016, pp. 133–138. Available online: http://bfyy.paperonce.org/oa/DArticle.aspx?type=view&id=201608037 (accessed on 12 November 2021).

MDPI
St. Alban-Anlage 66
4052 Basel
Switzerland
www.mdpi.com

Agronomy Editorial Office
E-mail: agronomy@mdpi.com
www.mdpi.com/journal/agronomy